THE FLIGHT OF THE MALFUNCTION

THE FLIGHT OF THE MALFUNCTION

SERGEANT WALTER BABINSKI - A B-24 BALL TURRET GUNNER IN WWII

Greg Babinski

The Flight of the Malfunction

Orzel Books
Copyright © 2017 by Gregory Babinski
Edmonds, Washington

Library of Congress Control Number 2017900998
ISBN 978-0-9986495-1-1

All rights reserved. Copyright under Berne Copyright Convention, Universal Copyright Convention, and Pan-American Copyright Convention. No part of this book may be reproduced, stored in a retrieval system, or transmitted in any form, or by any means, electronic, mechanical, photocopying, recording or otherwise, without the prior written permission of the author.

Printed by CreateSpace, An Amazon.com Company

Address inquiries to the author at:

Greg Babinski
Orzel Books
201 Third Ave. N., Suite 316
Edmonds, WA 98020 USA
gbabinski@gmail.com

DEDICATION

This book is dedicated to the memory of the crew of the Malfunction and to all those who flew with them, supported them, and helped bring them back home.

TABLE OF CONTENTS

Maps, Chronology and Illustrations		VII
Preface		XV
Acknowledgments		XIX
Introduction		XXI
Chapter 1:	Beginnings: We Love This Splendid Land	1
Chapter 2:	The World in the 1920s and 1930s: Era of Isms	19
Chapter 3:	At Home 1939-1942: A Tremendous Shock	34
Chapter 4:	Induction and Basic Training 1943: Confident Faith	44
Chapter 5:	Airplane Mechanics School 1943: The Mop on the B-24	63
Chapter 6:	Aerial Gunnery Training 1943: Faith, Trust, and My Duty	99
Chapter 7:	Bomber Crew Phase Training 1944: Goodbye Golden Gate	123
Chapter 8:	Southwest Pacific 1941-1944: How Long and What Cost?	160
Chapter 9:	Over the Isles We Fly: Guadalcanal & the 13th Air Force	165
Chapter 10:	Los Negros: The 307th Bombardment Group (Heavy)	179
Chapter 11:	Wakde: Thanks for the Memories?	210
Chapter 12:	Noemfoor: The Damned 13th Air Force Strikes Again	222
Chapter 13:	Back Home 1943-1945: Beyond a Wide and Spacious Sea	251
Chapter 14:	Morotai: With Courage and Devotion to Duty	257
Chapter 15:	1945: The Old Malfunction and the Return Home	314
Chapter 16:	New Beginnings: Dreams of Life	333
Chapter 17:	Afterthought: Lines of Life	336
Appendix A:	The Ball Turret	343
Appendix B:	Selected Military Records	351
Bibliography		363
Notes		378

Chronology, Maps and Illustrations

S/Sgt. Walter A. Babinski Wartime Chronology — VIII

Maps:

    S/Sgt. Walter A. Babinski Travels Across the U.S. 1943-1945 — X

    S/Sgt. Walter A. Babinski Travels Across the Pacific 1944-1945 — XI

    S/Sgt. Walter A. Babinski in the Southwest Pacific 1944-1945 — XII

    The map 'Southeast Asia and Pacific Islands from the Indies and the Philippines to the Solomons' (Washington, D.C.: National Geographic Society, October 1944) is used under license from the National Geographic Society.

Illustrations:

    All photos and other images are from the Babinski/Kurzyniec Family collection, unless noted otherwise.

    Images from the 307th Bombardment Group (Heavy) Association and from the Missing Aircrew Project are used with permission. Other sources of images include Wikipedia Commons (https://commons.wikimedia.org), the Dr. Francis E. Fronczak Digital Collection at Buffalo State University (http://library.buffalostate.edu/archives/francis-fronczak), the Bretherton Family, Cliff Llewellyn, Biloxi Public Library, the St. Petersburg Historical Society, Willard Public Library, and the author's private collection.

Title page B-24 sketch by S/Sgt Walter Babinski

Cover photos from Babinski Family collections

Beginnings, Early Years, and Basic Training — 21-30

Keesler Air Force Base and Aircraft Mechanics Training — 71-80

Laredo Air Force Base and Aerial Gunnery Training — 111-118

Muroc Air Force Base and Bomber Crew Phase Training — 139-148

Island Hoping Across the Southwest Pacific — 183-194

S/Sgt. Robert A. Laessig Paintings — 231-240

    The watercolor paintings by S/Sgt Robert A. Laessig are used courtesy of the United States Air Force Art Collection. Laessig was an armorer in the Thirteenth Air Force who had an aptitude for art. He was commissioned to document life in the Thirteenth Air Force for the book 'From Fiji Through the Philippines with the 13th Air Force.' More than 80 paintings were produced for this project. Laessig painted from real life or from photos after consulting with eyewitnesses. It was intended that the paintings would all be printed in color, but this was found to be too expensive for a book that would be within the means of most veterans. Only 16 of the photos were printed in color, the others being reproduced in sepia tone. The 20 Laessig paintings reproduced here have never before been published in color. They can be viewed on my companion Map Story website, *The Flight of the Malfunction* (see: http://arcg.is/2mWOzse). They provide an enhanced sense of the colors and ambience experienced by the veterans of the Jungle Air Force that black and white photos and written descriptions cannot convey.

    Morotai — 265-278

    Lines of Life — 325-332

    The Ball Turret — 347-350

1943 - January / February

1	2	3	4	5	6	7	8	9	10	11	12	13	14	15	16	17	18	19	20	21	22	23	24	25	26	27	28	29	30	31	1	2	3	4	5	6	7	8	9	10	11	12	13	14	15	16	17
F	Sa	Su	M	Tu	W	Th	F	Sa	Su	M	Tu	W	Th	F	Sa	Su	M	Tu	W	Th	F	Sa	Su	M	Tu	W	Th	F	Sa	Su	M	Tu	W	Th	F	Sa	Su	M	Tu	W	Th	F	Sa	Su	M	Tu

Segments: a | Active Duty Leave | b | Processing | rail 2 | c-1 | d | c-2 | d | (e) | d | f | Training | f-1

- Detroit >>> Rail 1 to Battle Creek >>> St. Petersburg, Fla. >>> Rail 3 to Clearwater Fla. >>>
- (a) Inducted into US Army
- (b) Arrived at Camp Custer
- (c-1) Arrived at St. Petersburg w/ weekend leave
- (c-2) Arrived @ St. Petersburg
- (d) At St. Petersburg
- (e) Assigned to US AAF

1943 - April / May

Training | (i) | rail 6 | j | Training | (k)

- >>> at Harvard AFB, Neb. >>> Keesler AFB, Miss. >>>
- (i) Promoted to PFC
- (j) Arrived via St. Louis & Montgomery, Ala. at Keesler Field, Biloxi, Miss.
- (k) Started Airplane Mechanics Course
- Assumed am class & days: 1 1 1 1   1 1 1 1 2 2   2 2

1943 - July / August

Training - Airplane Mechanics Course

- >>> at Keesler AFB, Biloxi, Miss. >>>
- WABBD
- 7 7 7   7 7 7 7 8 8   8 8 8 8 8 8   9 9 9 9 9   10 10 10 10 10 10   10 10 10 10 10 10   11 11 11 11 11 11   11 11

1943 - October / November

Training | rail 7 | m | m1 | Training | (n) | Training - Aerial Gunnery Course

- >>> at Keesler AFB, Biloxi, Miss. >>> Laredo AFB, Texas >>>
- (m) stop enroute in Houston, Tex.
- (m1) Arrived at Laredo AFB, Texas
- (n) Started Aerial Gunnery Course

1944 - January / February

rail 11 | (s) | Training | r12 | (t) | u | u1 | u2 | Overseas training course | u3 | u4 | u5 | v | u6

- Hammer AFB, Cal. >>>   Muroc AFB >>>   L.A.   Muroc AFB >>>
- (s) Arrived at Hammer AFB, Fresno Cal.
- (t) Arrive at Muroc AFB, Cal.
- (u) Began Overseas Training Course ('u' subscripts indicate flights)
- r12 By train to Muroc AFB
- (v) @ Los Angeles on Furlough

1944 - April / May   4/9 = Easter

O/S Prep | y | z | a | TH | b | c | Waiting for crew | c1 | XIII Bomber Command Training | d | XIII BC Training | d

- F-S AFB Hamilton AFB Guadalcanal >>>
- (y) b14 bus back to Hamilton AFB
- (z) 16-Leave Hamilton AFB via ATC for Overseas
- (a) Arrived at Hickam AFB, TH
- (b) Arrived at Canton Is.
- (c) Arrived at Guadalcanal via Nandi on Fiji
- (c1) 8 members of crew 430 arrive @ G'canal
- (d) Training Flights

1944 - July / August

e13 | Combat | e14 | Combat >>> | e15 | Combat >>> | e16 | e17

- Los Negros, Mokerang Strip >>>
- WABBD
- (e13) Yap (e14) Yap (e15) Yap (e16) Yap (e17) Yap

1944 - October / November

e27 | Combat >>> | e28 | e29 | e30 | e31 | e32 | e33 | e34 | e35 | e36

- Noemfoor >>>   Morotai >>>
- (e27) Balikpapan TB
- (e28) Balikpapan
- (e29) Balikpapan
- (e30) Shipping Prowl-Celebes
- (e31) ? Noemfoor-Morotai & Ret?
- (e32) Shipping Prowl-Mindanao Sea
- (e33) Los Negros - Alicante AS
- (e34) Fly to Morotai (new base)
- (e35) Los Negros - Fa
- (e36) Los Ne

1945 - January / February

e54 | e55 | e56 | e57

- Morotai >>>
- (e54) Luzon - Grace Park AS
- (e55) Luzon - Nielson Field
- (e56) Los Negros - Fabrica AS
- (e57) ??

1945 - April / May   4/1 = Easter

L7-USS Puebl | n | 18 rail | o | p | Furlough in Detroit | 19 rail | q | r | process

- San Francisco   Detroit >>>   Santa Ana AFB, Cal. >>>
- (n) Arrive at San Francisco (Angel Island)
- (o) Arrive at Fort Sheridan, Chicago, Ill.
- (p) Arrived at Detroit on 21 day furlough
- (q) Arrived at Santa AFB, Calif.
- (r) WAB applies for and receives c

VIII

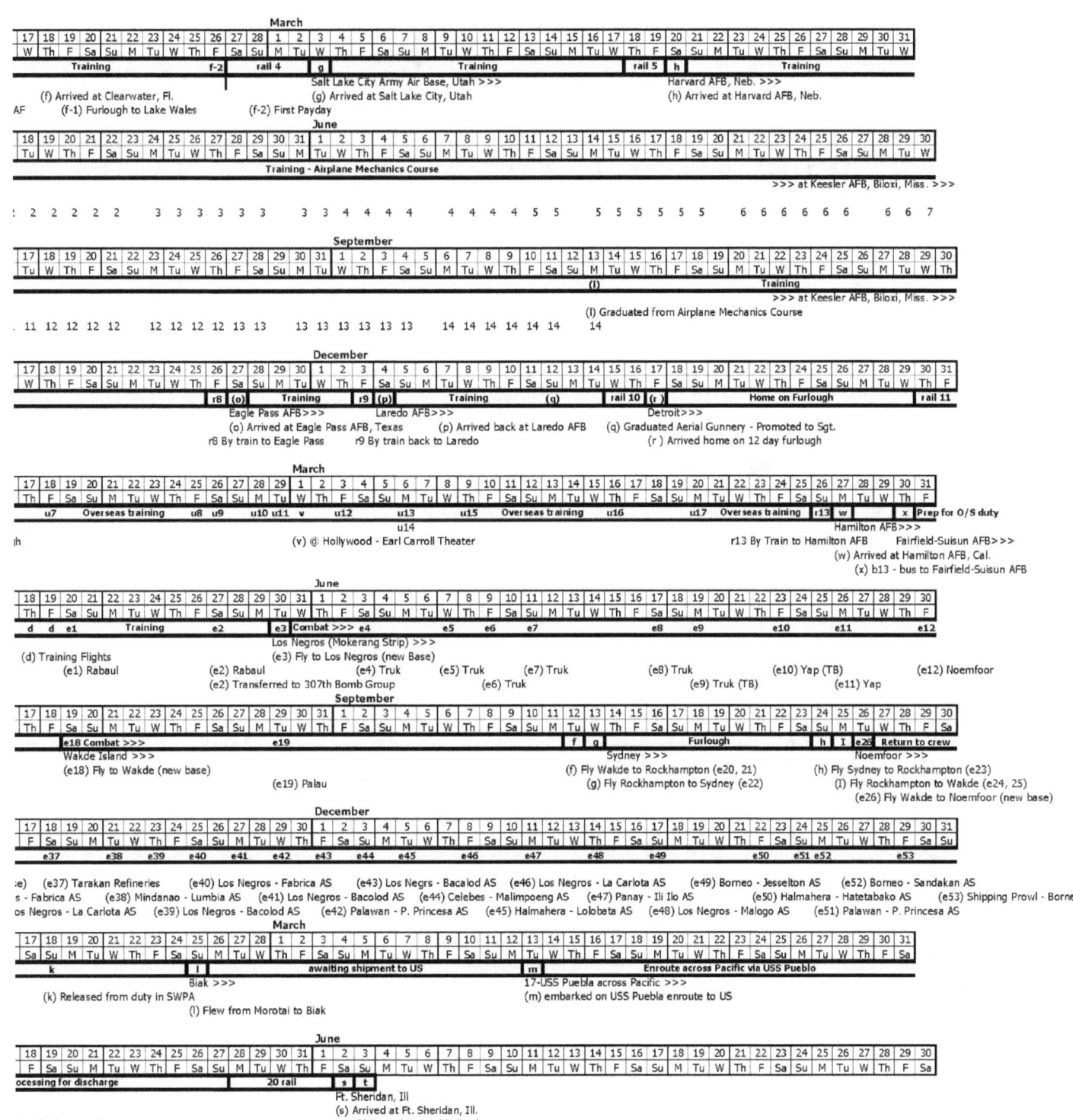

Flight of the Malfunction
Walter Babinski Chronology 1943-1945

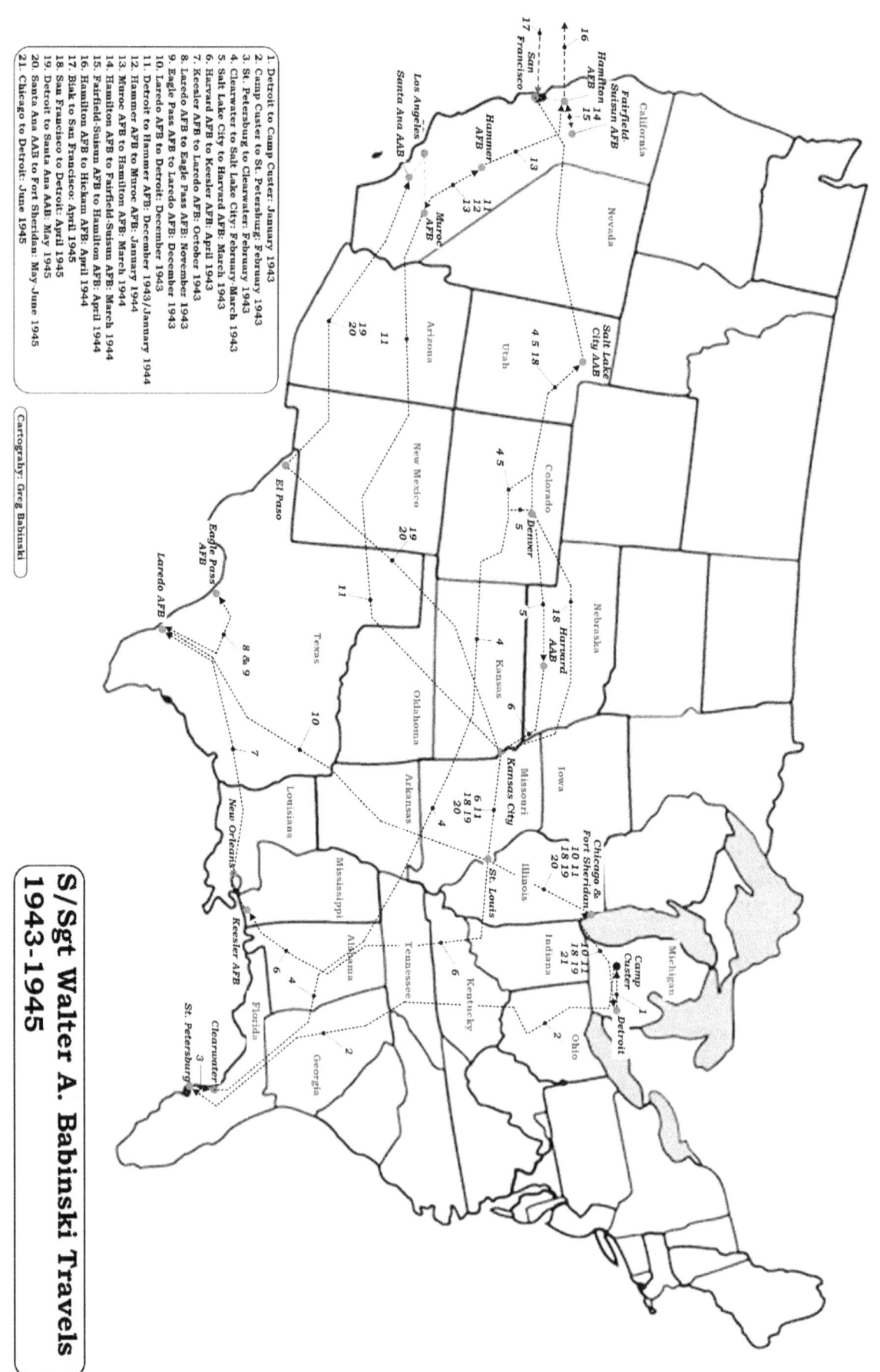

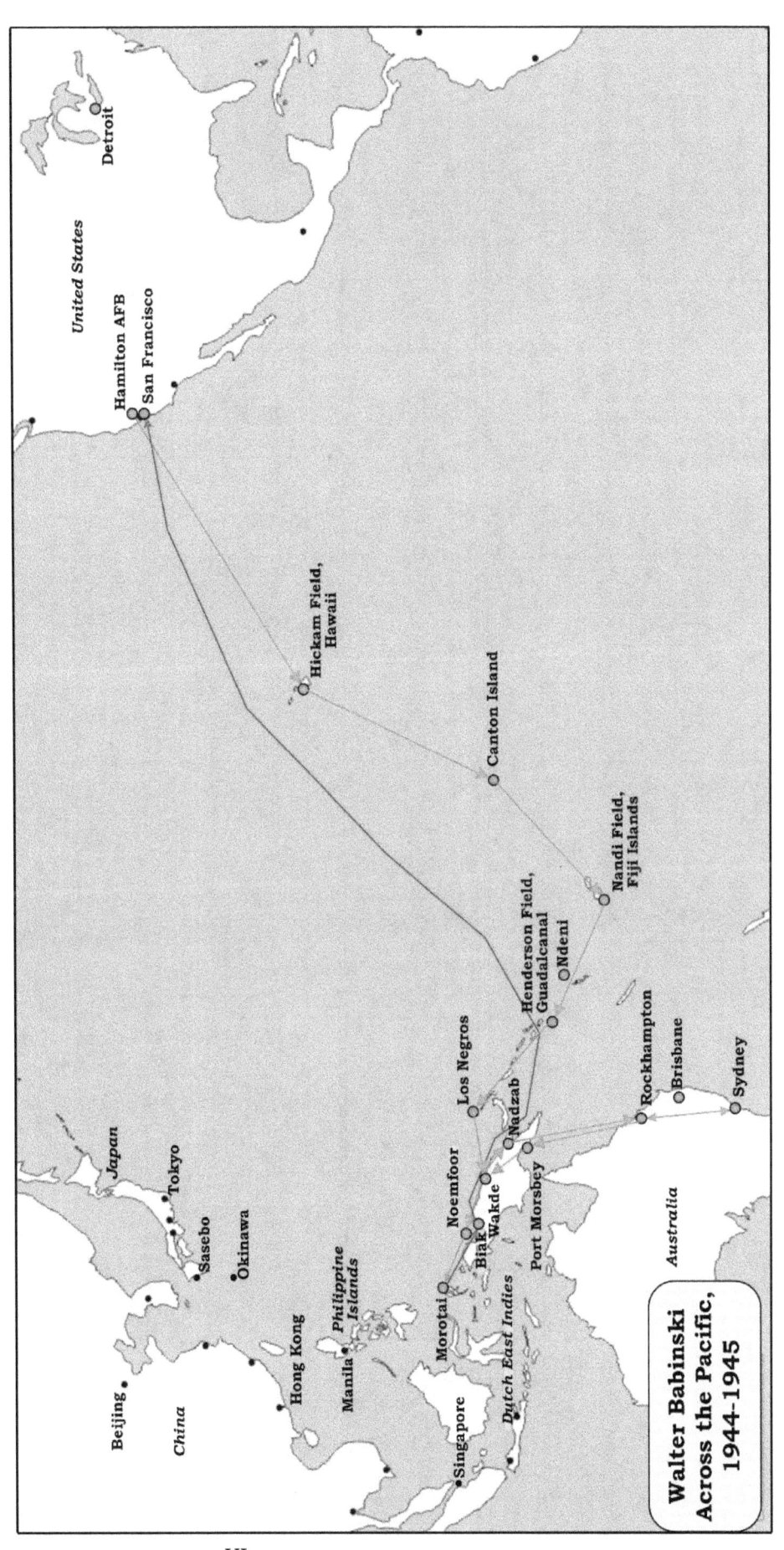

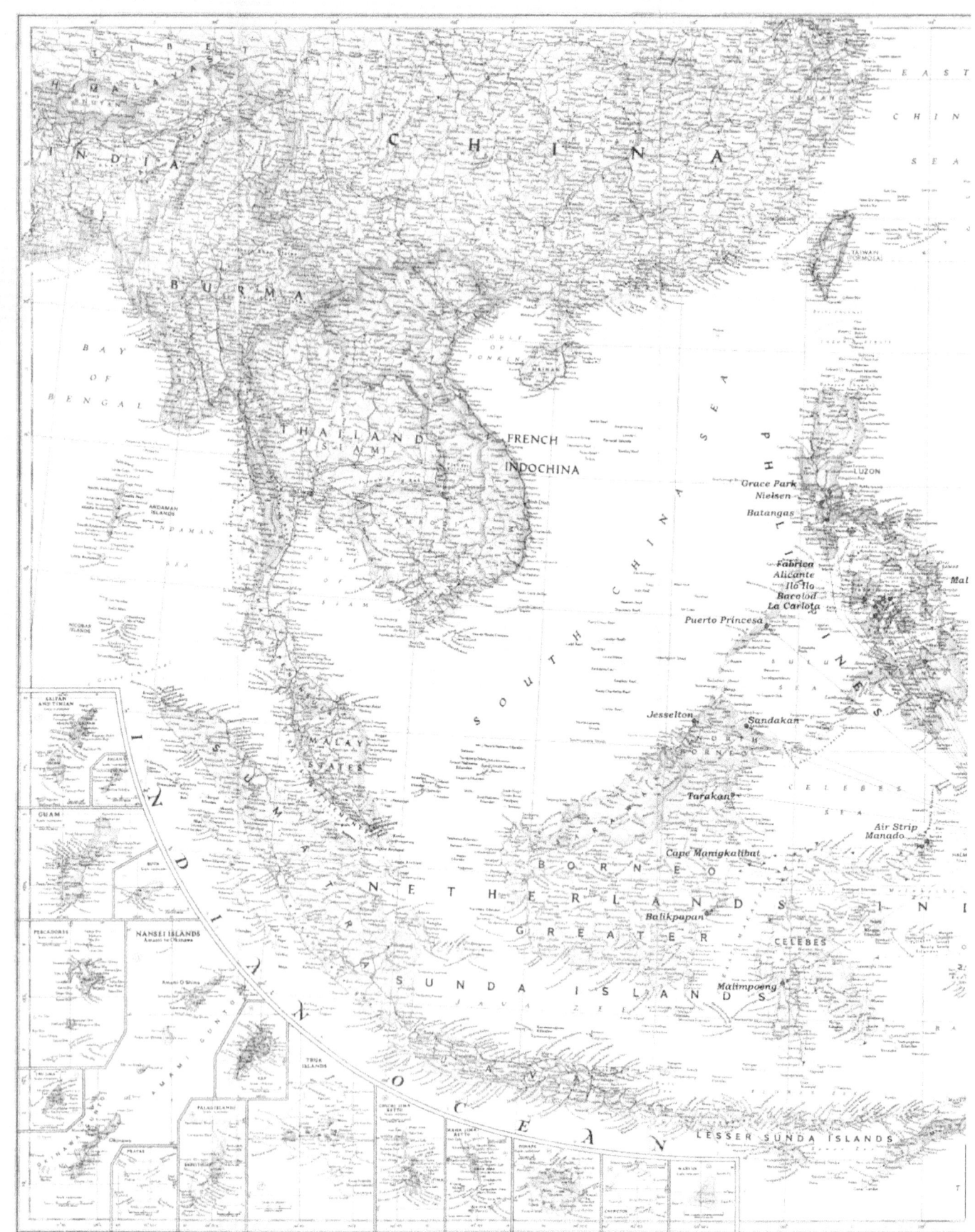

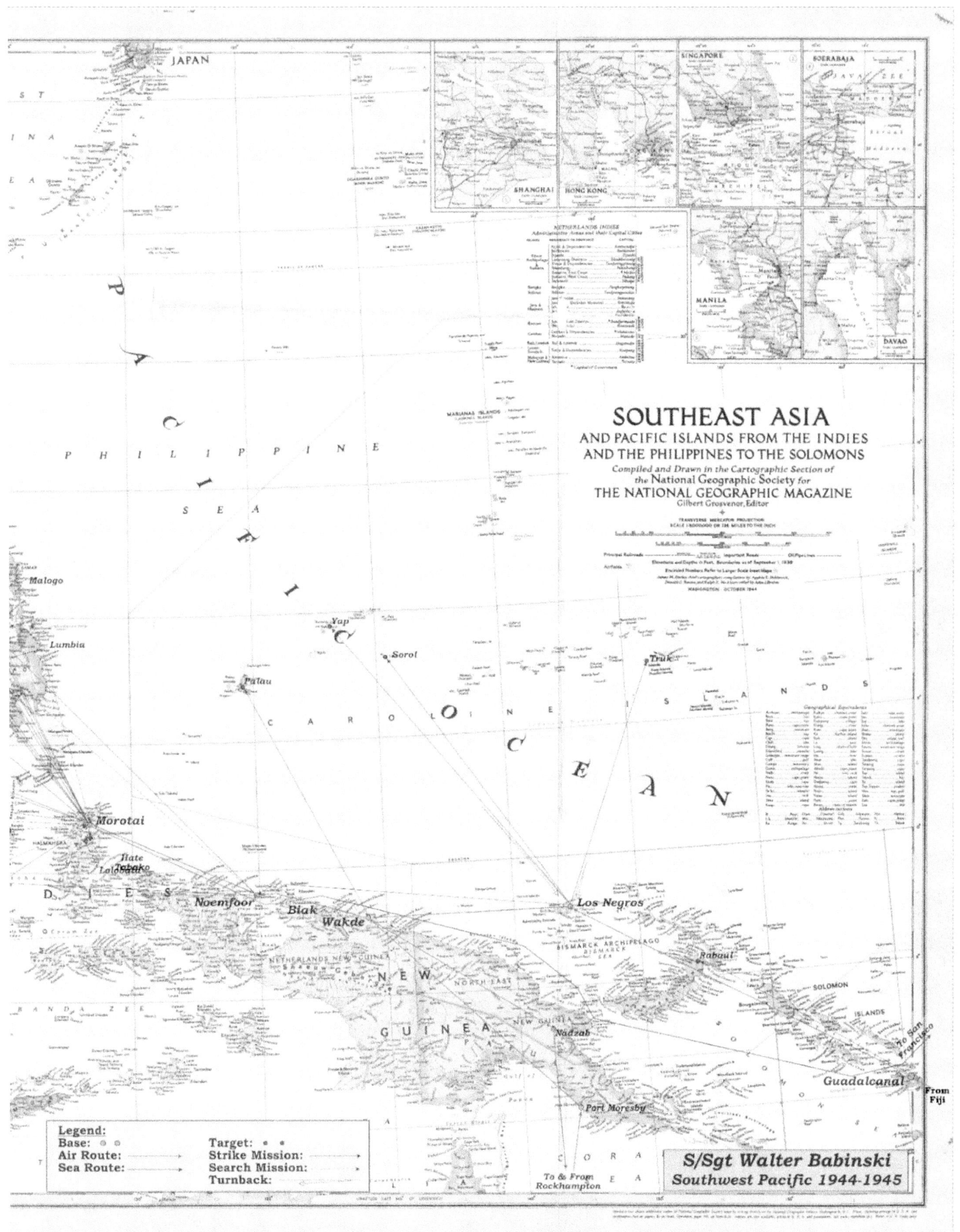

PREFACE

Pulaski Street in the Delray neighborhood of Detroit, where three generations of Babinskis and Kurzyniecs called home is gone, except for the boarded up and forlorn shell of what used to be St. John Cantius Church. Many families launched the lines of their lives along that quiet tree-lined street. However, memories fade fast. Soon the witnesses to that place and time will all be gone.

The inspiration for this book came from several sources and evolved over many years. About three or four years after my father, Walter A. Babinski, died in 1992 my mother asked me to look through and take some of the military records and photos that he kept after his years in the service. As a child, I had seen some of this material, stored in a box in the attic of our family home in Dearborn, Michigan. There were yellowed military orders and records, many small black and white photos, tattered magazine articles, mysterious maps printed on silk, a set of dog tags, old post cards – some blank, but many with brief messages addressed to names I did not recognize, a moth eaten felt pennant with military ribbons pinned to it, a couple of books about something called the 'Thirteenth Army Air Force,' and a handful of metal 'wings.'

My brothers and I would play 'war' with various items that our dad gave to us: canvas webbing belts and metal canteens. Dad gave my sister Christine a penny from Australia. My father made no secret of his service in the Air Force during World War Two – but neither did he brag or even talk about it much. I remember just a handful of vague stories or incidents related to his military service that he recounted. I knew he was a ball turret gunner and that he served in a B-24 bomber in the South Pacific. I remember that once he gave me a plastic model kit of a B-24 bomber. After I finished assembling the model, he talked about what it was like flying in a B-24 bomber across thousands of miles of ocean. Once, when I was about eight or nine years old, my mom drove me to an airshow at Willow Run outside of Detroit. I remember it was a great day with just mom and me, and I knew that my dad had talked about the significance of Willow Run and the role it played in the War. Other times Dad talked of jungles and humid heat that would rot leather boots in a matter of days. He talked of crewmates who were his friends, and of danger, and of crews that did not return. Most of what he told me I now know I have forgotten.

I am sure I was interested in his stories, but as a young boy in a large and growing family, there were many distractions. Dad worked hard alongside my mother – making a living and raising a family, so there was not a lot of time for quiet reflection and reminiscences. He lived for today with an eye to the future welfare of my mom and our family.

For much of my youth I was in awe of my dad. He always had an air of self-confidence and optimism that things would work out and that life's problems could be solved – not by providence or good luck, but by hard work and 'Yankee-ingenuity'. It seemed to me later in life that this self-confidence was a requirement to raise a family of ten children with the degree of joy, humor, and love that mom and dad always showed us kids.

Mom and dad reminisced often about their childhood and early years in the Delray neighborhood. Mom told stories about working at Fort Wayne during World War Two, about the Kurzyniec Family store across from St. John Cantius, collecting coal during the Great Depression along the tracks that ran nearby, and about trips to the Skrabut farm 'up-north' in Standish during the summer.

Dad told stories about playing baseball near a freight yard with right-field stretching across the tracks, riding Wabash and Pere Marquette Railroad car ferries across the Detroit River, and selling newspapers on East Jefferson Avenue. Sometimes he told stories about big bombers flying vast distances over the Pacific. But because Fort Wayne, St. John Cantius, baseball, Delray, and selling newspapers were things I could see myself, they made a more concrete impression in my mind.

As he and I grew older, my subconscious sense that he would be there if I needed to talk to him was a given. I moved out of the family home in Dearborn during college, then away from Detroit to the San Francisco Bay area, and soon I had a family of my own. Trips to Detroit to visit the family home, or his trips with my mom to California were a joy. Playing tour guide when mom and dad would visit California provided a curious environment where we could talk freely and there were many distractions to keep things from getting stale. We crisscrossed San Francisco, Marin County, the Peninsula, and down to Cupertino, Monterey, and Carmel, as well as east to Yosemite.

During visits to the Bay Area I do remember dad would sometime point out places that he had visited while in the Air Force during World War Two: Chinatown in San Francisco, Angel Island in the Bay, and a stretch of city street in Sausalito he remembered from when he had traveled by bus into San

Francisco on leave. Once we drove with mom and dad to a Scottish Highland Games in Santa Rosa up in Sonoma County, along with our Scottie dog, Mrs. McGillicudie. I remember dad remarking on the signs in northern Marin County for Hamilton Air Force Base, then closed down, but remembered as a location he had passed through on his military journeys.

On another occasion when my oldest son Danny was about 2 years old we drove to Sacramento to visit the California Rail Road Museum. En-route, driving east on I-80 past Fairfield and Travis Air Force Base, dad mentioned that he had passed through Travis on his way to the South Pacific. During the Rail Museum visit, dad also reminisced about his many rail journeys back and forth across the U.S. while in the Air Force.

One Bay area landmark that mom, but especially dad focused on was the Golden Gate Bridge. He talked about crossing the bridge en-route to San Francisco while on leave, but it seemed to fascinate him beyond just a structure to cross the Bay decades before to raise some Cain in 'Frisco.' I may have thought then that his interest in the bridge was only as an outstanding engineering structure, or for its esthetic qualities, or perhaps as a symbol of America building its way out of the Great Depression. I now believe I know the special symbolic meaning it had for him then.

In 1991, my family and I attended my brother Walter's wedding at Big Bay, in the Upper Peninsula of Michigan. The weekend of the wedding and the day or two afterwards was the last 'normal' family event that I experienced with my dad. His older brother John (my uncle 'Bob') was there also.

Within months, John had died and it was apparent that dad was very ill. His health declined rapidly, and in January 1992 I made a rushed visit back to Michigan. Seeing him in Oakwood Hospital in Dearborn, it was evident that he was wasting away, but by his demeanor it was clear that he was fighting hard and trying to enjoy every remaining minute. After a few days, I had to return to California. I said my good-byes knowing I would never see him again. On May 1, he died, in the home on Anthony Street in Dearborn in which mom and dad had raised a family of 10. It seemed like just a few years before that I was confident he would always be there to answer the questions I was too busy to ask, and then he was gone.

A year later in 1993, having learned about the U.S. Military Personnel Records Center in St. Louis and motivated by curiosity, I wrote requesting copies of all the military records they had related to my dad's service career. A few months later I received a bulging envelop from the Center, but my initial optimism at what I might find inside faded when I looked through dozens of copies of burned documents. I could deduce that these were the partial remnants of military medical, pay, rank, and service records, as well as fragmentary copies of orders and service awards. The cover sheet from the Records Center included with the copies indicated that the files had burned in a fire at the Center in 1973. The charred copies I looked thru reminded me of pictures I had seen of the Dead Sea scrolls when they were first discovered. Going thru that envelop I remember two distinct emotions: first, anger that the government which had asked millions of men (and women) to risk their lives in World War Two was so careless with the records of their individual contributions, and second, real sadness that an opportunity to learn something new about my dad – a chance to look back into the past, even to the time before my birth – was seemingly lost.

Then in 1995, during a stop-over in Detroit while traveling to a conference in Baltimore, mom asked me to look through the box of my dad's old military records and photos that she had stored in the basement of her apartment building in Dearborn and to take what I was interested in. The same curiosity that led me to contact the U.S. Military Personnel Records Center reasserted itself and I took about two-thirds of the material in the box.

Back home in California, when I looked through this material, I discovered hints of events, locations, people, and activities that my dad had experienced. Names like Aubrey, Muroc, Riley, Tarakan, Hidalgo, Los Negros, Keesler, and many others appeared and reappeared. An old railroad map of the U.S., showing routes darkened over with pencil and all leading to or from Detroit, hinted that there might be geographic and chronological patterns that could be discovered. But curiosity itself does not necessarily provide the inspiration to progress onward in a quest for knowledge.

About a year after that stopover, I received a box of old family photos from my mom in the mail. Included in the box was an original 8 by 10 black and white photo showing the nose of what I recognized as a B-24 bomber in a tropical setting, with ten confident looking crewmen posed in front. One image in the back row I recognized as my dad at about 22 years of age. On the back of the photo were the signatures of the ten men and a key identifying each man depicted. Interestingly, none of the men signed the photo with his rank, just their names: Jack B. Riley, Jupe Bretherton, Carl D. Appling,

Donald W. Aubrey, Ed Dunne, my dad, and four others. I had the photo framed and mounted on a wall with other family photos.

In April 2001, my curiosity was rekindled by a school project assigned to my son Micah. He had to research and report on some aspect of his family's past related to world history. He chose to write about World War Two air combat in the Pacific and his grandfather Walt's experience in the Air Force. I pulled out the box of dad's old military records, maps, and photos and paged through the items one by one with Micah, trying to explain what they might mean. We selected about 20 or 30 items, which I agreed to Xerox, so that he could use them for research and to try to describe what they told him about the grandfather that he just barely remembered. To try to create a historical context for these scraps though, we needed to do some outside research.

I knew that dad served with the 13th Army Air Force in the South Pacific. I remembered a history book of the 13th Air Force, illustrated with watercolor painting, but it had not been included with the material that my mom had showed me back in Michigan. It would have been the perfect background information resource for his assignment. Micah and I did have Internet access though, and I turned his assignment into a research learning experience for him (and for me as it turned out). After just a little coaching on how to search Yahoo, Lycos, and Google, I left Micah to his own devices.

I was surprised when in just a few hours Micah had found two 13th Air Force veterans' groups with web sites. Within a day, he was corresponding with a World War Two 13th Air Force vet, Mr. Mauro J. Messina, who provided Micah with firsthand information about the 13th Air Force, as well as his personal opinions about how the accomplishments of the 13th had been 'covered-up' by the government after the war for 'political' reasons. Micah's assignment was quickly and successfully completed and he moved on to other interests.

Shortly after his assignment was completed, I started browsing through the two 13th Air Force veterans' web sites, and I noticed that one included an on-line membership list. Curiosity reasserted itself, and I took the B-24 bomber crew photo down off the wall and compared the names on the back with the veterans' group membership list. I was disappointed that none of the names corresponded to my dad's old bomber crew. The photo went back on the wall and my curiosity faded.

Then in early autumn 2001, my cousin John Besek in Michigan sent me an interesting email. He had been thinking about how he had heard bits and pieces of my dad's 'war stories' from various past family gatherings. John had recently graduated from the University of Michigan and was starting a career teaching at a Catholic School in suburban Detroit. To try to make world history more immediate for his students, he had tried recounting some of the stories he recalled from my dad, but soon realized how little he remembered and how difficult it was to make them meaningful without knowing the wider historical context.

John had heard from my mom that I had the box of photos and military records, and he asked if I could send copies to help him. I told him about the copies I had made for Micah, which I still had, and we agreed that they would meet his need. As I mailed the copies away to my cousin, I reflected on them as relics - relics which helped illuminate human experiences that could be thought of as having some minor significance for a major historical event of the Twentieth Century, but which must also have had major significance, both for the father I had known for 43 years and for the other nine men whose faces stared at me from that picture on my wall back home.

Early the following year, 2002, I began reflecting on the approach of the tenth anniversary of my dad's death. I became motivated at least to organize the records and photos I had. I began looking at the material from multiple dimensions: geographical, chronological, events, activities, and people. It dawned on me that these relics represented about two and a half years of my dad's life. He was ten years gone, but if I could really learn something about him from these paper items, perhaps I could add some new understanding to the 43 years that he and I shared together.

While I made preliminary efforts to organize the material, I remained frustrated about the lack of historical context. This frustration led me back to the Internet to do additional research. The 13th Air Force veteran's web sites Micah had found seemed to offer no real help. I found another web site that provided a number of suggestions for researching 'your dad's time in the military.' Suggestions included library research, contacting veterans' groups, and the Military Personnel Records Center in St. Louis (with the caution about the large number of individual records that had been destroyed or damaged in the 1973 fire).

Then I found a web site devoted to veterans who had served in B-24 bombers during World War Two. A feature of the web site was a 'discussion page' where people could post public questions. With nothing to lose, I posted a message requesting information anyone might have about my dad, or about any of the other men I knew he served with from the photo. I also added other information I had learned from studying my dad's records, including bases that he was stationed at, the targets of missions he had flown on, and the specific 13th Army Air Force unit he had been assigned to: the 307th Bombardment Group (Heavy), 424th Squadron. I hoped this additional detail might help in my search. Within just a couple of days I received a number of sympathetic replies, plus a suggestion to contact the 307th Bombardment Group (Heavy) Association. In all my prior research, it had never dawned on me that there might be a veteran's groups for just a portion of the 13th Army Air Force.

I quickly sent letters to James Kendall, 307th Bomb Group Association Historian, and Cena Marsh, 307th Association Secretary, inquiring if they could help with my search for information. Both of these kind people suggested that I join the Association, pointing out that without the sons, daughters, wives, and grandchildren of the 307th veterans, their membership would only decline in numbers. After I submitted my membership, the association sent me three of their very informative bi-annual reunion books plus a membership roster. The reunion books provided a wealth of information, both about life in the 307th during World War Two and about the surviving vets who attended the reunions. Here were photos of hundreds of men who my dad had likely worked with, fought with, played with, and lived with. I saw dozens of B-24 crew photos similar to the photo of S/Sgt. Babinski and his nine crewmates that I had on my wall. Others photos and articles described life on the tiny island bases that served as home for these men in 1943, 44, and 45.

I wondered if I could find out more. I compared the names of the nine B-24 crewmen pictured with my dad, with the 307th membership roster, but I found no matches. I then scanned through the other military records and photos and found more names than just the nine. There were names of pilots for each mission my dad flew on, names of some other crewmen pictured along with my dad during training, names of still other crewmen on various Army Air Force records. At last I was able to find the names of five men my dad had served with who appeared to be 307th Association members.

But then what? I could try to contact these men – but should I? Would they remember Wally Babinski after 57 years? Were their memories of my dad something I wanted to know? Where their memories of that time too dim to recall or too painful to share? Then it occurred to me that while I wanted information, or memories, or just a connection to a time I did not know, there was something I felt I owed these men as well. As combat comrades of my dad, I knew they had faced danger and possible death together. My dad survived 42 combat missions. That was 9/10th due to the jobs they had done together as a crew. I felt I had to thank them for that. In addition, they had also played a tiny part together to help defeat evil when their country called. They had helped keep the world a relatively peaceful place for more than fifty years. I knew I needed to thank them for that as well.

I drafted a simple one-page letter that introduced myself, told them about my dad's later life, family, and death in 1992, requested any recollections or memories they could share (but told them I understood if they cared not to reply), and stated my thanks to them. It was with some reservations that I mailed these first five letters in April 2002. About one week later, on a pleasant Sunday afternoon, I was mowing the lawn when my son Micah called me into the house for a phone call. The gentleman on the line introduced himself as Boris Hidalgo, one of the five I had written to. Mr. Hidalgo proceeded to talk to me for about an hour about my dad, about his life, and about how he thought many times over the past 57 years about the whereabouts of many of his old comrades. He provided suggestions on others to contact from the 307th who knew my dad. He also told me he had already contacted some of these men about Wally Babinski, and his son, Greg.

Boris Hidalgo's phone call that Sunday convinced me that it would be possible to find out more about these 2-1/2 years of my dad's life, plus the lives of some of the men he shared that time with. I knew then that because I could discover more that I had to make the effort to learn all I could before it was too late. I also felt that because I could find out more that I had an obligation to all who had inspired or helped me. That obligation is to preserve and make sense of all the memories and information I have collected. I have since met many of the vets my dad served with. We have swapped stories and pictures, shared beers, shed tears, and I believe we have become friends. To a very large degree, this story is about them too, and I believe that is what my dad would want.

I was never disinterested in what my dad had experienced during World War Two or at any time in his life really. I just felt that he would always be there to answer my questions, tell his stories, and thereby enrich my life. This book is my feeble effort to try to understand what I should have asked about and share it with you.

ACKNOWLEDGEMENTS

The inspiration for this work originated with the curiosity and the questions of my son, Micah, and of my cousin, John Besek. Many family members cultivated this inspiration. First and foremost of these was my mother, Mrs. Dorothy F. Babinski. For six and a half decades now she has provided her love, support and comfort. If she had not thought this work was worthwhile, it would not have been done. Her memories and other help along the way made the work easier and kept me focused on the last chapter.

Several relatives provided direct help with information and personal reminiscences. My sisters Christine Babinski and Mary McDonald told me what they remembered. My uncle Henry Babinski and Aunt Genevieve Mendryga provided insights into early life in the Babinski family. Additional information came from my cousins Edward Mendryga, Diane Smith Poirier, Karen Smith, and James Hefferan. Other siblings provided their support, including brothers David, Mark, Matthew, and Walter, and sister Susan Babinski. For my other siblings, Cathy and Gerard, all I can say is that Dad loved you both so very much.

For my sons Daniel and Micah, thank you for your support and I hope we did not miss too much time together because of this work. To a great degree, this book is intended for you. I wish that I knew more of your grandfather Dave Wolin's story too.

For information about the 13th Jungle Air Force, I appreciate information and encouragement from Mauro J. Messina, Editor and Historian of the 13th Air Force Veterans Association, as well as Harlan Price and Bess Wening. I garnered useful information from people on the B-24 Web Board and from the USAAF Web Forum, including Steve Burris, Sanders B. Walker, Harold Bolce, William (Bill) E. Correll, Bob Black, and M/Sgt, USAF (Ret) James S. Peters, Sr.

For more than 40 years now, the members and leadership of the 307th Bombardment Group Association have kept the memory of the Long Rangers alive and provided a place where we can show our appreciation for the portion of their lives that they devoted to us. In particular, Cena Marsh, 307th BG Secretary gave me my first encouragement. James Kendal, 307th BG Historian and his wife Dottie welcomed me into their Birmingham, Michigan home. Jim McCabe, current Historian has also helped search the archives whenever I asked. Jim Walsh, current 307th BG President, and his wife Greta have become friends. E. Stanley Batten, Lewis Smith, and Ann Quallia are other 307th BG Association members who provided aid and advice along the way.

I feel privileged to have met both Dr. Pat Scannon of the Bent Prop Project and Mr. Pat Ranfranz of the Missing Aircrew Project. I think that we share the same general goals as we pursue our own unique objectives. I am happy to have aided Laura Hildenbrand and Wil S. Hylton on their projects related to the 307th Bomb Group. I benefited from the information and insights that they have contributed to our common understanding of the saga of the Long Rangers.

I am proud of the achievement of two young people who contributed to this work and to preserving our common understanding of the Long Ranger saga. Ms. Emily Mount, a recent graduate of the University of Washington in 2006 with an interest in military history, and Mr. Derek Poppe, the son of a friend who need an Eagle Scout project, conducted and videotaped veteran interviews at the 2006 Reunion of the 307th Bomb Group. Assisted by two Air Force ROTC Cadets from the University of Washington, they interviewed 10 of the 307th Bomb Group vets: S.J. 'Ike' Ayala, Harry E. Coggins, Wayne Cooper, William D. Holston, Edward A. Jurkens, William Manley, Glenn Norwood, Thomas W. Pelle, Robert Robinson, and James V. Walsh. Later Derek coordinated the careful transcription of each videotaped interview by volunteers from Mercer Island (Washington) High School. Their work preserving the story of these 10 vets now resides in the Library of Congress Veterans Oral History Program in Washington, DC.

I am so very grateful for the information provided to me by the following veterans or their relatives who knew my dad in the 307th Bomb Group: Donald W. Aubrey, Dennis and Bill Bretherton (sons of Ed Bretherton), Charles Dowdy, Edward T. Dunne, William B. Fawcett, James Fielding (and his wife Nancy), Alan C. Guild, Boris V. Hidalgo, Clifford Llewellyn (and his wife Betty), Mrs. Virgie Mansir (wife of Frank Mansir), Harold Mitchell, William R. Pruett, Richard J. Reis (and his wife Mina and daughter Kitty Reis), Mrs. Nancy Riley and Chris Riley (Wife and son of Jack Riley), Raymond (Gene) Schreiner and his son Roger Schreiner, and John Vanderpoel.

Others who provided assistance include Donald R. Aubrey of Norwich, CT. Thank you for helping me track down your namesake in Florida. Thomas S. Berkey, Colonel, USAF (Ret) from the USAF Historical Office provided a treasure trove of annotated 424th Bomb Squadron Mission Reports. Aaron Bittner helped me acquire a Keesler Air Force Base B-24 mechanic's training manual and other student material. Dr. Raymond Puffer, Historian at Edwards Air Force Base, California provided valuable information and a video about Muroc Air Force Base before and during WWII. Other USAF historians who responded to my questions and provide useful material were Don May, Historian, Keesler Air Force Base, Biloxi, Mississippi and Dr. Gary Leiser, Director, Travis Air Museum, California.

The Babinski and Kurzyniec family photo collections include many images that were a mystery to me. I appreciate many patient hours looking at some of these photos with my mother, and with Cliff Llewellyn, Gene Schreiner, Dick Reis, Jim Fielding, and John Vanderpoel. They helped me to solves some of the mysteries and identify people, places, and events.

I am particularly grateful to U.S. Senator Patty Murray and her staff who helped me connect (after many failed attempts) with the USAF Art Collection. Russel D. Kirk in the Directorate of Operations at the Pentagon helped me to obtain permission to reproduce never-before published color images of some of the original watercolor paintings by S/Sgt Robert A. Laessig, which are held in the USAF Art Collection. Many librarians and other researchers helped, including: Ellen J. Babb, Historian of the Pinellas County Heritage Village, Michael Daisy and George Ellenwood from the Detroit Water and Sewerage Department Public Affairs Office, Kathleen Leles DiGiovanni from the Oakland History Room at the Oakland (California) Public Library, Erik Larsen at the Detroit Public Library Map Room, and Joe Moreno of the Laredo (Texas) Public Library.

Tara Makin from Riverton Utah, who was researching her grandpa's service, made copies for me of pages from the "Gunnery Information File" that I was missing. Information about the history of St. John Cantius Church and School was provided by Roman Godzak, Archivist of the Archdiocese of Detroit. When I sought information about life in Biloxi during 1943, Murella H. Powell, Local History & Genealogy Librarian at the Biloxi Public Library not only provided copies of contemporary newspaper clippings, but also wrote a letter describing her own personal memories of life in Biloxi during that era as a young girl.

Information about other locations that would be stops along the way in Walt's story was provided to me by Eleanor Scharf, Reference Librarian at the Clearwater (Florida) Public Library System, Alice Hoeft, Library Assistant from the St. Petersburg (Florida) Public Library, Gladys Frekas, of the Costa Mesa (California) Historical Society, and Catherine Renschler, of the Adams County (Nebraska) Historical Society.

The U.S. Maritime Service Veterans Association in Berkeley, California provided me with information about troop ships, and Mr. Edwin Drechsel, ship historian, provided me with information about early 20th Century immigrant ships. Velma Bolden of the Samsonite Corporation, in Denver, Colorado provided information about the Shwayder Company in Detroit. Sarah Lange of the Downriver Genealogy Society in Wyandotte, Michigan, and Dennis Reigel of Orcas Island, Washington also assisted. The Collings Foundation work preserving the last flying B-24 provided me the opportunity to feel, hear, see, smell, and touch a Liberator bomber in flight above Mount Vernon, Washington. I need to see if I can get into that ball turret though. The staff of the Australian National Archives at the Australia War Museum in Canberra, Australia quickly certified me as an archival researcher and provided access to rare contemporary files about Morotai in 1944-1945. To my Indonesian guide Haarrees, thank you for getting me from Ternate, across Halmahera, and onto the Indonesian Airforce Base on Morotai.

I met two people in the Seattle area who have embarked on a journey of discovery similar to mine. Although their approach to discovering and documenting their relatives' stories is different from mine, I appreciate the time shared talking about our common quests with Daks Hanson and Chris Avenius. Very late in this project an old high school friend, Jim Gilleran, provided me useful information about his father's service in the 8th AAF during WWII. My work is much richer for the help and support of all the people listed above. However, I alone am responsible for any inaccuracies or omissions.

To my wife Audrey, I express my sincere thanks for ample time, support, and space to finish this work. I wish that your father and mine could have met to share a few beers with us at their sides.

Let us gratefully lift a glass and drink to the memory of all of the men who served together in the Thirteenth Jungle Air Force and in the Long Rangers during the Second World War.

INTRODUCTION

This work started out to be only a narrow history of my father's time in the U.S. Army Air Corps during World War Two. Because of a variety of limitations on the one hand and opportunities on the other hand, it has grown into something beyond that.

The limitations on what this work could be are related to the small amount of firsthand information that I had when I began and that I have been able to compile subsequently. Nevertheless, this firsthand information represents the foundation and framework on which the entire work is built.

To make this work as rich and meaningful as possible, not only for the reader, but for me as well, I have found it useful to organize all the first hand sources available on the basis of a variety of dimensions. For example, a single military record in my dad's collection may have a date, location, and an indication of some activity that it represented. Another example might be an interview with someone who knew my dad during his service years and who could provide information about some activity and other people involved, but with less certainty as to date and location.

All of the sources used have been organized on this multi-dimensional basis. These 'dimensions' include geographic (where an activity or event took place), chronologic (time and date), people (who was there), events (during training, or a mission, on leave, etc.), and activities (KP duty, or in the barracks, or cleaning guns, etc.).

My narrative is mostly built on a chronological framework. At various points though, I pause where it makes sense and expand on the other dimensions to enhance the picture of what Walt and other young men like him experienced during those years.

The opportunities I have pursued have focused on what I consider direct and indirect secondary sources. I am a direct secondary source, for there are a few events and stories that I remember my dad telling me related to his time in the service. Other direct secondary sources include my mother, sisters and brothers, as well as uncles and aunts. They have provided recollections about the time before, during, and after World War Two by a variety of letters or by conversations with me that I have tried to record in writing shortly after hearing them.

Other direct secondary sources resulted from my association with the 307th Bombardment Group (Heavy) Association. I have either met, corresponded with, or spoken on the phone to several of the men my dad served with in the South Pacific, and they have provided some valuable recollections of their experiences with him. Some of these men knew my dad, but many others did not, though they served at the same locations and same times as he did.

In one case, a narrative written by one of my dad's comrades (Clifford Llewellyn) provides a unique insight into the life of their crew during formation at Muroc Field in California, and later at several of the South Pacific island bases, and the return to the U. S. by ship. Cliff referred by name to my dad in his account of the World War Two years. I have also been able to contact other men or the families of men my dad served with via searches of various web sites and directories. Most of these have provided some additional direct secondary information.

Indirect secondary sources include a variety of books, articles, correspondence, and other material that describes the experiences of people who had a career or experiences in the military similar to my father's. Where I have been able to determine with reasonable certainty from these indirect secondary sources that a certain set of experiences was fairly typical of the Air Corps during World War Two, I have incorporated portions of them into the narrative. Many contemporary articles were intended to describe for the 'home-front' what service men were experiencing. Although some of these sources may have been glorified or 'sanitized' to meet the requirements of wartime propaganda (maintaining morale and fighting spirit) they do provide some useful descriptive information.

I have relied on two major sources to describe what Walt likely experienced as he travelled back and forth across the U.S. in 1943, 1944 and 1945. My dad had traced in pencil the routes that he had traveled during the war years on a large-scale railroad map of the United States. These routes correlated with the various bases he was assigned to, based on his official military records. First, I used copies of the *Official Guide of the Railways of the United States* to make educated guesses about the routes he took, the likely trains he rode, and his travel schedules for his crisscrossing of the country. Second, I used my almost complete collection of state guidebooks from the *American Guide*

Series. These were written between 1937 and 1941 by about 6,000 individual authors as part of the WPA Federal Writers Project. They provide roughly contemporary information about every city and most sizable towns in the country, including history, culture, economy, customs, tourist attractions, and maps.

There is also a body of books and articles written after the Second World War by some of the men who were in the Air Corps. Some are based on journals or correspondence written during the authors' service career (although journals or diaries were evidently forbidden in combat areas). Others are narratives, based on the authors' recollections of events during their service career. What these works lack in the rigid chronological accuracy of books based on journals or correspondence, they make up with a more impressionistic image of key events or defining experiences, based on years of reflection by the author.

My narrative methodology attempts to distinguish reliable known facts from reasonable conjecture. Known events or experiences are footnoted to indicate the source. References to items from Sgt. Babinski's military records or to other firsthand sources are made clearly. References to second hand sources are also clearly noted. In addition, my text will often describe events with terms like: 'it is likely that...' or 'Walt and his buddies must have...' or 'did they wonder what...' These phrases serve as textual indications that the actual experience is unknown, but that they are very likely accurate for many servicemen in similar situations. Again, my goal has been to provide a reasonable impression of what my dad and the millions of others like him experienced for their country, their families, and their comrades-in-arms.

This work started out to preserve the history of a small portion of my dad's life. To the degree that this work is an amalgam of the specifics of my father's military career, as well as the common experience of the men he served with and the families and friends at home that supported and prayed for them, it has grown into a tribute to them all.

- 1 -
BEGINNINGS: WE LOVE THIS SPLENDID LAND

Early Years

It is usually very hot and humid in southeastern Michigan in July, and that was likely the case when a midwife arrived at a little single story white frame house at 9050 Pulaski Street in the Delray neighborhood of Detroit on Wednesday, July 5, 1922. Waiting anxiously to meet the midwife was a wiry and gruff 38-year-old man along with his three children: a girl age 11, a boy age nine, and another girl age six. The man greeted the midwife in Polish, and then quickly led her back to a bedroom off the dining room, where she found a 22-year-old woman beginning to go into labor. At about the same time a few other neighbor women arrived, to help the midwife and also to help feed and care for the children. The bedroom window looked out over a narrow walkway between the adjacent houses, so it was likely open to allow in as much of the shady cool air as possible. The pregnant woman was young and strong though, and after much work and help from the midwife, she delivered her first child, a healthy baby boy. The following Sunday, at St. John Cantius Catholic Church, a few doors down Pulaski at Anson Street, the boy was christened by the pastor, Father Walczyk, as Walter Anthony Babinski.[1]

Young Walter was the son of Walter Babinski Senior and his wife Kunegunda. Walter Senior had been born as Wladyslaw or Ladislaus on March 20, 1884 in the little hill community of Roczyny, near Andrychow and Wadowice in the predominantly Polish province of Galicia, in the old Austro-Hungarian Empire. The Babinskis originated in the Beshkid Mountains of southern Poland, near 5,659-foot-high Babia Gora (Babiagorski Mountain). The name Babinski likely stems from the family roots near Babia Gora, although rural overpopulation in 19th century Poland had resulted in movement of many families and individuals away from their traditional homes.

Wadowice is located midway between Krakow and the Polish brewing center of Zywiec. Young Ladislaus worked as a farm hand, although it is not clear if this was on his own family's farm or as a hired hand on other farms in the region. The population of Galicia expanded rapidly in the 19th Century and many rural residents lived on tiny plots of land that were too small to support an entire family, so many young men had to hire themselves out to work as hands on the farms of more prosperous landowners.

What was Galicia anyway? The name originated in 1772 when Austria participated with Prussia and Russia in the first of the three partitions of Poland. Within 20 years, Poland was extinguished as a separate political entity, not to arise again as an independent nation until 1918. Austria needed a name for this artificial region, and a few years earlier the Spanish province Galicia had ceased to be a separate entity. The region that Austria acquired from Poland had never been a separate entity. Parts of it roughly coincided with an ancient region of Kievan Rus called Halych. An Austrian bureaucrat chose the name Galicia which was confirmed by Empress Maria Theresa, the Austrian ruler at the time. Galicia was populated by about 45% Orthodox Ruthenians, 45% Catholic Poles, 9% Jews, and small numbers of Gypsies, Tartars, and other minorities. By the mid Nineteenth Century, Galicia had its own elected parliament (Sejm) in which Polish consciousness was maintained and a new Ukrainian consciousness developed from the Ruthenian population, all within a framework of loyalty to the Hapsburg Monarchy.[2]

Male subjects of the Austro-Hungarian Empire were liable for military service on their nineteenth birthday, so in 1903 Ladislaus was drafted into the Austrian Army. His military records show that Ladislaus was 170 centimeters tall, about 5 foot 7 inches. On April 7, 1903, he was assigned to the K.u.K. Infanterie-Regiment Graf Daun Nr. 56, 12. Feldkompagnie. This was the Imperial and Royal Infantry Regiment No. 56, known as the Count Daun Regiment, 12th Company, which had its depot in Wadowice, near Ladislaus' family home.[3] This regiment was comprised of over 4,000 men, and about 100 officers, with all the men of the regiment being Polish and most of the officers being German speakers from Austria or other provinces of the Empire. Austria-Hungary was comprised of 12 different nationalities, but the enlisted men of the Army were all taught to speak and understand at least eighty words of command in German, the common language of the Empire. While Poles and other ethnic groups were treated better in the Austrian Empire than in Germany or Russia, class and ethnic distinctions still weighed down on most enlisted men.

Life of a recruit in the Austrian Army was hard but most of the men were used to hard work and little in the way of comfort. The typical Army Regimental Depot like Wadowice had little entertainment for the men, perhaps just a little nearby tavern and a skittle alley. There was not much money to spend though, with privates getting just 16 Heller (about a penny) per day. Payday was every tenth day. Men were expected to march 100 paces to the minute, 24 minutes to the mile. While men could be expected to march 31 miles in a single day, 12 miles per day was the standard distance carrying more than 60 pounds of gear and clothing.[4] Ladislaus must have done well enough in this environment though, for by 1906 he had been promoted to Corporal with an increase in pay to 24 Heller per day. On September 8, 1906 Ladislaus was discharged from active duty with the 56th Regiment and assigned to the Army reserves until 1913, after which he would be again assigned to the Landwehr (inactive reserves) until he reached his 42nd birthday.[5]

Ladislaus had other plans though that did not include the Austrian Army reserves. Economic conditions in Austrian Poland were bad. In addition, most Poles chaffed under Austrian (or German or Russian) rule and Ladislaus' time in the 56th Regiment had not given him a positive view of continued life in Austria. Ladislaus had decided to join the growing flood of migration from the poorer regions of the Austrian Empire to America. Wasting no time, he said goodbye to family and friends, packed up a few clothes and belongings, carefully hid the money he had managed to save as a corporal in the Army and made his way to Antwerp in Belgium. There, on Friday October 26, he boarded a ship named the Samland for the crossing to New York City in America.[6] The passage aboard the Samland was slow and very crowded. The food was poor and even personal hygiene was difficult. Finally, on Tuesday, November 6, the Samland arrived in New York and the immigrant passengers were transferred to Ellis Island. After two additional days, Ladislaus cleared through U.S. Immigration on Ellis Island on November 8, 1906, just two months after he had been discharged from active duty with the Austrian Army.[7]

The Ellis Island Immigration officials Americanized Ladislaus' name to Walter. Before the end of the year, he made his way to the Delray area in downriver Detroit, where there was both a sizable Polish born population and a large and growing number of industries to provide employment for unskilled workers. As early as 1900 there were more than 14,000 Poles among Detroit's foreign-born residents. Ladislaus, now Walter Babinski, settled within the bounds of St. John Cantius Parish in Delray. Father John Walczyk, an Austrian-Polish priest, had started St. John Cantius Parish in 1902. The parish began with just 62 families, but $10,000 was raised quickly to build a church, school building, and rectory. The church, located at the corner of Pulaski and Anson, was completed and consecrated by Bishop Foley on October 26, 1902. By the time Walter arrived in Detroit in 1906, the Parish had expanded to more than 400 families and was beginning to outgrow its original small church. Once again, $75,000 was raised from the parishioners and from local businesses and a new combination church and school building was erected adjacent to the original church at the corner of Pulaski and Harbaugh Streets. This building was 77' by 165' with classrooms on the first floor and a large open hall on the second floor to serve as the place or worship with a seating capacity of 1,058. Perhaps Walter was present on Saturday, August 20, 1910 when Father Walczyk spoke at the laying of the cornerstone for the new church:

> *"We look on the United States as our second mother. Here the Poles can get ahead and show what is in them, and no class distinctions keep them down. We are ardent Americans as well as good Poles and we love this splendid land."*[8]

When he arrived in Detroit, Walter first lived at a boarding house located at 30 Copland Avenue in Delray. Delray in 1910 was primarily a residential community, but it was surrounded on four sides by various heavy industries – steel, chemicals, rendering and soap, pulp and paper, machinery of all kinds. Walter found it easy to find well-paying though hard work. His occupation was listed in the 1910 Polk City Directory as a laborer. He must have done well though, for within just a short time Walter was able to marry a woman named Mary Jarosz.[9] Walter and Mary prospered and began a family the following year with the birth of a girl, Anne, on June 18, 1911. Two year later, on June 5, 1913, a son named John was born, and two years after that, on July 17, 1915 a girl named Helen was born.

After a number of short-term jobs[10], Walter went to work for the Detroit Sulphite, Pulp & Paper Company. Their factory complex was located right in Delray, south of West Jefferson Avenue in the bend that the Rouge River makes in its course from the southeast to the north, before turning one last time to the east and into the Detroit River. The industrial economy of Detroit boomed during the years of World War One and Walter and Mary were able to buy the modest frame house at 9050 Pulaski Street, barely a half mile from where Walter worked. This house was small by modern standards. Built on a small 50-foot-wide by 100-foot-deep lot, it had no basement, just a crawl space (which was unusual for houses in Detroit). Unlike most of the houses on Pulaski, it had just a single floor of living space, with an attic above under a steeply pitched roof. On the ground floor, the small front porch led into a parlor with a small bedroom off to the left. Behind the parlor was a large dining room with a second larger bedroom off to the left. This bedroom was curious, because it had two separate doors, each opening onto the dining room. Beyond the dining room was the kitchen, with a wood stove and a third small bedroom to the left. Originally, there was no indoor bathroom, but an outhouse at the back of the yard outside the large back porch. Stairs at one end of the back porch led up to an attic that extended the length of the house. There was never central heat in the house. Warmth in the winter was provided solely by the kitchen stove and a small coal stove located in one corner of the dining room.

Walter and Mary had joined St. John Cantius Parish and little Anne and John were able to attend the parochial school run at St. John Cantius by the Felician Sisters. This is an indication that the young Babinski family was prospering, for the children could have gone to nearby Morley School – the new public elementary school built by the City of Detroit to serve the Delray community. Public school was free, but Walter and Mary chose to pay to send Anne and John, and later other children to St. John Cantius School. Catholic education in an environment that preserved their Polish culture was important to the young family though, so it was worth the small price.

Shortly after the end of World War One, Walter's wife Mary died. Being a strong woman of Polish peasant stock, she had been lifting a railroad tie while pregnant! The strain caused a miscarriage and she died of complications.[11] This was an extreme hardship on Walter, not just losing his wife, but also losing the mother of his three little children. Relatives, friends, and the church were relied upon for support to help care for the three children and keep house while Walter put in long hard days at his job at Detroit Sulphite.

At some point in late 1920 or early 1921, Walter proposed marriage by mail to a young Polish woman named Kunegunda Gebolys. Kunda, as she was known for short in Polish, or Cora as she would later be known by the Americanized version of her name, had been born Kunegunda Malgorzata (Margaret) Gebolys on April 12, 1900 at Brzezinka, Oswiecim, in Austrian Poland. Kunegunda's father was Joseph Gebolys and her mother was Anna Gawelek. They were married about 1896 and had eleven children. Kunda's oldest sister, Mary, died at the age of three. Another sister, Rose, was two years older than Kunda. The Gebolys family appears to have been relatively well off, with one family member who was an Alderman of the City of Krakow and another who was educated in Rome and worked as a Reverend Professor teaching at the Krakow Gymnasium.

There must have been some family or business connection between the Babinski and Gebolys families, for Kunda accepted Walter's proposal. In March 1921, speaking not a word of English, Kunda said goodbye to her family in Poland forever, made her way to Antwerp, Belgium, where she boarded the Belgian Royal and U.S. Mail Line Steamship SS Lapland.[12] The Lapland stopped at Southampton, England, and then made the Atlantic crossing to New York City, landing on March 31, 1921. Kunda was processed through at Ellis Island, then made her way by rail to Detroit. Kunda had an aunt who lived in Delray, and this is whom she lived with when she first arrived. Kunda and Walter met soon after she arrived and their engagement was confirmed. They were married on June 7, 1921 at St. John Cantius Church, then after the typical Polish wedding celebration, Kunda moved into the family house down the street with Walter and his children Anne, John, and Helen.[13]

Despite their differences in age, Walter and Kunda had a successful and long married life. Her three stepchildren, Anne, John, and Helen, loved Kunda. In later years Kunda's grandchildren (who knew her as Babka – or Grandma, in Polish) had no idea that she was not the natural mother of Anne, John, and Helen – such was the love and respect that they felt and showed for each other. So this was the community and family that young Walter was born into in July 1922.

The Babinski family prospered as the Delray district boomed. Less than two years after Walter was born, Kunda gave birth to a girl, named Genevieve, and two years after that another boy, Henry was born into the family. Walter worked hard at Detroit Sulphite and Kunda worked hard to raise her family. She also found time to take classes to learn to speak English, which she mastered quickly, although her Polish accent never left her. All the Babinski family children spoke English as their first language, but they all spoke Polish fluently too. Walter and Kunda were 'ma' and 'pa' to their children.

More and more east European immigrants found their way to Delray, as well as significant numbers of Negroes and Mexicans. St. John Cantius Church was bursting at its seams by the time Walter Junior was born, because of all the people flocking to Delray and its plentiful jobs. The following year, in November, the Parish laid the cornerstone for a new St. John Cantius Church. This structure was begun adjacent to the previous church, which would then be used as an enlarged school and parish hall. Soon the parish would be home to 2,000 families. On November 15, 1925, Bishop Gallagher from the Archdiocese of Detroit dedicated the parish's new church.

At around 1924 or 1925, Walt's pa changed jobs, going to work for the Ford Motor Company, at the new Rouge Manufacturing plant in nearby Dearborn.[14] Henry Ford had been born in Dearborn, and as his Ford Motor Company gained financial success with its innovative large-scale assembly line manufacturing processes, he began to buy land along the lower Rouge River in Dearborn. By World War One he owned over 2,000 acres, which he considered turning into a bird sanctuary. After the U.S. entered the war though, Ford was approached by Franklin D. Roosevelt, Secretary of the Navy, who asked him take on the mass production of 'Eagle Boats' – small anti-submarine ships designed to find and destroy German U-boats. Ford agreed and started developing the Rouge site as a manufacturing plant. After the war, Ford converted the plant to mass-production of the Fordson brand tractor. At the same time, he began expanding the facility to manufacture parts for the Ford Model-T, which was still assembled at his Highland Park factory.

Throughout the 1920's, the Ford Rouge Manufacturing complex expanded, and as it did so, it provided more and more jobs for people in nearby Delray and throughout southeastern Michigan. The Rouge River itself had been widened and deepened by the Federal government during World War One to allow the Eagle Boats to make their way from the Ford Plant to the Detroit River and then to the sea. This work included cutting a new channel of the Rouge River at its southern bend adjacent to the Detroit Sulphite plant, straight into the Detroit River, thereby creating the new industrial Zug Island. Soon Ford was bringing iron ore, coal, and limestone via boat to the Rouge Plant, where he had built new coke ovens and blast furnaces. Molten steel from the furnaces went directly to new foundries, the largest on earth at the time, where it was poured into molds before cooling, to make engine blocks, cylinder heads, manifolds, and other forged machinery parts. Ford employed the renowned architect Albert Kahn to design over 200 buildings at the Rouge complex. Kahn designed buildings that were not only functional, but also tried to provide air and light to humanize the manufacturing process.

Just after Walter Babinski went to work at the Rouge Plant, as an overhead crane operator, Ford added a steel rolling mill and glass-manufacturing complex. In 1927 the Rouge complex gained the distinction of being the first fully integrated 'ore to assembly' plant with the introduction of the Model A. Raw materials poured into the plant by ship and rail, and completed Model A's were shipped out, manufactured from parts all made at the Rouge Plant. Later, during the early 1930's 100,000 people would work at the Ford Rouge plant.

For most of the 1920's Kunda cared for the three littlest children at home, Wally, Genevieve (or Gene), and Henry, while Anne, John, and Helen went to school at St. John Cantius. By 1925, Anne had completed the eight years of education offered at St. John Cantius, so she started going to nearby Southwestern High School, about a mile and a half away on Fort Street. Two years later John started going to Southwestern also, but he dropped out of High School after just one year and went to work with a series of odd jobs.

A big event that young Wally remembered from May 1927 was the tension and excitement while Charles Lindbergh was attempting to make the first solo flight across the Atlantic. Although his plane was named the 'Spirit of St. Louis,' Detroiters looked on Lindbergh as a native son, because he had been born in Detroit. The Babinski family listened to coverage of the flight by radio and when news came in that Lindbergh had landed safely in Paris, Wally got his first concepts of the potential for long-range flight. The following year Wally joined his sister Helen at St. John Cantius School, although she too left for Southwestern High School the following year. In 1929 older brother John went to work along with hi pa at the Ford Rouge plant.

In October 1929 the great crash on Wall Street was big news, but it did not immediately affect the Babinski family or Delray. By the following year, 1930, Walter Senior was still working as a crane operator for Ford Motor. John had been laid-off from Ford, but he had gotten another job pumping gas at a local gas station. Anne, now 18, was nearing graduation from Southwestern High, but she also had a part time job as a packer at a nearby wholesale supplier. Both John and Anne contributed a substantial portion of their earnings to their ma and pa, to help make life better for all the family.

The Babinski's were doing well in 1930. They owned a modest house valued at $6,000 – smaller than most of their neighbors on Pulaski, but they were not renters.

By the end of 1930, young Wally was eight years old and in the second grade. St. John Cantius School at that time was staffed by 13 Felician nuns, ages 18 through 45, who all spoke Polish, but taught in English.[15] Wally had begun as an altar boy at the beginning of the second grade, assisting old Father Walczyk, the man who had baptized him. Once a week early in the morning, young Wally would make his way down Pulaski Street before breakfast to assist Father Walczyk with morning mass, then return back home for breakfast, before heading back to school. On Sunday's he would assist with mass too, but not always so early. He would continue as an altar boy as long as he was a student at St. John Cantius.

Other Delray Beginnings

When young Wally and the Babinski family would walk down their street to St. John Cantius, six doors down from their house they would pass a little grocery store off to the left at 9014 Pulaski. This store was a two-story structure with plate glass windows facing the street, a small retail store area on the ground floor in the front with sales counter and a storeroom in the back. Stairs in the back led up to a small residential flat on the second floor.

This store was owned by Frank Kurzyniec. Frank had been born in about 1867 in Austrian Poland, as had his wife Mary, who was eight years younger. Frank worked for many years as a laborer at the Semet-Solvay Plant located at 6995 West Jefferson, just past Post Street in Delray. This was hard work, manufacturing industrial soda ash, caustic soda, calcium chloride, and baking soda, but Frank and Mary saved their money and were able to buy the small store on Pulaski. By 1930 they had four sons – Joseph aged 24 also worked at Semet-Solvay, Jacob (Jack) aged 20 had worked as a clerk but was now unemployed, Anthony aged 18 worked as a machine operator in an auto plant, and Gustave 17 years old. Mary tended the store during the day while Frank and the boys were at work, and the whole family tended store on Saturdays and Sundays.[16]

There were many other little shops like the Kurzyniec Store in the neighborhood. Right across the street from the Kurzyniec Store was the Tyznia's Store and there was one other in that block across from the Babinski house. Each block in the neighborhood had one or two similar little shops of its own. Families would shop for meat, produce, and baked goods at specialty shops on Jefferson Avenue or Dearborn Street, but the little neighborhood stores were handy for canned goods, cigarettes, candy, newspapers, and various sundries. The Kurzyniec store was the most convenient for the Babinski's, so they shopped there often. Another reason they liked stopping there was that Frank Kurzyniec had come from Zywiec, in the same southern hill country of Austrian Poland where both Walter Senior and Kunegunda had been born.

At some point in the early to mid-1930's, Frank Kurzyniec retired from Semet-Solvay. He turned 65 years old in 1932 and his sons had begun to move out to start families of their own. Frank and Mary decided to move on also and to supplement his small pension they decided to rent the store on Pulaski to Frank's cousin Thomas Kurzyniec and his wife Martha.

Thomas Kurzyniec was born on August 15, 1887 in Gilowice in the Zywiec District of Austrian Poland. He was the firstborn child of John Kurzyniec. John Kurzyniec had been born in 1865 in Gilowice (as had his parents before him), which was then little more than a tiny village. The only non-farm related business in the Zywiec district was a little commercial beer brewing. John Kurzyniec had learned the blacksmithing trade in Gilowice and in 1886 he married a young woman from the village named Mary Jendrucek. Thomas was the first of John and Mary's nine children. John and Mary did reasonably well for the time, living on a small piece of land with a house, small barn, and the blacksmith shop where John practiced his trade. They had a little land to grow some vegetables on and to keep a few geese, chickens, and pigs.

Despite their very modest prosperity, John Kurzyniec had learned of Polish history and knew that Polish freedom had been stamped out more than a century before by Austria, Germany, and Russia. He chaffed under the petty Austrian bureaucracy. For example, the Austrian government imposed a new tax on each separate building used for business purposes that a person owned. To avoid paying this tax on his blacksmith shop, John rebuilt it as an addition to the family house. John also held secret political meetings at night in his barn, to help support the Polish underground and keep the hope for freedom alive. Conditions were so bad and so many Poles wanted to leave that the Austrian government would soon be forced to establish laws to try to slow the drain of people to America. Austrian Law 46, of 1903 stated: "Anyone who encourages emigration at a public meeting by speeches, or by distributing printed matter or pamphlets, or by exhibiting these publicly, shall be punished with imprisonment for not more than two months, and by a fine not exceeding 600 crowns."

In 1902 John's wife Mary died. Three years later he remarried – to a woman named Anna, also from Gilowice. Anna must have been quite a woman – to marry a man who already had nine children. But then John and Anna had ten more children. Despite having 18 other children to care for, Thomas was very close to his stepmother Anna. [17]

Thomas had some schooling in the village and learned to read, write, and do arithmetic. As the oldest child though, Thomas started working at a young age. He worked as an assistant to his father in the blacksmith shop, and began to learn to be handy around heavy tools. He also helped his mother and siblings with the farm chores on the small plot of land they owned. His father did not pay Thomas much money for working in the shop with him, for he felt it was his son's duty to help his family.

One time, when Thomas was young, some of the other children in the village wanted to fight him. One of these boys went to his father's blacksmith shop, but Thomas was not there. The boy told his father to tell Thomas that they wanted to fight him. When Thomas came to the blacksmith shop, his father told him what had happened. Thomas got a pair of red hot metal pokers from the forge in his father's shop and took them with him to where the other boys had gathered. He said, "Here I am, I am ready to fight you." The other kids looked at the glowing red-hot pokers in Thomas' hands and they all ran away.[18]

In 1906, Thomas turned 19, so it was probably then that he was drafted into the Austrian Army for his three-year tour of duty. He was assigned to the cavalry, probably because of his blacksmithing skills. During his three years of service, he was posted to various parts of the Austro-Hungarian Empire, which exposed him to an externally well-functioning multi-ethnic state. For a while, he was posted to guard the long border with Russia. He witnessed many cases of cross-border smuggling, which seemed to occur with little problem as long as the guards were paid the appropriate bribes. For a while his regiment was posted to Vienna, the Imperial capital, and he experienced what was one of the most cosmopolitan and exciting cities during the early 20th Century.[19]

By 1909 Thomas' term of duty was up, although he was still required to serve in the reserves for many more years. He returned to his family home in Gilowice and resumed working in his father's blacksmith shop. John still could not afford to pay Thomas much money though, so to earn more money for himself, he started crossing the border into Germany, to work for a few months at a time in the Silesian coalfields. After he had saved enough money, say to buy a new pair of shoes, he would then return home. Another time he left home to work at a shoe factory in Budapest, Hungary, but again returned home after several months.

Thomas slowly developed a reputation as a young man who had more money to spend than most of the other poor farmers in Gilowice. Thomas held the record in Gilowice for being the best man at over 20 weddings. In Polish tradition, the wedding band would only dedicate a song to the bride and groom if they received a little extra money to do so. The best man was the one responsible for paying for these dedications. And since Thomas was the son of the village blacksmith, who was one of the more well-to-do villagers, he was a popular choice to be best man.[20]

For four years Thomas continued in this pattern – working in his father's blacksmith shop, then working for a few months in the German coalmines, then back home again. He also had to report each summer for his annual training duty in the Austrian Army reserve. By 1913, Thomas had decided that his prospects could be brighter. He had become used to living and working in different cities and with people of different ethnic backgrounds, but everywhere he went in Austria or Germany, he was treated as an inferior with limited opportunity, because of his Polish ethnicity.

Thomas decided that he wanted to try his fortunes in America. He had heard that this was a land where people from many ethnic backgrounds lived together, like in Austria, but in America, all men had equal opportunity. Many of his neighbors in Gilowice, and the men he met in the German coalfields had heard of friends and relatives who had made the journey to America and sent back reports of the good life there. Thomas had cousins who had immigrated to America and they were doing well. He learned from one of his cousins that the cost of the fare to travel from Poland to America would be the equivalent of 36 dollars, and Thomas also learned that the American immigration officials required each immigrant to have an additional 20 dollars in cash once they arrived, so they had money to live on until they found employment. Thomas went to his father and stepmother and told them his plan. He promised that he would send money back from America to his family. His father and mother approved Thomas' decision, and gave him the money for his fare and more than the 20 dollars required to help him start his new life. Many Poles living in and near Gilowice had heard from friends and relatives who had already left that there was ready and good paying work to be had in the industries of Detroit. Thomas' cousin had written that there was a growing Polish community in Detroit, so this is where he decided he would go.

In June 1913, Thomas said goodbye to his family. The Austrian border guards had started to look for young men – Army reservists – who wanted to emigrate from Austria. Any reservist who left was considered by Austrian law to be a deserter and liable to be arrested. Thomas was worried but prepared for this though. He had carefully hidden most of his money and bought a ticket only as far as Opole, in Germany, near the coalfields. When he was questioned at the border as to why he was traveling into Germany, he showed the Austrian guards his ticket and told them that he was going again to work in the Silesian coalmines, and he was allowed to pass. Once in Opole, he bought another train ticket, this time for Bremen, a major German port on the North Sea where he hoped to arrange passage to New York on a steamer called the Kaiser Wilhelm.

Once he reached Bremen though, Thomas learned that the Kaiser Wilhelm had already sailed. Thomas went to a reliable emigration agent he had learned about back in Zywiec. This was the F. Missler Agency, located near the Bremen train station at Bahnhofstrasse 30. The Missler Agency specialized in arranging all details for Poles and other central European nationalities traveling to America, including steamship passage and train tickets straight through to the immigrant's final destination in America. Thomas told the agent that he wanted to go to Detroit, in America. The agent issued Thomas a steamship ticket to New York, and arranged for a train ticket from New York to Detroit. The steamship ticket cost 140 German marks and the train ticket was an additional 57 marks. The agent also told Thomas that he would have to pay the American immigration officials a four-dollar head tax when he arrived.

On June 13, 1913, Thomas sailed from Bremen aboard the North German Line steamship SS Lutzow.[21] The Lutzow was an immigrant ship, so conditions aboard were crude and the Atlantic crossing was slow. During the passage, many people became ill and some died. Once in New York, the immigrants were transferred to Ellis Island where each person was given a tag to wear that showed their final destination. Thomas and most of the other newly arrived immigrants did not speak English, but there were immigration officials who spoke a variety of central European languages to help process them. Each person was interviewed and checked to ensure they had the required money and a destination with someone they could stay with. Thomas and the others had to have a physical exam to ensure that they did not have any disease. If the doctors found any problem with a person, their coat was marked with chalk and they were taken out of line. Thomas passed through without problem though, and he was allowed to enter New York and proceed to Detroit, where his cousin met him. His total journey time from Zywiec to Detroit had taken three and a half weeks.

Once in Detroit, Thomas lived with his cousin for a while, then he began work at a variety of jobs for the next seven years. Because he did not speak English, people laughed at him at first. Thomas knew from his experiences in Austria, and from working in Germany and in Hungary, that learning the local language was important, so he went to night school to learn to speak English as soon as he had some money. One distinctive memory that Thomas would recount many years later was a trip to Belle Isle, the large City of Detroit Municipal Park in the Detroit River. Thomas was there the day in 1915 that the original 2,000-foot-long wooden bridge connecting Belle Isle to East Jefferson Avenue burned down.[22]

During and after World War One, work was plentiful in America. Thomas worked a variety of factory jobs, he worked in a bakery for a time, and he took farm jobs as well.[23] He lived alone in boarding houses and while he made good money, he missed his family back in Poland, now an independent country after the end of the First World War. At some point in 1920, Thomas learned that his father had died and that his stepmother Anna was gravely ill. Thomas decided to return to Poland. He had made all his arrangements to return, but on the day he was scheduled to leave Detroit on the six o'clock train for New York, he received a letter that his stepmother had died. Thomas cancelled his plans and never returned to Poland, although he continued to send money and packages to his siblings for many decades.[24]

Thomas decided that the way for him to get ahead was to become an American citizen. While he continued to work a variety of jobs as a general laborer, he went back to night school to learn American history and government. At some point during this time, Thomas considered going to Mexico where he had heard that there were very good paying jobs available for hard workers with mechanical skills like his. He decided against this though, because he knew that if he left the U.S., it could jeopardize his chances for American citizenship. In 1924, Thomas was living in a boarding house at 9610 Herkimer Street in Delray. It was then that he submitted his application for American citizenship. He had to take an exam to prove his knowledge of American history, laws, and government. Many years later Thomas could remember each question from this exam, which he passed. On April 15, 1925, Thomas was sworn in as an American Citizen at the Wayne County Court House in Detroit. Thomas was always proud of the Polish background and heritage he had been born into, but he was even more proud of the American citizenship that he had earned by struggle and hard work.

Thomas' citizenship papers described him in 1925 as being 37 years old, five feet eight inches tall, with a fair complexion, blue eyes, and brown hair. 1925 was a very eventful year for Thomas, for in addition to gaining his citizenship, he also went to work as a laborer for the Ternstedt Steel Division of the General Motors Company. It is possible that he was able to get a job with General Motors, only because he had become an American citizen. The Ternstedt Division plant was located at Fort Street and Livernois Avenue, about a mile and a half east of Delray, but an easy streetcar ride from his Herkimer Street address. Ternstedt specialized in the manufacture of steel automobile body parts and stampings. General Motors was starting to overtake Ford by the mid-1920's as the leader of the motor industry, so a job with GM provided not only good pay (for very hard work) but also more stability in employment than Thomas had known in his dozen years in Detroit so far.

The third significant event for Thomas in 1925 was his marriage to Martha Skrabut on February 3 at St. John Cantius Church. Father Walczyk performed the ceremony, with his cousin Joseph Kurzyniec as best man and Martha's friend Frances Halat as bride's maid.

Martha Josephine Skrabut was born on July 29, 1907 in Kensington, Illinois, which is now a part of Chicago, west of Lake Calumet. Little is known about Martha's father, John Skrabut other than that he had been born in the predominantly Polish province of Galicia, in the old Austrian Empire. At the time of Martha's birth, his occupation was listed as laborer.

Martha's mother was Bertha Niktacki. Bertha was born on December 30, 1875 in Wisconsin. John and Bertha married in 1895 and had four other children, including an oldest son Joseph, daughters Sophie (or Genevieve?) and Florence, and a younger son Tony, before Martha was born in 1907.[25] A family photo taken about 1910 showed John to have a broad, stern face with a full head of light colored hair and a wide mustache. Although both John and Martha are seated in this photo, John appears to be no taller than his wife. Bertha had a thinner face than her husband, with a fine nose, thin lips, and darker hair done up on top of her head. John and Bertha lived near the intersection of East 117th Street and South Indiana Avenue in Kensington (now Chicago) in 1907 at the time of Martha's birth.

In 1910 Martha's father John died in Chicago. Shortly after her husband's death, Bertha Skrabut moved with her five children to Detroit. At some point afterwards she married again to a man named Paramba and moved to the small northern Michigan town of Sterling. This marriage failed after a few years and Bertha divorced her husband then moved with her five children to nearby Standish, where she bought a small 80-acre farm. Bertha and her children worked hard to make a living from the farm and Bertha also worked in a bakery in Standish to supplement their income.

At some point in the early 1920's, Martha's daughter Sophie (Genevieve) married Joseph Kurzyniec, Thomas' cousin. Thomas came along with his cousin Joseph to Standish for the wedding and there he met Genevieve's young sister Martha. The Skrabut family at that time was so poor that young Martha did not even own a pair of shoes. Nevertheless, Thomas fell in love with Martha and soon proposed marriage. At the time they were married in 1925, Thomas was 37 and Martha was 17.[26]

Thomas and Martha's wedding photo tells us something about them and their circle of family and friends. Martha appears a beautiful young woman, with fine features like her mother and haunting eyes. Though they are both seated, Thomas appears a couple inches taller than his wife, with a stocky build. To Thomas' left sits Martha's mother, Bertha. Now more than a dozen years after her husband's death, her face is more worn with age and her hands show years of hard work from raising her children as a single mother. Martha's bridesmaid, Frances Halat, is a stout woman but with a twinkle in her eye. Thomas' cousin Joseph Kurzyniec, his wife Sophie (Martha's sister) and their three children, Joseph, Leo, and babe-in-arms George are there, as well as Martha's brothers Joseph and Tony. A wedding custom brought from Poland to Delray was that the bride and groom would feed each other slices of bread and salt, as the guests would say *'We greet you with bread and salt so that your home might always enjoy abundance.'*[27]

A favorite summertime recreational activity for Thomas, and later Martha after their marriage, was a weekend excursion by steamship from downtown Detroit to one of the many nearby resorts on the Great Lakes waterway system. Favorite destinations were Bob-lo Island (actually in Canada, opposite Amherstburg on Bois Blanc Island near the mouth of the Detroit River), Put-In Bay on Bass Island out in Lake Erie, and the Tashmoo Resort on Harsen's Island in the St. Claire Flats across Lake St. Claire. A day trip to any one of these resorts started with a morning boat ride, then a day of sports, games, amusement rides, drinking, dining, and dancing at the resort, followed by a late evening boat trip back to Detroit, with more food, drink, and dancing aboard.[28] These were popular trips and well within reach of many immigrants during the prosperous 1920's. Thomas (and later Martha) both began to develop as sports fans around this time. Baseball, boxing, and hockey were favorite spectator sports for the two.

From 1925 until about 1928, Thomas and Martha lived at 9928 Herkimer.[29] Three children were born at home in this house to Thomas and Martha – Martha Frances in 1926, Dorothy Florence on March 13, 1927, and Stella Rose in 1928.[30]

At some point around 1928, Thomas and Martha moved into a small rented house on Gates Street in the Carbonworks[31] area just north of Delray. In 1928, the family moved yet again, about a block away to another house, a duplex located at 149 Dey Street, also in Carbonworks. Thomas and Martha paid $18 per month rent for their flat, while the Trevlia family, Samuel and Stella and their two children, Samuel Junior and Sophia, rented the other flat in the house for $20 per month. Samuel, born in Italy and Stella, born in Bohemia, were both three years older than Martha. Samuel Junior was the same age as Tom and Martha's daughter Martha, while little Sophia was the same age as Stella. Samuel Trevlia worked as a laborer in a bearing factory. In 1928 Thomas was still working as just a laborer, but by 1930 he had advanced to a foreman's position with Ternstedt's and life seemed to be going well, despite the stock market crash of the previous October.

The Kurzyniec house on Dey Street was near the intersection of several major railroad lines known as Delray Junction. A spur line of the Wabash Railroad left Delray Junction and cut right across one-block long Dey Street. The tracks ran right next to the Kurzyniec house. The three girls, Martha, Dorothy, and Stella would wave at the engineers and firemen and they would wave back as the many passenger and freight trains rolled though. Their mother though would despair of keeping her house clean and keeping the family laundry white that she would hang on the outside line to dry.

Thomas and Martha had joined St. John Cantius Parish, but they did not send the three girls to the parish school. Instead the girls went to Gersham School, a small Detroit Public Elementary School located in Carbonworks that had classes just up to the third grade. Martha started school around 1932, Dorothy a year later, and Stella a year after Dorothy. When Martha finished third grade, the family had to move so that they would be close to a school with a fourth grade for Martha.[32]

Thomas and Martha were hoping to buy a house of their own, but the depression that followed the stock market crash made a major purchase like this too risky for a while. Ford and General Motors both felt the effects of what would be known as the 'Great Depression'. Layoffs, short workweeks, and reduced pay made life hard for everyone in Delray. By 1932, Thomas had managed to retain his foreman position, but he had seen many of his coworkers either laid-off or fired for very trivial 'infractions.' Even a man who merely yawned at work was asked if he was tired the first time, then sent home for the day or for good the second time he yawned. Confidence in the future was very low for everyone during this period, though life did go on.

Good Times and Bad: Delray in the 1930's

The first three decades of the twentieth century had been good times for immigrants who had come to America. The 1920's in the U.S. in general and in Detroit in particular had seen years of unprecedented growth and prosperity. The 1930's though would be a great challenge to every family in Delray. Month after month, and then year after year, the Great Depression would unfold new hardships, new problems, and new tensions. People were going to be facing a major life-test, but most did not realize it until it was over.

In 1930, the population of Delray was estimated to be about 24,000. The vast majority of the population consisted of foreign-born people or those born to the foreign born. No single ethnic group dominated in Delray, but the two largest groups by far were Hungarians and Poles. Armenians, Gypsies, Germans, Mexicans, and Blacks were present in large numbers also.

In the immediate vicinity of the Babinski house on Pulaski Street were found families with immigrants or ancestors from Poland, Lithuania, Ireland, Germany, Czechoslovakia, Mexico,

Hungary, Italy, and Russia. There were also Negroes – some born in Michigan and others who had moved from Louisiana, Georgia, Illinois, and Alabama. Families living on Dey Street near the Kurzyniecs came from Poland, Italy, and Czechoslovakia.

The Polish families living in Delray preserved many old-world customs. Easter Saturday, families would take their basket of Easter food to St. John Cantius to be blessed, with kielbasa, ham, jars of sauerkraut, rye bread, and butter molded into the shape of a paschal lamb, with peppercorn eyes. 'Bidingus Day' was the day after Easter. The Polish Bidingus Day custom was for girls to dump a pail of water on their unexpected beau walking down the street, then chase after them with willow branches yelling the nonsense words 'Smigus Dingus'. The water was a symbol of fertility.[33] Walt would have participated in the St. John Cantius May Day procession. Led by the Felician nuns, all the children from the school would march through Delray by class, led by Walt and the other altar boys, singing religious songs, before returning down Pulaski Street for a church service.[34]

St. John Cantius was a predominantly Polish parish, while Holy Cross Church on South Street was predominantly Hungarian. Nearby on South Street there was also Saint John the Baptist Hungarian Byzantine Rite Catholic Church. There was also a Hungarian Lutheran Church at 8151 Thaddeus Street, a Hungarian Baptist Church at 400 Vanderbilt Street, a German Evangelical Church at 8156 Burdeno Street, with the Hebrew Congregation of Delray located at 8124 Burdeno.

The main commercial streets in Delray were West Jefferson Avenue to the south and Fort Street to the North. Both led northeast to downtown Detroit, paralleling the Detroit River, while to the southwest they crossed the Rouge River toward a string of industrial suburbs: River Rouge, Ecorse, Wyandotte, and Lincoln Park. Major streets that connected Fort Street and West Jefferson Avenue were West End to the northeast, and closer to St. John Cantius, Dearborn Street. Delray was a highly industrialized area; so these major commercial thoroughfares, and even many of the residential side streets, had various factories, warehouses, coal yards, docks, and shipping platforms along their length.

And then there were the tracks. Two major railroad lines crossed right through the heart of the area, with Delray Junction being one of the busiest railroad intersections in the state of Michigan. One line ran from Fort Street Union Depot in downtown Detroit and was used by four railroads: The Pennsylvania, Baltimore & Ohio, Wabash, and Pere Marquette Railroads. In addition to many passenger trains, the lines also served the many industries and warehouses located between Delray and downtown Detroit. For example, this line served the Ternstedt steel plant on Livernois, bringing in coal, coke, and iron, and carrying out castings and other auto parts. The Wabash and Pere Marquette also both used a spur off this main line that ran to docks on the Detroit River where they could transfer freight cars and even entire passenger trains on to rail ferryboats that carried the cars across the river to their Canadian rail lines. The other major rail route crossing at Delray Junction was used by both the New York Central and Grand Trunk Western lines, to carry their trains between Detroit and Toledo, Ohio. In addition to these major rail routes, a bewildering number of freight only switching lines threaded Delray. One such line passed right behind the Babinski house on Pulaski Street.

The 50-foot-wide front of the small lot at 9050 Pulaski was enclosed by a short wire fence, with a gate on the right side, which was all kept painted white. Behind this fence there was a narrow patch of grass, perhaps six feet deep, crossed by a short paved walk that led to the doorsteps leading up to the front porch. This porch had been enclosed at some point in the late 20's or early 30's, creating more living space for the growing family, at least in the summer, and helping to insulate the front of the house. A neatly painted wooden fence, perhaps three and a half feet high, enclosed the other three sides of the lot. The narrow paved walkway that led to the front door continued back, between the house and the fence on the right side, leading into the back yard. There were tall two story houses to the right and left, so it was always shady on the sides of the house and nothing much grew there in the sandy soil. Once at the back of the house though, the walkway turned left to lead to the stairs up into the back porch. This porch had been enclosed also, and an inside bathroom added allowing the family to abandon the back yard privy, although there was never a bathtub in the house. Saturday 'bath' was provided in a large galvanized tub that was brought into the kitchen, and then filled with hot water heated on the wood-burning stove. Pa and ma would take turns bathing first, and then the children by age. The water would be reused several times before it was drained out and replaced by clean water again.[35]

Behind the house there was more light than along the sides, with a good sized yard with a grassy lawn and neat flowerbeds along the fence to the left. On the right side of the yard, the grass grew right to the fence except for a large rose bush closer to the house. In the left side corner at the back of the yard where the outhouse used to be was the barn – really just a shed or small garage. A paved walkway led from the back porch across the yard to the shed door and to a gate in the back fence. If you were to step out the wooden gate at the back of the yard, to the right was an alleyway that led behind other houses on Pulaski towards Harbaugh, while off to the left, not five feet away, ran one of the many railroad spur lines that served local industries.[36]

Beyond this railroad spur line was a large open area of vacant lots, threaded by more, seemingly random, rail lines. There were many jobs available in the area, but one thing that both Carbonworks and Delray lacked was a designated safe play area for children. There was a small playground behind Morley Public School, but it tended to be used mostly by younger children. The vacant lots out the back gate of the Babinski House and across the tracks came to be a prime play area for young Wally and older friends from the neighborhood. These lots were adjacent to the Wayne Soap Factory and a cardboard box factory. In the summer of 1930, Wally was eight years old and he had begun his lifelong love of baseball above all other sports. The neighborhood boys would fashion a baseball diamond out of the empty lots every summer, clearing away debris and tall weeds. The lot had very little grass, if any at all, so fielding was both skill and a large degree of luck. Bases were made of pieces of wood, perhaps squared off, or pieces of cardboard, whatever would do. The lots were enough room for a game of 'sandlot baseball' when the boys were young, but as the years passed and they could hit the ball further, the outfield came to include stretches of the railroad track that crossed the area. Wally played outfield often, and he came to be expert at navigating the tracks safely as he was moving back or to the side to field a fly ball or grounder.

Sometimes brothers Johnnie and later young Henry would play baseball with him as well. Once, young Henry was playing second base. A player hit the ball, went racing around the bases and slid into Henry, breaking his ankle. Henry screamed in pain. Walt rushed over and carried Henry home to be tended by his ma and taken to the local doctor. Later as Henry was bed-ridden and recovering, Walt would amuse him by drawing pictures and making copies from the Sunday comics. Later, Walt would carry Henry out to his little red wagon and pull him around the neighborhood with a stop for a treat from the Kurzyniec store.[37]

The industrial area out the back gate of the Babinski house and across the railroad tracks held other adventures for Walt and the kids in the neighborhood. The field that they used for baseball in summer could be flooded in winter to make a skating and hockey rink. Between the box factory and a meat packing plant, there was a secluded area where, one summer, Wally and his friends built an underground clubhouse. First, they dug a big deep pit and lined the sides with wood. The built a ladder that led down into the pit and covered the top with log rafters and boards. Inside the pit-clubhouse, they built shelves and crude furniture. The boys were sad though when workers from the two adjacent plants discovered what they had done, and filled in the pit.[38]

As the depression ground on, every family had to tighten their belts. Walter's Pa usually managed to hang on to some work during those hard years, but there were periods of layoffs with very little money coming in. Toys for the children were rare, but Wally remembered that he did get a baseball mitt from his pa during those years, which he cherished and cared for lovingly. Other toys were improvised, just like the playground between the tracks. Cars and trucks were fashioned out of wood. Soapbox racers were made from wheels salvaged off junked baby buggies. In the fall, boys and girls in the neighborhood would make kites, fashioning frames from boxwood carefully slit into thin strips, and covered with newspaper, glued into place. A favorite summertime toy for the boys was a piece of wood fashioned into the shape of a pistol or rifle, which 'fired' strong rubber loops, made by cutting up discarded auto tire inner tubes for the rifles, and bicycle inner tubes for the pistols. A favorite game with these guns was cowboys and Indians.

A few people in the neighborhood did very well during the 1920's and early depression years of the 1930's though. The prohibition amendment of 1917 had created a huge unmet demand for beer and liquor across the U.S., despite its illegal status. In many parts of the country after 1917, this unmet demand was filled by illegal 'moonshine' stills, bathtub gin, and secret breweries set up in warehouses and barns. Along the borders with Canada and Mexico there was an additional option to meet the nations thirst for 'booze,' by smuggling it into the U.S. Detroit became a hotbed for 'rum-running' from Canada and Delray and the downriver area had its share of bootleggers. The warehouses, docks, and even houseboats along the Detroit River and River Rouge provided a multitude of spots to land boatloads of Canadian whisky and beer snuck across the river on dark nights. Both the Babinski and Kurzyniec families knew young men with fast motor boats who did not seem to have day jobs, but always seemed able to afford a nice car and new clothes, with plenty of spare spending money. Most of the Delray bootleggers were relatively 'small-time' operators who had surprisingly good reputations with most of their friends and neighbors. Many were generous with their money through the 1920's. After the stock market crash and the growing depression, they often helped families in need. There were stories too of Delray rumrunners who would buy baseballs, bats, and mitts for kids in the neighborhood whose families could not afford them due to extended job loss. After prohibition ended in 1933, many bootleggers moved on to other illegal activities, including running 'numbers.'

Little Wally's love for playing 'cowboys and Indians' with homemade toy guns was further nurtured by Hollywood. The Saturday matinee movie, with a double feature after cartoons and newsreels, was a popular pastime and relatively inexpensive. Western movies were a standard fare out of Hollywood. During the late 1920's young Walt and his friends saw movie after movie with 'Western' stars like Buck Jones, Hoot Gibson, William Boyd, and Wally's favorite cowboy, Tom Mix. Into the 1930's stars like Gene Autrey, Tex Ritter, and Roy Rogers kept the Western movie popular. Delray was served by two movie theaters on West Jefferson near Westend Street: The Grande and the Delray, which were right next to each other.

Even though the depression brought hard times for the Babinski's, they always managed to keep some money coming in to the family. Walter senior managed to keep a car through the depression years, but belts had to be tightened. Depression era Christmas celebrations always included a special dinner for the family and a Christmas tree – decorated with ornaments and lighted – very briefly and very carefully – on Christmas Eve night with candles.[39] Presents for the Babinski children were often very modest though – Wally would remember a couple oranges and a few pencils being typically all the gift he received on some of the early 1930's Christmas mornings.

Wally and other children who attended St. John Cantius School in the 1920's and 30's learned the three R's plus one more 'R' under the strict tutelage of the Felician Nuns. Although the nuns were almost all of Polish origin, all instruction was in English. There was never any doubt that the children were in an all-American school. Reading ability was mastered early on at St. John Cantius and Wally developed good reading skills, which he used all through his life. Writing and penmanship were other skills taught at St. John's and Wally developed beautiful handwriting ability. 'Rithmatic was the third of the three traditional 'R's' and here again Wally seemed to have good aptitude. He was extremely systematic and neat in his computational practices, skills that he would try to hand down to his children decades in the future. The additional 'R' at St. John Cantius was religion, and Wally had religion instruction every day, with a curriculum focused around the Baltimore Catechism of the American Roman Catholic Church. St. John's was an American school...but it did serve a predominantly Polish parish, so another class Wally took was Polish. Even though his ma and pa spoke Polish at home (in addition to English) this class taught Wally some of the grammar and structure of the language, as well as reading and writing ability.

In 1932, as the depression deepened, John had been laid off from his job at Ford Motor. He decided shortly after though to turn this loss of a job into an opportunity by going back to school. John had learned during his years at Ford that the way to get ahead in the future would be by furthering his education. He enrolled in an industrial arts curriculum at Henry Ford Trade School. This school had been founded by Henry Ford in Highland Park before 1920, to train young men to be skilled industrial workers. By the mid 1920's the school had been moved to a three-acre facility within the ground of the Ford Rouge plant. At its height the school had more than 100 instructors and over a thousand students. During the next three years, John worked part time jobs when he could, while pursuing his technical training at the Trade School. He studied tool and die making for two and a half years. He took machine shop courses for the full three years, sheet metal layout for eight months, and algebra and geometry for a year and a half, followed by a half year of trigonometry. He also took blue print reading classes for a year followed by two years of mechanical drafting.

About the time John was laid off, Anne and Helen were both working at least part time when they could to add to the family income. Walt began doing odd jobs when he could too. For several years young Walt earned some money by selling newspapers. He did not have a paper route, but instead sold the evening edition of the *Detroit Free Press* to passers-by on busy Fort Street or Jefferson Avenue. Sometimes his young brother Henry would help him.[40] He developed a 'marketing technique' where he would look for an obscure story somewhere behind the front page of that day's paper, that he could use to try to attract customers to buy papers from him rather than from the news stand or other paperboys. For example, if the day's headlines read something like "Mayor Announces New

Budget" – he might find an article on page six about a fire in a meat market, and then go down the street yelling: "Read all about it! Terrible tragedy - 200 dogs burned in big fire! Read all about it!"

One news story that made an impression on 10-year old Wally was about U.S. Army Air Corps maneuvers that brought the 32nd Pursuit Squadron to Detroit. A few dozen all-metal Curtis Hawk biplanes flew from Selfridge Army Airfield near Mt. Clemens, to Detroit City Airport. Here they were on display to the public and also preformed impressive massed fly-overs above Detroit.[41]

Most of Walt's earnings went to his family, especially during the hard years of the depression. He was able to keep some though, which he used for Saturday matinees, candy and pop, and later for his interest in baseball and other sports. Walt had been only four years old in 1926 when the great Ty Cobb played his last year with the Tigers, as a player-manager. The New York Yankees dominated the late 1920's, and Detroit hovered around a 500 average each year, finishing no higher than fourth place. Harry Heilmann, who played outfield and first base for the Tigers, was Walt's favorite player until he was traded to Cincinnati in 1930. The early 1930's were lackluster years for the Tigers too, finishing in fifth place three years and as low as seventh place in 1931.

As a young boy, Walt enhanced his interest in baseball and hockey in several ways. It was relatively expensive to attend a game, so those were rare experiences until Walt began to have regular income of his own. Radio was the primary medium to follow along with the game during the season. Interest in sports enhanced Walt's habit of reading the newspaper on a regular basis also. Sports section articles helped him keep informed of happenings with the Tigers and Red Wings, as well as other teams across the leagues. Player and team statistics motivated young Wally to master math concepts; with the result that math was often his best subject at school. From a very early age Wally became an avid baseball (and later hockey) card collector. Year after year, from the late 1920's through the 30's and into the early 40's Walt would amass a large collection of all the Tigers and most of the other players in the major leagues. Money earned from selling newspapers and other odd jobs funded his interest in sports and card collecting and motivated him to always maintain some source of regular income.

At age nine, Wally made his 'First Communion' at St. John Cantius Church. His First Communion photo (taken at the S.A. Rochowiak Studio at 8714 West Jefferson) shows him looking solemn in a dark suit with nickered pants, white tie, and corsage. His light brown hair is combed back, his face full as he stares at the camera, clutching his first communion rosary and prayer book. In 1934, when Wally was 12 years old, his parents gave birth to another child, Leona. Wally would refer to her as 'little sister' in later years.

In 1934, the Detroit Tigers acquired a catcher from the Philadelphia Athletics named Mickey Cochrane, who would become Walt's favorite baseball player. Cochrane was hired not only to catch, but to manage the Tigers as well. A lot started coming together for the Tigers after Cochrane's arrival. They had three solid hitters now, not just Cochrane, but also Charlie Gehringer at second base and Hank Greenberg in outfield. The pitching star of Detroit that year was a chunky, second-year right-hander named Lynwood 'Schoolboy' Rowe. Schoolboy Rowe had a 24-8 record for 1934, at one point winning 16 games in a row. From fifth place the year before, Detroit jumped to first, winning the American League pennant with a 101-53 win/loss record. They played the St. Louis Cards in the World Series, going the whole seven games, but finally losing the last game.

In 1934 or 1935, when Thomas and Martha Kurzyniec's oldest daughter Martha was ready for fourth grade, the family moved to Pulaski Street, because Gersham School on Dey Street only went to the third grade. Thomas and Martha had decided to rent the little Kurzyniec Store at 9014 Pulaski from Thomas' cousin Frank. Frank had retired from Semet-Solvay a few years earlier and his sons had moved out of the house. The small rent payment from Thomas would supplement his income. Thomas, Martha, and the three girls, Martha, Dorothy, and Stella, moved into the flat above the store.

Martha ran the store with the help of her three girls. The store never made much money...there was another similar store right across Pulaski and others in adjacent blocks. The girls would help with the window displays – stacking cans and boxes of products like oxydol soap, fels naptha soap chips, and Campbell soups. The store had a Coca-Cola sign below the display windows and above, signs advertised 'cigars, confectionary, & cigarettes'. On occasion, after the girls were a little older, they would have to walk to a warehouse on West Jefferson to pick up stock for the store. The girls remembered that cigarettes costs just $1 per carton back in those days.

Behind the store ran the same railroad tracks that passed behind the Babinski house. During the worst years of the depression, men and boys would sneak aboard slow passing trains and push coal off the coal cars. They would then come back and scoop up the coal into sacks, which they would sell in the neighborhood. Once Thomas Kurzyniec bought some of this purloined coal and put it in the barn behind the Pulaski store. The next morning though, the Kurzyniec family discovered that someone had made a hole in the side of the barn and stole the coal.

The little store never made much money, but every little bit helped. One sideline for the store was 'running numbers' for a local small time bookie. The store served as a distribution point for betting slips that would get deposited by locals and picked up a few times during the week by the bookie. [42]

By 1935 the three Kurzyniec girls were enrolled in Morley Elementary School, a Detroit Public Elementary School located at Portland and Beaumont Streets. To get to school, every morning they would walk down Pulaski, past the Babinski house, as Wally and the other Babinski children were walking to school at St. John Cantius.

Unlike St. John Cantius School, which was attended mostly by children of Polish origin, Morley Elementary School was a melting pot for the Delray neighborhood. Dorothy Kurzyniec remembered being bothered by bullies when she first started school there. She had a friend, a little black boy, who later went to the bullies and told them to leave Dorothy alone or he would take care of them. The bullies never bothered her again. The Kurzyniec girls left a legacy at Morley School – one year in April, on Arbor Day, on the front lawn of the school they planted a tree that grew as long as the school remained.

Early in 1935, Wally's brother John was hired back at Ford Motor Company. He had completed more than two and a half years at Henry Ford Trade School while he worked odd jobs part time. The national economy was picking up enough so that Ford was hiring again for workers at the Rouge Plant in nearby Dearborn where Walter senior still worked on a steady basis. Initially John work on an assembly line, assembling refrigeration equipment. Later, based on his trade school education at Henry Ford, John was hired into the Fabricating Department at Ford as a layout man. This was a skilled job and paid relatively well. John 'drew and developed exact patterns and forms on sheet metal according to designs specified on blue prints.' He used various drafting tools – scales, compass, dividers, etc., to mark bend points, cut lines, hole locations, and other operations on sheet metal before the pieces would go to the stamping plant for forming into auto body parts. John would eventually supervise 10 men in the fabricating department by the end of the 1930's. [43]

During the summer of 1935, Walt travelled to northern Michigan, for some relaxation, fishing, and taking in the scenery and clean air. [44] On July 30 he sent a postcard showing the Pere Marquette Depot on the beach at Petosky, Michigan to his brother Johnnie back in Detroit:

Hello Bob,
We're at the same place we were last year. Hope that you're working and enjoying yourself.
-Walter[45]

A week later Walt sent Johnnie another postcard from Alanson, Michigan:

Hello Bob,
Having a swell time. The fishing is fair and the weather is very wet. Have been having a lot of rain here.
-Walter[46]

The following week Walt took a side trip to Sault Ste. Marie:

Hello Bob,
Came over to the Soo to spend the day. Having a swell time.
-Walter[47]

1935 was an exciting, never to be forgotten year for sports fans in Detroit. Under Mickey Cochrane's management, the Tigers played well all year long. Cochrane, Charlie Gehringer, and Hank Greenberg all had great years – Greenberg hitting 36 home runs during the season. The Tigers also had great pitching, led again by Schoolboy Rowe, but Tommy Bridges and Elden Auker also won their share of games. The Tigers finished the season in first place in the American League with 93 wins and 58 losses, well ahead of the Yankees. Early in September, as Wally went back to school, it must have been difficult for the nuns to keep him and his classmates focused on their schoolwork because of the World Series excitement. The Tigers played well though and Detroit defeated the Chicago Cubs four games to two. Then following two years Detroit slipped back to second place behind the Yankees, but many Tiger faithful, including Wally remembered well the 34 and 35 seasons.

People in Delray did manage to get away from the industrial hustle and bustle of Detroit on occasion. Thomas and his cousin Joe had both married Skrabuts, and Bertha Skrabut still worked her farm near Standish, a few miles west of Saginaw Bay. Each summer Martha and her daughter's Martha, Dorothy, and Stella went by train via the Michigan Central Railroad to Standish to see Grandma Bertha. Thomas had to work through the summer, except for his two-week vacation, when he would come north too by train to join his family on the Standish farm.

One summer Martha's brother, Joe Skrabut, drove his car with the family up to Standish. Joe's car was a little two-door coupe. This was quite an adventure. Joe and his sister Martha rode in the front with Martha's daughter Dorothy on the seat between the two. Stella, the smallest of the Kurzyniec girls would lie on the little shelf behind the seat (there was no back seat) in front of the small rear window. The third girl - young Martha Kurzyniec and Joe's son would ride in the open trunk behind, with their luggage shoved in around them. They would ride this way all the way from Detroit to Standish. There were no freeways of course, so even though it is only 150 miles or so from Delray to Standish, it would take the better part of a day to make the drive. Their route would take them up Woodward Avenue, right through the center of Pontiac, Flint, Saginaw, and Bay City. What an adventure for the little kids!

Johnnie Babinski was working steady into 1936 and a friend of his, Walter, would travel each summer to northern Michigan to hunt and fish. By 1937 Johnnie had started vacationing 'up north' when he could also. For example, in May 1937 he wrote to ma and pa Babinski from Omer, just north of Standish on the Rifle River while on a fishing trip of his own:

> *Hello Folks...*
> *This is where I spend my weekends.*
> *-John[48]*

Later in July, Walt took a vacation in Northern Michigan. First, he rode the Pere Marquette Railway to Alanson, just north of BayView, Michigan, to spend time fishing on the Crooked River and nearby Crooked Lake. This was an area where a young Ernest Hemingway had fished, camped, and hunted as a young man on family vacations from his home in Chicago. Walt wrote a post card with a scene of a canoe on the river to Johnnie from Alanson:

> *Hello Bob...*
> *The fishing is swell out here. But we haven't taken the trouble to find out for ourselves yet.*
> *-Walter[49]*

Then Walt continued on the Pennsylvania Railroad up to Mackinaw City where he took the ferry to Mackinac Island and spent a night at the Grand Hotel. Walt wrote another postcard with a view of the Grand Hotel to Johnnie from Mackinac Island:

> *Hi Bob...*
> *Nice place to come for a rest. The scenery is beautiful and the weather is fine. Wish you were here.*
> *-Walter[50]*

In 1937, John went back to school again, enrolling in a two-year night school class in metal-smithing at Wilbur Wright High School in Detroit. He studied blue print reading, as well as layout and drafting for six hours per week.

Young Walter's father and his older brother Johnnie were witnesses to an event in 1937 that electrified Detroit and had ripple effects across the country. The Ford Rouge plant, where Walter senior and John Babinski worked, was not unionized at the time. Henry Ford, although he had been a pioneer in paying workers a living wage in the past, had become stridently anti-union by the hard depression years of the 1930's. To keep all union organizing activity out of his factory, Ford had hired a thug named Harry Bennett to run his 2000 man strong 'Service Department' which was his private internal security force. By 1937, the United Auto Workers Union, under the leadership of the Reuther brothers, Roy, Victor, and Walter, had succeeded in getting union recognition from General Motors, Studebaker, Chrysler, and many other auto industry manufacturers. Union recognition and labor agreements with these companies brought union members a decent minimum wage, the end of piecework pay, grievance committees, and seniority rights.

In early 1937, both Walter Senior and Johnnie Babinski were approached at various points by UAW men secretly organizing for the union within the Rouge plant. Walter Reuther and the UAW had decided that the time had come to take on Ford, the most stridently anti-union company in America. Secrecy was critical for the UAW, because Harry Bennett's Service Department included many spies, and anyone discovered handing out or receiving union information in the plant would be fired at best, if not beaten up. The UAW also began a public relations campaign, putting up billboards around Detroit saying "Fordism is Fascism" and "Unionism is American." The next step planned by the UAW was to begin handing out union literature at the gates to the Ford plant as workers left for home at the end of their shift.

Miller Road in Dearborn separated the main gates to the Rouge plant from the large employee parking lots where workers with cars parked. Buses and streetcars also ran up and down busy Miller Road with stops at each gate. To carry workers across Miller Road to the parking lots, Ford had built elevated overpasses from a number of the gates. On May 26, 1937, Walter Reuther, along with Richard Frankensteen, Robert Kanter, and J. J. Kennedy, three other top UAW organizers, had arranged for about 100 women from Local 174 to hand out leaflets to workers leaving the Rouge during the afternoon shift change. Because he knew that Bennett's security department could create trouble, Reuther also invited various clergymen, reporters, photographers, and even staffers from a US Senate Civil Liberties Committee.

As the UAW organizers and journalists arrived at about 2:00 p.m. on the 26th, they saw about 25 parked cars on Miller filled with Bennett's men. Bennett's thugs began telling the journalists to get out and threatened the photographers. One photographer, James Kilpatrick from the Detroit News, avoided Bennett's men though. He saw an opportunity for a great photo, so he asked Reuther, Kanter, Frankensteen, and Kennedy for a shot on the overpass at Gate 4, with the main 'Ford Motor Company' sign in the background. The men agreed and climbed the two flights of stairs up to the overpass. As the UAW men stood with their backs to the gate for the photos, a group of Bennett's thugs came up from behind and began brutally assaulting them as Kilpatrick took photos. Bennett's men kicked and punched the men repeatedly. Two men held Frankensteen's legs apart while another kicked him several times in the groin. Kanter was pushed off the overpass, falling 30 feet to the street below. Reuther was punched to the ground, then dragged by his feet to the stairs where he was thrown down the first flight, then followed by Bennett's men who picked him up again and threw him down the second flight of stairs to the ground, where they beat and kicked him still more.

At about this time, as the day shift was leaving the plant and streaming across the overpasses, streetcars carrying the union women with the leaflets began to arrive. Bennett's men met them at the doors and either pushed them back in or pulled them out and beat them up. One lone Dearborn policeman arrived and pleaded with Bennett's thugs to stop beating the women, but did nothing else. Then the thugs began going after the reporters and photographers, grabbing their notebooks and cameras. Kilpatrick at this point had gone back to his car, where he was stopped by a couple of Bennett's men who demanded his photographic plates. Kilpatrick gave the men some plates he had on his front seat and was then allowed to leave. He had hidden most of the photos he had taken of the melee in the back seat though.

Back at the Detroit News offices, Kilpatrick's photos were quickly developed and the News editors immediately knew that they had a major scoop. Headlines called it the Battle of the Miller Road Overpass, and many other newspapers across the country carried the story and Kilpatrick's photos. Walter senior and Johnnie saw the chaos and aftermath of the battle as they left work that day. Ford and Bennett had won the battle, but they lost the war to keep out the union. Walter Junior and the rest of the Babinski family listened with awe to the stories about that day that Walter senior and Johnnie told. The Babinski family became staunch union supporters from that day forward, as did many other Ford families. The Federal National Labor Relations Board castigated Ford and Bennett for their conduct on May 26. Later that year, labor candidates swept elections in Detroit and throughout much of urban Michigan. By 1940, Ford signed a contract with the UAW.

Despite the conflict, work was mostly good and steady at Ford, at least from the viewpoint of those trying to make a life in nearby Delray. Both Johnnie and Walter senior were bringing more and more income home. John could indulge his interest in hunting and fishing in northern Michigan, and short trips to regional resorts.

Later, near the end of spring 1937, young Wally was confirmed at St. John Cantius. His confirmation photo shows him at age 14 posing in the backyard of the Pulaski Street house in a dapper, double-breasted suit, wearing white shoes and holding a fedora. His light brown hair is combed back, and he smiles wryly as he stares at the camera. Wally had started High School now at Southwestern on Fort Street. He showed top aptitude in math and had also developed beautiful handwriting skills. He enjoyed history but he was indifferent to most other subjects. He also had a slightly irreverent streak – the pages of his text books were often covered with typical school boy doodling, or funny poems or sayings – like 'He who steals this book of knowledge, will end up in Sing-Sing College.' Around 1938, after completing 10th grade at Southwestern, Wally dropped out of school and began working a variety of part-time or short-term jobs.

Relatively steady income allowed Walt to begin to indulge more in his interests. Of course baseball was a passion. In 1937 Charlie Gehringer and Hank Greenberg led the Tigers into second place with an 89 – 65 record, but still 13 games behind the Yankees. The following year the Tigers slid further to 4th place with 84 wins and 70 losses, but Hank Greenberg excited Detroit by challenging Babe Ruth's record by hitting 58 home runs.

One extravagance that Walt indulged in after he quit school he never told his mom about. Flying had always fascinated him since he had heard about Charles Lindbergh and other pioneers of flying. One summer he learned that a barnstorming pilot was offering biplane rides at Ford Airfield out in Dearborn. Walt knew his mom would not approve such a 'foolish and risky' thing so he made his way out Michigan Avenue to Ford Airfield in secret. For $5.00, the pilot took Walt up in an open biplane for a tour of the city from the air. Walt loved the thrill of being in a plane – a thrill that would perhaps influence him a few years in the future.

In June 1938, a young Walt took an excursion to Michigan City, Indiana. He visited a new waterfront attraction: The Rock Garden, constructed a couple years earlier by the WPA as a tourist attraction. He wrote to his brother:

Hi Tornado John,
Having a nice time. This is the place for you boy. Lots of beautiful women. Plenty of altes lager for you to get tight on, but no nerds to drive you home.
-Stella & Walter[51]

The Babinski's were a very tight-knit family. Wally's pa, Walter Senior was remembered as the clear head of the family. He was hard working and a good provider for his wife and children, but stern as well. Wally would recall many years later the shaving strop that his father used on the back porch or back in the barn to administer 'corporal punishment' if one of the boys missed his chores, was late coming home, or somehow otherwise crossed him. Walter Senior smoked all his life and clearly enjoyed a beer or two on the weekends or even a shot of the hard stuff on special occasions. There was always enough food for the family though, and clothes, and a little money for the movies and a radio for the house. Walter Senior and Cora were a loving couple and she could moderate the harsh extremes of her husband's personality. As was common during that era, a woman's place was primarily cooking and cleaning, and Cora kept their small house neat and tidy and the family very well fed. She helped keep many Polish customs alive in their household and most of the family meals were built around traditional Polish dishes like kielbasa, kieshka, pierogis, guomkies, and stuffed cabbage. Holidays, birthdays, weddings, and funerals would also bring out Cora's cooking prowess, with a wide variety of crullers, poppyseed cake, and ornate cookies.

St. John Cantius continued to play a huge role in the lives of the Babinski and Kurzyniec families. Christenings, weddings, funerals, and the annual cycle of the Catholic holy days all centered around the church. Celebrations also had a secular aspect. For example, weddings and christening often were followed by a reception or dinner at a local hall, where there would be food, drinking, and dancing late into the night. Celebration of 'Polish Weddings' in Delray evolved into an elaborate ritual, based on old-world customs supplemented with the bounties of the new world. They usually included a live band, playing a variety of popular music, as well as traditional Polkas, Mazurkas, and other old-country dances. Everyone danced, although Wally's ma, Cora, was remembered as being an especially lively and skilled dancer.

As the 1930's neared an end, the Babinski's, the Kurzyniec's, and most of the other families in Delray had survived the worst of the Great Depression. Many had to bend with the pressures exerted during those years, but few broke. The first generation immigrants did not give up on the promise of America. Their children, the second generation Americans like Walter, Genevieve, Henry, Leona, Martha, Dorothy, and Stella, did not have extensive memories of the good times before 1929 to compare to, so each year of improving conditions in the 1930's must have seemed even more dramatic. Prosperity had still not reached the levels of the mid to late 1920's, but people sensed that the country under Roosevelt was making economic progress and moving in the right direction. For most people who made it through those years, the experiences tempered them and made them stronger. Few though realized that in not many more months new challenges would arise in the far corners of the world that would confront America head on. Those challenges would reach every city in America, right down into Detroit, and into Delray, and into every house on Pulaski Street and on every street across the country. The strengths developed during the 1930's would have to carry them further still.

- 2 -
The World in the 1920's and 1930's: Era of Isms

The Growth of Competing 'Isms'

It is human nature to spend our lives usually focused most intently on only our small, local circle of family and friends, within the context of where we live, work, worship, educate, and entertain ourselves. A person who is happy at home and content with his or her lot in life tends to stay put. There is something reassuring about a familiar environment – knowing what you will find down the streets near where you live, or a friendly greeting from the folks in your neighborhood, or where you work or shop. This is not to say that there are not some restless souls who live in a happy environment, but who still feel driven to experience something new – to try their luck in a new town, or country, or continent. However, this type of person has clearly been in the minority throughout world history. Sometimes though just the idea that there is a much better life possible somewhere else can motivate a person to pull up stakes and move thousands of miles away to start a new life.

This process seems to be related to a mental image that a person develops of what a far distant local environment might be like. Perhaps a person reads an article in a newspaper, or talks to a traveler, or meets a visitor from the distant environment who paints a glowing picture of what life in this distant land might be like. The age of exploration and various gold rushes fall into this category, although even more mundane events can produce similar attractions. For much of the 19th century in Europe, images of cheap and plentiful land or steady well-paying industrial work in America, Canada, Argentina, Australia, and dozens of minor colonies attracted millions of people to migrate just as much as harsh local conditions pushed people away.

Local conditions around the turn of the Nineteenth Century had motivated individuals and families like the Babinskis, Kurzyniecs, Gebolys, Skrabuts, and millions of others to leave Europe and move to the United States. External forces often create local conditions (both good and bad) though, and this was the case with the family stories described in the preceding chapter. A myriad of external forces made Detroit and Delray just one of thousands of attractive environments for migrants from around the world. A relatively benign political environment, plentiful raw materials, cheap transportation by rail and water, a skilled industrial workforce, and hundreds of capitalists with money to invest led to the formation and expansion of dozens of industrial ventures in Detroit by the late 1890's.[1] New technology played a role too. Steam power was being replaced by electrical, gas, and diesel powered motors and Detroit developed quickly as a center of this new technology, especially related to the new auto industry. This process of industrialism in Michigan, which in its early phases results in many new and expanded industrial ventures, soon exhausted the local supply of skilled labor, so working conditions and pay began to rise quickly as industries attracted workers to fill the needs of its factories. All of these external factors were creating local environments like Delray and Carbonworks that looked very good to people seeking a better life.

Local conditions during the Nineteenth and early Twentieth Centuries in many parts of Europe (as well as in China, Japan, Mexico, and other countries) had further motivated large portions of the population to think about leaving their homes for new lands. We have seen in the preceding chapter how laws and conditions originating in Vienna, capital of the Austrian Empire, affected typical Polish families living in the province of Galicia under foreign rule. This is not to say that foreign rule by itself caused people to emigrate. Huge proportions of the populations of Sweden and Denmark emigrated to the U.S. during the same years, and they came from countries with democratically elected national parliaments. Economic conditions were bad during the same years in Portugal and Spain, but few people from those countries migrated to the new world.

What was called the 'Great War' or the just the 'World War' of 1914-1918 was a watershed of history that had both worldwide consequences and a profound impact on Delray in Detroit and thousands of similar immigrant communities across the country. Had they not emigrated from Poland to the U.S. but stayed in Poland instead, Walter Babinski Senior and Thomas Kurzyniec would surely have been called back to active duty in the Austrian Army. Their chances of survival would have been very low indeed – the Austrian Army suffered millions of casualties during the war, including hundreds of thousands of Poles, Czechs, Serbs, and other minorities. Poland itself was a constant battlefield during the war, with the armies of Austria, Germany, and Russia marching east and west across Galicia dozens of times as the front lines shifted back and forth.

The end of the war led to the end of the three multi-national empires that had ruled Poland and in 1918, many nations remerged on the map of Europe: Poland, Lithuania, Latvia, Estonia, Finland, Czechoslovakia, Hungary, and Yugoslavia. Nationalism and national self-determination was a key feature of Woodrow Wilson's 14 points, the Versailles Treaty that ended the War, and of the League of Nations set up to resolve future international disputes. Life in these newly independent countries was not easy, but at least they were free to set their own destinies. As Poland was trying to make a nation out of lands that had been ruled from three separate capitals for more than a century, she also had to fight off the Communist Russian Red Army that contested Poland's eastern border and hoped to bring revolution across Europe and into Germany. There were border disputes with Lithuania, Czechoslovakia, and ominously for the future, with Germany herself. Nevertheless, independence eliminated a key motive for immigration. Indeed, we have seen that Thomas Kurzyniec was very close to returning to Poland in the early 1920's, just before he learned of his father's death.

The birth of Communism in Russia and its transition to the Soviet Union in the mid-1920's created a new force and a new fear in the world. The Soviet Union was something new – a state built upon the foundation of an ideology. In addition, this Communist ideology was very attractive to many workers (and large numbers of intellectuals as well) in many countries. Communist ideology promised better conditions and a fair share of the economic pie to those who did the work of the industrialized economies of the world. This ideology also promoted the concept of a world government, where wars like the senseless slaughter of 1914 – 1918 would be a thing of the past. Communism would be a major force throughout almost all the 20th century. Even though the anti-religious nature of Communism made it abhorrent to most European immigrants living in the U.S., during the 1920's and 1930's there was a suspicion and fear that immigrants (especially Slavic immigrants with a supposed sympathy to Soviet Russia) were a fertile breeding ground for Communism in America. The 'Red Scare' would be a feature of American life throughout the 1920's, 1930's, 1940's, and even into the 1950's. Saco and Vanzetti were just one example of innocent immigrants who were tragic victims of America's fear of foreign communism. Thomas Kurzyniec was able to overcome this challenge by acquiring American citizenship in 1925, which opened up doors to a better and more secure job with General Motors.

Out of nationalism and in violent opposition to Communism, a Fascism sprang up in Italy in the 1920's. Fascism was an attractive political option to many Europeans. Nominally working within the structure of parliamentary democracy, it offered a strong opposition to godless Communism. It claimed to offer an alternative to the apparent chaos of the multitude of political parties that sprang up in most countries after the War. Under Benito Mussolini, Fascism appeared to deliver on its promise in Italy. The economy improved, political chaos and industrial strikes declined, and many Italians felt good about themselves and their country. Of course, these benefits came with a price – while the trappings of democracy were maintained, all political opposition was ruthlessly stamped out, via laws if possible or by violent crackdowns by brutal Fascist paramilitary groups if necessary.

The birth and steady growth of Fascism in Italy in the 1920's was never earth-shattering news in the U.S. Nor was its spread to Germany and other European countries. The story of how Hitler came to power in Germany is well known now, but from the perspective of the typical American in the 1920's or 1930's there was little knowledge, interest, or concern with what was happening in Germany. Hitler had actually managed to forge together two ideologies – Socialism (the more benign relative to Communism) and Nationalism, into his National-Socialist Party, or the Nazi's. The combination of Nationalism and Socialism appealed to a very wide base of Germany society and allowed Hitler, step by little step, to impose ruthless anti-democratic Fascism on Germany. After his Nazi movement had a false start in the 1920's, the worldwide economic depression of the 1930's provided the fertile environment Hitler needed to bring Fascism to full flower in Germany.

Versions of Fascism would grow elsewhere in Europe in the 1930's – in Spain and Portugal, in Greece and Turkey, and outside Europe – in Argentina and Japan. Other countries struggled with Fascism – Hungary, Poland, and Romania had strong Fascist political forces and even the liberal democracies (Britain, France, and the U.S.) had many prominent admirers of Fascism. In many of these countries, the U.S. included, many questioned whether existing democratic institutions could keep Communism at bay, and suggested that Fascism might be needed as a defense against Red revolution. We forget how bad worldwide economic conditions were in the 1930's and the degree to which those on the far right and on the far left both thought that worldwide revolution was at hand.[2]

The nationalism that led to the emergence of a dozen or more new states in Europe after the great 'World War' of 1914-1918 was also not without its problems. The geographic distribution of ethnic

NPRC Records: Typical copies of S/Sgt. Walter Babinski records as received by the author in 1993 from the National Personnel Records Center.

Names from Crew Photo: Some of the photos that the author found with Walt's military records included names on the back. These names are from a photo of crewmen with a B-24 named 'Rose-o-Day'. Evidence like this began to strip away the anonymity of Walt's Air Force comrades.

Names visible in handwriting:
- Leonard Sherman, Francis Mansir, Clifford Llewellyn
- Myself, Jack Riley
- Edward Bretherton, Jimmie Clark, Ed Doone
- Don Aubrey, Charles McRae

Cpl. Wladislaw Babinski: Pages from Cpl Ladislaus Babinski's military record book from the K.u.K. Infanterie-Regiment Graf Daun Nr. 56, 12. Feldkompagnie, showing his mustering in date in 1903 and his discharge into the reserves in 1906.

SS Samland: Ladislaus Babinski took passage on the Samland in 1906 from Antwerp, Belgium to New York City

SS Lutzow: Thomas Kurzyniec crossed from Bremen, Germany to New York City aboard the SS Lutzow in 1913

SS Lapland: Kunegunda Gebolys travelled from Antwerp, Belgium to New York City in 1921 on the SS Lapland

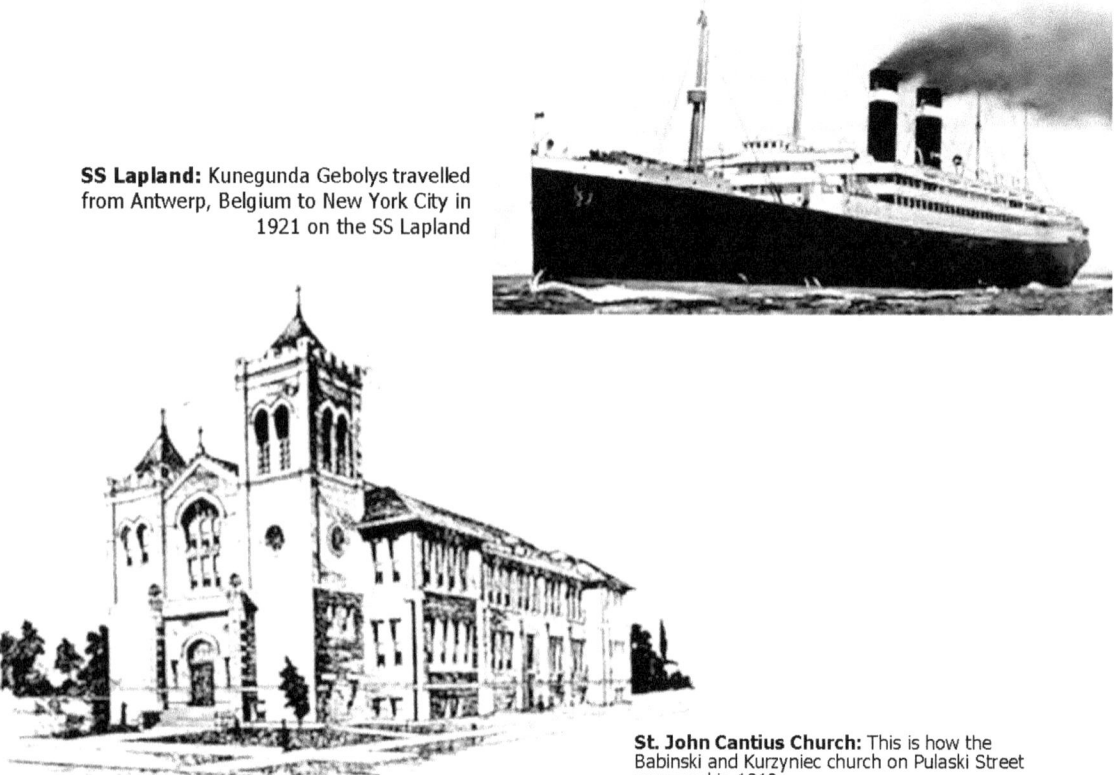

St. John Cantius Church: This is how the Babinski and Kurzyniec church on Pulaski Street appeared in 1910.

Wedding of Thomas Kurzyniec and Martha Skrabut, February 3, 1925:
Back row, L to R: Unknown male, Joseph Skrabut (brother), Sophie Skrabut (sister), Joseph Kurzyniec (brother-in-law and cousin of Thomas K.), George Kurzyniec (baby), Mr. & Mrs. Walter Zielinski (cousins of Thomas K.), Tony Skrabut (brother)
Front row, L to R: Unknown female, Martha Skrabut Kurzynieec, Thomas Kurzyniec, Bertha Skrabut (mother), Joseph and Leo Kurzyniec (sons of Sophie and Joseph Kurzyniec)

Ford Motor Company Rouge Industrial Complex 1927:
The wind blows smoke from the Ford Rouge plant over Miller Road and towards Delray in Detroit.
Detroit Publishing Co. Collection, U.S. Library of Congress Prints and Photographs collection.

Ford Tool and Die Room May 1941: Walter Senior worked here as an overhead crane operator and Johnnie worked as a layout man.
U.S. Office of War Information photograph in the collection of the U.S. Library of Congress Prints and Photographs collection

Father and Son: Walt and his pa pose in front of the back fence in the yard of the home at 9015 Pulaski in the late 1930's.

Ma, Walt, Henry and Leona: Walt poses with his Ma, Hen and Leona in front of the back door of their home at 9015 Pulaski in the late 1930's.

Walt 1931: Walt poses in his First Communion suit with prayer book and rosary.

Walt 1937: Walt (or Lize to family and friends) poses in his Confirmation suit in the back yard of the Babinski Family home.

Walt and Younger Siblings: Kunegunda's four children: Walt with Genevieve, Leona, and Henry, about 1939

Three Friends: Buzz Kaszyka, Walt, and Robert Symborski on an outing to nearby Belle Isle, 1942

Seaman John J. Babinski, USN:
Taken in 1942 or 1943.
Babinski Family Collection.

Private Walter A. Babinski:, USAAF:
Taken in February, 1943 at Bergeson's Photo Shop in Clearwater, Florida.
Babinski Family Collection.

Seaman John J. Babinski, USN @ Great Lakes Naval Training Station, 1942:
In August 1942 Walt's brother John sent him this post card from his basic training base, saying: "Hello Walter Jr. This is how we get our (chow) or food- it's all cafeteria style. It's very clean but no beer. -Bob"
Babinski Family Collection.

Aerial View of Camp Custer 1942: Fort Custer was a busy recruit reception center in 1942-45 as it had been during WWI. When Walt was at Camp Custer, the trees were bare and the base snow-covered. *Willard Public Library, Battle Creek, Michigan.*

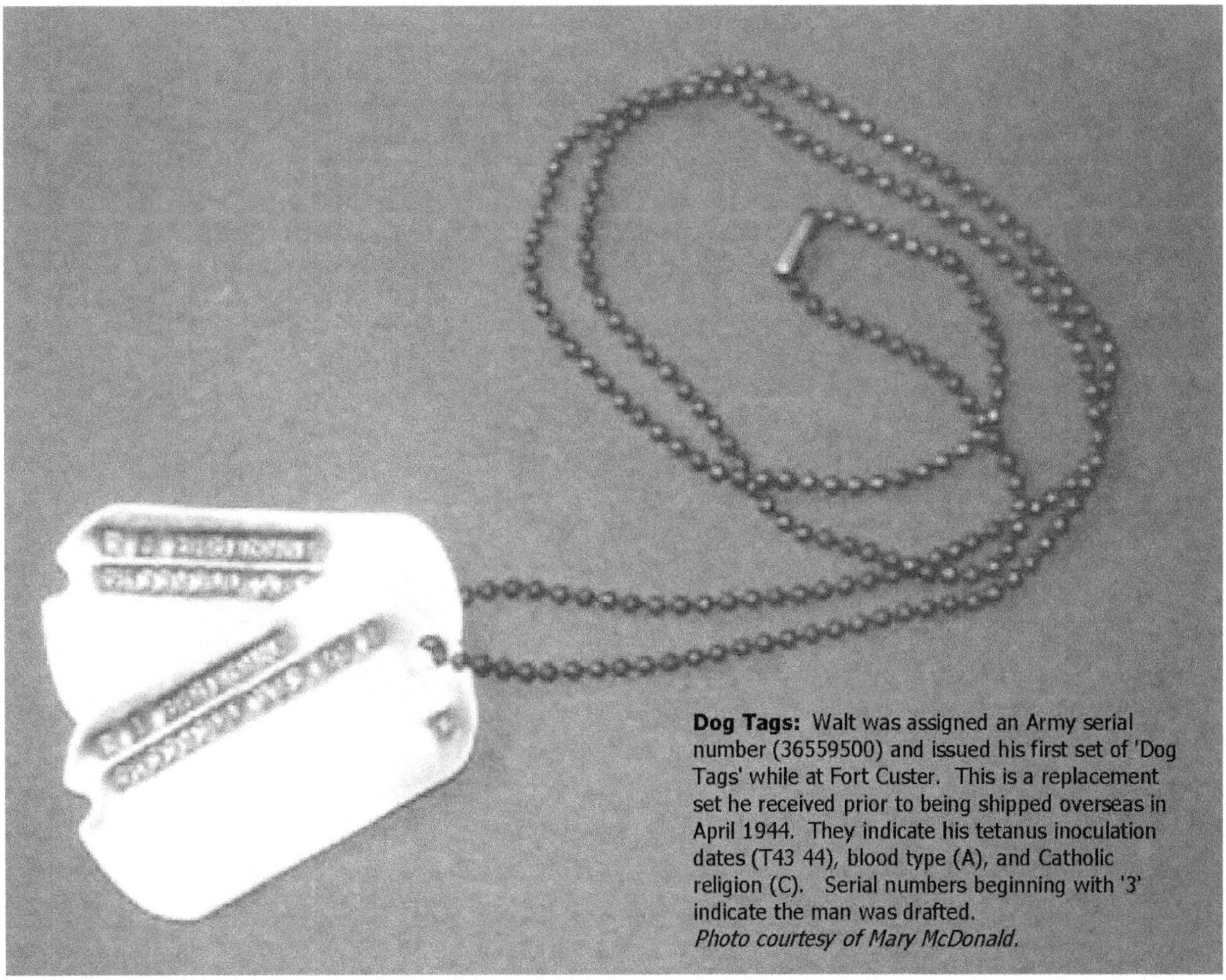

Dog Tags: Walt was assigned an Army serial number (36559500) and issued his first set of 'Dog Tags' while at Fort Custer. This is a replacement set he received prior to being shipped overseas in April 1944. They indicate his tetanus inoculation dates (T43 44), blood type (A), and Catholic religion (C). Serial numbers beginning with '3' indicate the man was drafted.
Photo courtesy of Mary McDonald.

Troop Train Travel:
Many of the rail journeys that Walt, his brother John, and other military personnel made during WW2 were in specially constructed troop train cars, similar to those shown in this New York Central Railroad ad in Life Magazine. His trip from Camp Custer in Michigan to St. Petersburg, Florida in January 1943 was likely in a special troop train. At other times Walt traveled in regular passenger train cars on scheduled passenger trains.
Greg Babinski Collection.

St. Petersburg, Florida, Hotel Beverley:
One of the many hotels taken over by the Air Corps BTC. Pvt. Babinski began his training and testing here. February, 1943.
St. Petersburg Public Library.

St. Petersburg, Florida, St. Mary's Catholic Church:
About five blocks from the Beverley Hotel, Catholic recruits attended Sunday mass here. February, 1943.
Babinski Family Collection.

Clearwater, Florida, Hotel Fort Harrison:
Walt continued testing and training while based here, where he was selected for aircraft mechanics school. February, 1943.
Babinski Family Collection.

St. Petersburg, Florida, the Municipal Pier:
The Hotel Beverley was located three blocks back from the beach, to the right of the street leading out to the pier. February, 1943.
Babinski Family Collection.

Clearwater, Florida, St. Cecilia's Catholic Church:
About a mile from the Hotel Fort Harrison, Catholic recruits attended Sunday mass here. February, 1943.
Babinski Family Collection.

Harvard Army Air Base: Pvt. Babinski looks glum with an unmilitary posture as he prepares for a stint on guard duty at bleak Harvard AAB.

Harvard Army Air Base: KP duty: Pvt. Babinski (L) and a friend take a break from 'kitchen patrol' chores.

Harvard Army Air Base: Pvt. Babinski (R) and friend pause in the chill April Nebraska air. Walt is dressed to go on leave, while his friend is dressed in his work fatigues.

Hastings, Nebraska: Aerial view of the commercial district in Hastings.

Harvard, Nebraska: Pvt. Babinski poses in front of the CB&Q RR Depot in Harvard in March or April 1943. Harvard, with just a few hundred residents, was about two miles south of the Air Base and about 15 miles east of Hastings.

Hastings, Nebraska: Pvt. Babinski pauses on a commercial street in Hastings, the liveliest city within reach of an airman from Harvard AAB with a day's leave to kill.

Harvard Army Air Base: Pvt. Babinski looks more at ease and up-beat in this pose. Has he recently received his orders to Aircraft Mechanics School at Keesler Field in Biloxi?

groups in Europe was not everywhere neat and tidy. Over vast quarters of the continent, various ethnic groups lived side-by-side or adjacent to each other. This was a consequence of the mixing of people that the large Russian, German, and Austrian empires had facilitated. And before them, the Polish Commonwealth, which in the 16th and 17th Centuries was the largest European state in area, was actually constituted as a multi-ethnic entity where Poles, Lithuanians, Germans, Jews, Cossacks, and others all had equal rights under the law. The result was that when the map of Europe was redrawn in the early 1920's, there were many Russians living in Estonia, Latvia, and Lithuania. Many Hungarians were left out of their state, to live as minorities in Czechoslovakia, Romania, and Yugoslavia. Poles found themselves living as minorities in Czechoslovakia, Lithuania, and Danzig, while Italian communities lived in Yugoslavia. And most ominously, almost every country of central Europe had large minorities of both Germans and Jews.

Outside of Europe and the Americas, most of the world was still controlled in the 1920's and 30's by the great European empires. Britain, France, Holland, Belgium, Spain, Portugal, and Italy controlled vast swaths of territory across Asia and Africa.[3] Among this club of European imperial powers, there were few disputes. By the 1920's most of them began to reduce their overseas military garrisons. They could always rely on the League of Nations to resolve any problems that arose (or so they thought), and so their budgets for navies and colonial army posts declined year by year.

However, simmering under the surface, imperialism still had the potential to upset world stability. There were four potential culprits. First, Germany had been stripped of its colonial empire in Africa and Asia by the Versailles Peace Treaty, which ended the war. Many Germans dreamed of regaining these possessions. Second, Italy, which had a small colonial empire in Africa, dreamed of adding possessions there if the opportunity arose. Some Italians and many Italian Fascists talked of a new Italian 'Roman Empire.' Third, there was a new player in the game of world imperialism: Japan. Japan had grown phenomenally as a world power since Commodore Perry's visit in 1853. It had acquired control of Korea in 1905 as a 'spoil' of its victorious war with Russia. In 1914-1918 Japan had been an ally of Britain, France, and the US, and acquired a string of island across the Pacific from Germany under the terms of the Versailles Treaty.[4] Japan saw itself as the equal of the US, Britain, France, and Russia, and coveted a large portion of the world imperial pie. In the 1920's and 30's Japan would look to China as ripe for exploitation and control. As the decades wore on, she would also consider the possibility of relieving the older imperial powers of some of their nearby possessions, if the conditions were right. Fourth was the growth of small but determined independence movements in many of the colonies themselves. These ranged from violent revolts of the sort that occurred in French Morocco or the Northwest Frontier area of India, to more peaceful and democratic independence movements led by then unknown activists like Gandhi in British India, Sukharno in the Dutch East Indies, or Ho Chi Min in French Indochina.

Racism was another problem during the era and one that most people did not substantially recognize at that time. Racism had a variety of manifestations. Many Americans now focus on the racism of the era that relegated blacks to second-class status in their own country. Within the U.S., racism had many regional flavors, certainly more pronounced against blacks in the South, against American Indians across much of the West, against Mexicans in Texas, and against Chinese and Japanese along the Pacific Coast. There was also American racism directed at European immigrant groups. In their turn, the Irish, Italians, Poles, Jews, and other newcomers were discriminated against by 'native' Americans or by recently assimilated groups.

Racism also included dislike and distrust between ethnic groups within Europe, which could quickly ignite into hate or violence under certain conditions. Here was a case where undemocratic leaders could manipulate nationalism to generate racist fears and hate to their political advantage. Racism of course was a key prerequisite to rationalize European, American, and Japanese imperial ambitions. Imperial racism could range from the relatively benign British version that could knight Indian princes or see Africans practice law in London, to the more blatant and exploitive racism of the Belgians and Japanese, who saw Congolese or Koreans as merely inferior 'under-races' who would benefit and learn by working for and serving their more civilized rulers. One curious aspect of European and American racism related to China and Japan. Americans, and to a lesser degree the British, had concluded that imperialism must not extend further into China. Modern Chinese leaders like Sun-yat-Sen and Chiang-Kai-Shek were idolized and admired by many Americans as bringing modernity and progress to China. Most Europeans and Americans though belittled and were suspicious of the progress that the Japanese had made in the previous three quarters of a century.

More importantly, their racism led them to deride and underestimate what Japan could accomplish and strive for. In many ways over the coming decades the world would pay dearly for its racism.

The dangers to world peace posed by nationalism, racism, and unchecked imperialism were recognized in the post-war years, leading to the establishment of the League of Nations as an international body that was intended to resolve disputes between nations peacefully. U.S. President Woodrow Wilson had conceived the League of Nations to help ensure that the needless slaughter of the Great War would never have to be repeated. The League was probably doomed from the start though, because the U.S. Senate refused to ratify the treaty and the Soviet Union was kept from joining. The Senate refusal stemmed primarily from another 'ism' – the isolationism that grew out of American disgust with the causes of the World War and the cost in American fortune and lives that had been required to end it. There was a very broad and deep sense in America in the 20's and 30's that never again should the country get involved in the problems and wars of the rest of the world.

The Looming Impact of World Events

However, one thing leads to another and just as the world would change, so too life in Delray and Carbonworks would change. The good economic times of the 1920's gave way to the crash of 1929 and the depression of the 1930's. American isolationism, coupled with denial of dangers to world peace in Britain, France, and the other major democracies, created a stage free of interference for Communism to solidify in Russia and for Fascism to take firm hold in Germany, Italy, and Japan. Within the Fascist states, dreams of imperialistic expansion grew. In 1931, Japan fabricated a provocation with China leading to the conquest of Manchuria. 14 years of continuous war between Japan and China began, but the first 10 years were substantially ignored by the rest of the world. America was the friend of modern China, but did not want to get involved and the League of Nations did nothing of substance in response. Four years later, in 1935, Italy fabricated a minor provocation and invaded Ethiopia, ending that ancient empire's independence after a thousand years. Once again, no effective response came from the League or from the U.S. By the end of the decade Italy would strike again, snuffing out Albanian independence and adding to its little 'New Roman Empire.'

The Fascists in Europe used Nationalism to further their international aims in the 1930's. Hitler used nationalism to violate the Versailles treaty in 1935, when he sent the (then) tiny German Army into the demilitarized Rhineland. Many outside Germany actually congratulated Hitler and Germany, for simply exercising their nationalist rights. The history of what happened over the next few years in Germany are well known now, but again the pattern of Hitler's plans and the looming danger were not universally well recognized at the time. After 1935, Germany began rearming (in violation of the Versailles Treaty) as rapidly as its industry and economy could manage. Then in 1938, German nationalism reasserted itself, first with the 'Anschluss' that united Austria with Germany. Later in 1938, after some mild protests from Britain, France, and Russia (and no involvement at all from the U.S.) Germany absorbed the Sudetenland region of Czechoslovakia (supposedly to protect its German population from Czech oppression). Within a year Germany not only completed the destruction of democratic Czechoslovakia when its troops marched into Prague, but also forced Lithuania to cede the port of Memel on the Baltic along with an adjacent strip of land with a small German population.

German propaganda then began to raise questions about reuniting the League of Nations mandated free city of Danzig with Germany. They also started to talk about the need for a German controlled road and rail corridor across Polish territory to unite East Prussia with Germany proper. In addition, concerns were raised about the supposed lack of fair treatment of the Germans who were left outside the borders of Germany when Poland had regained its independence in 1918.

The first real international mobilization of effort in opposition to Fascism during this period came not from nations, but from individuals. In the late 1930's, civil war broke out in Spain when Fascist army officers revolted against the democratically elected but socialist leaning government. The revolt, led by General Francisco Franco quickly developed into a stalemate. Franco's right-wing 'Nationalists' had support from the bulk of the Army, the business community, land-owners, and most of the Catholic church, while the left wing 'Loyalists' (so named because of their loyalty to the democratically elected government) had broad support amongst industrial workers, trade unionists, impoverished farm workers, and most students. To help their Fascist ally Franco break the stalemate, Hitler and Mussolini facilitated tens of thousands of German and Italian 'volunteers' who were sent (and equipped with the latest German and Italian tanks, artillery, and warplanes), to fight alongside their Spanish Fascist brethren. Slowly the tide of the civil war began to turn against the loyalists. Mild international protests against the German and Italian intervention had no effect, but individuals in France, Russia, Britain, the U.S., and dozens of other countries began to volunteer to fight alongside the Loyalists. Eventually more than 10,000 would serve in Spain. The thousands of Americans (including Ernest Hemingway) were organized into the Abraham Lincoln Brigade. This effort on the part of thousands of individuals helped show the world that opposition to Fascism was an acceptable moral position. In the case of Spain, it was a losing position, but the surviving veterans who returned in 1939 began to spread the word of how brutal Fascist military tactics could be.

Delray, like thousands of other local American communities had continued in isolation to most world events through the 1920's and 30's. Nationalism, Imperialism, Fascism, Communism, Racism, each had little direct meaning. Then as summer neared its end in 1939, Hitler struck again, invading Poland without warning on Friday, the First of September. Headlines in the Detroit News read: "Nazis Attack Poles by Land and Air; Warsaw Invokes Pact; Italy is Aloof." A large map on the front page of the paper depicted German attacks aimed around Cracow, which the Babinski and Kurzyniec families both realized would bring the fighting very close to their families left behind in Poland. Most Americans though took comfort from what Detroiters read off to one corner of the front page that afternoon: "Roosevelt Pledges Policy of Neutrality."[5]

Everyone in Delray had relatives in Poland or at least had friends with relatives in the 'old country.' A distant relative of the Babinski family, Zbigniew Babinski, from Sasnowiec, was a fighter pilot in the Polish Air Force.[6] One of Kunegunda's cousins back in Poland, Julian Gebolys, was also in the Polish Air Force[7]. The Babinski family prayed for them and the parishioners of St. John Cantius prayed for all of their compatriots back in Poland. The Poles and others in Delray now also had a new sympathy for their Czech neighbors who had seen their ancestral country snuffed out by Germany a few months earlier. Two days later, Britain and France declared war on Germany and the Second Great World War had begun. By the end of the month, Poland had been attacked by the Soviet Union also and was partitioned once more, ceasing to exist, except in the hopes and dreams of the Poles themselves. The Detroit newspapers still carried European war news, but more mundane local news was again finding a place on the front page, with stories about new auto industry union contracts, major public housing projects, remodeling of the Detroit Coliseum, and the election of a new president of the American Legion.[8]

Isolationism and a wide variety of political factors would keep America out of the war for more than two more years. Indeed, many felt very strongly that the U.S. should just not get involved in any way on the world stage again, no matter what the threat on the other side of the Atlantic or Pacific. Even in Delray, most people continued to focus only on family and friends, work and worship, and making the most of the American dream. How many on the friendly and familiar streets of Delray in September 1939 knew the profound impact world events would soon have on their hardworking and peaceful little community in general and on most of them very personally during the coming half dozen years? Very few, if any.

- 3 -
AT HOME 1939 – 1942: A TREMENDOUS SHOCK

Normal Life in Abnormal Times

Despite the troubling news from Europe in September 1939, life pretty much went on as normal in Detroit and Delray through the end of 1939, 1940, and most of 1941. Walter's dad and his brother John continued to work at Ford's Rouge plant. Walter senior still worked as a crane operator . In 1939 he averaged 40 hours per week, with one week off for vacation. He earned $1,530 for the year. During the same year, John also averaged a steady 40 hours per week. He also got a week off for vacation and earned $1,530 that year. The work at Ford was steady and the pay and hours were getting better as America finally began to pull itself out of the great depression. The economy had already begun to pick up when the war in Europe broke out. Soon Britain and France began placing orders for a wide variety of American products – from war supplies, to grain, meat, oil, chemicals, and hundreds of other key items. Ford and other major Detroit manufactures got many of these orders directly and also benefited from increased demand for autos and trucks as the rest of the American economy prospered.

Sisters Anne and Helen found steady work too, adding to the growing prosperity of the Babinski household. Anne started working at the Ford plant as an inspector and in 1939 worked 13 weeks at an average of 30 hours per week, for which she earned a total of $260 for the year. Helen found work as a housekeeper for the owners of a nearby Delray restaurant and nightclub, called Joey's Stables.[1] The owners lived in a flat above the restaurant and in 1939 Helen worked there an average of 60 hours per week and for 52 weeks she earned a total of $416.[2] Every son or daughter was expected to contribute some income, so the Babinski household had a reputation for big dinners and well-dressed family members. The Babinski family house at 9050 Pulaski was well maintained and in 1940 is was valued at $3,000.[3]

Young Walter, who had dropped out of school after the tenth grade at Southwestern High, also began a series of part time and full time jobs. Eventually Walt settled down working at the Shwayder Brothers Inc. manufacturing plant in nearby Ecorse. Shwayder was a Denver company that got its start during the 1910's manufacturing trunks and suitcases under the brand name 'Sampson.' By the 1920's, Shwayder claimed that they "…build luggage like Ford builds automobiles. Conveyors, assembling systems and automatic machinery, especially designed for us, enable us to produce quality luggage at prices previously considered impossible."[4]

Nevertheless, by 1927 demand for Shwayder's Sampson brand trunks and suitcases had outgrown the capacity of their Denver plant and the company opened a new 85,000 square foot factory in Ecorse to supply the East Coast market. Two years later though, Shwayder's business declined by half due to the 1929 stock market crash and production was cut back at both the Denver and Detroit plants. The company struggled through most of the 1930's and looked for ways to diversify and increase utilization of their manufacturing capacity. In 1931 they began experimenting with manufacturing and marketing folding card tables. Many of the card table manufacturing processes were similar to how Shwayder built trunks and suitcases, and they were able to produce a low cost quality product, also under the brand name Sampson. Originally of all wood construction, by 1936 tubular steel replaced wood for the table legs and a tube mill was put into operation at the Detroit plant. By 1938 card table sales had passed luggage sales for the first time and Shwayder also began manufacturing folding chairs to go with their card tables. In 1939 the brand name was changed to Samsonite.[5]

Walt was able to get work at Shwayder's as they expanded their operations to meet the growing demand for folding card tables and chairs. He did a variety of manual jobs at Shwayder, mostly related to the card-table line of products. In 1939, he worked a total of eight weeks part time for Shwayder's, earning a total of $75. Walt's commute to Shwayder's was easy. He would walk down to West Jefferson Avenue in the morning and catch a DSR bus headed south. The trip was about three miles, crossing the Rouge River through the suburb of Rouge River, then into Ecorse. Shwayder's was on High Street, just a block west of Jefferson, and backing on to a New York Central RR freight siding. In good weather, Shwayder's was close enough that Wally could walk home if he wanted to. Over the next several years, Shwayder's provided Walt with more and more work and he was able to contribute to the Babinski household and also keep a sufficient amount of disposable income to accommodate his own interests in sports and music.[6]

Wally cultivated two major spectator sports interests during these years, as well as two active sports interests as a participant. We have seen that he was a lifelong baseball fan and continued to enjoy playing when he could through these years. The Tigers in 1939 had had a good, but not great season. Led by manager Del Baker, the team finished the year with an 81-73 won-lost record, but just in fifth place. Hank Greenberg continued to lead the offense with 33 home runs and 112 RBI's, along with Charlie Gehringer and rookie Barney McCuskey in outfield. Tommy Bridges and Bobo Newsom led the pitching squad, but Schoolboy Rowe had a disappointing year with just 10 wins and a 4.99 ERA.

Walt's other favorite spectator sport was hockey. He had experienced the hometown Red Wings winning the Stanley Cup Championship in 1936 and 1937. The following three years the team had losing seasons, but Wally continued to cultivate a following for the sport and his steady income allowed him to take in the occasional game at the Olympia out on Grand River Avenue. For the 1940-41 season, the Red Wings improved to 21-16 and qualified for the playoffs, with star players like Sid Able and 'Black' Jack Stewart. However, Boston's star goalie, Frank 'Mr. Zero' Brimsek, allowed the Red Wings only six goals during the playoffs as the Bruins took the Stanley Cup in four straight games.

The other sport Walt participated in regularly was bowling. Now that Walt was working regularly, there was less time during the day for fun, and bowling allowed him to socialize with his friends in the evening and compete as well. Walt started to play on bowling teams and compete in various downriver leagues.

Big band swing music was another interest of Walt's. He did not have any particular favorite bands, but he enjoyed listening to music and occasionally going to downtown Detroit theaters when famous bands came into town. Hanging out with a circle of friends as a young man, Walt and everyone else began to have a nickname applied to him or her. At some point in the late 1930's or early 40's, Walt was christened with the nickname 'Lize' by his friends. Lize was a name taken from a character in an old Steppin Fetchit movie. Some of his other Detroit friends included Joe Kaszyca, known as 'Clark,' 'Pee Wee' Drabek (who was born the same year as Walt and lived in the house just to the west of the Babinski's on Pulaski), Edward 'Eddie' Nycz, and Joseph 'Babe' Lafata. Babe was one of the best baseball players in the Delray neighborhood.[7]

Two of the three Kurzyniec girls that Walt would see in the Pulaski Street store were given nicknames. The oldest, Martha, was nicknamed 'Jerk' while Dorothy was 'Butch.' Stella was too young for a nickname. Wally's brother John was known by some of his friends as 'Tornado John.'[8] Another nickname that Johnnie was known by was 'Bob' – from an Americanized pronunciation of the name Babinski – often pronounced more like 'Bobinski' then shortened in Johnnie's case to just Bob. The Babinski and Kurzyniec families continued to be very friendly to each other and the Babinski children would always be welcomed into the store at 9014 Pulaski.

Thomas Kurzyniec continued his job as a working foreman at the Ternstedt Division of GM. In 1939, he worked an average of 37 hours per week with two weeks off for vacation, earning a total of $1,700 for the year. His wife Martha continued to tend the store, working an average of 49 hours per week in 1939. The store was owned by Thomas' cousin Anthony. Anthony (age 29 in 1939) and his brother Jacob (a year older than Anthony) lived in the upstairs flat in the store, which was valued at $500. Anthony worked as an agent for a nearby coke factory, while Jacob worked only part time as a stockman in a bearing plant.[9] Thomas, Martha and the three girls lived in a small flat on the first floor, behind the store.[10]

Around 1939 Martha Kurzyniec completed Eighth Grade at Morley School, and then started at Southwestern High School. The following year Dorothy Kurzyniec started at Southwestern also, as did Wally's sister Genevieve. Dorothy and Gennie knew each other as neighbors on Pulaski, and continued as friends through their four years at Southwestern High. Often they would walk to school together. It was a mile and a half walk from Pulaski Street to Southwestern, on Fort Street near Post. Martha, Gennie, Dorothy (and Walter for two years before them) made this walk to and from school every day, rain or shine, autumn, winter, or springtime. A particularly harrowing part of the trip was crossing the New York Central Railroad tracks to get to Fort Street. The crossing was at one end of the major NYC freight yard in Detroit, and involved crossing 16 sets of tracks, with freights, passenger trains, and switch engines shuttling to and fro. As the girls got older, they would occasionally get a ride home from school in car owned by one of their friends. It was typical for six or seven kids to pile into the car for the ride back home to Delray.[11]

Wally's brother Johnnie continued his periodic fishing and hunting trips up north. In June 1940, he spent a week at a cabin owned by John Tobias at Norway Cove on Lake Avalon in Montmorency County, near Hillman, Michigan. He wrote his ma and pa:

Hello: Ma & Pa & the Family...
The weather is bad, it's raining, but it's nice out here and quiet. Having a nice time.
-Johnny[12]

As war raged in Europe and threatened in Asia, life went on in the U.S., seemingly unaffected. The German blitzkrieg had struck again in Europe in April and May 1940. Denmark and Norway were quickly conquered by the Germans, followed by the Netherlands and Belgium. The Detroit News reported: "Nazis Occupy Copenhagen and Oslo; Danes Submit Without Fight."[13] Later headlines announced: "Dutch Cities Bombed: Battle Rages in Rotterdam: Germans Enter City by Air and Seaplanes."[14] Then the German armies poured into France, quickly overwhelming the British and French. All seemed lost for the Allies, then there was a miracle at Dunquirk, as a third of a million British and French troops were evacuated by sea to England in the face of the German army and air force. Winston Churchill came to power as Prime Minister and the British will seemed to stiffen, even after France surrendered. A few months later, the Battle of Britain demonstrated that the Germans were not invincible. Americans listened to radio reports broadcast from London by Edward R. Morrow, who described both the horrors of the German terror bombing of civilians, as well as the continued resolve and success of the British people, their leader Churchill, and the Royal Air Force. In October, Italy attacked Greece from Albania, but amazingly, the Greeks stopped them dead in the rugged mountains of their country.

The little news that the Babinski and Kurzyniec families got from their relatives back in German and Soviet occupied Poland was not good. Life under the Nazis was not easy. However, they knew that one relative had escaped the Nazis and made his way to England. Lt. Julian Gebolys, one of Cora's cousins, had been a pilot and instructor in the Polish Air Force. He was one of thousands of Poles who refused to accept defeat and vowed to fight on against the fascist and Soviet invaders of their homeland. But there was no news at all of Capt. Zbigniew Babinski, another relative captured by the Red Army shortly after the attack on Poland in mid-September 1939.

Still America was nowhere near participating in the war. To be sure, there was a steady stream of news about defensive preparations Roosevelt and the Congress were making. At the end of May, the newspaper reported: "U.S. Arms Program in Full Swing,"[15] "Ford Studies Army Plane,"[16] "Billions Voted by Congress: Tax Rates Up,"[17] and "Treasury Backs New War Tax."[18] But what was the big news in Detroit during 1940? The Tigers improved to 90-64, finishing in first place in the American League! Wally's steady income allowed him to attend Saturday or Sunday games at Brigg's Stadium on Michigan and Trumble, usually in the bleachers. That summer of 1940 he saw Hank Greenberg lead the league in home runs with 41 and win the MVP award. Infielder Rudy York was not far behind with 33 homers and 186 hits, while Barny McClosky slammed 200 hits also. Bobo Newsom led the pitching crew with 21 wins, while Schoolboy Rowe improved again to 16 wins. The Tigers went seven games in the World Series, but lost the last game to the Cincinnati Reds.

In September, a year after war began in Europe, reports continued of Luftwaffe attacks: 'London Raided All Day; Dover Shelled.' The *Detroit News* also reported that FDR has signed a bill reestablishing the military draft, with October 16 set as the date for the first call-ups.[19] However, life in America seemed safe from war and Americans were not taking chances as they re-elected Franklin D. Roosevelt to an unprecedented third term as President in November 1940. Defensive preparations continued, as National Guardsmen were called up for training all across the country, and by a new peacetime draft that saw Detroiters being called into the service beginning in January 1941.

In late 1940, as the British Royal Air Force defeated Hitler's Luftwaffe in what came to be known as the Battle of Britain, many Polish-Americans in Delray thrilled at the exploits of the handful of trained Polish Pilots who had escaped to England and helped fly RAF Hurricane fighters in the battle. Walt may have heard his mother talk about her cousin, Julian, who was rumored to be among the Poles fighting with the English in Britain.

In 1941, Americans and the rest of the free world watched the weekly newsreels as Britain continued to hold out against Germany and Italy, winning a series of victories in North Africa. Hitler countered with a move of his own into Yugoslavia and Greece in April, ending the independence of two more countries. In May, the powerful German battleship Bismarck broke into the North Atlantic after sinking the British battleship HMS Hood. The Bismarck threatened the convoys bringing vital war materials from the U.S. and Canada to Britain. After a desperate search though, the Royal Navy tracked the Bismarck down and sank it before it could return to a German controlled port.

That month Detroiters read "U.S. in State of Full Emergency as Roosevelt Warns of Armed Action: Freedom of Sea is Proclaimed: Capital and Labor Told to Put Defense Work First."[20] The following month Hitler again surprised the world with a massive sneak attack against his ally Stalin and the Soviet Union: "German Troops Smash Into Russia As Hitler Proclaims War on Ex-Ally."[21]

A little known connection between Michigan and Poland at this time was the re-training of Polish non-commissioned officers (NCO's) at Camp Custer. These men fought in France for the British in 1940 and were veterans of the Dunkirk evacuation. Later they came to the U.S. After their training was complete, they paraded in Detroit with their commander, Captain George Ciepielowski, for inspection by Major General C. R. Powell of the U.S. Army in front of the Federal Building in downtown Detroit.[22]

However, with the U.S. still at peace, dramatic world news still often took second place to local happenings. On May 22, Father Walczak, pastor of St. John Cantius Church since 1902, died after a long illness. He was greatly loved by his parishioners, having come from Austrian controlled Poland like the Babinskis, Kurzyniecs and many others in the neighborhood. Father Vincent Anuskiewicz replaced Walczak as pastor on July 8, 1941. This appointment actually created some conflict as many parishioners had hoped that Father Tompor, a Pole, rather than Father Anuskiewicz, a Lithuanian, would be made pastor. There were actually a number of petitions circulated amongst the parishioners, requesting that the Archdiocese of Detroit replace Father Anuskiewicz, although it does not appear that the Babinski's or Kurzyniec's participated.[23]

In mid-August, 1941 the newspapers and newsreels reported that President Roosevelt and Prime Minister Churchill had met at sea in secret and agreed on 'War Aims' even though the U.S. was not at war. They signed a document known as the 'Eight Points' that envisioned a world that abandoned force as a tool to settle disputes and promoted fair trade, access to raw materials, social security, and fair labor standards for all. They also agreed that the end of Nazism was a joint objective.[24] Most Americans must have wondered, reading news such as this, how long it would be before they were drawn into war with Germany.

That summer the Tigers dropped back to fourth place, with a 75-79 won-lost record, although Wally continued as a faithful fan, attending games when he could or listening on the radio other times. Into the fall, the Red Wings had a poor start to the hockey season, with a 4-7 record by the beginning of December. All summer and fall, the *Detroit News* talked about the German Army driving deeper into Russia. Now at the beginning of December, they were within sight of the Kremlin, 20 miles outside of the city. The world held its breath…was Russia about to fall as had so many of Hitler's other victims?

The World Intrudes: December 7, 1941

Sunday, December 7, was cold in Detroit. Most every family in Delray attended mass that Sunday morning in their local parish church. The Babinski and Kurzyniec families attended mass at St. John Cantius church and then headed back home to get out of the cold and to enjoy a day of rest and good food. Young Wally read the Sunday paper, and then tuned the radio to listen to the Red Wings game that afternoon at the Olympia against their rivals the Montreal Canadians.[25] The Kurzyniec family was gathered in their little living room at the back of their store, having finished their Sunday supper and cleaned up the dishes. At some point early in the afternoon, the bells of St. John Cantius began to ring. Living in the shadow of St. John's, the Kurzyniec family noticed the unusual ringing of the bells right away and wondered aloud what was going on. Thomas got up and turned the knob on their console radio. The vacuum tubes behind the grill began to glow as the radio warmed up, then they heard a bit of static followed by a reporter's voice. They heard the announcer state:

> *"The Japanese have attacked Pearl Harbor, Hawaii, by air, President Roosevelt has just announced. The attack also was made on all naval and military activities on the principal island of Oahu. We take you now to Washington. The details are not available. They will be in a few minutes…"*[26]

At first they were all confused. What was, where was, Pearl Harbor? Then, within a minute or two it became apparent to everyone in the room what the reporter was announcing: the American naval base in Pearl Harbor, Hawaii had been attacked by aircraft of the Japanese Navy just after dawn...and the reporter made it clear that the attack was still going on as he spoke, with follow up waves of planes.[27] By mid-afternoon, live reports were coming in from Pearl Harbor. One NBC reporter climbed to the roof of the Advertiser Building in Honolulu with a microphone and broadcast: "We have witnessed this morning the attack of Pearl Harbor and a severe bombing of Pearl Harbor by army planes, undoubtedly Japanese. The city of Honolulu has also been attacked and considerable damage done. This battle has been going on for nearly three hours....It's no joke, it's a real war."

The reactions of people recorded around the country that Sunday were likely typical of most people in Delray:

> *Reporter: "How do you feel about, when you first heard about the Japanese war?" Man on street: "Well, uh, I felt...I'll be called into the draft pretty soon. I'm eligible, I'm in the 1-A classification and it hit me pretty bad. I wasn't expecting it so soon."*
>
> *Man on street: "Well I was surprised that it was so sudden, I think Japan has an awful lot to lose by doing such a thing. "Are you eligible for the draft too?" "That's right. I expect to be called any week now."*
>
> *"My name is Pete Seeger. It was Sunday here....somebody burst in saying, 'Hey, the Japs have bombed Pearl Harbor.' I didn't even know where Pearl Harbor was. He said, 'Don't you know? It's the big navy base in Hawaii.' Ah-ha. We were at war, finally. We could see it coming but now there was no turning back."*
>
> *"My name is Helen Thomas. I was home in Detroit on December 7th. I was just a schoolgirl. It was a tremendous shock. At the same time it was not on the mainland here, and so still had the sense of being remote, but not really remote because every family felt it at that time. The draft had been going on since 1940..."*[28]

Many people gathered that Sunday and talked for hours. What did it mean? Would the Japanese bomb the American mainland? Or invade? People were in a daze.[29]

The following Monday Wally, his pa, and brothers and sisters went to work or school as normal, but everyone was talking about what the attack would mean...war had come again, but how would the U.S. respond? The newspapers that morning reported news from many fronts, and none of the headlines sounded good:

> *"3,000 Casualties in Raid on Hawaii"*
> *"Battleship Capsizes; Destroyer Blown Up"*
> *"Philippine Army Base is Bombed"*
> *"Japs Claim Supremacy in Pacific"*
> *"Lights Dimmed in Capital as War Flares"*
> *"City on War Basis; Plants Under Guard"*[30]

During the middle of the day, in most schools and factories across the country, workers and students paused and gathered around radios to hear an address by President Roosevelt to a joint session of Congress:

> *"Yesterday, December 7, 1941 - a date which will live in infamy - the United States of America was suddenly and deliberately attacked by naval and air forces of the Empire of Japan. The United States was at peace with that nation and, at the solicitation of Japan, was still in conversation with its Government and its Emperor looking toward the maintenance of peace in the Pacific. Indeed, one hour after Japanese air squadrons had commenced bombing in Oahu, the Japanese Ambassador to the United States and his colleague delivered to the Secretary of State a formal reply to a recent American message. While this reply stated that it seemed useless to continue the existing diplomatic negotiations, it contained no threat or hint of war or armed attack.*
>
> *"It will be recorded that the distance of Hawaii from Japan makes it obvious that the attack was deliberately planned many days or even weeks ago. During the intervening time the Japanese Government has deliberately sought to deceive the United States by false statements and expressions of hope for continued peace. The attack yesterday on the*

Hawaiian Islands has caused severe damage to American naval and military forces. Very many American lives have been lost. In addition American ships have been reported torpedoed on the high seas between San Francisco and Honolulu.

"Yesterday the Japanese Government also launched an attack against Malaya. Last night Japanese forces attacked Hong Kong. Last night Japanese forces attacked Guam. Last night Japanese forces attacked the Philippine Islands. Last night the Japanese attacked Wake Island. This morning the Japanese attacked Midway Island. Japan has, therefore, undertaken a surprise offensive extending throughout the Pacific area. The facts of yesterday speak for themselves. The people of the United States have already formed their opinions and well understand the implications to the very life and safety of our nation.

"As Commander-in-Chief of the Army and Navy, I have directed that all measures be taken for our defense. Always will we remember the character of the onslaught against us. No matter how long it may take us to overcome this premeditated invasion, the American people in their righteous might will win through to absolute victory.

"I believe I interpret the will of the Congress and of the people when I assert that we will not only defend ourselves to the uttermost but will make very certain that this form of treachery shall never endanger us again.

"Hostilities exist. There is no blinking at the fact that our people, our territory and our interests are in grave danger. With confidence in our armed forces - with the unbounded determination of our people - we will gain the inevitable triumph - so help us God.

"I ask that the Congress declare that since the unprovoked and dastardly attack by Japan on Sunday, December seventh, a state of war has existed between the United States and the Japanese Empire."

Later that day, Congress declared war against Japan and before the week was over, Germany and Italy declared war on the U.S.

The remaining days of December 1941 in Detroit saw a strange combination of normalcy and frantic reactions to the outbreak of war. The *Detroit News* counted down the remaining shopping days until Christmas prominently on the front page each day, next to reports of war news from the Pacific and war preparations at home. The News reported mundane front-page stories throughout December, covering topics like new left turn driving restrictions on Grand River Avenue, a farmer in South Carolina rescued after three days stuck in a well, and the theft of a beer truck in Spokane, Washington. War news close to home included the arrest of 38 Germans and two Italians in Detroit, seized by the FBI as 'dangerous aliens.' At the same time, the *News* reported that the FBI interviewed each of the 100 or so Japanese living in the Detroit area, but they were all allowed to return to their homes. Army troops were seen guarding the Ambassador Bridge and the Detroit-Windsor Tunnel, as well as major industrial plants around the city. The war news was bad – two British battleships sunk by the Japanese off Malaya, Manila under air attack and Japanese troops landing on Luzon Island in the Philippines. Wake Island and Hong Kong were both reported to be holding out desperately against Japanese attacks. Ominously for young men and their families across the country, there were reports that the U.S. Selective Service Administration was considering expanding the draft to cover all males between 18 and 64 years of age, from the current range of 21 to 27 (with registration between 20 and 35).[31]

The weather in Detroit turned mild just before Christmas, with highs in the low 50's. There was plenty to eat and seemingly good economic prospects for the Babinski and Kurzyniec families that Christmas season. Both Thomas Kurzyniec and Walter Babinski had reliable work and plenty of overtime at Ternstedt and the Rouge Plant, where Johnnie worked also. Young Wally's prospects seemed good too. Shwayder's was doing well with steady work, and there were rumors that the plant might begin converting to war production soon.

Detroit greeted the New Year and 1942 with a blizzard that brought snow and high winds. Detroiters enduring the blizzard read the war news that reported that the Nazis had been stopped by the Red Army outside Moscow in deep snow and subzero temperatures. This good news contrasted with the continued bad news coming out of the Pacific, as Japan continued to push back the British, Dutch, Americans, and Philippinos in Malaya, the Dutch East Indies, and the Philippine Islands. The steady conversion of the American economy in general, and the Detroit area in particular into the "Arsenal of Democracy' is reflected in typical *Detroit News* headlines as 1942 progressed. Restrictions were announced on the civilian economy and consumer goods were curtailed:

> *"Auto Ban Clears Plants for War"*
> *"Tire Regulations are Released"*
> *"No More Girdles Girls; The Rubber Won't Stretch"*

At the same time, ambitious plans were announced by the government to begin producing the arms needed to win the war:

> *"12 Billion Asked for 33,000 Army Planes"*
> *"Navy Bill asks over 17 Billion; Program Described as Staggering"*[32]

In mid-February, Johnnie was required to register for the draft, as were millions of other young men across the country and an estimated 100,000 in Detroit. War news included German submarine attacks against allied ships off the East Coast and in the Caribbean, and black outs against night time bomber raids in east coast cities, California, and even Detroit. By March, Singapore and Batavia had fallen, as had most of the Dutch East Indies. In the Philippines, only tiny Corregidor Island in Manila Bay still held out. Further west, the Japanese had invaded Burma and naval attacks against India were feared.

At home in Detroit, the bad war news was tempered for some by familiar routines and diversions. Wally followed the progress of the Red Wings, after they qualified for the playoffs by mid-March. They beat Montreal and Boston in semifinals, but lost in the Stanley Cup championship series to Toronto, 4-3.

Finally, after more than four months of unending Japanese victories, Detroiters read some encouraging news on Saturday, April 18th: "Tokyo Bombed by American Fliers; 3 other Japanese Cities Also Raided"[33] In Britain, the RAF had been launching bombing raids against Germany for two and a half years. But this was the first blow by the U.S. against the homeland of the enemy. Details released over the coming days made it clear that this raid was by 16 Army Air Force Bombers, led by Gen. Jimmie Doolittle, launched from an American Navy aircraft carrier. For the first time, the U.S. could demonstrate the will and ability to strike right at the heart of its foe. This was a big morale boost to the American public, and many young men began to think about serving their country as a bomber crewman, when their time to be drafted came.

Americans on the home front went on with their lives mostly as normal. However, one by one, the young men in Delray and thousands of other communities across the country began to get their draft notices or in some cases, they made the decision to volunteer for service on their own. Departure of these men for training in the army, air corps, or navy was usually preceded by a couple of weeks of preparations and last chances to hang out with friends and say goodbye to family members. This all occurred within familiar patterns...work with likely overtime in some nearby factory or war industry plant, then a double feature at the movies on Friday or Saturday night, or a late evening spent at a bar or pool hall or bowling alley. Sunday morning would bring church and dinner with the family, then perhaps taking in a ball game if the Tigers were in town, or listening to the game on the radio if they were away. Many of the star Tigers from 1940 were gone, Hank Greenberg and Charlie Gehringer were typical of many, volunteering for the Army and turning their backs on their baseball career at the peak of their games. For the Tigers in 1942, Rudy York moved from catcher to first base and was the batting leader along with Barney McClosky in outfield. Dizzy Trout and Virgil Trucks were the journeyman pitchers and Hal Newhouser the outstanding rookie pitcher for the season. Still, the Tigers struggled from the beginning of the season.

As baseball season opened in the spring, Japan continued its advance in the Pacific and Indian Oceans. Corregidor fell in early May. After a major naval raid against India, the Japanese began moving to the southeast, along the north coast of New Guinea, then into the Solomon Islands. It was clear that their goal was to cut the supply lines between the U.S. and Australia by occupying a string of island bases across the Pacific. Rabaul and its harbor in the Solomons fell to Japan, then one by one additional island to the south were seized. In early May, the *News* reported a major naval battle between the U.S. and Japan in the Coral Sea, with both sides loosing major ships, but still the Japanese advance continued. Then Alaska was attacked by Japan, first with carrier air raids, then landings in the Aleutian Islands. Finally in June came news of a major victory at Midway in the

Pacific. The *Detroit News* headline "8 to 16 Japanese Warships Damaged in Crushing Defeat in Mid-Pacific"[34] was incorrect in its numbers, but correct in its assessment.

In early July, just before Walt's 20th birthday, war news was still bad, with little to cheer the allies. In Europe and Africa, the Germans were on the move again. The British had been driven from Libya and back deep into Egypt where they began to dig in to try to stop the Nazis and Italians one last time at a place called El Alamein, just 30 miles from their naval base at Alexandria. In Russia, the Nazis, with their Italian, Rumanian, and Hungarian allies were attacking to the east again, not towards Moscow, but this time towards Leningrad in the north and to the Caucasus Mountains and the Russian oil fields in the south.[35]

On Tuesday, June 30th, Wally left work early at Shwayder's and reported to the Selective Service Office downtown on the eighth floor of the Owen Building at 250 Lafayette between Shelby and Washington Boulevard, where he registered for the draft. Many of his friends had already done so or would soon. Beulah Saurtrey, the registrar who processed his registration recorded that Wally was 5'-10-1/2" tall, weighed 165 pounds, with blue eyes, brown hair, and light complexion.[36]

Walt turned 20 on Sunday, July 5, 1942. After mass and Sunday dinner with his family, he drove out to Belle Isle Park with a couple of his Delray neighborhood friends, Joseph 'Clark' Kaszyca and Robert Symborski. Pictures taken of Walt at the Belle Isle Zoo, lounging on the Cass Fountain, and on the steps leading up to the conservatory, show him looking relaxed, confident, and fit. His hair is cut short, with a slight 'flat-top' look to it. A picture taken of Walt and his two friends on another day off that summer shows them in a field with a number of cars parked off in the distance. One of them has driven his car out onto the grass and they're posed, Clark with a cigarette in his mouth, Robert with a mug of beer in his hand, and Walt with a beer in one hand and a cigar in the other. They clearly look like very young men trying to appear older, but perhaps realizing that they would be growing up quickly soon. They each knew the meaning of the draft cards they now carried in their wallets, and the bureaucratic wheels that were turning that would eventually generate a call to each one to serve his country.

In early August, news began to trickle in on the pages of the *Detroit News* about a battle developing on Guadalcanal, the capital island of the Solomons, and the site of the southernmost Japanese air base in their drive to cut off shipping lines to Australia. The *News* reported that the Marines had landed on the island and were making progress against the Japanese defenders. Few Americans had any idea how long the U.S. and Japan would struggle on Guadalcanal. Wally may have read about this and also about the progress at Ford's Willow Run bomber plant, where the new B-24 long-range bombers were now slowly rolling off the assembly line. At the same time, the paper reported the first big USAAF bomber raid against the key Japanese naval and air base at Rabaul at the northern end of the Solomon chain.[37]

As the summer wore on, Wally, his brother Johnnie, and many of their friends talked more and more of the draft and the alternatives. The war was not going well, not in Europe, nor in the Pacific, and they all knew that the military would be looking for millions of young men to fill its manpower needs. They had all seen some of their friends get their draft notices already and leave for training, usually in the Army. Knowing the history of the bloody toll taken on the infantry during the First World War (as it was now known) as well as the news of horrific infantry battles raging in Russia, China, and Africa, the infantry was not a desirable way to serve your country in the eyes of many Americans. Especially for many of the young men of Delray, who had lived their whole lives in the heart of industrial America, there was a sense that the application of industrial technology to modern warfare was going to be a key factor in winning the war. Because most of them had worked in or around factories, many felt they could make an important contribution by serving in the branches that relied most heavily on technology: the Navy and the Army Air Force.

There was another option too, and that was continuing to work in war industry. In 1942, the supply of weapons, uniforms, tanks, planes, and the thousands of other items needed by the millions was nowhere near sufficient to equip the needs of the military. It was relatively easy to get a draft deferment for war industry work and many took advantage. Wages were increasing too, so the money was good and the financial incentive to continue working was great. Most Americans were motivated though to do what they could to best serve their country. Many enlisted, and in the summer Johnnie began to think about this more and more. Finally, he made his decision after talking to a recruiter – he would join the Navy, with the promise that he would have a good chance of advancing his technical skills aboard ship in the Navy. By the beginning of August, he gave notice at Ford, ending seven and a half years as a sheet metal layout man. His pay rate when he left Ford was $60 per week.[38]

On Thursday, August 13, 1942, Johnnie reported to the Broadhead Naval Armory on East Jefferson Avenue where he was sworn into the Navy for a three-year enlistment. Instead of the $60 per week salary he earned at Ford Motor, he was to be paid $78 per month as a seaman first class. He also received an enlistment uniform allowance of $133.81. His military records show that when he enlisted, he was 29 years old, weighed 186 pounds, stood 5'-7-1/2" tall; with blue eyes and light brown hair.[39] A photo taken sometime shortly after his enlistment shows him staring confidently at the camera in his Navy uniform and with his hat cocked slightly to his right. This photo was sent home and sat in a frame in the Babinski family living room for the duration of the war.

The following morning, Friday, August 14th, Johnnie said goodbye to Wally, Anne, and Helen, as they left for work early in the morning. They each wished him luck, told him to take care, and each hoped in their heart and mind for his safe return. Walter Senior, who had arranged to take the morning off from work, helped Johnnie load the small bag he was limited to taking into the family car, then they got in along with Johnnie's ma, and his siblings Gennie, Henry, and Leona. They drove over to the Michigan Central Station on Michigan Avenue, where they said their goodbyes. Johnnie received hugs from his three younger siblings, then a firm handshake from his pa, then a hug and tearful kiss from his ma, who got Johnnie to promise to be very careful and not take chances. Then he turned and walked into the station, to find the USO lounge where he was to find the rest of the party of Michigan Navy recruits he had been ordered to travel with to basic training.

He quickly found the officer in charge, reported in and soon the group was aboard a NYC train headed west. By early afternoon they had arrived in Chicago. They had time for lunch before changing over to a CNW train for the short trip north to the Great Lakes Naval Training Center, just past Lake Forest Park, Illinois. Here Johnnie and groups of recruits were assigned to an NCO who announced that they would be under his command for the duration of their eight-week basic training that would begin as of today.[40] Johnnie was assigned to the 4th Regiment, 23rd Battalion, Co. 750, USNTS Great Lakes. His civilian clothes were exchanged for poorly fitting military fatigues. Later Johnnie and the other recruits would use their allowances to purchase tailored dress uniforms.

Within a week and a half or so, Johnnie was able to start writing back to his family, to let them know how things were going. Wally received a postcard showing hundreds of young recruits seated in a mess hall at Great Lakes Naval Training Station, dated August 24, 1942:

Hello Walter Jr.
This is how we get our (chow) or food. Its all cafeteria style. Its very clean but no beer.
-Bob[41]

By September, the Tigers ended a disappointing season with a 73-81 win-loss record. During the third week in September, the family received news from Johnnie that he had completed his eight weeks of basic training at Great Lakes Naval Training Center. A few days later a letter came saying that he had been selected for additional training at Great Lakes – he was to start a 16 week long Mariners Mechanics Course.[42]

Early in October Wally's pa received a postcard from Johnnie at GLNTS, showing hundreds of sailors seated in a huge drill hall. Johnnie wrote:

Hello:
Well everything is allright yet. The weather is fine, nice and warm, but the mornings are cool. This is a drill hall where we have our mass.
-Johnny[43]

That fall, the war news focused on the German siege of Leningrad and the Nazis attempt to capture Stalingrad from the Soviets. In North Africa, the British and Axis forces remained deadlocked in the dessert, but the threat to Egypt and the Suez Canal seemed to have passed as the Brits were now holding their own.

On Saturday, November 1, Wally might have gone out to the Olympia to see the Red Wings open their 1942-43 season with a game against the Boston Bruins. The Red Wings would win 3-0. Sid Abel, who had joined the Red Wings in 1938, was now team captain. Under his leadership the team had an outstanding season. Four days after their opener, the Red Wings demolished the New York Rangers with a 12-5 win at home.

On Sunday, November 8, the *Detroit News* reported a major surprise attack by American and British forces: "U.S. Forces Invade French Africa; This is 2nd Front, Says Roosevelt." Except for New Guinea and Guadalcanal in the South Pacific, eleven months after the U.S. entered the war, this was the first major commitment of American ground forces in combat with the Axis. Indeed, the paper included stories about the continuing fighting on Guadalcanal and throughout the Solomon Islands. There was also encouraging news about 100,000 Axis troops being surrounded by the British under Montgomery, attacking westward out of Egypt at the other end of North Africa. The paper also included news about gas and food rationing, which affected everyone on the home front.[44]

A week later the paper was reporting the rapid progress of American and British forces in North Africa, as they swept across Morocco and Algeria, toward the Germans and Italians in their last stronghold in Tunisia. The first photos of the fighting and of American troops in North Africa brought the reality of the war home to Americans, and especially to the young men waiting for their call in the draft. The news also referred to continued bitter fighting with the Japs in the South Pacific. Other prominent front-page articles included more rationing news. The *Detroit News* also featured front page cooking advice, with suggestions on how to use the limited cooking oils, meat, and sugar available through ration allotments most effectively for home meals.[45]

The last few weeks of the year passed by quickly, with the war news from Russia, North Africa, and the South Pacific as the backdrop. Again, normal routines of life continued. Thanksgiving, Christmas, and New Year's Eve were celebrated in their turn.

However, for Walt, his world was definitely changing. His brother Johnnie was off in the Navy. His circle of friends was getting smaller week by week. Joe 'Clark' Kazyka, Buzz, Joe (Pudgy) Prus, Stanley Prus, Ed Czopeto, Joe (Fuzzy) Sklarzek, Joe Suchyta, Joe (Jo-Jo) Jekielek, Joe Kurzyniec, Joe Krupka, Adam Sordyl, Theodore Sordyl, Joseph Stoklosa, and Eddie Chopek were just some of Walt's friends who had entered the service or knew their time would come very soon. At work, he watched the other men his age trickle away one by one too, as their draft call came up. Walt managed to finish the year at Shwayder's, having earned a total of $1,782.69.[46]

By the end of the year the Red Wings were having a good but not great season. On New Year's Eve they won their tenth game of the season against the Rangers in New York to end the year with a 10-6-5 W-L-T record. As 1942 came to an end on that cold Thursday in Detroit, Walt and millions of other Americans wondered what changes and challenges the New Year would bring.

- 4 -
INDUCTION AND BASIC TRAINING 1943: CONFIDENT FAITH

Induction

Walt, his pa, and sisters Anne and Helen returned from work on Tuesday evening, January 5, 1943, looking forward to the family dinner that his ma and sister Genie had been preparing for them and the little ones Henry and Leona. Shortly after walking thru the door into the warmth of the house on Pulaski Street, his ma nervously handed Walt an official looking letter that had arrived in the mail earlier that day. By the return address they all guessed what it likely meant, for many of the young men they knew in their neighborhood, from school, church, and work, had already received similar letters.

Glancing at the return address from Wayne County Local Board 43 of the Selective Service System, Walt opened the envelope and read:

ORDER TO REPORT FOR INDUCTION
The President of the United States,
To Walter Anthony Babinski Jr.
Order No. 13551
Greetings:
Having submitted yourself to a local board composed of your neighbors for the purpose of determining your availability for training and service in the armed forces of the United States, you are hereby notified that you have now been selected for training and service in the Army.
You will, therefore, report to the local board named at 3162 East Jefferson Ave. Detroit, Michigan at 7:00 A.M., on the 14th day of January, 1943.[1]

Reading further, they learned that when Walt reported to the draft board office he would be examined to determine his suitability for military service. He was advised to keep this in mind when arranging his affairs over the remaining few days, including letting his employer know he had gotten his induction notice, and that if accepted for service he would need to be replaced at work.

Dinner conversation that day and into the evening must have focused on what awaited Walt. At twenty years old and healthy, his induction notice was no surprise. Walt's older brother Johnnie had enlisted in the Navy the previous year and was still completing advanced mariner mechanics training at the Great Lakes Naval Training Center in Illinois.[2] While the U.S. had been at war for more than a year now the path to victory was far from certain. Indeed, as they contemplated the future for their son and brother that January evening, the Babinski family was well aware from the news that epic battles were raging at that time in three corners of the world. In Russia, vast Soviet forces were locked in battle with the German Army at Stalingrad, and Hitler was boasting to the world that he would smash the Red Army. In North Africa, American and British forces had driven across Morocco and Algeria from the west and British General Montgomery had driven out of Egypt and across Libya from the east, but the Germans and Italians in Tunisia had stopped the allies and were receiving reinforcement every day. In the southwest Pacific, American Marines and the Japanese seemed to be locked in a never-ending battle of attrition at a place called Guadalcanal Island. As Walt went to bed that night his mind must have raced, wondering what the future would hold for him.

The following day at the Samsonite plant Walt informed his boss of his induction notice. Once again, it came as no real surprise – every employer had experienced the same many times over during the past year. It was part of doing business during a time of total war. He just asked Walt to let him know the day after he reported for induction whether he was accepted or rejected, although he must have been quite certain that he was losing another good worker to military duty.

The weekend after receiving the induction notice Walt was busy letting his friends know he had been called up. They talked about who had left for service already and what they knew of the process. He likely sent a letter off to his brother John in the Navy, letting him know the news as well.

Very early on Thursday, January 14 Walt woke up and ate a hearty breakfast with his ma and pa. He then walked in the freezing early morning weather the few blocks to Fort Street, where he boarded an eastbound streetcar, paying his six cents fare and an extra penny for a transfer. Once downtown, he transferred onto an East Jefferson streetcar getting off about a mile and a half east, just past McDougall Street. The induction center was in a low, modern, brick industrial building that the Army had converted less than a year earlier. Arriving just before the appointed time of 7:00 a.m., Walt found that dozens of other young men were waiting for processing also. He quickly checked in with a clerk at the center, presented his draft card and was told where to report next. Later in the morning he went through a series of standardized physical exams, where to no surprise he was found to be fit for service. More time was spent waiting, and then he and a group of other inductees were called by name, one by one, and assembled in a room where an officer read them the "Articles of War" then swore them in as members of the U.S. Army.

The "Articles of War" was the legal document that defined the obligations of an enlisted man while in the U.S. military. Article 110 stated: "Articles one, two, and twenty-nine, fifty-two to ninety-six, inclusive, and one hundred and four to one hundred and nine, inclusive, shall be read and explained to every soldier at the time of his enlistment or muster in, or within six days thereafter, and shall be read and explained once in every six months to every garrison, regiment, or company in the service of the United States." The individual articles the men were read included various legal definitions, a clarification of who is subject to military law, the meaning of their enlistment and the terms of discharge, their obligations related to muster, absence without leave, insubordination, confinement, various war offenses, and other miscellaneous crimes. The power of officers to enforce discipline and to arrest deserters was also included. The officer then told the men to stand and take the oath as prescribed by Article 109. Walt stood shoulder to shoulder with the others and recited the oath:

> *"I, Walter Babinski, do solemnly swear that I will bear true faith and allegiance to the United States of America, that I will serve them honestly and faithfully against all their enemies whomsoever, and that I will obey the orders of the United States and the orders of the officers appointed over me, according to the Rules and Articles of War."*[3]

The officer then addressed the men with instructions on what to do next. Most of them, including Walt, were placed on inactive duty for one week, to allow them to finalize getting their personal affairs in order. The following Thursday, in one week, they were instructed to report for active duty and transportation to their initial processing center at Camp Custer, near Battle Creek. They were told what to bring and, more importantly what not to bring. Then they were again each called by name and handed their own individual orders....the first of many they would receive during their time in the service. Walt glanced at his, noticing that he had been assigned serial number 365 59 500.[4]

After just a few hours on 'active' duty, Walt had been released into the 'Enlisted Reserve Corps,' prior to reporting for active duty the following week. As a member of the Enlisted Reserve Corps he was not entitled to pay, allowances, or government medical care. Walt and a few other men who had completed processing waited for the westbound streetcar to carry them back to their homes, and talked quietly in the cold afternoon air about what was to come. The following day, Friday, he informed his boss at Shwayder's that he had been OK'd for active duty and that he would have to report the following Thursday. Over the last few days of work he said his good byes to the new friends he had made, wondering how many of the young ones would soon follow his path into the military.

That last weekend before induction Walt was busy hanging out with his friends and relatives. He stopped in at the Kurzyniec store just down Pulaski Street to say goodbye to Mr. & Mrs. Kurzyniec, and their daughters Martha, Dorothy, and Stella. Saturday night he likely took the streetcar with some of his friends over to Olympia Stadium to see the Red Wings play the Black Hawks as they battled to a 1 to 1 tie. Sunday after mass at St. John Cantius there was surely a special dinner for Walt and all the family. At some point early the following week Walt packed away some of his personal belongings up in the attic. He carefully packed away his extensive collection of baseball cards and hockey cards, each year tied together with string and placed in a box, which he put in a corner of the attic along with his other things.

Early in the morning on Thursday, January 21, after another big breakfast at home, Walt said goodbye to Anne and Helen as they left for work and to Genie, Leona and Henry as they left for school. Walter senior had managed to keep a car through the depression, and gas was hard to come by now, but it is likely that Wally's father and mother drove him the few short miles to the Michigan Central Station at 14th and Michigan Avenue, where he was to report. Packing had been easy. His instructions were to not bring a change of clothes – he would be issued new clothes at Fort Custer that afternoon. His civilian clothes would then be sent home. He was not even supposed to bring shaving equipment or a toothbrush – nothing that he could not carry in his pockets, for everything he needed would be new Government Issue.

The lack of baggage made saying goodbye to mom and dad easy physically, but surely not emotionally. Walt's dad was fun loving, but stern, so he surely got a firm handshake, perhaps a brief hug and a wish of good luck. Walt's ma was not stern though…she knew she would worry and miss him a lot. Walt kissed and hugged his ma, promised to write home often, then said one last good bye, turned and walked into the station. There were very likely a number of tears shed as they parted their ways that cold January morning.

Walt walked into the warmth of the Michigan Central Station, beneath the massive 15-story office tower above, and into the waiting room, itself 100 feet wide by 233 feet across. It was still early, but the station was crowded already with people beginning travel by the easiest means during those wartime years. Walt had been instructed to report to the USO Lounge, so he walked over to the information booth, opposite the ticket windows, to find out where the lounge was located. Just at 7:45 a.m., his appointed time, he walked into the USO Lounge, gave a clerk at the entrance his name and was told to take a seat and wait for instructions.

A few minutes later, a man introduced himself to the group of waiting men, stating that his name was Edwin Wloszek, the designated leader for their group on their short journey to Camp Custer. Wloszek told the men that they were now under his command and that they would follow him for transportation via rail to Camp Custer.[5] He would also provide them with a meal ticket, for lunch at Augusta. Shortly before 9:00 a.m., Wloszek assembled the men and instructed them to follow him across the station concourse, down the ramp leading over the station platforms, and on to a waiting train. The group of men took seats in a single car on the local train. The men had been introducing themselves to each other back in the USO Lounge, and here on the train they had more time to talk, find out about the backgrounds of their new comrades, and speculate on what would come next. Then a short whistle was heard and slowly the train began to move west, carry Walt on the first of many thousands of miles he would travel over the next few years.

Fort Custer, Michigan

At around noon on the 21st, Walt's train pulled into the small community of Augusta, 130 miles west of Detroit, and about a dozen miles west of Battle Creek. Wloszek ensured that the men all detrained at that point, and then assembled them on the platform. He handed each man a meal ticket and told them they had a half hour to get lunch on their own, and then they were to report back to the depot. Augusta was only a village of 700 or so people, but activity at nearby Fort Custer[6] for the past three years ensured that the local cafes and diners were well prepared to get the men fed and back to the appointed assembly point in plenty of time.[7]

Early that afternoon the men were driven by bus into the base and met by a non-commissioned officer who introduced himself as their squad leader while at Fort Custer. He told them that they were now part of the 1609th Squadron Unit. He explained that Fort Custer was only an Army Reception Center where the recruits would go through just a few days of basic training, processing, and paperwork before being sent elsewhere for further testing and training. One recruit described his first day at the reception center: "Boy! Did we catch heck when we got here. 'All right, you low down son of a -----, you -------, sit down! Don't go out of the barracks; no smoking; no running around; shut up,' etc. Believe me, I got down in the dumps, but then they fed us."[8]

That first day at Fort Custer each man was issued new "GI" clothes, shoes, shaving kit, toothbrush, and other items by the camp quartermaster, as promised when they were first inducted. Some of the clothes fit well, but for other men they did not…some too big, some too tight. But there was no going back to exchange any ill-fitting clothes. The men were told that when they got to their next base, they were responsible to have them altered at a tailor, at their own expense.[9]

They were then marched to their barracks, assigned bunks, and instructed to change into their new clothes. Shortly after they reported back to the quartermaster where their civilian clothes, shoes, and other items were packaged up to be shipped back to their home address. On the way back to barracks each man also reported to the camp barber for a "GI" haircut – cut 'down to the nub.'[10]

Most of the officers and NCOs (non-commissioned officers) at Fort Custer were older men with previous military experience, but too old for active combat duty or even assignment as a drill instructor. The next few days were very busy as the men started to learn army routine. Their squad leader worked hard to ensure that they had little free time to themselves. They awoke early, dressed, and arranged their bunks ready for inspection. They marched as a group to the mess hall for each of their meals (if they were not already there for the first of many KP or 'kitchen-patrol' duties – assisting with meal preparation and clean up). Part of each day was spent with their squad leader, learning subjects like military courtesy (deportment, saluting, etc.), marching and drill, or in PT – physical training. They learned the importance of keeping their barracks spotless for inspection. The men would scrub the floors on hands and knees – hard work, but the consequence for an unsatisfactory inspection could be assignment to KP duty, which typically began at 4:00 a.m. and lasted until after 9:00 p.m. at night.[11]

Most of their brief time at Fort Custer was spent in processing and paperwork though. For example, on Friday the 22nd, Walt met with Lt. S. F. Brower, an elderly officer from the Adjutant's Office who still signed his name with the title: 1st Lt., Cavalry. Brower explained to Pvt. Babinski how he could have a portion of his pay set aside each month to buy War Bonds. Walt agreed and completed an application to have $1.25 allocated from his pay each month, to be saved in an account to buy $25 war bonds. He designated his mom at their Detroit home address as beneficiary.[12] Then Brower outlined the National Service Life Insurance that Pvt. Babinski could also acquire. Walt must have thought, 'Hmmm – I'm in the Army now…I guess the insurance will take care of ma and pa if anything happens to me.' He signed up for $10,000 of insurance to be paid to his parents in case of his death, with a monthly premium of $6.50 to be deducted from his pay.[13]

Later that morning Walt and the rest of his squad had a series of physical exams and received various inoculations, including a smallpox vaccine. They were tested for blood type. Walt was found to be type A. His height was recorded as 5 feet 11-1/2 inches and his weight was 184 pounds.[14]

On Friday afternoon, each man was issued a copy of the Army's "Basic Field Manual: Soldier's Handbook." While it outlined the basic knowledge, skills, and abilities that each man would need to master in order to perform the basic function of a soldier, it also made it clear that the men were not expected to be merely unthinking tools in the hands of their officers. Paging through the manual that day, Walt read:

> *"You are now a member of the Army of the United States. That Army is made up of free citizens chosen from among a free people. The American people of their own free will…have declared that…we will defend our right to live in our American way and continue to enjoy the benefits and privileges which are granted to the citizens of no other nation. It is upon you and your comrades….that our country has placed its confident faith that this defense will succeed should it ever be challenged.*
>
> *"Making good as a soldier is no different from making good in civilian life. The rule is the same and that is - know your own job and be ready to step into the job of the man ahead of you. Promotion is going to be very rapid in this Army. Be ready for it. You will have little time to learn the duties of a noncommissioned officer after you become one. You will be expected to know those duties and show that you know them. You want to know what is expected of you and be ready to do it.*
>
> *"The things that a trained soldier must know, and the way in which they are done, will be taught to you as rapidly as you can absorb them. The basic military information is described and explained in this handbook…by mastering the contents your future progress will be much more rapid."*[15]

So here was an outline of what was expected for the first part of Walt's military service. Walt must have reflected though that what he was being introduced to was presented in a surprising way. For over three years most Americans had seen newsreels that depicted their seemingly unstoppable German and Japanese foes. Germans seemed to fight out of obedience to their 'Fuehrer' and to impose their racial dominance on Europe and North Africa. The Japanese seemed to be motivated by devotion to their god-emperor and again to impose their racial domination over eastern Asia. With both foes, the image of the soldier was a man who was an unthinking tool in the hands of their leaders and officers.

However, what Walt read was that he was defending the American way of life. There was no talk of domination and the only superiority implied was that of the benefits and liberties that come from being a free American. And the image of the ideal soldier was not that of an unthinking tool, but of a man who was motivated and capable of not only doing his own job, but also the job of the comrade to his left or right. He was even encouraged to prepare himself to take over the job of his superior. Walt had been on active duty for not a hundred hours yet and he was being told to prepare for his own advancement!

Saturday the routine included some basic testing and interviews to determine which branch of the Army each man would be assigned to. Walt made clear his interest in the Army Air Corps and described his experience working at the Samsonite plant in Detroit. He also took some basic intelligence and aptitude tests. Shortly after he was notified that he had done well on his tests and that he would be assigned to the Air Corps for basic training and additional evaluation.

On Sunday, the 24th, Walt attended Catholic Mass at the chapel on the base...the first of many times he would see most of the men in his barracks attest to their religious faith. At many bases separate service were available for Catholics (both Roman Catholic and Orthodox), various Protestant denominations, and on Saturdays synagogue service for the many Jewish men he would serve with.

There was probably some free time too, when the men had an opportunity to relax in the enlisted men's day room. The recruits were encouraged by their squad leaders to write home to their family and friends. One minor benefit of military service was the ability to send letters and postcards without postage. To use this benefit, the men had to include their military unit and base on the return address, then just write the word 'Free,' where a postage stamp would normally be applied. Walt wrote home often during his time in the military. His letters and postcards were written in the beautiful penmanship that he had learned from the nuns at St. John Cantius School. We can assume that Walt wrote from Camp Custer to his ma and pa in Detroit, and to his brother Johnnie at Great Lakes, letting them know that things seemed to be starting out OK and that he was being assigned to the Air Corps.

On Monday the 25th each man received orders for their next assignment. Walt was assigned to a new squad, with a new squad leader, and ordered to proceed by rail to St. Petersburg, Florida, where he would receive additional testing for possible training in the Army Air Force. He received a daily meal allowance and was told that his squad leader would provide his transportation. They would leave the following day.

St. Petersburg and Clearwater, Florida

Sometime after lunch on Tuesday, January 26, Pvt. Babinski and the other members of the squad traveling to Florida assembled with their squad leader. Each man was wearing his new GI issue uniform and had with him a duffel bag with his spare clothes, toiletries, and other possessions. When ordered, each man shouldered his duffel and marched to a waiting bus to be carried back to Augusta to board the afternoon eastbound local train headed for Detroit. Once back in Detroit the men had just a couple hours of spare time before they boarded their next train. They were confined to Michigan Central Station, but found time for a fast meal. Walt and the other men from the Detroit area were able to make a fast phone call home to let their families know they were OK and on the move.

Late that evening the men boarded an old Pullman sleeper car and began a long slow journey south. As the train headed west, then south, if Walt was not too distracted by his new companions or too tired, he may have gazed outside his car window into the dark and cold January night to notice the lights of Delray Junction, where the rails crossed Dearborn Avenue. Just a few blocks down Dearborn was Pulaski Street, with his home and family. How long, if ever, would it be before he saw that house, ma and pa, his sisters and brothers, and his neighborhood friends again? But the train moved on, sleep came, and the miles passed.

Wednesday morning the men woke up in southern Ohio, having passed through Toledo in the early hours of the morning. Over the following two days or so their train wound its way south over a crazy collection of rail routes. Normally a passenger train bound from Detroit to St. Petersburg would make the journey in about 38 hours, traveling over the rails of just four lines (Wabash, Pennsylvania, Louisville & Nashville, and the Atlantic Coast Line). Even top passenger trains during World War Two did not have top priority on the rails. Often over the next two and a half years, Walt would experience being on a train that made good time from city to city, then move into a remote siding to wait for a heavy freight to pass by. Trains with munitions, weapons, equipment, and supplies had priority to keep the troops already fighting well supplied and the wheels of war industry turning.[16]

The troop train Walt traveled on though would take at least 72 hours and would change from one railroad to another eleven times! From Detroit to Fostoria, Ohio they rode the rails of the New York Central; from Fostoria they rode on the Nickel Plate Railroad to Lima, Ohio, where they changed to the Detroit, Toledo & Ironton, which pulled their train to Springfield. In Springfield, the New York Central took over again, hauling them into Cincinnati. Crossing the Ohio River, the Louisville & Nashville pulled their train across Kentucky and Tennessee into Atlanta. From Atlanta, they rode on the Atlanta & West Point Railroad into Newnan, Georgia, where they changed to the Central of Georgia. The C. of G. carried them into Macon, where their train was switched to the Southern Railroad. The Southern turned the train over in Cordele, Georgia to the little Albany & Northern Railway, which carried them into their namesake city. From Albany, the Atlantic Coast Line took over, pulling their train through Thomasville to Quitman, Georgia. In Quitman, the South Georgia railway carried them across the state line into Perry, Florida. From Perry they rode along the Atlantic Coast Line for the rest of their journey into St. Petersburg.[17]

Tired and dirty, their train pulled into St. Petersburg, Florida on Friday, January 29, arriving at the Atlantic Coast Line Railway station at First Ave. South and Third Street. They had left a cold and snowy Michigan three day ago, but now it seemed as if they had arrived in a different world, for stepping out into the street, they were dazzled by the sunshine. Whereas the average high January temperature in Detroit was just 31 degrees (with lows below zero not uncommon), the average low in Florida was 61 and the average high 74 degrees, with generally sunny weather![18] Strange trees, varieties of palms, Australian pines, and subtropical shrubs lined the city streets in every direction. They breathed warm, floral scented air, wafting gently on the breeze. Regaining a semblance of order, their squad leader assembled the men and marched them off with their duffels. They walked just two blocks north (crossing Central Avenue, the city's main thoroughfare) then one block east to come to First Avenue and Second Street North and a small, graceful three-story hotel that would be their new home. This was the Beverly Hotel, with just 77 rooms, old-fashioned gabbled roof, attic windows, second story verandas, and spacious ground floor porches. Built around 1906, the Beverly for many years had been a popular tourist hotel. It had been taken over by the Army Air Force in 1942 as a testing, training, and processing center.[19]

The men were assigned to rooms that had been stripped of their normal resort hotel furnishings. Instead, each room was furnished with several army cots and footlockers. The rooms were bare and crowded, but much, much better than the Fort Custer barracks they had first encountered and certainly better than any Army accommodations they could have imagined. But it got better still. After being assigned to rooms, the men were told they would be fed in the first floor hotel dining room each day, plus their testing and processing would not begin until the following Monday! They had been assigned to the U.S. Army Air Force 604th Technical Service School, Flight 296. But they had a whole weekend to explore the new tropical paradise they had been deposited in.

After cleaning up, changing into summer uniforms, and eating lunch, the men struck out in groups of two, three, four, or more to explore the city. St. Petersburg in the early 1940's was Florida's fourth largest city, with a population just over 40,000. Narvaez, the Spanish governor of Florida, had explored the region as early as 1528. The peninsula that the city occupies was home to Indians into the 19th century, and old Indian mounds were still to be seen on the outskirts. The first recorded white settler did not arrive until 1843 and the first permanent settler did not appear until 1856. Fishing, cattle ranching, and early citrus groves were the first economic activities, mostly with an eye to markets in Tampa, across Tampa Bay.

In 1876, John C. Williams, a promoter from Detroit arrived on the scene and acquired extensive acreage in what was to become the downtown area. He struck a deal with Piotr Alexeitch Dementieff, a Russian exile of noble birth, to build a railroad line into the area from a rail junction in the north, in exchange for a share in his land holdings. The two men tossed a coin to decide upon a name for the new town. Dementieff won and named the settlement for his birthplace in Russia: St. Petersburg. Later Williams built the first large resort hotel in the city and named it for his hometown: The Detroit. In 1885, the American Medical Association endorsed St. Petersburg as a health resort, based on a scientific study that claimed the location was one of the sunniest in the United States. Thereafter St. Petersburg grew by spurts and periodic land booms, as it capitalized on its attraction as a health resort.[20]

The war that began in 1941 and the rapid industrialization that began many months before that had brought jobs and prosperity to most communities across the country, as factories were converted from civilian to war related needs. But St. Petersburg, Clearwater to the north, and many other communities across Florida did not have significant industry to convert. Their major economic basis before the war, tourism, was severely impacted by wartime travel restrictions. Its chief business had been and was providing housing to transients. The St. Petersburg Chamber of Commerce and Florida Congressman J. Harold Peterson had waged a campaign beginning in 1942 to use Florida tourist accommodations as quarters for military recruits for basic training. By the spring of 1942, St. Petersburg and other Florida cities were notified that the area had been selected as a basic training center for the U.S. Army Air Corps Technical Services. By fall of 1942, 20,000 air corps recruits were being processed in St. Petersburg at any given time. Every large hotel in the city, except for the Suwannee Hotel (left for civilian use) was taken over by the Army. For most of the rest of 1943, St. Petersburg would be filled with air corpsmen. Walt and his mates saw them everywhere as they wandered through St. Petersburg that first weekend. They marched and drilled in the streets and parks. They had classes in the open, along the waterfront, and anywhere else that could be found.[21]

Walt and his buddies were drawn east to the waterfront on that first afternoon on the town. Walking a block north from their hotel, then turning right onto Second Avenue North, they walked along a broad tree lined, four lane road, which carried a streetcar line as well. Within two blocks, Second Avenue extended into the bay, first along landscaped fill and then a further 3,000 feet onto the steel and concrete municipal pier, still carrying a four-lane highway and streetcar tracks. On the north side of the pier was the municipal spa and beach. The spa was an indoor heated fresh water pool that cost 35 cents admission, while access to the adjacent white sand beach was free.

Further out, on either side of the pier was found the municipal boat slips, which catered to both pleasure craft and fishing boats. Walt and his buddies would have gaped as large sharks, dolphins, or whiprays were unloaded from the fishing boats for sale in the city markets. Around the perimeter of the pier itself was a wide pedestrian promenade, with park benches every few feet and fishing balconies spaced farther apart to cater to anglers without boats. At the far eastern end of the pier was the Casino, a stuccoed, arcaded structure with red roof tiles and low towers in each corner. The Casino was a popular tourist and entertainment spot, with souvenir booths, restaurants, snack bars, and shops that rented fishing poles and bait. On the second floor of the Casino was a large ballroom for dances, as well as the studios of NBC radio station WSUN. Surrounding the Casino on the second floor was an open balcony that provided a view of the St. Petersburg skyline to the west and of Port Tampa across the bay, with its towering oil tanks and phosphate elevators busy with wartime activity.[22] The Casino was also the location of USO sponsored dances (with local girls provided companionship as "bomb-a-dears") and entertainment like a beauty contest to choose "Miss Personality" with the finalist to compete for the title of "Queen of the Air Force." Order was kept by patrolling squads of MP's (military police) who outnumbered the local police by far.[23]

That evening, sitting on one of the park benches on the pier, Walt wrote out and a couple of postcards he had purchased at one of the Casino souvenir shops. The first, with a picture of St. Mary's Catholic Church in St. Petersburg, he sent to his sister Genevieve:

> *Hello Jenny*
> *Just walking in this town is like being a millionaire, that's how rich this place looks. I'm glad I got in the Army Air Corp. Tech Sq. School.*
> *-Walt*[24]

The second card, with a picture of the St. Petersburg Municipal Pier and Casino, he sent to his youngest sister, Leona:

> *Hello Leona,*
> *How are you my little sister? I hope you are allright. Boy you could swim a lot here.*
> *Take care of yourself.*
> *-Walt*[25]

Walt dropped the cards in a mailbox that evening as he returned to the Beverly Hotel by curfew. Sleep must have come easy after three days on the train and the excitement of being in what must have seemed like an exotic city for the first time. The following days, Saturday & Sunday, Walt and the other members of Flight 296 were largely at liberty to explore the city more. They would have walked or taken a street car north or south along the broad drive that bordered Tampa Bay, with large, stately houses or imposing apartment blocks to the west and landscaped parks or the white sand beach down the water's edge on the east. Many of the residences were built in a pseudo-Spanish style that must have appeared opulent to boys from grimy northern industrial cities.

Later, they would have walked up Central Avenue, St. Pete's 100-foot-wide 'White Way,' past the faded brick and frame buildings that made up the original business district on the east, to the more modern commercial core, with taller hotels, business blocks, and office buildings of light brick and stucco that rose above the palms to give the city its urban profile. Interspersed with the modern buildings, they found old houses, false-fronted for business, that were used as fruit stands and fruit-juice dispensaries, and curio shops selling live alligators, turtles, and doll-sized green benches, among other novelties. The green benches were a trademark of St. Petersburg, for the city had installed over 5,000 along major streets and in parks. As Walt and his buddies walked along and observed who occupied the benches most often, they realized another economic fact of life of St. Petersburg. Most of the bench sitters were in their 60's, 70's, or older. In addition to being a tourist destination, St. Pete was one of the first Florida retirement centers, and the benches were installed as resting stops for the elderly on their daily rounds in the sunshine and mild air.

Further west along Central Avenue, past 9th Street, they came to Methodist Town, the Negro section of St. Petersburg. While the housing was mostly dilapidated, there were a few substantial brick and frame houses of the more economically successful Negros in business and in the professions. A couple blocks south of Central Avenue, along Second Avenue and adjacent to the Atlantic Coast Line tracks, was the nightlife center of Methodist Town, with upstairs dance halls, jazz clubs, barbeque stands, and small shops.[26]

Sunday morning, January 30, Walt and the many other Catholics in Flight 296, made their way south several blocks to St. Mary's Roman Catholic Church at 515 5th Avenue South. Three priests from Cuba were the beginning of St. Petersburg's Catholic community in 1892. By 1906, there were still only 12 Catholic families, but as St. Pete experienced its progressive real estate booms, the parish grew along in size. The St. Mary's Church that Walt attended was built in 1925, having replaced three earlier churches of the same name. St. Mary's was designed by Henry L. Taylor. The stained glass windows were created by Heimer & Co., and installed in 1930.[27] Most of the rest of Sunday was spent exploring more of St. Petersburg and relaxing in the warm Florida sunshine. Walt had noted exotic sounds as well, including the quite call of the whippoorwill, most common around dawn and dusk. He probably picked up a Sunday paper to review the war news and scan the sport section to see how the Red Wings were doing, learning that they had recently beat the Rangers at home 7-0 and the Bruins in Boston 5-3.

Monday, February 1 was a new routine for the men. Training and testing began early that morning – Walt and his comrades began doing the work of the U.S. Army.[28] The typical morning routine began with a whistle which sounded at 5:50 am. By 6:00 am, the men were supposed to be shaved, with their beds made, bathroom cleaned and dry, and lined up outside the hotel with their dog tags hanging outside their fatigue shirts in the cool morning sun.[29] Most of that first Monday morning was spent in orientation briefings with the flight officer. There was to be no more leave for some time, the men were confined to the Hotel, where they would be subjected to a series of interviews and tests. The interviews were designed to identify the men's education level, any previous military experience, their peacetime work history, and any other specialized skills they already possessed. They were also given a series of intelligence and aptitude tests.

Based on these tests and interviews, the men were placed into one of three classifications. Some were determined to have sufficient skill and experience to be termed specialists and to be put to work without further training. These men would leave soon after. A second group consisted of men whose tests showed they had the aptitude for specialized jobs, but lacked the specific knowledge to be put to work without further training. These men would be sent to various Air Corps technical schools around the country. The third group consisted of men whose test scores were too low to warrant further specialized training. These men might be transferred for training in the infantry, or if they were lucky, they would be retained in the Air Corps but assigned to non-specialized duties at a variety of air bases either in the U.S. or overseas.[30] Later that morning and well into the following evening Walt and the rest of flight 296 took these physical and mental aptitude tests, then more physical exams. The following day the test results were posted...a few of the men had indeed been found to have specialized skills and others washed out and were to be transferred early the following day. Walt did not 'wash-out' though. Later that afternoon he wrote a postcard to his family back home with the news:

2-2-43

Hello Ma & Pa
Will someone explain this to them. I had an examination and I past. I'm going to school to be an airplane mechanic or armorer for six months. After that I've got a chance to be a gunner on an airplane.
Well so long

God bless you   -Walt[31]

Before beginning his training as an airplane mechanic though, Walt and the other men assigned to various specializations had to complete basic training. Over the next month or so, the men truly began the process of transformation from civilian to soldier. Some of this process occurred in the hotel, some in and around town. This phase of training included a minimum of six hours' instruction in 'military courtesy,' four hours of lectures on the 'Articles of War,' 12 hours of classroom and field work on personal hygiene and first aid, eight hours of class work on the proper wearing of a military uniform, two more hours of math and alphabet testing, three hours of lecture on government insurance, and 127 hours of soldiering school. Soldiering school included physical fitness drills, squad drills, platoon drills, and lots of marching. They would also spend at least six hours on interior guard duty.[32]

Even though they were living in a resort hotel, Walt and his friends were not on a vacation schedule. During basic training, Walt arose at 0430 (that's 4:30 a.m., in the morning!), swept and mopped under his bunk, attended reveille, then ate breakfast. He then got his equipment in order and reported to his squadron by 0645, where his flight leader would inspect the men for proper haircut, clothes neat, shoes shined, and general personal cleanliness. Morning training began at 0730 and continued until 1130. From 1130 until 1245 was lunch and mail call. At 1245 he reported back to his flight leader who formed up the men and marched them to the afternoon training site. Afternoon was generally field-training which lasted until 1730. If the men had any equipment to return to the supply room, they did that on their free time. They were also expected to study the training manual on their own time. The men did their own laundry during basic training – at least once a week and again on their own time, but there were often long lines to use the laundry tubs, so this task could last well into the evening.

Once or twice a week during basic training, each recruit was selected for KP duty. Those on KP were awakened at 0145 to complete their normal barracks duty first. Then at 0245, they met formation to be assigned to specific mess hall duties. KP duty lasted until 1800, with breaks only to eat their own meals. At the end of the day, the men on KP still had to report for mail call and other flight duty, resulting in a 19-hour day. On average, recruits worked 100 hours during their six-day training week. Sunday at least was an official day of rest.[33]

As the men drilled then marched through town and out into the countryside to commence their 'soldiering school', they must have thought fondly of the first weekend on their own in St. Petersburg. As basic training progressed, strict discipline was established. The drill instructors pushed the men hard to conform to military discipline and to march with precision. The motivation of the drill instructors was to have their squad picked by the top brass as the best in the group during mass reviews and parades. The men did tend to take pride in the performance of their unit, especially when there was a band to play march music, which made it easier to endure the hours and hours of practice.[34]

A variety of trivial infractions could keep Walt confined to the hotel in the evening after training and chow. Typical 'infractions' might include: 'toilet bowl dirty or unflushed, bed improperly made, shoes unlaced or unpolished, dust on the window sill, ashtrays not emptied, button on shirt not buttoned, floor or wall in room dirty, lint on rug,' and many more.[35]

The basic training that commenced that week in St. Petersburg started with 'military courtesy.' The men were taught that the lessons they had learned about politeness and consideration to others while at home and in school extended to the military as well. The concept was that a soldier was expected to have pride in his profession and respect for his comrades and that displays of military courtesy reflected that pride and respect. To show proper military courtesy the men were trained on the correct way to salute and address superiors.[36]

The men were also given detailed instructions on the wearing of their uniforms and the proper way to care for and clean them. They were given training on the various insignia and badges of rank and honor that they might see being worn on other uniforms. Recognizing the rank of another soldier by these insignia was a key part of military courtesy as well. A lot of attention was placed on keeping boots and shoes clean and in good condition.[37]

Early the following Monday the men were ordered to clean out their rooms, pack their duffels, and report for a short trip to a new base. After just a week in St. Petersburg, they were on the move again, although this time not far. Their destination was Clearwater, just 18 miles north by rail....less than an hour on the ACL Railroad passenger train. St. Petersburg had been primarily a processing and testing center run by the USAAF Basic Training Command. The Fourth BTC, with headquarters in Miami Beach, had taken over 326 hotels throughout Florida to house men going through the first steps in testing and training. Clearwater, they were briefed, would be the site of additional weeks of basic training.[38]

On Monday February 8, Walt and the other men stepped off their train and onto the streets of Clearwater.[39] From the small Clearwater station they had a short walk with their duffels to the large Hotel Fort Harrison[40]. They went through additional processing, including another medical exam and any shots they required – Walt received a triple typhoid inoculation.[41]

They could tell that Clearwater was not as large or bustling as St. Petersburg, but the warm weather and glimpses of salt water just a few short blocks to the west suited the men just fine. In the early 1940's Clearwater was a small city of about 8,000 people. It was the county seat, occupying a slight ridge running parallel to the coast. Like St. Petersburg, it had been founded in the 19th century and grew in the early 20th century as tourists and retirees discovered the healthy and pleasant climate of the Florida coast. Clearwater still served as something of a service center for the many citrus groves on the outskirts of town. They had noticed some of the 11 fruit packing houses along the tracks, as well as several fruit-juice-canning plants. But Clearwater had been built on wealthy tourists as well. The major attraction was the exclusive Belleview-Biltmore Casino and Hotel, an imposing white structure of over 600 rooms about a mile south of the center of town. The Belleview-Biltmore had been begun in 1896 and included two 18-hole golf courses on 275 acres of beautifully landscaped grounds.[42]

The Hotel Fort Harrison, located at 118 North Fort Harrison Avenue in Clearwater was just two blocks from the waterfront. While not in the same league as the Belleview-Biltmore (also commandeered by the 4th BTC), the Fort Harrison was still a luxury hotel and the best that downtown Clearwater had to offer prewar guests. It boasted 253 rooms with baths on a dozen floors. It was a modern building, finished in 1927. Like all BTC commandeered hotels, the luxury furnishings of the Fort Harrison had been put into storage and replaced with olive drab army cots and footlockers. The Hotel was the location of the Army Air Force's 588th Training Squadron.

The men of Flight 296 were told that they had been transferred from the 604th to the 588th Training Service Squadron and would be based for additional basic training at the Fort Harrison for the next several weeks. Life in the hotel was a curious blend of tourist luxury with bustle, routine, and discipline of Army life. Revile and taps were played to rouse and call the men to rest by an army bugler from atop the Hotel roof. Meals were taken in the elegant Hotel dining room, but the men took turns on KP duty in the Hotel kitchen. Hi-jinx did occur – the story was told of one private from the 588th on dishwashing duty in the Fort Harrison kitchen who snuck under one of the sinks to take a nap. While he slept, some of his KP mates piled stacks and stacks of dirty dishes and pans around him so he awoke trapped by the very work he had tried to hide from.[43]

On Tuesday morning the 9th the men were given a little free time to write home and give their families and friends their new address. Walt wrote out several postcards, including one to his mom and dad in Polish[44] and another to his sister Genie (with a color picture of the Hotel Fort Harrison):

Hello Jenny,
  How are you? I received your letter and am glad you wrote, and I see a lot of things
  here I never seen. That's my hotel I live in on the picture – show it to the family. So long.
    -Walt[45]

A few days after arriving in Clearwater, Walt got a postcard from his brother in the Navy. John put the card in the mail in Albuquerque on Sunday, February 7, addressed to Walt in St. Petersburg. From there it had been forwarded from the 604th to the 588th TSS in Clearwater. John wrote:

John J. Babinski, U.S. Navy
Hello Walt.
  Well I am heading west hoping you are fine. I have seen some nice country. Still going. I
  will write you when I arrive.
    -Bob[46]

Walt had learned from other letters sent from his family in Detroit that his brother John had graduated from his 16 weeks of Mariner Mechanics training at the Great Lakes Naval Training Center. John had completed a basic mechanics course, and then advanced mechanics training, including math, boiler making, mechanical drawing, welding, shop theory, machine tools, ship fitting, steam engines, and carpentry. Finally, he completed Metal Workers School, with an overall score of 84.84%. Now he was headed west (via Albuquerque) to San Francisco where he would be assigned overseas, somewhere in the Pacific.[47]

Just as they had in St. Petersburg, the men of the 588th TSS continued their training and marching in and around Clearwater. As the days of February wore on the men became used to the daytime temperatures in the 70's and lows at night that rarely went below 60 degrees. Walt and the other Catholic men in Flight 296 attended mass on Sundays at St. Cecilia's Catholic Church. St. Cecilia's was about a mile walk to the southeast, located in a quiet residential neighborhood at Jasmine Way and South Prospect Avenue.[48]

On Saturday, February 13, the men were able to hop on a bus while on leave and travel about 90 miles (via downtown Tampa) to nearby Lake Wales, Florida. Tampa was a bustling industrial port city that had little attraction to the men on leave compared to Clearwater or St. Petersburg, but Lake Wales was a well-known small resort center. Lake Wales was a community of about 3,400 surrounded by several small lakes. Crystal Lake Park contained a tourist club building, a bathing beach, and an open-air theater. The main attraction in Lake Wales was the 'Singing Tower' – located in landscaped grounds next to a small lake. The Singing Tower was built at the highest spot in Florida, constructed of pink Georgia marble and containing one of the largest and finest carillons in the world, consisting of 71 bells. Mr. Edward W. Bok had built it as a symbol of beauty and as thanks for the American Flag.[49] On his return from Lake Wales Walt wrote brief postcards to his brother Hank and sister Leona:

2-14-43

Hello Hank,
I received your letters and don't have much time to write, I will write more later on. So long.

-Walt[50]

2-14-43

Hello Leona,
How are you my little sister, I hope you are taking good care of your self and I hope you like that emblem I sent you and Henry. Goodbye.

-Your Brother, Walt[51]

Walt's days began early with two hours of physical training (PT) early in the morning, followed by breakfast. Marching and drill continued later in the afternoon and after lunch. Training sessions were held in subjects such as camouflage, mapping and map reading, first aid, manual of arms, and how to 'stand guard duty.'[52]

The second weekend on leave, Walt put on his best uniform and hat and walked over to Bergeson's Photo Shop in Clearwater. There he had his photo taken to send to his family back in Detroit. He looked thinner in this photo than the 180 pounds he weighed when he was first processed at Camp Custer. His hair is still short from his military haircut and his hat is characteristically cocked slightly to the side. He looks at ease in his uniform and his smile suggests that he had adjusted well to Army life and the men he lived and trained with.

Walt would have also gone to the local Clearwater USO where he was happy to read in the local Clearwater newspaper that the Red Wings had continued their winning season, beating the Rangers in New York 5-4 the day before. USO stood for the United Service Organizations, a fund-raising and coordinating group that grew out of the YMCA, YWCA, National Catholic Community Services, National Jewish Welfare Board, Traveler's Aid Association, and the Salvation Army in January 1941. These six groups each began their own uncoordinated efforts the year before to provide facilities across the country for rest, relaxation, and civilian contact for the hundreds of thousands of young men (and later women) now in the service, as the military geared up for war. The goal was to maintain both the morale of the troops, while minimizing the negative impacts that millions of service personnel would have on local communities. Initially the Army and Navy wanted to maintain direct control over the facilities and programs begun by the six agencies. However, President Roosevelt himself decided:

> *"This is the way I want it done. I want these private organizations to handle the on-leave recreation of the men in the armed forces. The government should put up the buildings and some name common to the organization should appear on the outside."*[53]

The USO was incorporated in New York State on February 4, 1941. Much of its operation was funded by private donations, with over $33 million contributed throughout the war. By 1943 USO centers had spread across the country and overseas. They were located near training centers and active bases, as well as bus, train, and shipping terminals to help service men in transit. They were located in barns, churches, museums, auditoriums, stores, and railroad cars. Many of course were built from the ground up. 3,000 locations worldwide had USO facilities.

The USO provided service men with a 'home away from home.' Young women from the local community volunteered as hostesses, providing civilian contact at dances, parties, and other events. They would help with mending clothes and writing letters for men who could not or would not write home. The men at a USO could wash up, relax, get a coffee or snack, and sit in a lounge and listen to music and talk with friends. The USO provided a bag check service and provided travel and sightseeing information. Local clergymen provided counseling.[54]

The following week Walt and Flight 296 marched out of Clearwater for several days. This was a standard part of basic training and included several nights of bivouac out in the Florida countryside. Each man had a forty-pound pack, plus a gas mask, canteen, rifle (another nine pounds), and mess kit. Even in February it was hot and dusty marching down the dirt rural roads to the north of Clearwater. The men were kept on their toes looking out for rattlesnakes and scorpions.[55]

While on the march out in the country and in bivouac, the men learned and practiced a variety of new skills. This exercise developed both self-reliance and teamwork. Men knew they were responsible for their own pack and equipment. This was true from the ground up, starting with ensuring good broken-in boots and clean socks, to knowing how to pack their gear quickly, to be ready to fall-in when the order was given. They learned how to march for long periods and rest efficiently when halts were made. They learned how to pitch their tents while on bivouac and ensure that they slept warm and dry under a variety of circumstances. They learned how to care for their feet – washing them carefully when they first arrived in camp and how to attend to blisters if they occurred.[56]

Friday, February 26 was Walt's first Army payday.[57] It was also his last day in Clearwater, for he and a few of the other men in flight 296 had received orders to report to Salt Lake City, Utah.

Walt's next two destinations are something of a mystery. He had been selected for training as an airplane mechanic on February 2, while quartered in the Beverly Hotel in St. Petersburg. He would arrive at Keesler Field in Biloxi, Mississippi by the middle of April to begin airplane mechanics training. Clearwater to Biloxi is about 600 miles by land. However, for some reason the Army Air Corps sent Walt from Clearwater to Biloxi by way of Salt Lake City, Utah, then Harvard, Nebraska.[58] None of Walt's military records, correspondence, or other information explains the purpose of the detour of an additional four thousand miles travel and six weeks' time. Walt would spend about two and a half weeks in both Salt Lake City and Harvard. We can only assume that the Air Corps sent him to these two destinations to continue the testing, processing, and basic training that he had begun in St. Petersburg and Clearwater. Some evidence suggests that Salt Lake City served as a major classification, processing, and reassignment center for the Air Corps in 1943 and 1944.

Salt Lake City, Utah

Walt began his long journey to Salt Lake City in the late afternoon of Friday, February 26. He made his way from the Hotel Fort Harrison back to the Atlantic Coast Line station a few blocks away to await the 6:00 p.m. train headed north that would begin him on his journey. ACL train 32 was a local which made its way slowly north in the Florida evening, finally crossing the state line around midnight, arriving in Thomasville, Georgia at 1:10 a.m. Thomasville was a southern Georgia city of about 14,000.[59] It was located at the junction of several important highways and was also a junction point on the ACL Railroad. Walt spent just a little time on the station platform in Thomasville, waiting for the ACL Railroad train from Jacksonville, bound for Montgomery, Alabama. After Train 57 arrived, Walt boarded, found a seat and resumed his quest for sleep as the train made its way northwest to Montgomery.

Train 57 arrived at Union Station in Montgomery, its terminal destination, at 8:00 a.m. Union Station was located adjacent to the Alabama River, at the junction of Lee and Water Streets. Walt checked at the ticket counter to verify that his next train did not leave until well after noon, so he had time to clean up, have breakfast and explore a little of Montgomery. As in most medium and large railroad stations across the country, a volunteer would have been available at the USO Lounge in Union Station to direct Walt and his pals to the nearby USO club where they could get a bite to eat and shower.⁶⁰ After he was refreshed, he likely walked up Commerce Street to Court Square, the heart of the downtown business district. It was a Saturday morning, so many offices and businesses were closed, but the city would have still been filled with shoppers at the stores radiating out from the Square.

At the center of the Square was the McMonnies Fountain, with 25 jets of water surrounding a life size female figure. It was erected in 1886 as the South was coming out of the reconstruction era. Looking east from the Square up Dexter Avenue, Walt could see the State Capitol building on Goat Hill, patterned after the National Capitol and topped with a 97-foot dome.

In addition to being the state capital, Montgomery had a long history as a center of the cotton industry. The decades prior to the war had also seen the development of diverse industry as well, including livestock and dairy industry, lumber mills, furniture plants, chemicals, and food packing plants. The city's population had more than doubled since the turn of the century to more than 78,000 by 1940.⁶¹

Despite this recent industrialization heightened by war business, Walt would have been struck by the poverty of Negro slums readily visible closer to the river. Montgomery was the heart of Alabama's "Black Belt" where blacks represented the vast majority of the population but wealthy white landowners still dominated all aspects of life and ensured the continuation of "Jim Crow" laws and discrimination. Walt and other servicemen from up North would have been surprised at the deference that "Black Belt" Negroes would have shown them as white men, removing their hats and stepping aside whenever they passed.⁶²

After exploring a little of Montgomery, Walt stopped at a lunch counter for some food, then made his way back to Union Station. There he boarded Louisville & Nashville Railroad train 10, leaving at 2:30 p.m., bound for Birmingham.

The L&N train made its way north and west, gradually moving into hilly country. The train reached the outskirts of Birmingham just as dusk began to fall, but Walt recognized that Birmingham appeared more like home in Detroit than any of the other Southern cities he had visited. The city's population was more than 267,000 in 1940, making it the largest in Alabama. It was also the major iron and steel-producing center of the South. Like Detroit, it was both a well-planned industrial city and a hurriedly built boomtown, combining plenty and poverty, beauty and ugliness. Located in narrow, 15-mile-long Jones Valley, Birmingham was plagued with the smog that marked mid-twentieth century industrial centers like Detroit and Pittsburgh.

The train pulled into the L&N Station at 20th Street and Morris Avenue right at 6:00 p.m. Walt determined that he needed to get to Terminal Station, located at 26th Street and Fifth Avenue North by 11:30 p.m. It was still Saturday, but most stores had already closed. As Walt made his way the ten blocks or so to Terminal Station, on the lookout for a likely spot for dinner, the night sky was lit up by flames from the Sloss Iron Foundry just a few hundred yards to the east. The location of other foundries up the slopes of either side of Jones Valley could be identified by the glow of their blast furnaces as well.⁶³

Much of this section of Birmingham must have reminded Walt of Delray and home. Another aspect of this city that set it apart from Montgomery and the rest of the South was the demeanor and status of the blacks. Most blacks with jobs worked in the steel mills and foundries. It was hard work and blacks tended to get only the dirtiest, most difficult jobs, but compared to other Alabama Negroes, they were very well paid. "Jim Crow" laws were still on the books in Birmingham, but they were resented and a prosperous Negro business section along Fourth Avenue was a center of nightlife for all of Birmingham. Here there would be none of the automatic deference by blacks that Walt had seen earlier that morning in Montgomery.⁶⁴ America was changing.

After dinner, Walt made his way to Terminal Station, where he confirmed the departure time for his train, then found a seat to rest up as much as possible. Just before its 11:30 p.m. departure, Walt boarded Saint Louis and San Francisco Railroad train 108, bound for Fort Scott, Kansas. As the train departed the station, Walt dozed off to sleep with the glow of the foundries still illuminating the Alabama night sky.

Just after 7:00 a.m., the train pulled into the Memphis, Tennessee Grand Central Station at Main Street and Calhoun Avenue, having crossed the northern corner of Mississippi during the night. After a pause of just 20 minutes or so, the SLSF train pulled out of the station and soon was crossing the Mississippi River via Harahan Bridge, the first bridge built across the river south of Cairo, Illinois.⁶⁵

Walt's train progressed during the rest of that Sunday, steadily west and north. It crossed the flat Arkansas plain of the Mississippi, covered by cotton fields and rice plantations. Past Jonesboro, then past Walnut Ridge it pierced the Ozark Escarpment following the winding Spring River then turned north into Missouri at Thayer. The route was not fast, but scenic. The major SLSF junction of Springfield was passed at around 3:00 p.m. At 6:00 p.m., the train paused in Fort Scott, Kansas. Walt got off at the SLSF depot at 623 East Wall Street, as the train continued north to Kansas City.

Fort Scott was a small city of 10,000 or so about five miles west of the Missouri state line. It was an important shipping point and manufacturing center for southeastern Kansas. On a Sunday evening at the end of February though it must have seemed 'dead' to any young servicemen like Walt waiting there for their next train due to depart nine hours later. The choice of eating establishments must have been very limited indeed. If they were lucky, there might have been a USO club for them to spend their time in. If not they would have eventually made their way to the Missouri Pacific Railroad depot at 219 North National Avenue, where they might have played cards for a couple hours, then settled into waiting rooms seats to doze.[66] Shortly before 3:00 a.m. on Monday morning, they would have been roused for the arrival of westbound MP Railroad train 19, which they boarded quickly in the cold Kansas night air. Then they settled into uncomfortable seats once more and struggled to sleep.

Next morning Walt awoke to a scene of gently rolling wheat and cornfields, with occasional stops at small Kansas farming communities. At 9:30 a.m., the train pulled into the MP depot at C Street and Main Street in Hutchinson, Kansas, where Walt got off to make his next train change. Once again checking with the ticket agent, he determined that his next train for his journey west would not leave until 5:50 p.m. Walt likely stopped at the USO for something to eat and, more importantly, a place to clean up. After he had rested he likely stepped outside to explore a little of Hutchinson. This was the fourth largest city in Kansas, with over 27,000 people. Hutchinson had broad, straight streets bordered by homes with large lawns and many parks. The commercial area extending north up Main Street was fronted by two and three story brick buildings, with a few office blocks of four, five, or even eight stories at key intersections. Hutchinson was not just a farming and ranching center, but also an industrial center. It was not surprising that flourmills and grain elevators supported the economy, but salt plants (dependent on underground salt mines within city limits) and refineries (that served oil fields surrounding the city) were also important. With several hours before his train left, Walt had time to explore the commercial district or perhaps take in a picture at one of the five movie houses within the city. As the afternoon wore on Walt made his way over to the Santa Fe Railway Station at Third Avenue and Walnut Street.[67] Just at 5:50 p.m., westbound ATSF Railway train 5 pulled into town, Walt boarded and he was off again to the west.

Hour after hour through the dark Kansas night, they moved steadily westward. Perhaps the passengers were awakened in the early hours of the following morning, when the train paused in La Junta, Colorado, a major cattle market and junction of the Santa Fe Railway, where the route split in two directions.[68] Most of the cars continued southeast, toward New Mexico and California. The car Walt traveled in and a few others were switched off the main train and left shortly afterwards towards the west and northwest. A couple hours later, at 5:40 a.m., the train paused in Pueblo, Colorado at the joint Santa Fe and Rio Grande depot and Walt walked off and into the station building.

It was now Tuesday, March 1, but Walt still had far to go to reach his destination. His next train left shortly after noon, so once again he had time to find the USO in Pueblo for some breakfast and to clean up and relax. Pueblo, like Birmingham was a center of the steel industry, relying on minerals and coal from the nearby Rocky Mountains and supplying much of the west with manufactured products. This was not the typical dusty western cow-town that Walt had expected to find.[69] However, Pikes Peak off to the northwest did provide Walt with his first glimpse of the Rocky Mountains.

Just after noon, Denver & Rio Grande Western Railway train 1 pulled into the station and Walt boarded. At 12:15 p.m. it pulled out and headed westbound for Salt Lake City. Almost all the trains Walt rode during the war were very crowded. Gas rationing had severely restricted the ability of people to travel by car or bus, and the highway network was not well developed in the 1940's (there were no interstate freeways). In many cases, Walt rode in normal passenger coaches, but on the occasions when he rode on top passenger trains, the railroads typically added two or three cars for military personnel with their own separate dining car.[70]

West of Pueblo the train traversed dry rolling hills covered with gramma grass and russian thistle. This was the image of the west that Walt had expected from the weekly western movies he had enjoyed as a boy back in Detroit. Near the town of Florence, the train passed irrigated apple orchards then began following the Arkansas River closely. Past Canon City, the train entered the "Royal Gorge" or the Grand Canyon of the Arkansas. The rail route here followed a canyon so narrow, only a single track rail line was feasible, while the vertical canyon walls towered over a thousand feet above on either side.[71] Emerging from the Royal Gorge the train gained elevation mile after mile. Past Salida, it turned north and climbed higher still, until just at dusk it crested the Continental Divide at Tennessee Pass, the highest railroad crossing of the Rocky Mountains at 10,427 feet. Once again, the now very weary travelers attempted to sleep. Early the following morning they awoke in Utah, just crossing over Soldier Summit as the train began to descend the western slope of the Wasatch Range. Soon the train emerged from the foothills and sped north across the Valley of Utah. Provo was passed, with Utah Lake to the west, then neat Mormon farms and orchards. Finally, at 10:00 a.m., the train stopped at the Rio Grande Station at Third Street South and Rio Grande Street in downtown Salt Lake City and Walt detrained.

Salt Lake was a city of about 150,000 people with broad tree lined streets, which were laid out by the original Mormon settlers foursquare with the compass. Down the streets, Walt could see mountains in every direction, despite the fact that the city itself was situated on the broad flat floor of ancient Lake Bonneville. From the doors of the Rio Grande station and looking northeast across Pioneer Park he could see the copper-domed State Capitol building on a bench above downtown. South of the Capitol and lower down was the six-spired Temple of the Church of Jesus Christ of Latter-day Saints – the Mormons.[72]

Shortly after arriving at Salt Lake City, on Wednesday, March 3,[73] Walt checked in at the Army transportation office in the station to report his arrival and find out about transportation out to Salt Lake City Army Air Base. While he was waiting at the station for the bus to take him to the base, Walt had time to write a couple fast postcards to his family back home in Detroit, which he mailed just before noon:

Hello Jenny,
Didn't write for four days because I traveled on a train. Busy now can't write much
but will write more tomorrow all about my trip here.
-Walt[74]

Hello Leona,
How are you. I'll bet you were lonesome because I didn't write for a while but I
couldn't because I was on a train. Will write more soon.
-Walt[75]

Early in the afternoon, a bus from the Air Base pulled up at the station, and Walt and the other service men destined there boarded for the 16-mile ride. Salt Lake City Army Air Base was less than a year old. After the Pearl Harbor attack, the Air Force decided that a large training base should be located far in the interior of the country, invulnerable to coastal attack. Utah was ideally suited, being roughly equal distance to the three major western ports of Seattle, San Francisco, and Los Angeles. It is also halfway between Canada and Mexico. In early 1942, the Army selected a 5,450-acre dry farming area in Kearns, a small community west of Salt Lake City. Base construction began in April 1942 with the installation of water mains, sewage systems, streets, tarpaper buildings, and electrical systems capable of supporting 30,000 to 70,000 people. By August 1942, the barracks had been completed at an estimated total cost of $17 million. The post included 926 tarpaper buildings, 16 mess halls, three theaters, two gyms, three fire stations, recreation fields, chapels, barbershops, a ten-wing hospital, a post office, a library, and a bank. 25,000 trees and shrubs and fields of grass were planted to keep the dirt in place while beautifying the base. While the tarpaper barracks Walt was assigned wasn't up to the standards of the Beverly Hotel or the Hotel Fort Harrison in Florida, at least it was relatively new and in good condition considering its cheap construction. The walls were flimsy though and it was cold, with just small stoves provided for heat. The men learned to wear both summer and winter underwear, with their summer uniform, and then their woolen olive drab uniform over that, to try to keep warm.[76]

It appears that at Salt Lake City, Walt was attached to the Air Corps 18th Replacement Wing.[77] Although we do not know for sure why the Air Corps transported Walt halfway across the country to

Salt Lake City Air Base for just two weeks, we can assume that he likely continued his basic training. The base had been officially designated a facility of the Army Air Force Training Command as a basic training center. It also appears to have been used as a classification center, where recruits were held, processed, sorted out, and then issued orders for additional training or active duty. The training facilities included a grenade-throwing ground, a gas demonstration area, a beachhead maneuvers area (lots of beach sand, but minus the water), and a mile-long obstacle course. It also had the nation's second largest rifle range. By the spring of 1943, the base had grown into Utah's third largest city and at its peak, stationed 40,000 troops and employed nearly 1,200 civilians.[78]

Walt spent two weekends during the time he was assigned to Salt Lake City AAB, so he had some time on leave to explore the attractions of the area. Buses between the base and down town Salt Lake City ran frequently and made the trip in 35 minutes.[79] Temple Square was the tourist center of Salt Lake City and Walt was surely drawn to the area, but the Mormon Temple itself was not open to non-Mormons. One attraction of Temple Square that made an impression on Walt was the Seagull Monument. Erected to commemorate the gulls that ended a plague of crickets in 1848 that threatened to destroy the crops of the early Salt Lake Pioneers, Walt would recount the story of the Salt Lake Seagulls to his children many times in later life. However, the Mormons did not allow smoking anywhere within Temple Square, so Walt (a longtime smoker already at age 20) probably wandered away soon. In any case, the daytime temperature in Salt Lake in early March would likely not be higher than the mid-40's with lows below freezing at night, so exploring on foot was not especially pleasant.[80] Souvenir shops and restaurants surrounded the Square, so perhaps these helped him while away his free time.[81]

The habitants of Salt Lake City presented a strange contrast. As a center of municipal, county, state, and Federal government activities, as well as the corporate headquarters of the Mormon Church, briefcases and suits were common garb and accessories. But just as common were real cowboys in ten-gallon hat, copper-riveted Levi's, and high-heeled boots.[82] Walt was surprised to find that residents of Salt Lake City were not necessarily hospitable to servicemen, but resisted complaining about the men on leave from the base because of the jobs the base provided and the beneficial effect on Utah's economy from men on leave spending some of their meager Army pay.[83]

One weekend Walt made his way to nearby Bingham, west of the Air Base and out in the desert mountains. Bingham was the site of the Utah Copper Mine – its 'glory hole' had been excavated more than 1,500 feet deep and must have made quite an impression as an example of the progress and industry of the Country he was prepared to serve and fight for.[84] Back at base he relaxed with the Sunday newspaper, reading that his Detroit Red Wings had just lost two back-to-back games against the Maple Leaf's – 1-3 in Toronto and 3-5 at home in the Olympia. He was relieved to read though that they would finish the regular season high enough in the standings to make the Stanley Cup Playoffs.

Time passed by quickly enough with continued training, as well as processing and classification. By Wednesday, March 17, Walt was given orders to depart the following day for a new destination and more training. His next base was to be Harvard Air Force Base in Nebraska.

Harvard AFB, Nebraska

Mid-day on Thursday, March 18, Walt packed his clothes and gear in his duffel, made sure he had his travel orders in his pocket and then took one of the base buses headed into Salt Lake City and the Rio Grande station downtown. After checking in to confirm his departure time, he had a few minutes to wander the city one last time and perhaps sit in Pioneer Park outside the station and view the mountains off to the east. He probably had an early dinner at around 5:00 p.m., and then boarded the eastbound Rio Grande train shortly before its late evening departure.

Sleeping on crowded passenger cars must have been somewhat routine to Walt by that point, so he slept while the train made its way east in the darkness through Utah and western Colorado. Dawn came up as the train approached Glenwood Springs. The rest of the day Walt and his fellow travelers were treated once more to the crossing of the Rocky Mountains. This was not the route via Tennessee Pass that Walt had taken coming west, but the new crossing of the Rockies completed in the 1930's via Gore Canyon and the Moffat Tunnel. 15 or 20 miles east of Glenwood Springs, the modern Moffat Tunnel route branches off to the northwest for a direct route to Denver. Just 50 miles west of Denver the train emerged from the east portal of the Moffat Tunnel and began its descent to Denver. Walt was impressed with the scenery, so he bought a post card showing the Gore Canyon route to send to his family later. Finally, just before dusk, the Rio Grande train pulled into Denver's Union Station at 17th and Wynkoop Streets and Walt got off.[85]

It was now late on Friday, March 19, but Walt did not have much time to explore Denver. He tried to clean up at the nearby USO, before returning to Union Station in time to board an eastbound Chicago Burlington & Quincy Railroad train for its early evening departure. The conductor awakened Walt the following morning as dawn was breaking. The train had crossed into Nebraska in the early morning hours and now they were traveling through a region of large rolling farms.[86] At about 7:00 a.m. on Saturday, March 20, the train arrived in Hastings, Nebraska at the CB&Q Rail Road station at 501 West First Street, and Walt got off.[87]

Walt's destination was not Hastings, but Harvard Army Air Base, about 20 miles east out in the prairie. Hastings though had become a major military center by 1943 and was the logical location for service men to arrive and depart. Harvard AAB was dwarfed in both size and number of personnel by the area's major military installation: Hastings Naval Weapons Center. Located about three miles east of Hastings, the Navy Weapons Center covered many thousands of acres served by a dense network of new rail spurs. The mission of the Weapons Center was to store reserve supplies of munitions and bombs for the U.S. Navy in the center of the county, out of the range of a possible enemy naval or air attack on coastal cities or bases. Many hundreds of Navy personnel worked at the Center, a large proportion of whom were Negro enlisted men.

When Walt detrained in Hastings, he checked in with the transport officer in the station and learned that he had a couple hours before the first bus out to Harvard AAB. This gave him time for a quick breakfast and also to stop at the Western Union office and send the following telegram back home:

HASTINGS NEBR 20
MRS CORA BABINSKE =
9050 PULASKE ST DET =

HELLO FOLKS ARRIVED AT MY DESTINATION SEVEN AM SATURDAY MARCH 20 AM FINE. HOPE TO HEAR FROM YOU SOON. MY ADDRESS 485TH ARMY AIR BASE SQD DS HARVARD NEBRASKA =

WALTER A BABINSKE[88]

The Hastings telegraph operator evidently was unfamiliar with Polish names ending in 'ski' but it was the message that was important. The telegram was not received in Detroit until near midnight on the 20th, so it was not delivered until early the following Sunday morning. His ma & pa, brothers and sisters likely got out a map to see where Walt was.

Hastings, with a population of 15,490 in 1939, was the fourth largest city in Nebraska. Located south of the Platte River in the heart of the great Kansas-Nebraska wheat belt, the city depended mostly on agricultural business supplemented with a few other industries and a college. When Walt first walked around near the depot, he saw a skyline dominated in the north by the tall smokestack of the city's municipal power plant. To the south were other tall grain elevators and the smokestacks of the local industries, including brick and tile, agricultural machinery and food processing plants.

Soon Walt boarded the bus bound for Harvard AAB. It headed north one block, then turned east down Second Street, Hastings' main commercial thoroughfare, which was flanked by blocks of buildings, some with elaborate cornices and others more modern stores.[89] At Elm Street the bus turned south, crossed the CB&Q rail line, then turned east once again on East South Street. This was US Highway 6, the main route through town. Soon the residential neighborhoods of Hastings were left behind. After just a few miles the main gate to Hastings Naval Weapons Depot was passed, followed by five miles of barbed wire fences and guard towers that surrounded the base to the south.

About 15 miles out of Hastings the bus turned north, crossed the CB&Q line once more, and turned right onto Oak Street, the main street of tiny Harvard. It then turned left and headed another mile north on Airport Road to the entrance gate to Harvard AAB. The location and appearance of this base must have been quite a shock to Walt and any other men arriving there for the first time. This was a small facility, covering just one square mile, with a single runway aligned north and south. On the west side of the base were located a few tarpaper covered barracks, a mess hall, shops, hangars, and a small PX. Mud was everywhere. Incredibly, the base had not existed four months earlier. Initial construction had begun in September 1942 and was completed by November.[90]

The first Air Force personnel did not arrived on the site until December 12, 1942 and civilian staff first appeared on January 1, 1943. Additional construction work continued. The PX was completed just 5 days before Walt arrived, but there was no recreation facility on the base until July, long after Walt had departed.[91]

The bus pulled through the Harvard gate and Walt got off and walked into the small base headquarters building to report for duty. The orders he had received in Salt Lake City assigned him to the 485th Army Air Base Squadron. Walt discovered when he reported that he had a couple days free, for his records show that his official duty at Harvard AAB began on Tuesday, March 23.[92] He was in good spirits though, writing a postcard with the picture of Gore Canyon in Colorado that he had crossed a few days earlier to his sister Jenny back home on the 23rd:

Hello Jenny,
How are you Jenny and how are you getting along in school, in work, and hope you have fun in your spare time, I do. See that the package gets here will you. Write more later.

-Walt[93]

For his first couple days at Harvard, Walt had frequent KP duty and he was assigned to help clean barracks and police the base. He even assisted with construction work on the recently completed but still rough base. After the 23rd, he may have continued some of the basic training he had begun in Clearwater and Salt Lake City. Harvard AAB was also the location of the AAF 426th Sub-depot, which was a maintenance group for Second Air Force heavy bombers. Second Air Force B-24 bombers were sent to Harvard a few at a time for one hundred hour mechanical inspections. Although Walt had not yet even begun airplane mechanics school, he was assigned there to work with experienced mechanics from the 426th Sub-depot to get familiar with specialized tools and aircraft maintenance procedures.[94]

Walt got day-passes to leave the base one or two times while he was stationed at Harvard. Even though the town of Harvard itself was just a couple miles outside the base, it was a rural farming community with a population of just 725 in 1943. The town did eventually establish two small USO facilities, one for whites and one for colored troops.[95] Walt and the other airmen did not stop in Harvard on their free time, but continued back to Hastings. Walt sent his family a number of post cards depicting some of the major sites in Hastings. The city had three motion picture houses to amuse the service men. By late March, average high temperatures in Hastings would be reaching the 50's, so Walt could comfortably walk around the city. He walked out to Hastings College, strolling through the 82 acre landscaped campus with 50 faculty and 1,000 students. Just north of the college was Heartwell Park, where a dam stored water from the city storm sewer system to create a small tree fringed lake used for skating in winter and as a site for picnics in summer. In the center of town the USO used the large Masonic Temple at Fourth Street and Hastings Avenue for various live entertainments for men from the Navy Weapons Center and from Harvard AAB.[96] Sunday April 4 he read in an Omaha newspaper that the Detroit Red Wings had had won their first round, beating Toronto four games to two, and winning their first game against Boston on April 1 for the Stanley Cup championship.

On Thursday, April 1, the men at Harvard were informed that they were no longer attached to the 485th AB Squadron, but now belonged to the 521st Base Headquarters and Air Base Squadron. They were told that the 485th was actually assigned to nearby Kearny Army Air Field, and that since January 28, the 521st was assigned to Harvard, but "due to error in correspondence channels" this information was not relayed to the base. The men must have rolled their eyes and wondered if the military was really capable of keeping track of the millions of men they were calling into service. As of April 1 the base had a strength of 18 officers and 419 enlisted men.[97]

The change from the 485th to the 521st is reflected by comparing the telegram Walt sent home when he first arrived at Harvard with the return address on the series of postcards depicting various scenes from Hastings that he sent back to his family on Thursday, April 8. This was near the end of his short stay at Harvard AAB and we get the sense that he continued to do well:

Hello Jenny,
How are you. Fine I hope and I want to thank you for them song sheets you sent me. When I asked for a couple I didn't know I would get that many. Write more later.
-Walt[98]

Hello Hank,

Well I am doing allright yet just thought I'd drop you a line. By the way how did the Red Wings make out? Did they beat the Boston Bruins for the championship? Well I will write more later.

-Walt[99]

Hello Leona,

How are you my little girl? I hope you are getting along fine and is the weather outside very nice. Its very nice here. I will send some more pictures home soon. Goodbye.

-Walt[100]

So the first week of April in the middle of Nebraska, Walt was pleased with the weather, happy to have received some song sheets from his sister, and interested in how the Red Wings had done in the Stanley Cup playoffs. Perhaps he was thinking back to that last Red Wings game he had seen with his friends back in January before reporting for active duty.

On Saturday, April 10, Walt and 45 other men stationed at Harvard AAB were summoned to base headquarters and given new orders. First, Walt and the others were told that they were promoted from Private to the grade of Private First Class. This would mean an increase in pay! As a PFC Walt would make $50 per month, with $6.50 taken out for laundry of clothes and bed sheets, $6.00 for the purchase of one-third of a war bond each month, and a small amount for life insurance.[101] Next the same 46 men received orders to report to the Army Air Force Technical Training School at Keesler Field, Mississippi for assignment as students in the Airplane Mechanics course. They were instructed to proceed to Keesler by rail, departing on Tuesday, April 13, to arrive no later than Thursday, April 15. The orders stated that they would receive breakfast at Harvard AAB on the 13th, then each man would be furnished with six meal tickets that would provide $1.00 for meals taken in a dining car while enroute, or $0.75 for each meal taken elsewhere. The base adjutant, 2nd Lt. Charles F. Brigance, signed the orders.[102]

- 5 -
AIRPLANE MECHANIC SCHOOL 1943: THE MOP ON THE B-24

Keesler Field, Mississippi

On the morning of Tuesday, April 13, after breakfast, Walt again packed his clothes and gear in his duffel; made sure he had his travel orders in his pocket and got onto the waiting bus along with four-dozen or so other men for the drive from Harvard AAB into Hastings. After the half-hour drive, they were deposited at the CB&Q Station in Hastings before eastbound train 2 arrived, just before 11:00 a.m. The men boarded, found seats as well as they could, and were off to the east. Train 2 pulled into the Lincoln, Nebraska Union Station located at Seventh and P Streets right at 1:30 p.m. and the men got off as the train continued east to Omaha. They did not have time to explore though, because shortly after they arrived, CB&Q train 42 from Billings, Montana arrived and they boarded to continue on to St. Louis. Walt did manage though to buy a couple postcards of Lincoln while crossing over from one train to the other.

Train 42 pulled into St. Joseph, Missouri at about half past six, and once more Walt was able to dash out and buy a handful of local postcards. At about a quarter till nine in the evening their train paused at the Kansas City Union Station and yet again Walt was able to buy a handful of postcards, which he put in a pocket until he had time to write out a few and mail them home. The men probably played cards into the evening or perhaps sang, aided by the song sheets Walt had gotten from home.

It was now late morning and Walt stopped in a nearby USO to clean up, rest, and write out some of the post cards he had been collecting. Pulling out a card showing the University of Nebraska in Lincoln, he wrote to his ma and pa:

> *Hello Folks,*
> *Am on my way to school. Having a lot of fun on the train. I may get there*
> *Wednesday night. Will say so long for now.*
> *-Walt*[1]

Another card showing downtown St. Joseph was intended for his younger brother:

> *Hello Hank,*
> *Am on my way to school. Hope you all are fine. Having quite a time here on the train.*
> *-Walt*[2]

A third card, depicting Kansas City Union Station at night, he addressed to his youngest sister:

> *Hello Leona,*
> *How are you little sister? I am on the train going to school. I am having a lot of fun here.*
> *-Walt*[3]

He put these cards back in his pockets, ready to be mailed the following morning. Military regulations prohibited him from saying where he was going, or even his route, while he was in transit. His choice of these postcards though would help his family trace out the route he was taking from Hastings towards the southeast. Soon the men tired. The singing, games, talking, and letter writing ended, and they settled back into their seats for sleep.

Wednesday, April 14, at about 7:25 in the morning, train 42 pulled into St. Louis Union Station at Market and 18th Streets. Walt and the other men were told that they had about eight hours on their own until their next train left for Nashville, Tennessee. They were told to report to the St. Louis Union Station Army Transportation Office by 3:30 p.m. The location of the nearby St. Louis USO Club was pointed out and they were reminded to be on their best behavior and not be late in returning. With those instructions in mind, the men left in groups of two, three, or four to explore the city. Walt mailed his postcards, and then he struck out along Market, the main commercial street, eastward toward the Mississippi River. Glancing back at Union Station, he saw a four-story structure built of Bedford, Indiana limestone, with a slender, 230-foot high clock tower. A few blocks east, he passed two more prominent buildings constructed of the distinctive Bedford limestone: the Municipal Auditorium on his right and on the left, a little further on, the Soldier Memorial Building. Further east approaching the Mississippi River he found a more dilapidated section of the city. This was the original commercial district along the river, when the life of St. Louis was tied to shipping heading to and from the north, south, east, and west, via the Mississippi, Missouri, and Ohio River systems. Now the river was lined with largely unused wharves bordered by empty warehouses, factories, and commercial buildings. Before the war, St. Louis had begun to clear the area for a major waterfront park, but those plans were now on hold[4]

Glancing briefly at the Mississippi, Walt turned back west along Market Street and found his way to the USO near Union Station. It was near lunchtime and Walt entered the USO to clean up, rest, and write out another post card, this one showing St. Louis' Old Courthouse:

> Hello Jenny,
> I am here in St. Louis for 8 hrs. We are having some swell fun here. We are in the USO right now. So long.
>
> -Walt[5]

After a couple pleasant hours spent at the USO, the time came to head back to Union Station. By 3:30 or so all the men from Harvard AAB had reassembled and were ready to board Louisville & Nashville Railway train 91, in time for its 4:22 p.m. departure for Nashville. Shortly after the train crossed the Mississippi into Illinois and headed southwest. Soon darkness came, the men ate dinner in the canteen on the car and then gathered for their various card games, jokes, or songs. At 9:05 p.m., the train pulled into Evansville, Indiana, on the Ohio River. There was a scheduled 15-minute stop here, so Walt and several other men walked over to the local USO in the nearby Evansville Red Cross Canteen, where they were able to pick up some snacks quickly and get back to their train in time.[6] Soon after they each drifted off to sleep. Early in the morning hours of Thursday the 15th, they reached Nashville, where they were awakened and led off the train. Train 91 continued on shortly after bound to Florida. Walt had about a 40-minute wait in the quiet Nashville Union Station until L&N train 1 pulled in, the men boarded shortly before 3:00 a.m., and they headed off south again towards Alabama.

Walt awoke and had a fast breakfast well before the train pulled into Birmingham at about a quarter past 8:00. Later in the morning L&N train 1 had a 20-minute stop at Montgomery, to allow a change of engines and crews. Walt had time to dash into the station, buy a postcard of the Alabama Governor's Mansion in Montgomery, and send off another of his very brief notes to his brother, writing:

> Hello Hank,
> Am here in Montgomery having a swell time. Will reach camp in the evening sometime. So long.
>
> -Walt[7]

Walt reboarded the train in time for its 11:05 a.m. departure. The route of train 1 trended off to the southwest, headed for its final destination of New Orleans. This was Deep South cotton country they were passing through, until the train crossed a broad swampy area and then entered Mobile. The industrial growth of Mobile was evident with many newly established war industries along the tracks. About 20 miles outside Mobile, they crossed into Mississippi and stopped briefly at Pascagoula, another small town on the Gulf of Mexico booming with war industry. The men knew they were nearing their destination and so became more interested in what the countryside looked like. The cotton fields had disappeared north of Mobile. The scenery now was lush and they could smell the Gulf salt air. Vegetation was a combination of palms, pines, magnolias, oaks, Spanish Bayonets, poinsettias, cape myrtles, and azaleas. 15 miles past Pascagoula, the train passed through the small town of Ocean Springs, then crossed Biloxi Bay on a mile and a half long bridge that brought them into Biloxi itself. At about 5:50 p.m. the train pulled into the L&N station at Reynoir and Railroad Streets. Walt and the other four dozen men got off with their gear, and the train continued on west to New Orleans.[8]

The Army was very efficient in Biloxi and a line of trucks were there waiting to take the men from the station into nearby Keesler Field, just a mile and a half to the west. The men boarded and then they were off. A few minutes later, the trucks turned left off Forest Avenue then paused at the base's Gate No. 1 to clear with the guard on duty. They then proceeded down "F" Street past row after row of barracks and classrooms. By mid-evening, the men had presented their orders and were assigned to platoon sergeants who marched the men with their gear to their barracks. It was Thursday, April

15, 1943 and Walt had arrived at the place that would be his home for almost the next six months – the longest period he would be at a single base during his Air Force career.[9]

Walt and the other men were assigned to one of hundreds of identical barracks buildings, each two stories high, 30 by 80 feet, and capable of housing 63 men on each floor. The construction of these building was crude and their unfinished interiors, without insulation, gave them a rustic feel. Sleeping was in double bunks and there was an open latrine on each floor that provided basic facilities but no privacy. Air conditioning was unheard of, despite the hot humid climate the men would endure through the summer. On the plus side, there were gas-fired heaters to keep them warm during any cold snaps and there were screens on the windows to control the mosquitoes. These were not the best living quarters that Walt would enjoy during his time with the Air Corps, but they were far from the worst.[10]

Keesler Field in 1943 was one of the largest military bases in the U.S., and yet it had not existed two years before. As early as 1938, President Roosevelt felt that the country might be involved in a future war. Germany's use of the Luftwaffe during the Spanish Civil war convinced FDR that American airpower was important to try to keep the country out of war, but critical to ensure success if it came to war. That year the Air Corps proposed an expansion to a force of 7,000 planes. Roosevelt went to Congress shortly thereafter to request funding for a force of 10,000 planes. As late as 1941, the U.S. had only three bases dedicated to the technical training of aircrews and aircraft mechanics. General Henry H. (Hap) Arnold, chief of the Air Corps, convinced the President that a force of 10,000 aircraft would require a vast expansion of air bases and trained men. The War Department decided by 1939 that two new training centers were required.[11]

Early in April 1939, a number of Biloxi civic and business leaders began discussing the possibility of their city being chosen as the site of one of the new Air Corps training facilities. Between 1935 and 1938, the WPA (Works Progress Administration) had constructed a small municipal airport with a 3,000-foot runway west of the city. It had been used for a short time in 1938 during Army training maneuvers, by an Air Corps observation squadron from Maxwell Field in Alabama. The economy of Biloxi had always been highly seasonal – dependent on tourism in the summer. Only small local seafood canneries provided significant year round activity. The Biloxi leaders reasoned that an Air Corps base and training center could provide a major boost to the local economy. Mississippi Senator Pat Harrison was enlisted to lobby the War Department for the Biloxi site, but the competition was daunting – at least 50 other cities were competing for the two new bases.

For almost two years, Biloxi city leaders and the Mississippi congressional delegation lobbied the Army Air Corps for their site. Biloxi started by offering the Air Corps the 164-acre municipal airport free of charge. For its part, the Air Corps began by specifying a base that could train 5,000 men at a time. Later this was increased to 6,000 men, then to 12,000 and finally 24,000 men. As the military demands were increased, the city added an adjacent 138-acre parcel, zoning to prohibit aerial wires or smokestacks that might obstruct the flight paths, the lease of a further 160 acres, local fire protection service, increased and expanded city bus service to the base, and more than 300 building lots for conversion to low-rent housing. The additional land allowed the proposed base to expand adjacent to the local Naval Reserve Park and a Veterans Hospital site so that the combined land available would be 832 acres. Finally, on March 6, 1941 Biloxi Mayor Braun was notified by the War Department that Wichita Falls, Texas and Biloxi had been selected as sites for the two bases, and that the president had appropriated $8 million for construction at Biloxi. The base was to be named in honor of Second Lieutenant Samuel Reeves Keesler, a native of Greenwood, Mississippi, who had been shot down and killed behind German lines in October 1918.[12]

Construction of Keesler Field began on June 13, 1941 with just 400 men. By August a peak of 11,471 men were on the base construction payroll. Initial plans were massive: 376 two-story barracks each housing 63 men per floor, ten mess halls each serving 2,000 men at a meal, 28 administration buildings including post headquarters and orderly rooms for each training squadron, a 1,200 bed hospital, five double hangars, six large academic buildings, an engine test block building, two motor repair shops, a service club, two theaters, seven post exchanges, four firehouses, a post office, and four chapels, as well as various warehouses, rail spurs, fuel depots, and an incinerator. Runways and base roads were constructed, gas lines run to the barracks and other buildings for heat, and water, sewer, and electric lines run. Training commenced in August 1941, even as work on the base continued. However, construction did not progress as quickly as initially hoped, despite the urgency that the Japanese attack on Pearl Harbor added in December. Finally, in September 1942, initial construction was completed and the base was deemed operational.[13]

Keesler had two primary training missions. In addition to its primary mission to train aircraft mechanics, it was later decided to provide basic training as well for new recruits. All trainees were assigned to a Technical School Squadron. Each TSS was either part of the Provisional Replacement Center Group A, responsible for basic training, or Provisional School Group C, responsible for aircraft mechanic training. On the day the Japanese attacked Pearl Harbor, Keesler Field had 7,051 recruits in basic training and approximately 4,000 in aircraft mechanics training.[14]

When the war began, Biloxi's population was just 14,850. It was the oldest permanent European settlement in the Mississippi valley, located on a low ridge at the eastern end of a narrow finger-like peninsula. The Gulf of Mexico was to the south and to the north, the Bay of Biloxi separated the town from the Mississippi mainland. The site was visited in 1682 by La Salle who claimed the entire Mississippi drainage for France. Biloxi was first settled by a group of Frenchmen led by Pierre le Moyne, Sieur d'Ibreville, in 1699, two years before Detroit was founded. It became the capital of the French claim to an area extending as far northwest as what is now Yellowstone National Park, and was soon after visited by hardy *voyageurs* from Canada who paddled down the Mississippi with loads of furs for trade. A few years later, the capital was moved to Mobile, then later still to New Orleans, and Biloxi settled into decade after decade as a small French commercial and fishing town.

After the U.S. acquired the area, Americans began to move into Biloxi, along with their black slaves, but the French influence remained dominant and the European flavor of the community was preserved. The modern growth of Biloxi began in 1869 when the railroad first reached the town, allowing seafood to be shipped to market in refrigerated cars. Later shrimp canning developed and Polish-Americans from Baltimore moved into town to work in the canning plants. In the 1890's the first resort hotels were developed and Biloxi's other major pre-war industry, tourism, began to develop. By 1940, 40 seafood-canning plants were in operation, employing 3,000. 800 sailboats with 2,000 boatmen supplied the factories with shrimp and oysters, as well as with fish such as sea trout, mullet, croaker, redfish, catfish, and pompano.[15]

Biloxi was booming in 1943 with all the new employment the base provided. By July 1943 there were 2,322 civilian employees working at Keesler Field. In April, when Walt arrived, Keesler was also home to 39,646 military personnel, dwarfing Biloxi itself in population. This number dipped to 31,078 in June as many early mechanics classes began to graduate, but rose again to 38,300 the following month with the arrival of even more recruits and trainees.[16] Biloxi and the surrounding Mississippi coastal communities were very busy indeed supplying food, supplies, and services to the base, as well as entertaining the troops when they were on leave.

The local Biloxi newspaper, the *Daily Herald* was useful for Walt to learn about Biloxi and for Keesler news as well. Walt had come to enjoy group singing as well, and he learned from the *Herald* that a new Army issue song book, or 'Hit Kit,' would be arriving at Keesler soon, with words to current popular songs such as "There's a Star Spangled Banner Waving Somewhere," "Coming in on a Wing and a Prayer," and "Roll out the Barrel." He also read in the *Herald* the day after he arrived at Keesler that men who did well in aircraft mechanics training would be promoted to corporal and assigned to more advanced training at aircraft factory schools run by Ford, Consolidated, and other manufacturing companies.[17] Walt wondered if he could make the grade, but before he started technical training his first two and a half weeks at Keesler were spent with more processing, medical checks, base orientation, additional drill, and duty assignments. Shortly after he arrived at Keesler, Walt was assigned to the 396th Technical School Squadron (396th TSS).[18]

Walt began to familiarize himself with the huge base. Religious service at Keesler was provided at a rustic wooden chapel named for Brig. General William 'Billy' Mitchell, head of the U.S. Army's Aviation Section during World War One.[19]

Jewish airmen had religious services on the base as well. Monday, April 19 was Passover, and Keesler provided passes to over 1,000 Jewish servicemen so they could attend a Seder supper at the Buena Vista Hotel in Biloxi, conducted by Chaplain Sidney Ballon.[20]

Sunday April 25 was Easter day so the religious services were well attended. Catholics and Protestants attended outdoors Easter services at an area north of the base hospital known as Oak Park. April in Biloxi was generally dry, with pleasant high temperatures in the 70's and lows in the 60's.[21] Seating for 6,000 men was provided, with Protestant services at 0700. Later, at 0900, Walt attended a Solemn High Mass and Benediction for Catholic servicemen, conducted from a large altar platform built for the service.[22] Later in the day, Walt picked up a copy of the local paper to see if he could find any baseball scores for the Major League season that began the week before. He read that the Detroit Tigers lost their opener in Cleveland 1-0, but beat the Indians the following day 4-0.

When Walt first arrived at Keesler, the Post Exchange, or PX, was located in a warehouse on the far south side of the base. It was completely remodeled and reopened for business in July 1943 as a small sized department store.[23] Whenever he could, Walt preferred to eat at mess hall number 4, because the sergeant in charge had rigged up a radio-juke box combination to the P.A. system. The jukebox had more than 200 records available to provide jive tunes to entertain the troops.[24]

Keesler was primarily a training facility, but it was also a major airfield. After several intermediate runways had been completed, then expanded, the final construction resulted in two major runways: one 6,600 feet long and 150 wide oriented southeast to northwest, and a second 5,000 feet oriented southwest to northeast. A 62-foot high steel radio-control tower coordinated flying operations and allowed the base to receive and fly off aircraft as large as the B-24 Liberator. The roar of aircraft on take offs and landings was a constant background noise that Walt quickly got used to. A variety of hangers located along the taxiways allowed student mechanics to test their newly acquired knowledge and skills on real first line Air Corps bombers.[25]

The post police, known as military police, or MP's, provided law and order at Keesler. The MP's numbered about 100, and in addition to manning the base gates and checking passes for those entering and leaving the base, they also provided a town patrol. The town patrol operated in Biloxi and other nearby communities to maintain order and discipline and enforce curfew and off-limits restrictions for Keesler personnel. There was also a Negro MP detail, responsible for patrolling the segregated black sections of Biloxi.[26]

Movies were shown at the Post Theater every evening at 1800 and 2000. However, if a trainee saw a movie at 1800, he would miss his evening meal, or he could see the 2000 showing if he could get by on just six hours sleep the following day.[27]

An Air Force band had been authorized at Keesler before Pearl Harbor, but its development was slow. Originally just a cadre of three men with an authorized strength of 28, as late as January 9, 1942 they were using waste paper baskets as drums. As more recruits arrived, qualified musicians were found, many of whom had worked professionally with big bands and orchestras. Band members stood reveille, cleaned their barracks, policed their squadron area, and performed all the duties of a soldier in addition to their band duties. Most of their days were spent in practice and working on arrangement in Recreation Hall No. 1. One of the musicians was chosen as drum major, based on his military bearing, knowledge of band formation and troop movements, skill with a baton, and in executing troop movements. By the end of 1942 the band organization at Keesler had grown to 79 men, with a separate 31 man drum and bugle corps. To provide entertainment for officers and men at Keesler, these men could be broken into five bands, three dance orchestras, a polka band, and a schottische band. These bands and individual musicians often performed for a daily 15-minute show called 'Free for All', which originated in Keesler's Theater No. 1 and was broadcast by radio station WWL in New Orleans. WWL was a clear channel a.m. radio station that could be heard across the entire country east of the Rocky Mountains.[28]

Trainees could obtain a Class B pass every night (unless the privilege had been temporarily withdrawn), which permitted them to leave the base and explore Biloxi. To obtain a pass though, they had to change from their coveralls into their Class A uniform, and then wait in a line to obtain the pass. At least once a month a man could obtain a weekend pass, which would allow him to leave the base from 1800 on Saturday until 2330 Sunday.[29]

Despite the focus on technical aircraft mechanic training however, regular military training continued as well. Because some mechanics would be assigned to airfields near front lines or other combat areas, defensive combat tactics were emphasized at Keesler. During these first two and a half weeks, Walt and other men from his flight of the 396th TSS participated in a 25 mile, 12 hour march. 200 to 300 men participated in these marches, carrying light packs, rations, gas masks, and rifles. The following week Walt and his flight went with a larger group of about 400 men on an overnight march and bivouac to the Keesler rifle range, which was about 8 miles north of Back Bay on the Mississippi mainland. They left Keesler by foot for the rifle range at 0730 and returned to base by 1800 the following day. The goal of these marches was to harden the men and get them used to field conditions. They learned reconnaissance and combat procedures, scouting, camp sanitation and hygiene, small arms firing, and camouflage techniques.[30]

A week after arriving at Keesler Walt had his first pass to explore Biloxi. He walked out of Keesler's Gate No. 1, turned right onto Forest Street, crossed over the L&N tracks, then turned left onto Howard Avenue, the main east-west commercial thoroughfare. Howard was relatively narrow, reflecting Biloxi's old-world French background. It was lined by one and two story stuccoed commercial buildings, which were mostly weathered to a soft cream color. Here Walt found various markets and department and drug stores. South of Howard was a newer residential and tourist section of Biloxi extending just a few blocks down to Beach Boulevard, which followed the water's edge for six miles to the west. Out along Beach Boulevard he found artificial beaches, resort hotels, and amusement parks. The free Biloxi beach and entertainment pier were just off Beach, between Nixon and Elmer Streets. The Boulevard was lined with tall longleaf pines, oaks, and camphor trees, which would catch the Gulf breeze. Spaced out along Beach, Walt also found older planter-type houses built by ante-bellum plantation owners who would come to Biloxi before the Civil War to escape the inland heat and fever. This section of Biloxi had some of the tourist gayety that Walt and his pals had experienced in St. Petersburg and Clearwater, but on a smaller, quieter scale.

North of Howard Avenue out to Back Bay, the arm of the Gulf of Mexico, which separated Biloxi from the mainland, was an older section of town. Most of Biloxi's population lived here in older and smaller steep roofed brick and stucco cottages, situated behind picket fences and shaded by elderly oaks. The area had a somewhat timeworn, dingy appearance. Most of the side streets were paved with finely crushed oyster shell that appeared grey when shaded by the great oaks, but dazzled white like snow in sunny stretches. At the far eastern end of Howard Avenue, where the peninsula ends at the 'Point' was located the shrimp and oyster packing and canning plants. The residential area around the 'Point' was comprised of box like wooden structures built in 1925 as temporary housing for immigrant laborers brought to work in the packing plants. This 'temporary housing' had never been replaced though and these tiny dilapidated cabins, some built on stilts out over the bay, gave this area a poverty-stricken appearance. However, the inhabitants of the Point, a mixture of Polish, Yugoslavian, Austrian, and Czech immigrants, with their heavy accents and old world customs, also gave the area a somewhat exotic feel. There were many dance halls in this area too, less formal than the dance pavilions found in the resorts hotels further west out Beach Boulevard. Perhaps in the area Walt was able to find familiar Polish music, sausages, and pastries while on leave.

North of the Point and Howard Avenue, in the northeast part of town, was the primary Negro residential section. Biloxi had the largest percentage of foreign born of any Mississippi city in the 1940's but the smallest proportion of blacks - about one sixth of the population. Biloxi's blacks were unusual for the state, because most were Roman Catholics, like the majority of the coastal white population. Although most Biloxi blacks worked as manual laborers, some owned their own fishing boats, and an unusually high percentage owned their own homes. Biloxi Catholics, both black and white, had a long history of worshipping together since emancipation. With the growth of Biloxi's black population though, the community built its own Catholic church, Our Mother of Sorrows, in 1914. The parish soon added a grade school and by the late 1930's had added a parochial high school to supplement the public school available to Biloxi Negroes.[31]

Biloxi could be a wild place on a Saturday night. With servicemen at Keesler outnumbering Biloxi citizens more than two to one, the city was hard pressed to keep them entertained. Before 1941, Biloxi was remembered by one resident as 'a sleepy fishing town.' By 1943, the town had boomed though, and on a Saturday or Sunday with thousands of airmen on leave, the sidewalks along Howard Street became so congested, you had to walk in the street to get from one shop to another. In the evening and late into the night, bars and gambling boomed. There had always been some gambling in Biloxi before the war, but now it was commonplace and open. Every bar, store, and restaurant had open slot machines, and it was not difficult to find poker, craps, and other dedicated gambling halls. Indeed, the city had actually started to tax gambling. The USO provided wholesome entertainment, supplemented by dances for servicemen staged by local hotels and lodges, but still a 'red-light' district developed, despite policing by the MP's and frequent education about the health dangers they presented. Things became so wild that the residents of nearby Gulfport would not allow their teen-aged children to visit Biloxi – it was 'off-limits' to them. All along the Gulf Coast during the war, Gulfport had a straight-laced reputation, while Biloxi was a 'wild' place.[32]

Walt was probably looking for just a movie or two, and then a good meal cooked off the base with a couple cold beers. There were at least four movie theaters for him to choose from in Biloxi. That first weekend the Saenger Theater was showing 'Hello Frisco, Hello' in Technicolor, with Alice Faye, John Payne, and Jack Oakie. The Saenger charged 31 cents admission. The Buck was showing a movie called 'Hitler's Children' along with a Donald Duck cartoon and sports shorts. The Roxy, which charged just 20 cents, featured 'Tarzan's New York Adventure' with Johnny Weissmuller and Maureen O'Sullivan, along with selected shorts, while the Meyer had a double feature of 'Moonlight Masquerade' with Dennis O'Keefe and Jane Frazee coupled with 'London Blackout Murders.'[33]

The men were all instructed by their squad leaders to keep on their best behavior because of a series of base inspections and tours that occurred during Walt's first week at Keesler. First on Monday, April 19, a group of journalists from the National Geographic Society, the Scripps-Howard Newspapers, and the Associated Press were given a tour of the base by Col. Goolrick. Three days later, Brig. Gen. John Lentz, G-3 of U.S. Army ground forces toured the base, again with Col. Goolrick.[34]

By Friday, April 23, Walt had settled into his routine. He had not yet started his airplane mechanics course - that was still a week or more in the future - but he did find time to write to his youngest brother and sister to give them and all his family some idea of what his future weeks would involve. To Leona he sent a post card showing a single engine trainer flying over Biloxi's Back Bay, with an aerial view of Biloxi and Keesler below:

Hello Leona,
Here is a postcard on which kind of airplanes I am going to learn to fix. Down below is the land where the camp is. Show this to Ma & Pa.
-Walt[35]

To Hank, he sent a postcard showing Keesler mechanics students swarming over a B-24:

Hello Hank,
Here is a postcard showing what I am going to learn to do. To be a mechanic on big bombers. Show this to Ma & Pa to know what I am going to do.
-Walt[36]

About a week later, Walt had started to receive mail from his family and friends at Keesler. His friend Joseph Kaszyca (from the Delray neighborhood in Detroit, and known to his friends as Clark), by then a Pfc. himself wrote Walt at Keesler from San Diego:

Hello Lize,
So you're a PFC now. Well I do say. You do get around fast and before I know it Lize will be a cpl. or sgt. Well keep up the good work & some day you'll be a "Lt." for I see your in a different place again. Well Lize...we sure had a swell time. As ever, your pal,
-Clark[37]

Walt wrote another handful of cards to his family in Detroit, each with the same basic message as this one to his sister Jenny:

Hello Jenny,
How are you Jenny? I am fine. Busy now, will write more later on. Hope you had a swell Easter because mine was swell. So long
-Walt[38]

On Saturday evening, May 1, Walt picked up a copy of the *Biloxi Daily Herald*. The newspaper provided ample coverage of activity at Keesler, as well as news of Biloxi and the world beyond. He read about the work he had seen starting on new 9,800 seat outdoor theater at Keesler. It would be completed by June and used for USO shows, weekly camp boxing matches, Keesler band concerts, and talent nights. Walt read about a Keesler airplane mechanic school graduate, S/Sgt. Augustine Mazzacaro, who had later completed aerial gunnery school and became a B-17 crewmember. Mazzacaro was being recommended for the Distinguished Service Medal for spotting a German U-boat, which was subsequently sunk by his B-17.[39] Walt also read that the Tigers had just swept a two game series in Detroit against the St. Louis Browns, 4-2 and 3-2, leaving them with four wins and three losses in the early days of the 1943 season.

On Saturday, Walt got a number of letters from home that brought him up to date on news of family and friends. He learned that his brother Johnnie was now in Brisbane, Australia. He had spent a few weeks in San Francisco waiting for orders and a ship, finally embarking on a vessel called the M/S Sommelsdyk, a Dutch passenger ship that had been taken over by the U.S. after the war broke out. He had 'crossed the equator' on March 13 aboard the Sommelsdyk, an event that included an elaborate and ritualistic shipboard ceremony where 'Neptune' greets anyone aboard who has never 'crossed the line' before. His crossing the line was verified by the signature of the ship's captain, Lt. Commander O. E. Hagberg on Johnnie's official Navy records! Johnnie arrived in Brisbane, at the US Navy Station Depot on March 25. On April 21, he was transferred to the US

Navy Seventh Fleet, 78th Task Force, in Brisbane, assigned to the Escort Ship and Mine Vessel Machine Shop – Refrigeration Repair Unit, responsible for repairs to refrigeration and water cooler units on small crafts. He was assigned two helpers and given a $3.00 per day subsistence allowance, in addition to his normal Navy pay as a Seaman, First Class.[40]

On Sunday, May 2, Walt wrote another postcard to his sister Jenny:

> *Hello Jenny,*
> *Did you get that letter I sent you in which I asked for that package? I hope you did cause I need the stuff. Write more soon.*
> *-Walt*[41]

What was the 'stuff' and why did Walt need it so desperately? The truth is that a PFC was not paid a lot of money and payday came just once a month. Perhaps the 'stuff' was cigarettes or some other supplies from home that would save Walt some money from payday to payday. Requests for cigarettes, food items, or even money were common in letters home from servicemen during the war. It is interesting to note the reference to a letter that Walt sent to his sister. We have only copies of postcards that Walt sent to his family, and the messages on these are generally very brief. How often did he write letters and what additional detail did he go in to?

A few days later, a postcard came from another of Walt's friends from Detroit, Edward Nycz. He wrote to Walt from Jefferson Barracks in St. Louis where he was beginning his processing in the Air Corps and basic training:

> *Hello Lize,*
> *How are you. Are you still kicking or are you lying down. Well it looks as though I'm in the same outfit your in. We do a lot of marching. We had a G.I. party Monday and boy did we dance with the G.I. brushes. I been here 4 days and I like it. We sleep in huts six men to a hut. Goodbye and a lot of luck to you.*
> *-Eddie*[42]

Aircraft Mechanic Training

Early the following week Walt and the other men in his flight of the 396th TSS were briefed on the technical training they were about to begin. The Technical School where aircraft mechanics training occurred was the reason for Keesler's development and existence. The commanding officer of Keesler Field was also commandant of the technical school, the running of which was his primary responsibility. When Walt was at Keesler during 1943, the base commander was Col. Robert E.M. Goolrick.

While attending the aircraft mechanics technical school, Walt and all the other trainees wore authorized work clothing consisting of a one-piece coverall fatigue uniform, with a belt at the waist and a piece of white tape above the left breast pocket with the man's name stenciled in black.

The Keesler Aircraft Mechanics Curriculum was divided into four major branches of instruction. The first branch was Maintenance Fundamentals. This introduced students to how the Air Corps organized aircraft maintenance functions and the policies and procedures that were used for inspection, maintenance, and repair tasks. Maintenance Fundamentals Branch classes were held in Keesler Academic Building 1. The second branch was Airplane Structures. This branch focused on the airplanes' main construction components and flight operation systems. Walt took his Airplane Structures classes in Hangar No. 3, which was designed for that use and allowed the men to inspect entire aircraft under the hangar roof.

Another branch was Airplane Engines, which trained men on the repair and replacement needed to maintain an airplane's power plant assembly. This branch also covered the hydraulic, electrical, fuel, and instrumentation systems that integrated the engines with the structure to complete an efficient, airworthy aircraft. The final branch was Aircraft Inspection, which taught students how to perform field aircraft inspection and repair. It was focused on training them how to apply the mechanical, hydraulic, and electrical systems training they received in the other branches on actual aircraft repair. This work was done initially on single engine training aircraft, progressing to B-24 inspection and maintenance. Special emphasis was placed on the correct and consistent use of aircraft manufacturers' maintenance and technical manuals. The goal was to standardize maintenance work so that men who were new to mechanical repair work could maintain complex aircraft operated under difficult conditions.[43]

Keesler AAF, Mississippi: The main gate to Keesler on the east side of the base.

Keesler AAF, Mississippi: New Aircraft Mechanics School students arriving at base. Pvt. Babinski arrived at Keesler on April 15, 1943. *Keesler Field, Mississippi* From Babinski Family Collection.

Keesler AAF, Mississippi: Newly arrived men being assigned to barracks. *Keesler Field, Mississippi* From Babinski Family Collection.

Keesler AAF, Mississippi: Standing in line for chow on a pleasant spring day in 1943. Later in the summer, the line would be in the shade to avoid the mid day heat..

Keesler AAF, Mississippi: "Mess Hall and Barracks" Oil on canvas by Sgt. Harry Dix. *Art and the Soldier* From Babinski Family Collection.

Keesler AAF, Mississippi: Evening classes at Keesler. The students studied in two shifts. They would switch shifts after each of the 14 eight-day courses was completed.

Keesler AAF: Walt relaxing in the shade with no school. Mechanic students usually attended classes wearing coveralls.

Keesler AAF, Mississippi: Keesler used 'war-weary' older B-24-D models for instruction purposes.

Keesler AAF, Mississippi: Classrooms were equipped with B-24 components and adequate tools to provide realistic hands-on training.

Keesler AAF: Walt with his hair carefully combed poses in the sun with a smoke, in front of a truck taking men on leave into Biloxi.

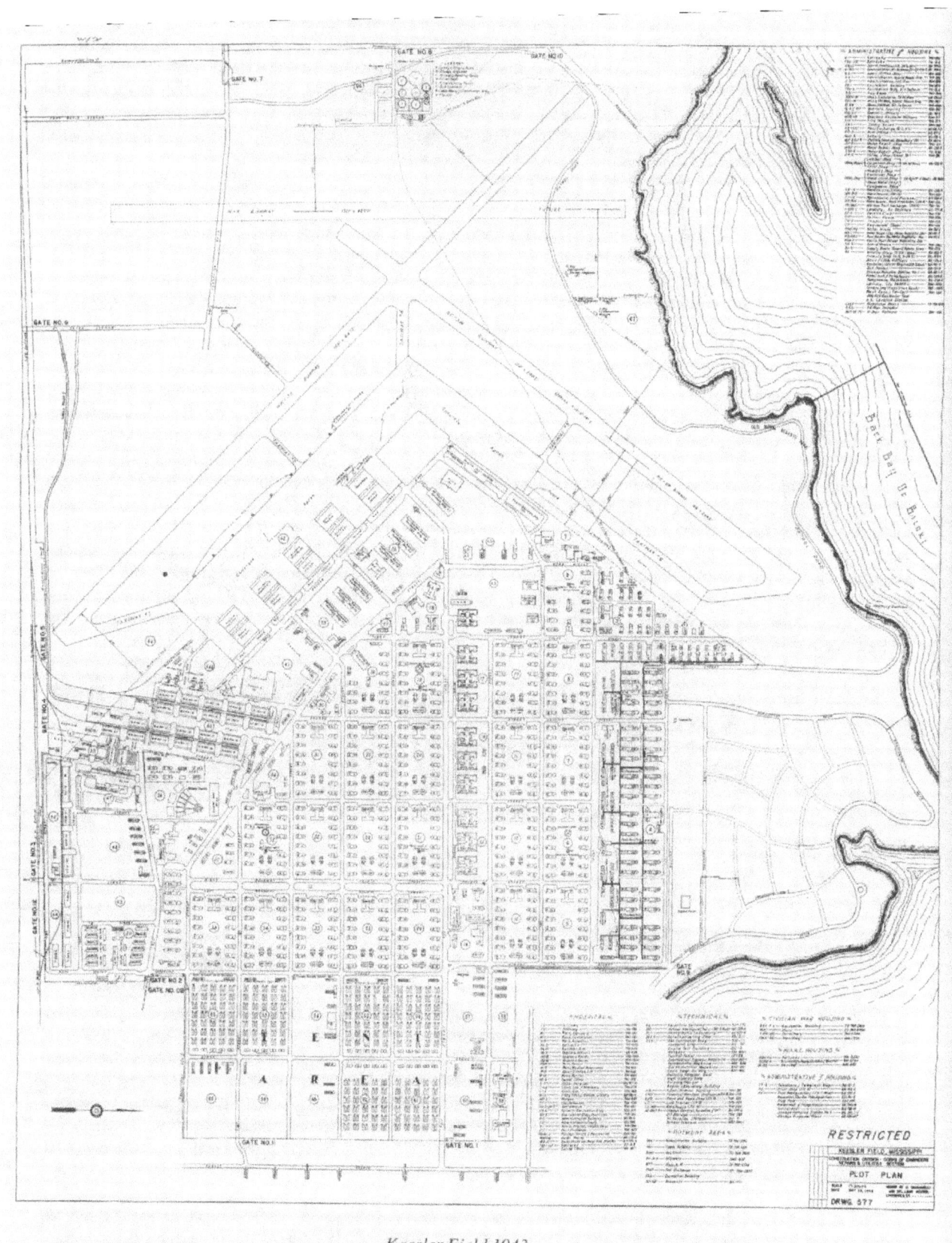

Keesler Field 1943

Keesler Air Force Base: Map of Keesler AFB in 1943, with the main gates to the east (bottom of the map) leading into Biloxi, Back Bay to the north, and two runways with hangars to the west.

Keesler AAF, Mississippi: Walt mailed this little cartoon-poster home to show Henry, Leona, and the rest of his family in Detroit what his daily routine was like as an aircraft mechanic student.

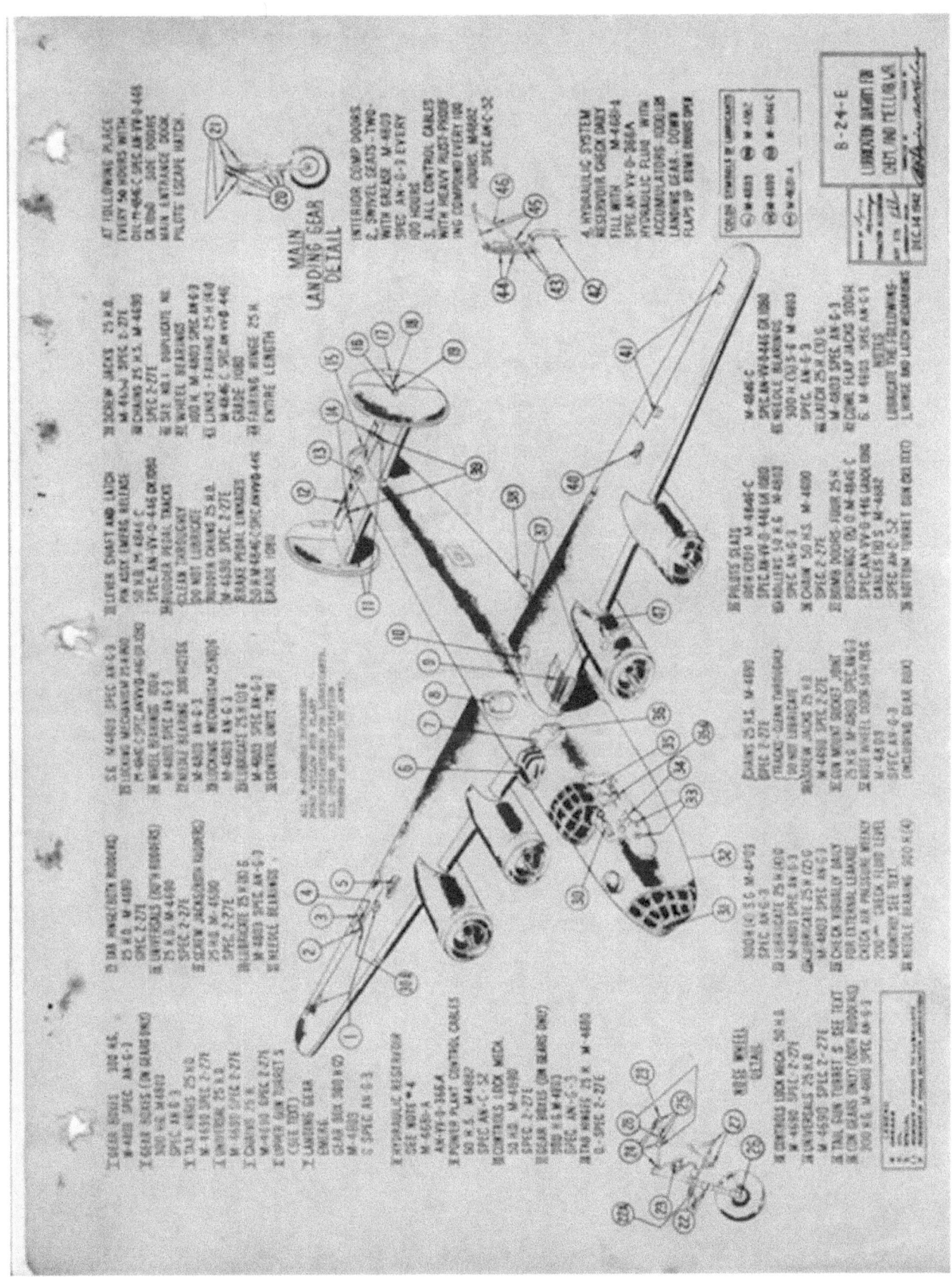

Keesler AAF, Mississippi:
B-24 lubrication diagram from the Keesler Aircraft Mechanics Student Workbook.

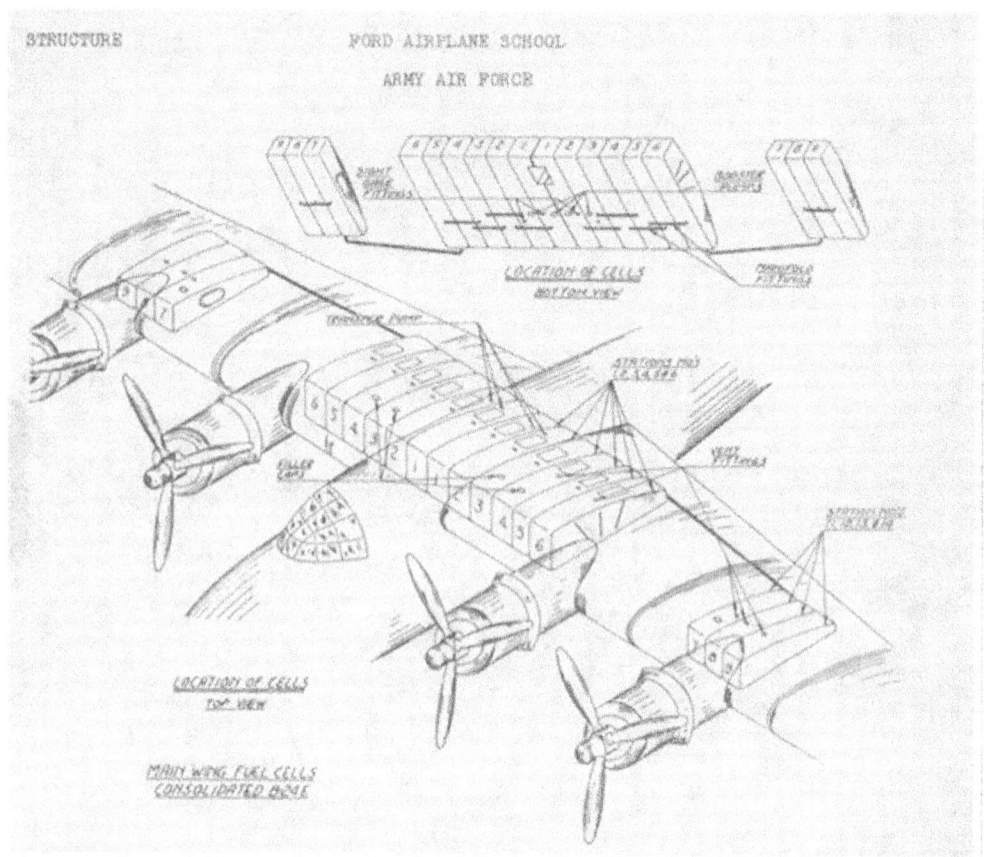

Keesler AAF: Aircraft Mechanic Student Workbook: B-24 Fuel Cells

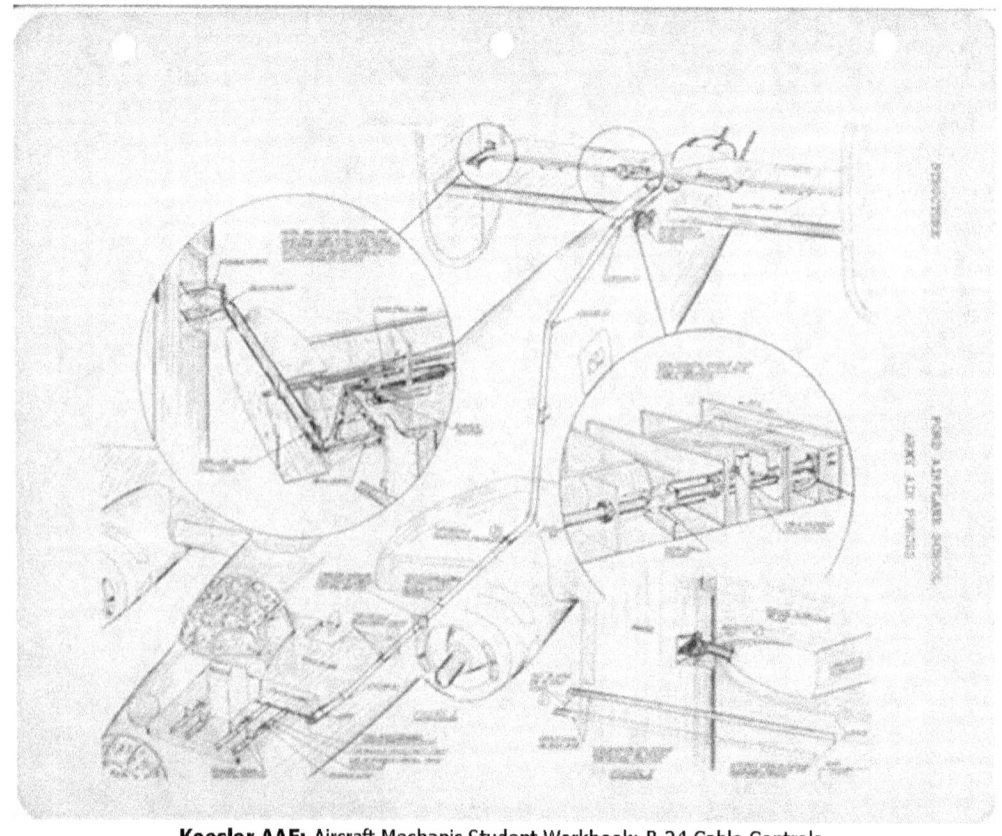

Keesler AAF: Aircraft Mechanic Student Workbook: B-24 Cable Controls

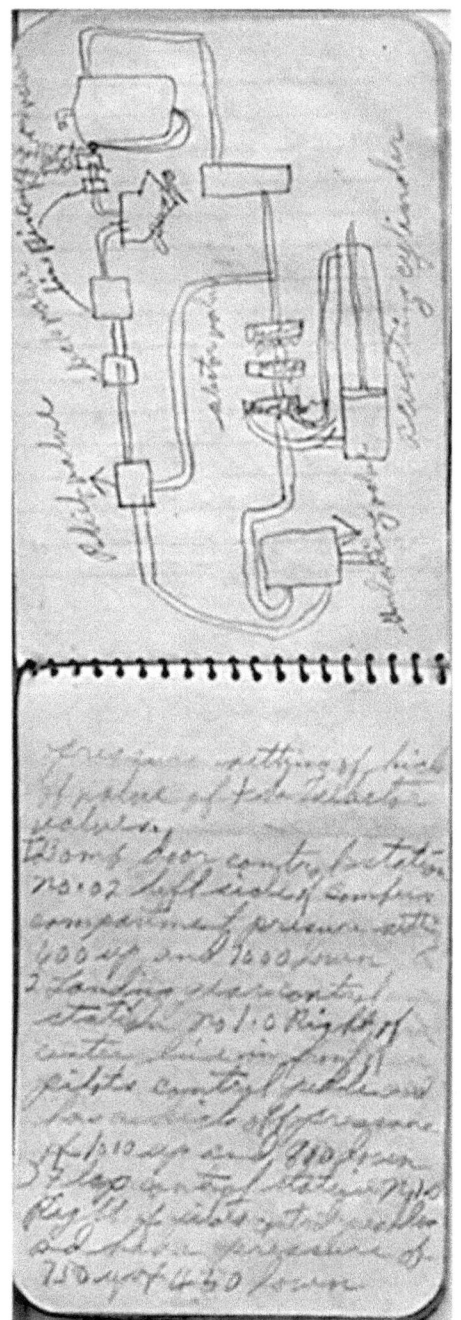
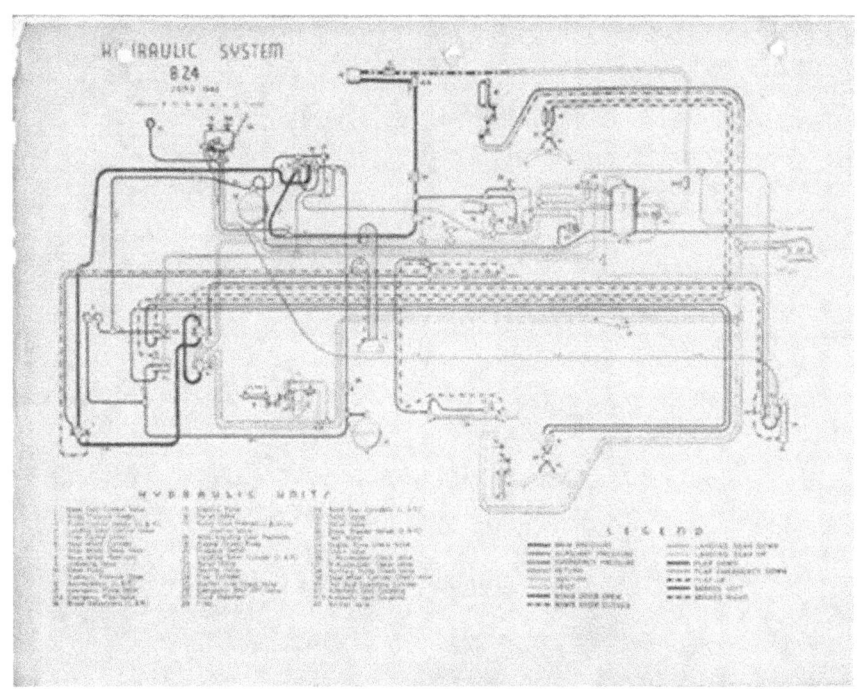

Keesler AAF: Aircraft Mechanic Student Workbook: Hydraulic Schematic (above) with Student Notebook (left) *Pearce Collection*

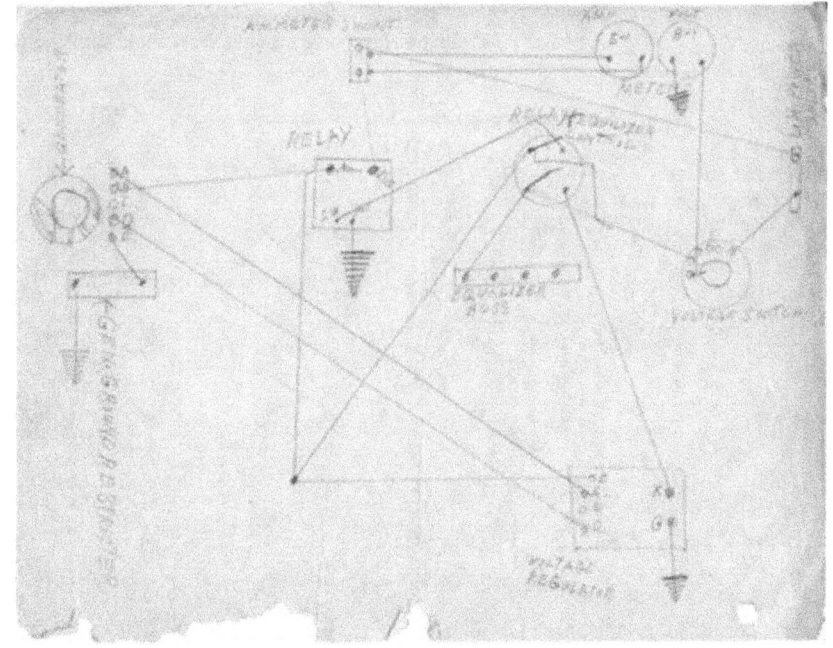

Keesler AAF: Aircraft Mechanic Student Workbook: Electrical Schematic Student Sketch *Pearce Collection*

Keesler AAF: Aircraft Mechanic Student Workbook: Airplane Engines Branch with Student Notes

Keesler AAF: Aircraft Mechanic Student Workbook: Electrical and Airplane Inspection Branches, with student notebook and student with a 'Mop on the B-24.'

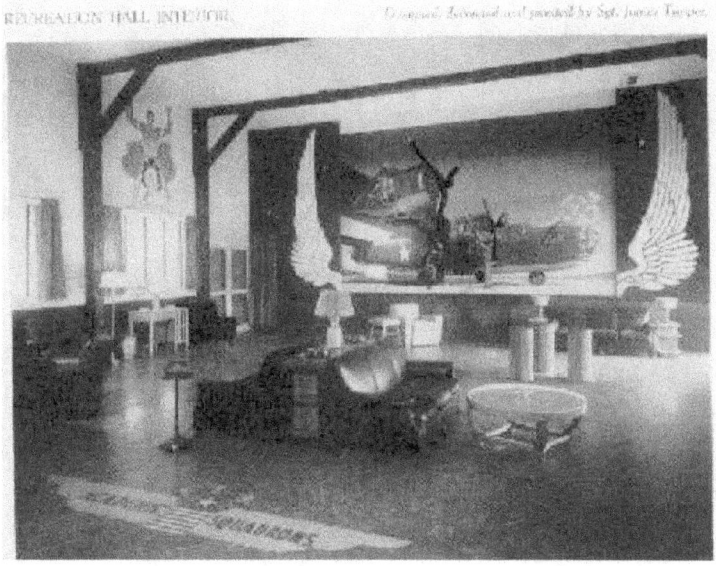

Keesler AAF: Work, relaxation, and inspiration at Keesler. 'Testing Block' - S/Sgt George W. Potts; 'Recreation Hall Interior' Designed, decorated, and painted by Sgt. James Tupper; 'Dayroom Interior' Designed and decorated by Cpl. Albert Bitters; 'Airplane Mechanic' photo by Sgt. Wallace Marley; 'Service Club Mural' T/Sgt. Samuel Bromberg; 'Night Bombers' watercolor by Sgt. Kenneth Gordon.

Biloxi, Mississippi: The Gulf of Mexico coast is in the foreground. In the distance is the site of Keesler AAF, and beyond that, Back Bay.

Getting Ready for Leave: Walt is looking handsome at Keesler AAF. His clean uniform and tie means that he has leave for at least part of the day to explore Biloxi.

Biloxi, Mississippi: Before the onslaught of thousands of men on day-pass from Keesler, Biloxi looks quiet and peaceful.

Keesler AAF: Walt takes time out from putting his laundry out to dry (above) to go on armed patrol to ensure that Keesler was free of German and Japanese attackers (below).

Keesler AAF: Walt taunts a friend who is not on the side of the fence with cold beers.

The final branch of aircraft mechanics training was the Graduation Field test. This curriculum was designed to simulate actual adverse working conditions in an airbase located within a combat zone. It provided students with a realistic environment in which to apply the mechanic knowledge, skills, and abilities they had learned over the previous three months. The Graduation Field test was conducted in a wooded area in the southwest corner of the base, across the NW-SE runway from the hanger line.[44]

When Keesler first commenced operation as an aircraft mechanics school, it focused on the development of general aircraft maintenance and repair training. After completing 110 days of training, graduates were sent to other training bases to specialize in maintenance of specific types of aircraft, such as the B-25, P-51, B-17, etc. By mid-1942, it was clear to the Air Corps that this was not producing adequate numbers of mechanics quickly enough. In May 1942, Keesler was informed by the Air Corps Training Command that it would specialize in complete maintenance training for heavy bombers, the Air Corps B-17 Flying Fortress and the B-24 Liberator.

Through the summer of 1942 Keesler worked to acquire aircraft for student training, as well as the specialized tools, repair and maintenance manuals, and spare parts needed to put this plan into operation. Difficulty in preparing to train men on maintenance and repair of two radically different aircraft types led the Air Corps once more to change the focus of Keesler. On August 27, 1942 the Air Corps ordered Keesler to specialize only on B-24 maintenance. By the time this change went into effect with the first class on October 19, the training command had been able to reduce the previous 110-day course to 96 days, a savings of 14 days.

The new schedule resulted in eight days of training for most subjects, with six hours of classroom or field instruction per day. Each of the instruction branches was assigned from 13 to 23 B-24's, which permitted class sizes to vary from 20 to 35 students. Most of the aircraft the men worked on were worn out older model bombers that had returned from active duty after they were replaced by more recent models with front line units in Europe and at training and patrol bases across the U.S.[45]

96 days of instruction was precisely the schedule that Walt completed. He started his aircraft mechanics training on Wednesday, May 5 and would complete it on Monday, September 13. This was 96 days, on a Monday through Saturday schedule, with only Sundays off.[46]

Walt reported very, very early on Wednesday May 5 for his first day of technical training. To maximize the use of training facilities at Keesler, the men were divided into two shifts. The first shift, shift 'A', reported in their classrooms ready to begin instructions at 0600 (6:00 a.m.). 'A' shift instruction ended at 1300 (1:00 p.m.) with an hour off for lunch. 'B' shift began at 1400 and ended at 2100, likewise with an hour off for dinner.[47] As each eight-day class session ended, the men switched from 'A' shift to 'B' shift. While the six hours of classroom instruction seems like a short day, the men were expected to do extensive study and review on their own during free time. They would be graded on completion of each class and a low grade could result in washout of the aircraft mechanic program, so this independent study could not be ignored. The men on 'A' shift would study through the afternoon and into the evening. Those on 'B' shift had to study in the morning before class time began. Men on both shifts had to develop self-discipline and time management habits, because they could also participate in recreational activities or just rest while not in class. Failure to prepare adequately though would quickly show up in their test scores.

On May 5 Walt began the Airplane Mechanics' Tools Branch course in Academic Building Number 1. The Keesler academic buildings were one-story classrooms, set in an area where many of the original golf course pine trees had been preserved. This gave the area a pleasant appearance and helped disguise that fact that most of the structures were less than a year old. The building were not air conditioned, so the trees added some shade and contributed to keeping them as cool as possible during afternoon classes.

As the men reported to begin this class, most likely thought that they were familiar with all the tools and tool handling techniques they would need to work as aircraft mechanics. Some may have felt impatient and wondered why they could not just start working on the various components of a trainer aircraft or bomber. They quickly learned though that there were many specialized tools and procedures that they were not familiar with. They had been all instructed to buy a number of small spiral notebooks to record notes from the instructors' lectures or demonstrations. The notebooks were to be small enough so that they could be placed in a coverall pocket while the men did hands-on training at various points during their day in the classroom.

The Airplane Mechanics' Tools Branch focused on tools used for the B-24 with a different topic each of its eight days. Day 1: Mechanic's Hand Tools; Day 2: Elementary Assembly Devices; Day 3: Precision Assembly Methods; Day 4: Fabrication of Materials; Day 5: Aircraft Plumbing Methods; Day 6: Safetying Methods and Devices; Day 7: Permanent Assembly Methods; and finally Day 8: Advanced Mechanical Projects.[48]

As the tool course began the men learned not only about individual tools, but also about their proper use with various aircraft systems. They began with bolts and nuts and the proper tools and techniques associated with their use. Almost all the men were sure they knew how to fasten and tighten a bolt and nut, but the instructors quickly educated them that in an aircraft with multiple vibrating forces acting on a bolted connection, the proper use of nut locking systems and the application of the proper torque was necessary to ensure that a bolted connection, like, say an engine bolted to a wing, did not fail. The men quickly grasped the significance and started writing in their little notebooks. It quickly dawned on them that an error on their part could result in damage or lives lost.

They recorded how to calculate the proper torque by measuring bolt diameter, number of threads in a nut, and applying figures from a calculations table. They moved on to pipe and tubing systems found in an aircraft. They started learning a completely new nomenclature, recording the names of the pieces in a tubing fitting (sleeve, coupling nut, and union) and the differences between a pipe union and a pipe nipple. They learned about repair problems they had never imagined. How to clean 'Flexiglass'?

> *"a – wet sandpaper in water and smooth around the scratch.*
> *"b – use 280 or no. 300 sandpaper in the process*
> *"c – then wash and clean with auto cleaner*
> *"d - after cleaning polish with auto polish"*

Many months later Walt would appreciate clean 'Plexiglas' and make sure the ground crew chief knew if the 'glass' in his turret needed cleaning.

As the days went by during this first course they learned about cables and methods of attaching them to mechanisms, and to each other. They studied electrical wiring joints and soldering, rivets and riveting, sheet metal repair (for holes in fuselage or wing skin – "…always select a gage higher than the skin is made of if the same cannot be had.").[49]

They learned the importance of obtaining all the proper tools for a given maintenance task before beginning work, to save time and ensure that no 'shortcuts' or improper tool use would be attempted. For example, they learned that to make a sheet metal repair using rivnuts, they needed to assemble the following tools:

1. *Layout tools*
2. *4-1/8" Cleco's*
3. *Cleco gun*
4. *Snips*
5. *½" drill*
6. *Half round file*
7. *No. 28 drill*
8. *Hand drill*
9. *11/64" reamer*
10. *Rivnut squeeze*
11. *Plain faces pryalene mallet*
12. *Dimpling set*
13. *Tinner's snips*
14. *Narrow lip screw driver*

The men soon realized how little of what the Air Corps expected of them as mechanics they really knew already. The little blue notebooks were filled by all the men with more and more information. The notebooks were consulted often through the day and studied in the evenings and on weekends. Finally, on Thursday, May 13 Walt and his classmates took their final mechanics' tools exam. A couple days' later grades were posted and Walt saw he had scored a 74.

All airplane mechanics' (AM) grades were bases on a 100-point system with the following categories:

95 - 100	*Superior*
90 - 94	*Excellent*
80 - 89	*Very Satisfactory*
70 – 79	*Satisfactory*
Below 70	*Unsatisfactory*[50]

So Walt's grade of 74 was in the middle of the 'satisfactory' range. That was OK, but he would have to try to do better if he wanted ensure that he did not wash out of the program.

Friday, May 14 Walt started the Air Forces Fundamentals course. This branch of instruction outlined Air Corps maintenance policies and procedures. The instructors also introduced the basic technical information that would be studied in more detail and applied to actual aircraft maintenance later.[51]

Air Force Fundamentals classes were held in classrooms located in the balcony of Hangar #1. Each day of the eight-day sequence they studied a different topic in a different classroom, custom equipped for the subject: Day 1: Technical Order Compliance Systems; Day 2: Visual Inspection Systems; Day 3: Elements of Fluids and Lubricants; Day 4: Elementary Hydraulic Systems; Day 5: Principles of Electricity and Magnetism; Day 6: Elementary Electric Motors, Generators, Transformers; Day 7: Electrical Symbols, Measuring Instruments; and Day 8: Application of Measuring Instruments to Electrical Circuits.[52]

They started with an introduction to reading and understanding blue prints and the various types of information they could provide to a mechanic (main assembly, sub-assemblies, details, and parts lists). They learned the different types of lines and symbols they would find on blue prints, including:

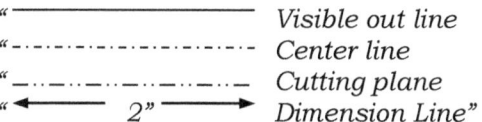

"——————— *Visible out line*
"-·-·-·-·-·-·-·-· *Center line*
"_··_··_··_··_ *Cutting plane*
"◄——— 2" ———► *Dimension Line*"

That Friday after class, Walt read in the *Biloxi Herald* that the Air Corps had announced a new promotion policy, for enlisted men who completed both a technical training course like radio operation, radar operation, or airplane mechanics, and then complete flexible gunnery training to qualify as a bomber crew member. Enlisted men who completed both technical training and gunnery training were to be promoted automatically to sergeant.[53] Walt had thought about applying for gunnery school, and this seemed like an added inducement. This would be further motivation for him to do well in the Airplane Mechanics Course at Keesler.

Air Force Fundamentals training continued on Monday the 17th with the study of technical orders, which included spare parts lists, lubrication data, and technical repair or maintenance information for main structures, engines, hardware, and instrumentation. Walt and his co-students were lectured on the importance of properly identifying the exact model and manufacturer of an aircraft that they were assigned to maintain or repair. This was critical, because even a basic Air Corps aircraft design, like a B-17 Flying Fortress or a B-24 Liberator, had periodic design changes that upgraded critical components, added new features or equipment, or changed the location of fittings, connections, or other construction features. Even within a given model, like the B-24 model D, there were many differences based on the specific contractor or manufacturing company that fabricated the components and assembled the finished aircraft. In short, there was no single set of operation, maintenance, and repair manuals for a given model, but potentially many different versions based on the specifics of the aircraft. A mechanic had to identify these specifics before he could consult the proper repair manuals, order the proper spare parts, and used the correct procedures for keeping a bomber airworthy.

Specific aircraft were identified by a combination of numbers and letters that indicated the basics needed for the mechanic to begin his work. First came an abbreviation for the type of airplane:

"*P = Pursuit*
"*A.T = Advanced trainer*

"C = Cargo
"B = Bombardment
"B.C. = Basic combat
"A = Attack"

This was followed by the basic aircraft type (B-24 for the four-engine Liberator Bomber) and a model suffix letter (for example 'D' for the models Walt and the Keesler students would begin working on). Then came a number that identified the manufacturer of the four engines:

"10 = Pratt & Whitney
"15 = Lycoming
"35 = Wright
"40 = Continental
"65 = De Havilland"

Finally came the identity number for the aircraft manufacturer:

"5 = Consolidated
"25 = Curtis
"60 = North American
"110 = Bell
"40 = Douglass
"70 = Boeing
"100 = Ryan"[54]

Once the men understood the proper identification of a given craft, they were instructed on how to order spare parts and lubrication supplies via the use of various forms:

"Store credits form no. 82
"Store charges form no. 81
"Store exchange form no. 249"

The proper entry and reading of information on aircraft inspection checklists was covered:

"Red X = indicates the plane is unsafe for use. Red (/) a minor defect. Red (-) inspection past due. The initial indicates it has been inspected (black). The black O around the initial indicates greased or oiled except in 46 which indicates water checked in battery."

As the days wore on the men learned the difference between preflight checks, daily aircraft checks, 25 hour, 50 hour, 100, 200, and 300-hour inspections, each one progressively more involved and detailed. They learned of the importance of visual and instrument checks to keep a repair and maintenance history of each individual aircraft.[55]

Walt paid attention, took good notes, and studied them diligently. He completed the fundamentals course and took his final exam on Saturday, May 22. Later that day he picked up a copy of the previous day's *Biloxi Herald*. Having spent many weeks now in the South, both in Florida and Mississippi, as well as traveling across Georgia, Alabama, Tennessee, and Missouri, he was aware that Southern attitudes to blacks and other minorities was different from what he had learned and experienced in Detroit. This was a common observation of many of the 'Northerners' Walt would serve with throughout his time in the service. However, perhaps the first slight changes in attitudes to race were beginning in the South. Walt read a number of articles in the *Herald* that presented minorities in a positive light. He read of a blackman in the 1170th Training Group at Keesler, Pvt. Julio Jose Gomez, who spoke and wrote 12 languages fluently, and who was then teaching himself Polish and Swedish as well. Another article was about Pfc. Doo Jam Ng, a Chinese-American enrolled in the Aircraft Mechanics Course in the 395th TSS. Ng, like Walt and many other Airplane Mechanics students, hoped to become an aerial gunner.[56]

Walt also checked up on the progress of the Tigers. They had lost both games of a recent doubleheader against the Washington Senators, 7-1 and 10-6, dropping Detroit dropped below 500 in the standings. He hoped they would do better against the Red Sox, in Detroit for a four game series.

The following day Walt took the bus into Biloxi to spend some free time and see the sights again. Students in the technical training branches were granted passes more liberally than those in basic training. Technical training students could be granted 24 hour passes each weekend. Three-day passes were also available, but their use was limited to ensure that the base operated at maximum efficiency. Later three-day passes were given as an incentive to students with a grade average of 90 or above. Attaining a grade this high would be a challenge for Walt.[57]

When grades for the Air Force Fundamentals course were posted on Monday, May 24, Walt saw that he had scored an 80 – bottom of 'Very Satisfactory.' Extra hard work and study had paid off.

That same day he began the Airplane Structures course. For this course, the men moved over to Hanger No. 3, just off the taxiway. The hanger buildings had specialized classrooms just like the academic buildings, but they also provided access to a large hanger space where the trainees could observe and inspect the specific aircraft components they were studying in a complete, operational aircraft. The Airplane Structures course familiarized the students with the general construction of the B-24 and its flight operation systems. The daily topics for B-24 Airplane Structures included: Day 1: Cleaning and Handling of a B-24, Rafts, Vests, and CO2 Cylinders; Day 2: Principal Structural Units and Riveting; Day 3: Cabins and Cabin Equipment; Day 4: Flight Control Surfaces; Day 5: Flight Control Mechanisms; Day 6: Main Landing Gear and Retractive Mechanisms; Day 7: Auxiliary Landing Gear and De-icing Equipment; and Day 8: Tires, Tubes, Wheels and Brakes.[58]

For the first time the men were taken in small groups to begin inspecting one of the B-24D's on the base for repair and maintenance work. The aircraft they inspected was probably manufactured by Consolidated, one of over 2,000 D-models built. The B-24D was 66 feet, 4 inches in length, with a wingspan of 110 feet and a height of 17 feet, 11 inches. With a maximum take-off weight of 64,000 pounds, it was powered by four 1,200 h.p. Pratt & Whitney R-1830-43 Twin Wasp radial engines. The B-24D had a maximum speed of 303 m.p.h. and an operational ceiling of 32,000 feet. Its maximum range was 4,600 miles and it could carry a maximum of 8,000 lbs. of bombs in internal bomb bays.[59]

The men were instructed on how to describe a location on the aircraft in a variety of orientations: longitudinal through the fuselage, lateral through the center of gravity of the plane, and vertical through the other two lines. They studied the design of the B-24, learning that is was a semi-monocoque construction. This meant that the framework of the fuselage was constructed of main lateral bulkheads and minor belt frames, connected by longitudinal structural members known as stringers and longerons. The metal skin of the aircraft was attached to this framework, as was the framework for the wings and tail.

They were likely somewhat crestfallen then to learn that their first 'maintenance' task was to learn how to properly wash the exterior of a B-24:

> "The mixture for washing: 1 – the soap solution is used with 50 pounds of soap to 20 gals. of water and let stand for 24 hours. Then put one pt. of the solution to 5 gals. of water and apply with rag or brushes and rinse with water. Washing of planes is done from the bottom."

One of the aircraft mechanics students over in Flight 305, Pfc. Julius E. Engram, composed a little song that reflected on the experiences that Walt and his classmates were going through in this class. His song referred to the detailed survey of the B-24 components that the men were making, but then their surprise at learning that they also had to wash the planes:

WHERE IS THE MOP ON THE B-24
(Tune – "Who Broke the Lock on the Henhouse Door)

(1)
In this school, we must know
All about planes
And what makes them go.
But there's one thing we don't know
Where they keep the mop on the B-24.

(2)
We mop the barracks.
We mop through a phase.
We mop latrines 'till we're in a daze
Still there's one thing we don't know
Where they keep the mop on the B-24

(3)
They have supercharger buckets
And prop wash too,
Generator brushes for the whole darn crew
Automatic pilot and bomb bay doors
But we can't find the mop on the B-24's.

(Chorus)
Where is the mop?
I don't know
I've looked high, and I've looked low.
Still there's one thing we don't know
Where the hell's the mop on the B-24.

This song and several more verses quickly became popular among the troops. The base print shop printed up the words and distributed copies to the men for 'sing-a-longs' before or after movies or other entertainment shows at the base theaters.[60]

The following day they returned to a sparkling clean plane and continued their detailed exploration of the interior and exterior, taking notes as they went. In addition to familiarizing themselves with the general structure of the plane, they identified accessories they would be responsible for inspecting and maintaining. They learned about the A-2 inflatable life raft and that it was included during the routine 50-hour inspection and also once every six months. The A-2 raft could hold up to 1000 pounds or five men. They learned they were responsible for inspecting the A-3 inflatable personal safety vests carried for each man on the crew. The pilot and copilot seat belts required periodic inspection also, as did the fire extinguishers located throughout the plane.

They learned about the location and meaning of exterior identification numbers and insignia. They would be responsible for keeping these visible and making changes as aircraft were assigned to new operational units. They learned about towing a B-24 by tow truck, with the maximum turning angles and the need for a mechanic in the cockpit to apply the aircraft's breaks when stopping. They were taught the proper method for securing and tying down a parked aircraft.

Airplane Structures training concluded on Tuesday, June 1. Again there was a final exam, with questions such as:

"Locate the outside release handles for the life rafts.
"What is the correct strut extension on a loaded airplane? Where should the measurement be made?
"Locate the six jacking points on this airplane.
"Locate the stowage rack for the tail turret gunner's parachute."
"Locate the manual crank for emergency operation of the landing gear system. What precaution would you take in re-setting this system?"

A few days later Walt saw that he had improved his score to an 82, again 'Very Satisfactory.'

The fourth course Walt began was Airplane Hydraulic Systems, on Wednesday, June 2. For this course, the men were in Academic Building #2, again with a custom equipped classroom designed for the daily topic. The daily topics for B-24 Hydraulic Systems included: Day 1: Open Center Systems; Day 2: Electric Auxiliary System; Day 3: Hydraulic Power System; Day 4: Flap and Bomb Door Hydraulic Systems; Day 5: Brake Systems; Day 6: Landing Gear Hydraulic Systems; Day 7: Assembly and Testing of Hydraulic Units; and Day 8: Complete B-24 Systems.[61]

Walt learned that the B-24 used a Vickers hydraulic pump to provide hydraulic pressure to automatically operate a variety of aircraft systems, including landing gear, wing flaps, bomb doors, fuel transfer pumps, and the bottom ball turret. The men learned that a key component of a hydraulic system is the adjustable pressure relief valve, with a maximum kick off pressure of 1250 psi on the B-24 to protect the individual lines from rupturing. Each man was given a schematic diagram of the B-24 hydraulic system to study.

As the June days wore on, Walt and his classmates learned of the various types of hydraulic tubing lines, including pressure supply lines, hydraulic fluid return lines, and alternating lines running from control valves to the object being controlled. A B-24 hydraulic system held 6-9/10 gallons of type 3580 hydraulic fluid in the reservoir and 18 gallons in the entire system. To understand the complexity of the hydraulic reservoir and the hydraulic system in general, most men made sketches of the systems, including the location of the air bleed, supply and return lines, oil fill port, hand pump, and emergency line connections. The function of the Parker three-way hydraulic valve was described, as was the importance of strategically located check valves to prevent fluid draining back and to keep hydraulic line pressure at proper levels.

More time was also spent on the operation of the Vickers piston hydraulic pump, with seven pistons driving three individual pumps via manifold and valving. The function of the relief valves was described, for instance, if a hydraulic line was smashed closed, or if a rudder was hung up, the relief valve would allow fluid to flow back to the reservoir, protecting the line from overpressure and rupture. The class learned of the importance to keep the hydraulic fluid clean and free of air bubbles. The Cuno filter was used on the fluid return line just before the reservoir. The reservoir itself included an air bleed off-valve at the top to remove air bubbles.

The men studied hydraulic actuators as well. These were the devices used to transfer hydraulic pressure into a mechanical force to operate a flap, the landing gear, door, or to lift or lower the ball turret. Time in the classroom was again supplemented with time in the base B-24's, to allow the men to become familiar with the location and operation of the hydraulic components. Details included: "Bomb door control station no. 02 left side of bombers compartment pressure setting 600 up and 1000 down....flaps control station no. 1 right of pilot's control pedals and has a pressure of 750 up and 450 down."

The men studied how to make hydraulic line connections via threaded connections in the classroom, using "13 parts of white lead 1 part of grafite 100 mesh or finer." They were also instructed on proper torquing of fittings to make a tight connection. They learned about the Bendix pressure switch, they learned about the alternate emergency hydraulic hand pump, they learned about the auxiliary hydraulic system (designed to provide redundant control during take-offs and landings, when flying 20,000 feet, etc.), they learned about the unloading valve, and they learned about methods for safety testing the various hydraulic components.

The proper setting for the adjustable relief valves was important. The main and auxiliary systems setting was 1250 psi, while the bomb door relief valve was set at 800 psi, wing flaps relief valve at 500 psi, and landing gear relief valve at 500 psi. The men practiced how to use a rag dampened with denatured alcohol to clean flexible hydraulic lines, and also to check for leakage while doing so. They learned about the need to drain the hydraulic system during major periodic maintenance cycles. The entire system was drained, and then flushed on the inside with denatured alcohol. The specific type of hydraulic fluid depended on the climactic conditions where the plane would operate. Heavy hydraulic fluid was specified where the temperature would always be above 50 degrees Fahrenheit. Medium fluid was used for conditions between 10 to 60 degrees, and light fluid used for operations below 20 degrees.[62]

Walt felt confident when he took the hydraulic system exam on Thursday, June 10, with questions like:

> "What is the fluid specification for the hydraulic system?"
> "Name the steps to check before reading the sight gauge on the hydraulic reservoir
> "Explain the use of the emergency suction line selector valve....and what position is it normally in?
> "Locate the hydraulic pressure switch."

Later that day Walt picked up a copy of the *Biloxi Herald* and read about the progress of the war and events at home. Key stories covered the anticipated invasion of Italy by Allied forces from Africa, appeals to Czechs to increase their guerilla war against their German occupiers and avenge the massacre of the residents of Lidice village, and counterattacks by Chinese troops against the Japanese near Owchihkow on the Yangtze River. Other articles described the huge bomber offensives that were being planned by the RAF and the US 8th Air Force from England against Germany later in the summer. Walt wondered if he would eventually contribute to this fight after his training was over. He also read about B-17 bomber attacks against Japanese positions at a place called Munda on New Georgia in the Solomon Islands. On the 'home front' he read about the 'Zoot-Suit' Riots in Los Angeles, striking coal miners in Pennsylvania, and the search for five German prisoners of war who had escaped from a POW camp at North Camp Hood, Texas.[63]

Walt also checked the paper for recent baseball scores, discovering that the Tigers continued to struggle with a lackluster season. The Senators, who had swept Detroit just a few weeks ago, had just completed beating them again three games in a row, followed by another loss to the White Sox. A few Tiger players were having decent seasons, including first baseman Rudy York, outfielder Dick Wakefield, and their star pitcher Dizzy Trout.

Late the following day Walt was shocked when he saw his score posted for the Hydraulics System exam. He had earned just a 70 – the very bottom of the 'Satisfactory;' range and just one point away from the potential to wash out altogether. He would have to redouble his efforts to succeed now, or he would not be going to gunnery school.

The Airplane Engines course began on Friday, June 11.[64] This course focused on the repair and maintenance replacement phases needed to keep a B-24's Pratt and Whitney R-1830-43 power plant operating properly. Most of this training moved back to the newly completed Academic Building #7, with occasional trips out to a hanger for engine inspection and disassembly. The daily topics for B-24 Airplane Engines included: Day 1: Engine Hardware and Tools; Cylinder Assemblies; Day 2: Removal and Replacement of Engine Units; Day 3: Nose Section; Day 4: Rear or Accessory Section; Day 5: Valve Clearance and Valve Timing; Day 6: Ignition Systems; Day 7: Ignition Timing; Fire Extinguishers; Engine Mounts; and Day 8: Maintenance Inspections; Preparation of Engines for Storage.[65]

As with other courses, the Engines instructor began with basics, talking about the four-stroke engine that powered the B-24. For any man with mechanical aptitude but no formal training, the instructor outlined the four cycles of the engine: "1 – down - intake of fuel mixture, 2 – up - compression stroke and just before it reaches top dead center sparks fires it, 3 – down – explosion or power stroke, 4 – up – exhaust stroke." The pistons and cylinders themselves were described, consisting of five rings – three compressor rings, a dual oil ring, and a scraper ring.

The B-24 power plant consisted of four radial engines designated as R-1830-43.[66] R designated 'radial design,' 1830 designated the total displacement of the cylinders in cubic inches, and 43 designated the series or make of the engine. The men also had to learn how to designate a location on the radial engine, such as B.C. – bottom center, T.C. – top center, A.T.C. – after top center, B.T.C. – before top center, etc.

The reduction gearing of the engine was described, along with illustrations explaining how power was transferred from engine to propellers. The men were lectured on proper compression ring clearance, for instance, compression ring #1 – gap clearance: 0.059" = 0.066"; side clearance: 0.003" – 0.005." The proper methods for checking valve clearance were outlined and also the importance to check other components, such as push rods and rocker arms, and the need to inspect for dirt in the ball and socket ends of push rods. As with earlier courses, classroom lecture was followed by hands-on work on R-1830-43 engines on the special test block designed for the purpose.

Valve and timing check procedures were explained next, and here the importance of knowing how to designate a location on a multidimensional mechanism like an engine became apparent. For example, the instructor might tell a student to "...turn the prop shaft in the direction of rotation. Exhaust valve should close 20 degrees A.T.C. Return to top dead center compression stroke #1 cylinder and reset cold clearance of 0.010". Repeat process on #8 cylinder."

After engine mechanical devices were covered, the instructor moved on to engine appurtenances. The R-1830-43 engine was supplied with electrical current via an American-Bosch Company magneto. This led into a lecture on trouble shooting engine electrical problems, such as spark missing - check for: "1 – Badly worn breaker points, 2 – Breaker points not adjusted, 3 – Oil on contact points, 4 – Breaker assy. installed incorrectly, 5 – Improperly assembled magneto," etc. A similar trouble-shooting checklist was provided the men if there was no spark at all. Other devices covered included spark plugs (with proper installation torque), oil pumps, fuel pumps, starter, generator, and vacuum pump. Although Keesler produced mechanics designated to work on B-24 bombers, the men were also instructed on the gun synchronization mechanism used on single engine fighter planes, designed to allow machine guns to fire through a rotating propeller without causing damage.

Saturday, June 19 came and Walt was nervous as he took his airplane engines exam. He did not want to repeat anything close to the grade of 70 he had gotten on the hydraulics course. Questions included:

> "Check all of the turbos for radial shake and end play. Check to see that the buckets are not cracked.
>
> "Locate the air filters on one inboard and one outboard engines.

"Locate the intercooler switches. Would you ever close them?

"Give the location of the primer, accelerating and mesh switches for engines #3 and #1. Should the switches be pushed up or down?"

Sunday, after Mass, Walt read in the *Biloxi Herald* that the Tigers had beat the St. Louis Browns the previous day 4-3, which improved Detroit's overall record to 24 wins and 24 losses. They would face the Browns again on Sunday in a doubleheader at Brigg's Stadium.

Walt was relieved to learn the following Monday that he had scored a 79 on the airplane engines exam. He knew the level of effort he needed to apply now to succeed. Never again would his score drop below the high 70's.

That Monday, June 21, Walt started the Airplane Electrical Systems course. The men learned basic electrical system concepts and how to read electrical diagrams. They studied the schematic diagrams for the B-24 electrical systems. They learned the concept of electrical circuits and how to test them for 'shorts.' They learned about and practiced using specialized tools for electrical maintenance and repair, such as soldering tools, wire stripers, circuit testers, etc.

This training section was in Academic Building #5. Daily topics for B-24 Airplane Electrical Systems included: Day 1: Aircraft Batteries and Lighting System; Day 2: Control Switches, Solenoids, Relays and Boosters; Day 3: Electrical Wiring System; Day 4: Starter Systems; Day 5: Electric Motors and Inverters; Day 6: Generator System Units and Auxiliary Unit; Day 7: Generator System Operation and Maintenance; and Day 8: Inspection and Maintenance of Electrical System.[67]

The previous course on engines had focused on the main B-24 power plant, the four big radial engines that not only applied power to the propellers to put the airplane in motion, but also supplied power for the generators that powered all the electrical systems (and indirectly the hydraulic systems) while in flight. However, an aircraft parked on the ground required electrical power even when the main engines where not operating. Therefore, the men were introduced to the operation and maintenance of what was referred to as the A.P.U., or the Auxiliary Power Unit. This was a small single stoke gasoline engine that drove a generator which could supply electrical power to the B-24 for inspection, testing, maintenance, or repair procedures. Those men who would later be assigned to B-24 combat units, either as flying crewmembers or as maintenance crews would refer to these small engines as the 'put-put.'

Wednesday, June 23, after classes, Walt was ordered to report for a medical checkup and to have his inoculations brought up to date. He continued to be healthy, but he needed his antitetanus inoculation renewed.[68]

Later in the week more detailed electrical systems topics included the study of storage batteries, solenoid switches, the ignition switch, and so on. The importance of fuses to protect the electrical system from overloads was covered. The men learned the purpose and location of each of scores of fuses. There were at least 15 fuse boxes on a B-24 and the men had to learn the location of each, the specific fuses located in each, and be able to find it quickly. Again, classroom instruction was supplemented with work on actual aircraft or on subsystems in special labs adjacent to the classes. Walt seemed to grasp the electrical concepts quite easily.

The Airplane Electrical Systems course ended on Tuesday, June 29 with the usual exam:

"Give the location of the master switch and the battery solenoid switches.

"Locate the voltage regulators. Who should adjust them?

"Locate the fuse for the heater solenoid on engine #3.

"Locate the battery neutralizing jar."

This time Walt scored an 84 – mid way in the 'Very Satisfactory' range.

The following day, Wednesday the 30th, Walt started the Airplane Fuel Systems course. In addition to learning the correct and safe procedures for fueling a B-24 with very flammable high-octane aviation gasoline, the men learned the detailed operation and maintenance of the various lines, pumps, and fuel cells (tanks) needed to deliver fuel to the engines during various phases of operation.

Fuel System training was held in Academic Building #6. Daily topics for B-24 Fuel Systems included: Day 1: Oil Systems; Day 2: Engine Control Systems; Day 3: Fuel System Units; Day 4: Fuel System Installation; Day 5: Combustion Heaters and Carburetor System4; Day 6: Pressure Injector Carburetors; Day 7: Supercharger Units; and Day 8: Supercharger Installation.[69]

A key aspect of fuel system training focused on safety. Smoking was a way of life for most of the men in the Air Corps, and Walt was certainly no exception. During fueling, standard procedure was for the men to leave any smokes, matches, or lighters in a special locker, to eliminate the possibility of an absent-minded light up. The instructors kept a careful eye out for any safety infractions and came down hard on anyone who slipped up. There was truly no opportunity for a second chance with fuel system errors.

Another obvious danger to a B-24 in combat was the potential for fire or explosion due to an enemy bullet or flak puncturing the fuel system. A key aspect of the B-24 design was the segregation of the fuel supply into 18 individual self-sealing fuel cells located in the wings. The cells were manifolded together to act as a unit, but any one could also be bypassed if it was punctured and all fuel drained out during flight. Each cell was made of a special treated rubberized multi-layer fabric. A key maintenance responsibility was to remove these cells periodically for inspection or replacement. To learn this difficult task, a wing section mock-up was installed in a classroom for teams of men to practice on.

As the days wore on, the men traced the fuel system lines from tanks to pumps to the carburetors located in each engine nacelle. The fuel system lines, transfer pumps, and valves were quite complex, to allow fuel transfer from one tank to another during flight. This allowed the aircraft weight load to be adjusted frequently, aiding fuel efficiency and speed optimization under various conditions.

Monday, July 5th was Walt's 21st birthday. The previous year he had celebrated his birthday at home, with his ma, pa, brothers, and sisters singing 'sto-lat' and his friends singing happy birthday. He almost surely received cards or letters at Keesler from his family back in Detroit wishing him a happy birthday, and also a package from home with small gifts and special food. He likely heard from his brother Johnnie, still based at Brisbane with the US 7th Fleet. He would have heard by this time that Johnnie had been promoted from Seaman First Class, to Mechanics Mate Second Class.[70] But this July 5 was like all the other school days at Keesler, with an early rise, many subjects to cover in class, and more study after class and chow. Any acknowledgement Walt's pals made of his birthday was very likely small.

Airplane Fuel Systems training concluded three days later, on Thursday, July 8, and again with a final exam:

> "Remove one of the gasoline filler caps. Re-safety it properly. Briefly explain the purpose of the yellow line.
> "Where are the fuel system "low point drains" located?
> "Explain the use of the wing drains in the bomb bay area. When should they be closed?
> "Locate the bomb-bay tank selector valve (fuel system) and the transfer pump.
> "Can fuel from the fuel booster pump bypass an engine fuel pump that is inoperative?"

A few days later Walt saw that he scored in the 'Very Satisfactory' range once more with an 81.

Thursday afternoon, after class at mail call, Walt got a postcard from Dayton, Ohio from a friend of his from Detroit, named Maggie:

> *Dear Walter:*
> *Myrle just left at 8:00 P.M. It's 9:00 now and Eva isn't back yet. Eric is trying to play the piano. Daddy & I picked 88 qt of raspberries Tuesday till 2:00 P.M. Were going today, but my arms were too sore from sunburn. We were here for supper. Had chicken & raspberries for supper. Very good. Am very busy, but hope to write again soon. – Rec'd card –*
>
> *Love, Maggie*[71]

Friday, July 9th was the first day of the Airplane Instruments course. The focus of this class was to master the location, function, and proper operation of the various gauges, indicators, warning lights, automatic switches, and position lights located throughout the plane to inform the crew of the status or operation of various pieces of equipment or components needed to operate the craft. Most of the instrumentation of course was located in the pilot and copilot compartment, but there were also many specialized or redundant system instruments located in the bombardier and flight engineer's stations as well.

Airplane Instrument training was held in Academic Building #4. Daily topics for B-24 Airplane Instrument Systems included: Day 1: Pressure Gages; Day 2: Thermometers; Day 3: Tachometers; Day 4: Remote Indicating Instruments; Day 5: Compasses; Day 6: Pitot-Static Instruments; Day 7: Gyro Instruments; and Day 8: Automatic Pilots.[72]

The men learned of basic instrumentation operation, testing, and repair. They took turns examining the pilot's compartment, and recording the type and location of instruments found there:

Altimeter

Airspeed

Directional Gyro

Rate of Climb

Bank and Turn Indicator

Radio Compass

Manifold Pressure

R.P.M.

Fuel Pressure Gauge

Flap Position

Oil Pressure

Oil Temperature

And many more as well. The men learned the importance of checking each instrument during preflight checks to ensure that the cover glass was intact, and how to replace it if loose or broken by cementing a new one in place with aircraft sealing compound Spec. 2-87.

Wednesday, July 14 Walt got a postcard from his cousin Angie from Detroit, writing from Hamilton, Ontario:

Hello Lize:

Believe it or not, but its me – honest. Sorry for not writing but will as soon as I get a chance. I'm on my vacation now and am having a wonderful time. Almost hate to go back but have to be back at work Monday morn. Phooey! Hope everything is okay with you. So long for now.

Yours, Ange[73]

Saturday, July 17th was the end of the Airplane Instruments course, with the usual exam:

"Give the location of the suction gauge. What should this read when the engines are all operating?

"List the instruments you would find in the bombardier's compartment.

"List the pressure shown on the brake gauges. What should the pressure be when the auxiliary system is being used?

"Where is the de-icer pressure gauge located?"

This exam must have been a little more difficult, for Walt's grade dropped down to a 78 – still high in the 'Satisfactory' range though.

On Sunday, July 18, Walt spent some time at the enlisted men's day room, reading the *Biloxi Herald* to catch up on war news and to learn what was happening around the country. He also read with interest that the Tigers had split a double header against the White Sox at Comisky Park the previous day, losing the first game 5-2, but shutting out the Sox 3-0 in the second. They had improved their record to 39 wins, 36 losses, and one tie (June 30 against the Red Sox). They were scheduled for a second double header against the White Sox that Sunday.

The radio playing in the background in the day room broadcast the hit songs of the day, including the Glenn Miller Orchestra's *Juke Box Saturday Night* and *Mission to Moscow*, Judy Garland singing *Zing! Went the Strings of my Heart*, and the Ink Spots version of *Don't Get Around Much Anymore*.

The next course was a short one, just six days rather than the normal eight. The Airplane Propellers class began on Monday, July 19. Airplane Propellers training was held in Academic Building #3. Daily topics for Airplane Instrument Systems included: Day 1: Removal and Installation of Hydromatic Propeller; Day 2: Disassembly and Assembly of Hydromatic Propeller; Day 3: Hydromatic Propeller Governor and Auxiliary Control Systems; Day 4: B-24 Anti-Icing Equipment; Checking of Blade Angle; Day 5: Removal and Installation of Curtiss Propeller; and Day 6: Etching, Inspection, Maintenance and Repair of Propellers.[74]

Topics covered included inspection of propellers and repair of defects. They also learned of the complexity of the propellers themselves. Most had assumed that a propeller was simply bolted in place to the engine drive shaft. This may have been how a propeller was attached for the simple biplanes that were used during the previous war, but this was not the case with a modern bomber like the B-24. They learned that installation and removal of a B-24 propeller involved 42 separate steps. Each propeller was comprised of two blades, attached by a barrel assembly. To disassemble, then reassemble the blades and barrel required 58 separate steps.[75]

The men also learned about the operation of the feathering controls, which allowed the pilot or copilot to adjust the pitch of the propellers while in flight, in order to maximize their efficiency depending on whether the aircraft was climbing, descending, trying to maximize speed, or minimize fuel consumption. In addition to working on propellers mounted on a test block in the classroom, the men also learned about the propeller controls in the pilot's compartment.

Saturday, July 24 the Propeller class ended with the usual exam:

> *"Give the location of the switches for the propeller governors.*
> *"When starting engines, should they be in the "INC" or the "DEC" position?*
> *"If oil were found leaking at:*
>> *"a. Dome Plug*
>> *"b. Barrel halves*
>> *"c. Blade shanks*
>> *"d. Rear cone*
>> *"e. Dome and barrel*
>> *"f. Spider and barrel*
> *"What oil seals would be damaged?*
> *"How would you know when the propeller governor had reached a maximum increase or decrease R.P.M. position?*
> *"Why are the propellers checked through their pitch range twice?"*

Walt achieved a grade of 78 again.

The weather in Biloxi had been seasonably very hot and humid through most of June and July. The average high temperature during these two months was 90 degrees, with an average low of 76 degrees. Relative humidity averaged 83% at 7:00 a.m. and 63% at noon. It rained an average of 6 inches each month in Biloxi, with rainfall occurring an average of half the days.[76] All in all, the weather was both oppressive and monotonous, but Walt was in for some meteorological excitement.

Late on Sunday, July 25, a tropical storm was reported by ships at sea about 125 miles south of Biloxi, in the Gulf of Mexico. The skies darkened, winds increased, and unusually hard driving rain came. Men with radios around the base on that Saturday listened to songs like Dooley Wilson singing *As Time Goes By* from the movie *Casablanca*, and Lena Horne singing *Stormy Weather*. Then radio reports came in from ships out in the Gulf that put the center of the storm the men could see out their windows at approximately 28 degrees north latitude and 87 degrees, 42 minutes west longitude and headed west with wind speeds of 60 m.p.h. The following day radio reports indicated that wind speeds were increasing as the storm moved slowly west, first to 75 m.p.h., then 80 m.p.h. by the end of the day. This was now a full-fledged hurricane. Stormy weather indeed![77]

Monday, July 26 was the start of the Airplane Engine Operation course, and as Walt and the other men of flight 396 made their way to chow and then to classes in the Engine Operation Building, they fought the increasing wind and arrived drenched by the rain. Walt's instructor tried to keep his focus and the focus of the trainees on the airplane engine curriculum, but the rain and wind beating on roof and windows made concentration difficult.

That evening, unknown to most of the 40,000 men on the base, two Army Air Corps pilots sat in the bar drinking beers at the Keesler Officers' Club. One of the pilots, Maj. Joe Duckworth, made a bet on a dare with the other pilot that he could fly a plane through the eye of the hurricane then passing south of Biloxi. No one had ever attempted such a feat of flying before, but Duckworth had confidence in his flying skill and in the strength and endurance of the single engine North American AT-6 trainer, of the type that Walt and his friends had begun to work on.

Early the following morning Duckworth got into an AT-6 with a volunteer navigator and took off, headed southwest for the assumed center of the storm. Although the ride was not smooth, Duckworth soon made his way into the eye of the storm, taking valuable readings as he went on wind speed, barometric pressure readings, and storm track heading. He found that maximum wind speed had increased to 85 m.p.h., and that the storm was headed for land in coastal Texas. Returning to Keesler later that morning, Maj. Duckworth was greeted by the base weather officer, who asked if Duckworth would return to the storm a second time with himself as a passenger. After a thorough check and refueling of the AT-6, Duckworth and the weather officer took off again and returned to the hurricane, confirming that winds continued to blow at 85 m.p.h., and that the storm remained on track for landfall in eastern Texas.

These two flights by Maj. Joe Duckworth on Tuesday, July 27 are considered to be the first attempts to gather information about a hurricane from the air. Weather officers up and down the Gulf Coast discussed Duckworth's two flights that day. His pioneering work led to the establishment the following year of the U.S. Army Air Corps 30th Weather Reconnaissance Squadron. Later the squadron was renamed the 53rd Weather Reconnaissance Squadron, nicknamed the 'Hurricane Hunters' and permanently based to this day at Keesler Field in Biloxi. The squadron is assigned 20 six-man aircrews and flies 10 special Lockheed WC-130-J aircraft.[78]

As the storm subsided and work crews from basic training cleaned up damage done by the brush with the hurricane, the men got back to focus on their airplane engine operations course. This class built on most of the proceeding courses to cover the integration of hydraulic, electrical, fuel, and instrumentation systems with the engine itself to permit controlled operation. Because of its highly critical and complex nature, this course took 12 days, two full weeks, to complete.

Engine Operation Training Branch topics included: Familiarization with R-1830 Power Plant Installation; Daily Inspection; Starting and Stopping of R-1830 Engine; Warm-up Procedure; Preflight Inspection of R-1830 Power Plant; Interpretation of Instrument Readings; Operation of Control Systems and Turbo Superchargers; Operation of Propeller and Electrical Systems; Operation of Fuel, Priming and Oil Diluting Systems; Operation of Oil and Coolant Systems; and Analysis and Correction of Troubles on R-1830 Power Plant Installation.[79]

Because a key function of an airplane mechanic was to test all components of the B-24 before and after a maintenance assignment, the men had to know how to both start and stop the engines. Starting the R-1830-43 engine involved 42 steps as the mechanic sat in the pilot's seat: "1. Remove pitot tube covers, 2. Check wheel chocks, 3. Open bomb doors, 4. Fuel selector valves in "ON" positions (Tank to Engine), 5. Hydraulic pump switch in "ON" position....40. Check switches for "OFF" position at 800 r.p.m. engines cold, 41. Check propeller governor operation at 1500 r.p.m., 42. TO CHECK MAGNETOS. Bring engine up to 30" Hg. with mixture control in "AUTO-LEAN". Drop r.p.m. to 2000, check mags from "BOTH" to "RIGHT" to "BOTH" to "LEFT". Maximum drop 75 r.p.m. Check should be made as quickly as possible."

On Wednesday, July 28, Walt got a postcard from the youngest of the three Kurzyniec girls, Stella, writing from Standish, Michigan where the Kurzyniec family was spending time on the Skrabut farm:

> *Hi Lize!*
> *Having a swell time here. Wish you were here too.*
> *As always, Stas*[80]

The Airplane Engine Operations final exam was given on Saturday, August 7:

> *"Give the location of the cowl flap switches. What is the position of the cowl flaps when starting engines?*
> *"Start the ship's A.P.U. by using the ship's batteries.*
> *"After the engine starts, how long should you wait for an indication of oil pressure? What should be done if this time limit is reached and there is no indication of oil pressure?*
> *"What would happen if the single magneto checks were made with the mixture control in "Automatic Lean"?"*

Walt scored a solid 82 to demonstrate that he had a good grasp of all the preceding training.

That Sunday he read that the Tigers were still struggling with a lack-luster season, having lost the day before to the White Sox 7-4 in Briggs Stadium. Their season record was 48 wins, 48 losses, with one tie, right at 500 where it had been most of the year. Individual players on the Tigers were having good years, but the team was not putting their talents together effectively.

Next followed a series of three Airplane Inspection Courses for Walt. Each one was designed to test the ability of individual trainees to use all of what they had learned, from mechanics' tools, air force maintenance fundamentals, and individual aircraft systems, in an environment of aircraft inspection, maintenance, and repair. The first Inspection Course began on Monday August 9. Each of these courses was handled differently, with the men reporting each morning for assignment to work on various maintenance tasks for single engine trainers or B-24's in different locations on the base. Some days they might work on an aircraft in one of the base hangers, on other days they might work on a plane parked just off the taxiway.

Airplane Inspection Branch-I covered 25-hour inspection procedures with classes in Hangar #1. Day 1: Inspection of Propellers, Fuel Systems, Fuel Tanks; Day 2: Inspection of Valves, Manifolds, Superchargers, Oil Systems, Engine Controls; Day 3: Inspection of Electrical Equipment, Night Flying Equipment, Batteries; Day 4: Inspection of Cooling System, Instruments; Day 5: Inspection of Cockpits, Cabins, Fuselage Hydraulic Systems; Day 6: Inspection of Fixed and Moveable Surfaces, Flight Control Mechanisms; Day 7: Inspection of Landing Gear, Nose Gear, Wheels, Brakes; and Day 8: 25-hour Inspection of Power Plant-General, Airplane-General, Oxygen Equipment.[81]

Later that Monday when Walt returned to barracks, he found that he had received a postcard from his friend 'Babe' Lafata from Detroit. Babe was writing from Niagara Falls, New York:

> *Hello Lize*
> *Here we are in N.Y. and I'm enjoying myself very much for the falls are really something to look at. We'll be back home Wednesday morning.*
> *As ever, 'Babe'*[82]

Walt was a decent 'sand-lot' baseball player back in Delray, but his friend Babe was really quite good. All that week Walt and his friends from the 396th were talking about entering various athletic events the following Sunday. This was the date of the third annual Keesler Field track and field meet, to be held at Commando Park, the base's main athletic facility. Walt was probably more interested in the baseball contest to be held during the track and field events. The contest was comprised of four separate events, including throwing for accuracy from home plate into a target at second base, distance throwing from home to center field, circling the bases, and fungo hitting from home plate. Everyone was eligible to enter, so Walt probably gave this a try.[83]

Airplane Inspection-I concentrated on single engine trainer types, like the AT-6. The men learned the correct use of technical publications and maintenance manuals to perform 25-hour inspections. They learned procedures for logging in problems found on the aircraft they were maintaining, as well as properly recording corrective actions taken. Because the field inspection test simulated real world conditions, the men trained on procedures and signals for directing an aircraft on taxiways and into an aircraft maintenance pad.

Once a plane was landed, taxied, stopped, and tied down, the men would be assigned various groups of systems to inspect and report on. For example, an instruments inspection would include the following specific jobs:

1. *Compensating the Compass*
2. *Visual Inspection of Panel*
3. *Inspection of Tachometer*
4. *Inspection of Oil Pressure Gages*
5. *Cylinder Head Temperature Gages*
6. *Inspection of Gyros*
7. *Inspection of Pitot-Static Assembly*
8. *Checking Circuits*
9. *Inspection Form Data*

Just a single 'job' would involve many detailed tasks. To 'compensate the compass' the following steps were performed in order:

1. *Put the compensating magnets in neutral (line up dots)*
2. *Head ship N, E, S, and W, enter readings under "Compensating Swing" columns (1), (2), and (1)-(2) on card (next page).*
3. *Using the recorded deviations, calculate the coefficient of C, B, and A on card (next page).*
4. *Note: All additions and subtractions are algebraic; that is +(+) = +, or +(-) = -, or -(+) = -, or -(-) =+.*
5. *With ship headed N, add Coefficient of C algebraically to Compass reading and adjust Compensating Screw N-S to make Compass read accordingly.*
6. *Head E, add Coefficient B algebraically to Compass reading and adjust Compensating Screw E-W to make compass read accordingly.*
7. *With ship on any heading add Coefficient of A algebraically and move Compass in mounting to read accordingly. (Do not make this adjustment.)*

> 8. *Swing the Compass on 45 degree headings. Fill in the "Residual Swing" columns (3) and (4) on card (next page).*
> 9. *Calculate and record Deviations in columns C to M and M to C.*

An accurate compass was a critical instrument for a B-24 travelling thousands of miles under all weather conditions over open ocean to find a tiny enemy target or friendly home base. As the 'items" of the Instrument Inspection were completed, the men would transfer their findings and/or actions to a standard inspection form:

> <u>Instructions</u>
> *On your 50 hour Inspection Form, page 5, enter inspection data obtained in performing Jobs 1 to 8. Use standard A.A.F. symbols listed below. The "remarks" column must be complete and accurate when the symbols indicate a defect.*
> <u>Symbols</u>
> 1. *Red Cross (X) – Indicates "Dangerous Condition."*
> 2. *Red Diagonal (/) – Indicates "Maintenance Work Necessary."*
> 3. *Red Dash (-) – Indicates "Inspection Not Made."*
> 4. *Black Last Name Initial – Indicates "Thoroughly Inspected – Condition Satisfactory."*
> 5. *Black Dash (-) – Indicates "Inspection Today Not Required."*
> 6. *Black Line (l) Drawn Vertically – Indicates "Not Applicable."*[84]

Walt completed Airplane Inspection-I on Tuesday, August 17. Just as for the earlier classroom instruction, he was given an end of course exam. He scored a grade of 80.

Walt started Airplane Inspection-II on Wednesday, August 18. The work was similar to the Inspection-I course, but from this point on Walt and the men from his flight worked on B-24's exclusively. They learned how to perform 50-hour inspections, which were more detailed and thorough than the work they had done during the proceeding course.

Airplane Inspection Branch-II focused on 50-hour inspection procedures and was held in Hangar #3. 50-hour Inspection Topics included: Day 1: of Propellers, Fuel Systems, Fuel Tanks; Day 2: Oil Systems, Fuselage, Cooling Systems; Day 3: Inspection of Engine Controls, Instruments, Oxygen Equipment; Day 4: Electrical Equipment, Batteries, Night Flying Equipment; Day 5: Cockpits, Valves, Manifolds, Superchargers; Day 6: Flight Control Mechanisms, Movable and Fixed Surfaces; Day 7: Landing Gear, Nose Gear, Wheels, Brakes; and Day 8: 50-hour Inspection of Hydraulic System, Power Plant-General, Airplane-General, and Preflight Inspection.[85]

A typical assignment might be to perform all oiling and lubrication required for a 25-hour inspection/maintenance for a B-24. This involved oiling and lubing dozens of points across the aircraft, including wing tab hinges (oil with M-4690 per spec 2-27E), rudder universal joints and screw jacks (oil with M-4690 per spec 2-27E), lubricate eight points on each main landing gear assembly (grease with M-4803 per spec AN-G-3), pilot and copilot rudder pedal tracks (clean thoroughly, but do not lubricate – but lubricate rudder chains with M-4690 per spec 2027E0, etc. The following day they would be assigned a new plane and ordered to perform the 50 hour lubrication, which included all the 25 hour points, plus additional points, such as the power plant control cables (lube with M4882 per spec AN-C-52), pilot and copilot brake pedal linkages (lube with M-4846-C per spec AN-VV-D-446, grade 1080), and so on.

Airplane Inspection-II ended on August 26. Walt's grade was 79.

Airplane Inspection-III started the next day, on Friday, August 27. A new topic was performing airplane preflight inspections. Now that the men were thoroughly familiar with almost all B-24 systems, they were capable of performing the critical preflight inspection on their own. Preflight Inspection Training was held in Hangar #2. Preflight Inspection topics included: Day 1: Engine Controls, Oil Systems, Flight Control Mechanisms; Day 2: Engine Instruments, Fuel Systems, Power Plant-General; Day 3: Preflight Inspection; Day 4: Inspection of Moveable Surfaces, Electrical, Wheels, Brakes; Day 5: Cockpits, Propellers, Fuel Tanks; Day 6: Fixed Surfaces, Cooling Systems, Valves; Day 7: Fuselage, Batteries, Hydraulic System, Landing Gear; and Day 8: Daily Inspection of Airplane-General, Navigation Instruments, Nose Gear.[86]

Preflight inspection comprised at least 91 distinct points and the men were required to complete a written checklist that tested their ability to quickly determine the condition of many routine components:

"1. Quantity of fuel in main tanks
"2. Fuel tank caps secure and saftied
"3. Wings free of defects....
"37. Landing, passing, and running lights operation
"38. Windows and windshield clean...
"64. Quantity of hydraulic fluid in reservoir
"65. Hydraulic accumulator pressure at unloading valve"

Walt completed Airplane Inspection III on Saturday, September 4. His grade was 78.

On Sunday, Walt checked up on the progress of the Tigers. He learned from the *Herald* that they had been walloped the day before in Briggs Stadium by the St. Louis Browns, 12 to 5. They had improved their overall record during the previous couple of weeks though, and now stood at seven games over 500, with a 66-59-1 record.

The final course for aircraft mechanics training was the Graduation Field Test. Walt started this session on Monday, September 6. It was held in the open – on the south side of the runway – to simulate real field conditions. The schedule included: Day 1: Repair of Simulated Combat Damage; Day 2: Adjustment and Replacement of Units: Mooring; Day 3: Engine Change – Removal; Day 4: Engine Change – Preparation of Engine for Installation; Day 5: Engine Change – Installation; Day 6: Engine Change – Completion of Installation, Final Checks; Day 7: Camouflage, Protection, Dispersal; Preflight Inspection; and Day 8: Servicing, Taxiing.[87]

The Graduation Field Test was designed to test the ability of the men to apply the skills they had learned under conditions that simulated maintenance work at a primitive airbase similar to what they would encounter when assigned to combat units. This last branch of the curriculum was added because many early Aircraft Mechanic School graduates performed poorly when they were first assigned to work under adverse conditions. The area devoted to this testing at Keesler was a low, wooded spot, with no buildings or paved roads, other than taxiways, at the southwest corner of the base, across the landing field opposite the line of hangars.

The men maintained gravel paths or boardwalks through the area, and all work was done in the open or in tents that the men erected themselves. They also dug latrines and slit trenches and erected camouflage. Meals were taken in a mess tent or in the open, and there was not even running water.[88]

The men worked independently, either on their own or in small teams, during graduation field test, but they were closely observed by mostly silent instructors the whole time. Once again, they were given a variety of inspection and repair assignments each day designed to test their mastery of a wide variety of subjects. Specific components tested included:

Exhaust system
Cowling, air ducts, turbo-superchargers, and regulators
Fuel systems
Electrical system
Oil system
Cylinder removal

For each of these systems, they were given test sheets with the following directions from their instructors:

"Use this sheet as you would an actual inspection sheet. Follow instructions explicitly. Do not initial after an inspection until you are satisfied it is correct to the smallest detail. If it will not pass inspection, write in the last column the trouble and see your instructor for repair parts. This is important as you will be checked in detail. Also write answers to the questions in the last column."[89]

Graduation Field Test ended after seven class days, on Monday, September 13. Walt completed this week of testing with a score of 88 – his highest score of any of the 14 individual classes he completed, and at the very top of the 'Very Satisfactory' range.

The following day, Tuesday, September 14, Walt and the other successful trainees from TSS 396 had finally completed the B-24 Airplane Mechanics Course. A few days' later results of the final course were posted and Walt's grade of 88 on the Graduation Field Test proved he could master complex systems. During the previous 19 intensive weeks, he had completed 678 hours of classroom training (along with many, many hours of independent study and practice) with an average grade of 79.6. It was a pleasant change for the men of flight 396 to be free of the tight schedule of classes and study that their training course had required.

A couple days later, Walt picked up a copy of the base newspaper, The *Keesler Field News*. Headline news and related stories focused on the Government's third War Bond Drive, with pictures of the parade down Howard Street the previous week, with the Keesler Band and hundreds of basic training troops. The paper also had news about entertainment and sports that Walt was interested in. One article spotlighted the new outdoors 'Commando Theater' just opened on the base that would seat 10,000 for USO shows and relieve the overcrowding in the indoors theaters. Walt read that the base movie theaters would be showing the new Warner Brothers production of the Irving Berlin hit show *This is the Army*, staring Ronald Reagan, George Murphy, and Joan Leslie and showcasing hits songs including *This is the Army, Mr. Jones* and *I Left my Heart at the Stage Door Canteen*. Other movies scheduled for the coming week at Theaters 1, 2, and 3 included *Destroyer* with Edward G. Robinson, *Johnny Come Lately* with Jimmy Cagney and Grace George, *The Phantom of the Opera* with Nelson Eddy, Susanna Foster, and Clause Rains, and *Frontier Bad Men* with Diana Barrymore.

Sports news at the base included the results of volleyball, archery, football, and bowling leagues. The previous week the 396th TSS volleyball team had been defeated by a team comprised of Chinese aircraft mechanics students. Perhaps Walt attended the Base Welterweight Title Bout that night between Cpl. Benny Montebana and Pfc. George Hays from the 306th TSS. Walt read about how the previous week, the 396th TSS won the base-wide singing program. Did Walt participate in this group? He was an avid singer – the songs assigned for the coming week included *Put Your Arms Around me Honey* and *East Side, West Side, All Around the Town*.

The *Keesler Field News* also demonstrated that the Army Air Force was a multi-ethnic force, with an article and photo about a father-son pair of black sergeants stationed at Keesler: First Sgt. Tom Gray and his son Sgt. Sam Gray. First Sgt. Gray Senior was completing his 20 year in the Army Air Force and preparing for retirement.

Presaging a future Walt did not yet know, he read about how Keesler technical training school graduates who had the necessary physical qualifications would likely be assigned to one of the A.A.F. Training Command's six flexible gunnery schools. He also read about how the men of the 13th Air Force in the South Pacific considered '13' their lucky number. He read how 13 gunners of the 13th Air Force had each shot down five or more Japanese planes. The article also reported that the 13th Air Force had been activated by Special Order No. 13, published at 1300 hours on January 13, 1943. Did Walt recall that this was the day before he had been inducted into the Army eight months earlier?

More somber was news about the 25,000 casualties suffered by the Allies during the recent conquest of Sicily in the European Theater of the war. This was a reminder that it was serious business that the Technical Training School students were preparing for.[90]

The following Saturday morning, September 18, dawned a little windy, with dark threatening clouds. Walt and the other successful graduates from Flight 396 were paraded to an outdoor graduation ceremony, with march music performed by the Keesler band. Col. Goolrick congratulated them and each man was given a Certificate of Proficiency, signed on behalf of the Army Air Forces Technical Training Command, Keesler Field Technical School, by Captain L. C. Eulberg.[91] At the conclusion of the ceremony, they marched back to barracks, again to the music of the Keesler Band, while the wind continued to increase and the sky turned darker still.

All through that weekend the weather was rainy and very windy. They were experiencing the effects of another hurricane, this one a couple hundred miles to the southwest in the Gulf of Mexico. Winds out in the Gulf reached 85 m.p.h. through most of Saturday. Into Sunday, the winds had decreased into the 70's, but the storm turned northeast towards coastal Louisiana and Mississippi. Men on leave that weekend dashed from bus to the shelter of various movie houses, restaurants, bars and stores in Biloxi. By Monday the 20th, the storm reached shore in Louisiana to the west, but the winds had decreased to just 40 m.p.h., with plenty of rain and wind in Biloxi.[92]

Graduates of Aircraft Mechanics Training had three options after their training was completed successfully. Most were assigned directly to work as mechanics in Army Air Force units, either in the U.S. or overseas. Some of the top students were selected for advanced post-graduate instruction at the various manufacturing plants where B-24 engines or planes were constructed. Some of Walt's pals would be going to the Ford plant at Willow Run to complete this post-graduate training.

Successful graduates who had the physical qualifications could apply for aerial gunnery school and the chance to train to be a bomber aircrew member. This is what Walt had wanted from the time he started basic training in St. Petersburg and at Clearwater. Airplane Mechanics Course graduates were desirable candidates for aerial gunnery crewmembers. The B-17, B-25, B-24, and later the B-29 bombers that were carrying the war into the heart of Germany, Italy, Japan, and other Axis powers were complex mechanisms. While they were all sturdy designs and well built by American war industry, in combat the complexity of their systems made them vulnerable to relatively minor damage.

For example, machine gun fire from an enemy fighter or flak damage from enemy anti-aircraft fire could leave a bomber crew and aircraft structure intact, but damage a relatively minor system that could still bring the plane down. Machine gun fire could cut a cable that allowed the pilot to operate the rudders, or flak damage could sever electrical, hydraulic, or fuel lines that could prevent a plane from returning to base. A crew of gunners also trained as airplane mechanics increased the likelihood that emergency, in-flight repairs could be made successfully.

Airplane mechanic school graduates also demonstrated mechanical aptitude that was of value to them as gunners as well. The .50 cal machine guns that defended American bombers, as well as the turrets that housed most of the guns, were complex mechanisms in themselves. A key responsibility of an aerial gunner was maintenance of his guns, and the ability to perform emergency repair and trouble-shooting while in the air.

Not all aerial gunners were trained as airplane mechanics, but this qualification helped Walt. As early as his first testing for aptitude in St. Petersburg and Clearwater, he had envisioned being an aerial gunner. During the second half of September, Walt went thru additional tests for physical fitness and aptitude. At some point near the end of the month, he was informed that he had been selected for aerial gunnery training and that he would receive orders to his new training base soon.

After graduation and while waiting for new orders, Walt tried to keep up with the progress of the Tigers through the end of their season. As late as September 17 they had been six wins above 500, but they had experienced a slow, steady slide after that. Late in the month they lost a four game series against the Yankees, 3-1, then traveled up to Boston where they lost a four game series against the Red Sox 3-1 again. Down to Philadelphia, they took a three game series against the A's 2-1, and then finished their season in Washington with a 12-5 win and a 4-1 win versus the Senators on October 3. The Tigers final record was 78-76-1, finishing in fifth place in the American League, 20 games behind the Yankees. As Walt analyzed their season though, he noted that the Tigers had lots of talent. Outfielder Dick Wakefield led the league hits and Rudy York led the league with 34 home runs, as well as RBI's. The Tigers as a team had the top American League batting average and total hits, while defensively their pitchers allowed the fewest hits by opponents and recorded the most shutouts and strikeouts. Their star pitcher had been Dizzy Trout, with five shutouts to his credit. Walt had a feeling the Tigers could do much better in 1944.

- 6 -
AERIAL GUNNERY SCHOOL 1943: FAITH, TRUST AND DUTY

Laredo Air Force Base, Texas and Aerial Gunnery Training

After chapel and breakfast on Sunday, October 10, PFC Babinski and dozens of other aircraft mechanic school graduates who were assigned to aerial gunnery training were assembled and ordered to pack their gear in preparation for transport to their next base – Laredo Air Force Base in Texas. Back at his barracks building, Walt quickly and carefully removed his clothing, gear, and belongings from his locker and packed it all in a large duffel bag, stenciled with his name on the outside.

There was a lot of speculation about what Laredo Army Air Base would be like, but no one really knew. The men's duffels were heavy as they walked the several hundred yards to the Keesler transport assembly room, trying to keep as cool as possible in the limited shade that the few live oak trees provided. Once they reported in at the transport office, they were issued their travel meal allowance and assigned to a designated lead enlisted man. They were told to report back to the transport office shortly after lunch.

Their meal allowance would cover three days of travel and their orders allowed them to travel at government expense via private transportation. Lunch in the Keesler mess hall was finished quickly, followed perhaps by a stop at the base PX for candy, gum, smokes, or a magazine for the journey. Walking back to the transport office Walt and his comrades looked down the base streets at the training halls and workshops that had been their home for the past half year. This would seem the most permanent and home-like base to them in retrospect, but they did not know that at the time and now it was time to move on.

The order came and the men retrieved their duffels, loaded them onto a truck, and boarded a bus for the short ride out past the base gate and through the streets of Biloxi to the Louisville and Nashville Station on Reynoir Street. Shortly after 3:00 p.m., the men boarded a local L&N train that had originated in Montgomery, Alabama, bound for New Orleans. As the train left Biloxi and made its way west at a leisurely pace, the men observed for the last time the Creole fishing villages and the resort homes that dotted this picturesque stretch of coast. Gulfport, the largest community between Biloxi and New Orleans was a modern city begun in the twentieth century, so it lacked the charm of the other gulf coast communities.[1] In addition, Gulfport was booming with wartime business, in particular the construction of thousands of military landing craft that would play a key part in the dozens of seaborne assaults that were still to come during almost two more years of war. 50 miles after leaving Biloxi, the train crossed the Pearl River and entered Louisiana. Angling away from the coast, the L&N route traversed a broad marshland laced with a maze of sluggish waterways with distant margins of pine and oak. After 40 or so more miles, shortly after 6:00 p.m., the train pulled in to the L&N Station at the foot of Canal Street in New Orleans.[2]

A night in New Orleans, with the French Quarter adjacent to the L&N station, provided an opportunity for the men to experience the unique Louisiana cuisine and local entertainment. New Orleans jazz in particular was a favorite of Walt's – and the city provided many clubs featuring the styles he liked, including Dixieland and the more modern 'hot' jazz typical of Louis Armstrong.[3]

The following morning the men made their way to Union Station on South Rampart Street to board a Missouri Pacific Railroad train around 10:00 a.m., bound for Laredo. Their route took them northwest along the left bank of the Mississippi to Baton Rouge, where the train crossed the river shortly after noon and headed more directly west. By 5:00 p.m., the train had crossed the Sabine River into Texas, and little more than three hours later arrived at Union Station on Crawford Street in Houston. Houston was a modern city in 1943, focused on the oil industry and shipping. Walt and his friends likely made use of rooms at the USO club for the night, and then made their way to the Southern Pacific Station on Franklin Street.[4] Before boarding their train shortly after 10:00 a.m. on Tuesday morning, Walt had time to send off a postcard to his parents back in Detroit:

> *Hello Folks*
> *Am here in Houston for a little layover will soon move on to where do not know yet.*
> *Am having fun on train.*
>
> *-Walt*[5]

Their train made all the stops between Houston and San Antonio. They traversed one of the most intensive agricultural districts in the state, with cotton fields, dairy farms, and cattle ranches. River valley bottoms were well wooded, in contrast to the wide vistas that increased as the train progressed on its journey.[6] The train arrived at the Southern Pacific's San Antonio depot on East Commerce Street in mid-afternoon. This allowed time for the men to have a leisurely dinner, and then make their way cross-town to the Missouri Pacific depot to catch their next train shortly before midnight.

As their train approached Laredo the following morning, it passed an arid brush country covered by numerous large cattle ranches. The area south of San Antonio is where the Texas cattle industry originated. Early ranchmen developed the methods by which the Texas cowboy evolved as a distinct type. A major problem for cattle ranchers in this part of Texas is the abundance of thorny vegetation. As Walt and his companions glanced out the train window they saw a country covered by prickly pear growing higher than a man, as well as other thorny plants, including catcaw, huajillo, rat tail cactus, and the dreaded junco, which Mexicans in the area thought had been the plant woven into Christ's crown of thorns. Leather chaps and jackets were developed as a necessary protective clothing to allow the successful working of cattle ranches in this area. Typical of the ranches as they approached Laredo was El Rancho Luz, 10 miles outside the town of Artesia Wells and covering 90,000 acres of rangeland. Closer to Laredo itself, irrigation water from artesian wells and from the Rio Grande had turned the land into productive fields of crops, mostly onion with some cotton.[7]

Laredo was an important commercial center for the upper Rio Grande valley, as well as a key center for trade between the U.S. and Mexico. Before the war over half a million tourists would pass through Laredo annually, enroute to and from Mexico City along the Pan American Highway. 60 percent of all freight between the U.S. and Mexico crossed the border at Laredo. In 1940 its population was around 35,000, approximately 85 percent of Mexican descent.

Tomas Sanchez, a Spanish ranchman, had first settled Laredo in 1755. By 1800 the town had steadily increased to a population of 1,000. In 1835-36, Santa Ana's army passed though Laredo on its way to attack the Texans at the Alamo in San Antonio. After the battle of San Jacinto, defeated Mexican troops passed through on their way back south to Mexico. After the establishment of the Republic of Texas in 1837, Laredo and the entire region between the Rio Grande and the Nueces River, was claimed by both Mexico and Texas. The Mexican American War firmly established Laredo as a part of Texas and later of the U.S.

This was the city that Private Babinski's train arrived at, pulling slowly into the Missouri Pacific Railway depot on Farragut Street early in the morning, just before reaching the international railway bridge across the Rio Grande leading into Mexico.[8]

The original U.S. military base in Laredo was Fort McIntosh, located west of the Missouri Pacific Rail Road yards, along a major bend in the Rio Grande. The fort had been established after the Mexican War as a star shaped earthwork structure. When the Civil War began it was occupied by the Confederacy. In 1863, Federal troops advanced up the Rio Grande from Brownsville to attack Fort McIntosh, but they were repulsed and retreated. During the border troubles with Mexico in 1916-17, the fort was reactivated and manned by troops from Maine, New Hampshire, Missouri, and Florida. The grounds of the fort that Walt and his comrades visited in 1943 covered 208 acres and contained mostly frame buildings, with some older ones of stone, brick, and adobe, but all painted yellow with a white trim. The grounds were covered with pleasant lawns and shrubs, as well as evergreens.[9] The fort was the headquarters of the 8th Engineers.[10]

However, this was not the destination for the gunnery school novices. Laredo Army Air Field began operation in November 1942.[11] It was located on the immediate outskirts of Laredo, north of Saunders Road. To the north and east of the base in 1943 were irrigated fields of onions and cotton. The primary mission of Laredo AAF was as a gunnery range and flexible gunnery school.[12]

Private Babinski arrived in Laredo on October 13, 1943, a Wednesday.[13] Typically a truck or bus would be waiting at the Missouri Pacific Depot to take the newly arrived trainees on the drive out to the Air Force Base. There they were processed into the 3rd Gunnery Service Squad, issued a one-piece flight suit and canvas helmet with holes for earphones, indoctrinated into camp regulations, assigned to a barracks building where they chose bunks, and told where the mess facilities were located.[14] Even though Walt and the other newly arrived trainees were used to the Army's habit of 'hurry up and wait' they were all surprised to learn that their actual training course would not begin for another week and a half – on Monday, October 25.[15] They had to report each day for an hour or two of general military and physical training. We do "physical exercise or as the army calls it calisthenics," he would write home. Some days they were also assigned to various work details, including keeping the grounds clean, minor repair work, or helping out on KP assignment.

Within a few days Walt had settled into this routine and (for the time being at least) was enjoying base life. Describing food at the base, he wrote of his camp mates: "Chowhounds all of them but I don't blame them – good food." For recreation, he discovered "Our swimming pool and very nice..." and "plenty of sports....we even have bowling alleys." He observed "plenty of good horseback riding around here," but it is not known if he had a chance to saddle up and go riding himself. For relaxation in the evening he discovered the base's small, but "very nice theater."[16] Walt would find the weather in Laredo pleasant and conducive to both outdoors recreation and (later) the frequent outdoors gunnery practice he would face. Typical highs through October and November would range from the 70's to mid-80's, with nighttime lows rarely falling below 55 degrees. What rain there was generally fell in downpours, then quickly dried as the sun reappeared.[17]

By Saturday, October 16, Walt had time to write a postcard to his older sister back in Delray:

Hello Genie,
Well here I am again since I wrote a couple days ago and I still will say this is the best camp I was in yet. The food is about the best I wrote to you about. We have steak or pork chops every day. God bless you.

-Lize[18]

Walt was an avid newspaper reader, so it is likely that he picked up a copy of the local *Laredo Times* the day he arrived, either at the Missouri Pacific Depot, or later at the base PX. He learned that Laredoans were interested in the war raging in the outside world and their part in it. Laredo's Catholic Bishop Garriga had made a Columbus Day address to the Lion's Club the day before, looking forward to the liberation of Catholic Italy, by the world's two great 'protestant' powers the U.S. and England. Other war news included an article about a Laredo native who had just graduated as an AAF B-24 bomber pilot, and another about the former commander of Laredo AAF, Col. William Kennedy, who had been shot down over Germany the previous August. The article mentioned that Kennedy's wife, still living at Laredo AAF, had just received her first letter from Col. Kennedy, from a POW camp in Germany. Walt would have also surely checked out the sporting events, movies, and other entertainment available in Laredo, for when he would have leave time to visit the city that was to be his home for the near future.[19]

Much of their time during the first 10 days in Laredo was free though, and Walt and his new buddies made the most of it. In addition to the recreation and morale building facilities on the base, there was also a small wood frame Catholic chapel, which Walt first visited on Sunday, the 17th.

That evening he attended a concert by the base's own jazz swing band, the LAAF "Jive Bombers," who performed on the base as well as at the USO club in Laredo.

Before starting the Aerial Gunnery Course, Walt likely took the bus into Laredo to the Rialto, one of the five movie theaters in town. (Others included the Royal and the Paramount.) The Rialto though was featuring the Warner Brothers movie "This is the Army" staring Ronald Reagan, Joan Leslie, and George Murphy. The film included 19 Irving Berlin tunes, including "Oh How I Hate to Get up in the Morning" and "This is the Army Mr. Jones." Locals and Laredo soldiers alike were also roused by Kate Smith singing "God Bless America" and Walt and any fellow Michigan natives would have also felt pride seeing Joe Louis in the film, punching a punching bag in beat with a band of Negro soldiers performing "What the Well Dressed Man in Harlem will Wear."[20]

On Monday, October 25, Pvt. Babinski was formally reassigned to the Laredo Army Air Field's 1027th Flexible Gunnery School, class 43-50, and began his gunnery-training course.[21] This class was comprised of more than 100 men, some of whom would not graduate for a variety of reasons.[22] It is likely that class 43-50 was addressed by the commanding officer of Laredo Army Air Field, Colonel Charles G. Pearcy. Colonel Pearcy described the origin and mission of Laredo Field and the Flexible Gunnery School in these terms:

"The gun totin,' lead slingin,' fancy booted bandito of the border is gone, but in his place and over the same ranges has grown up a new and deadlier aristocracy of death dispensers. For a long time after the disappearance of the "old days," nothing was to be found on the wild areas around Laredo. Mesquite, chaparral, sagebrush, and plenty of rattlers and jackrabbits. However, when the United States became hard-pressed for aerial

gunners; and it was evident that more schools would have to be opened, the Laredo area came into favor because of its exceptionally good flying weather.

"When the site was selected in July of 1942, there was nothing on the barren ground, yet a few short months later, a factory had sprung up. A factory unique among a land of surprises. For here [we take] raw material, from all corners of our land, all manners of men, trained in every conceivable trade; and in a few short weeks, they [are] turned into a homogenous group of deadly gunners and air crew members."[23]

The men were told that the Flexible Gunnery Training course was comprised of a standardized curriculum to verify their aptitude and their ability to master the skills necessary to serve as an aircrew gunner. Aerial gunnery training typically consisted of at least 320 hours of instruction, to be completed in about seven weeks. Specific topics with required hours of instruction included:[24]

Weapons (56 hours): Nomenclature, stripping & assembling, operation, and malfunctions
Sights & sighting (38 hours): Theory, range estimation, synthetic trainers
Turrets (50 hours): Maintenance and manipulation
Aircraft Recognition (20 hours)
Ground firing (48 hours): Skeet & moving base, turret mounted shotgun, hand held & turret mounted machineguns - both .30 and .50 caliber
Air to air firing (48 hours): Flexible & turret gunnery, camera & air to ground
Medical training (10 hours): Malaria control and first aid course #2
Pool period (10 hours): Orientation, oxygen equipment, and pressure chamber
Military physical training (6 hours per week)

In addition to their classroom instruction, aerial gunnery trainees were expected to study and practice during their free time as well. 320 hours of instruction over seven weeks represented about 9 hours per day six days a week in the classroom. With other normal military duties and study and practice, this did not leave much free time at all, but these were all volunteers motivated to succeed and avoid 'washing out.'

The first three weeks of the flexible gunnery course were spent in the classroom only.[25] That first week of aerial gunnery training, Walt and his comrades were introduced to a variety of weapons and also to the theory and practice of gun sights and sighting. Classes began as early as 5:30 a.m. (but usually at 7:00 a.m.) and went to about 4:30 p.m., followed by physical training before dinner. Sometimes there would be further classroom work after chow too.[26]

The classroom instructor introduced Walt and his fellow trainees to the material they would be covering:

"The Caliber .50 machine gun has proven its reliability in combat. It has worked in even the worst kind of field conditions.

"But its reliability depends on the gunner. The gun is only as good as the gunner. Even the most skilled do not know everything about a machine gun. Only by constant study, attention to every detail, and careful care and maintenance of his gun, will the gunner be doing his job, defending his airplane and members of the crew."[27]

Their classroom training commenced with learning about the specifications, definitions, and nomenclature for each individual part of the .50 caliber machine gun. The function of each part was explained to the class, followed by demonstrations of the adjustments they would be required to make. Just like during aircraft mechanics training, there were exams to pass:

"Name the parts of the oil buffer body.
"What is the purpose of the oil buffer tube locking spring?
"What is the total recoil of the piston rod?"[28]

They also studied the smaller .30 caliber machine gun. They were instructed on the proper care and maintenance of their machine gun, learning a checklist of thing to do with their gun before and after each flight:

"Points to be observed before flight:
1. Wipe the bore and chamber of the gun barrel.

  2. *See that the adjusting screws are screwed in tight against the buffer discs in the back plate.*
  3. *Test functioning by hand, using dummy cartridges. Particular attention should be given to headspace adjustment.*
  4. *Oil the gun carefully.*
  5. *See that the sights are clamped securely in place.*
  6. *Make sure that the ammunition belt is in good condition and loaded properly.*
  7. *Load gun partially or completely as directed.*
"*Points to be observed after flight:*
  1. *Unload gun completely and remove the ammunition belt.*
  2. *Clean bore and all working parts. If this cannot be done at once, oil gun to prevent rust.*
  3. *Release the firing pin spring.*
  4. *The armorer must get a detailed account of the gun's behavior in the air. If stoppages have occurred, their cause must be determined and corrected immediately.*
  5. *At the first opportunity, dismount the gun. Clean, oil, and inspect all parts. Stone off all burrs. Make needed repairs and replacements.*
  6. *After assembly, check operation with dummy cartridges and release the firing pin spring after insuring that functioning and adjustments are correct.*"[29]

  They learned about the different types of ammunition, appropriate use, and how to recognize markings on ammunition to determine the type. Ammunition types included ball, blank, dummy, armor piercing, tracer, and incendiary. There were also combination types, for example armor piercing with tracer. The studied the banding on individual cartridges that identified the type, as well as the markings to be found on boxes of ammunition to indicate the type.

  Still in the classroom, they studied the operating principles of the .50 caliber machine gun, including firing, recoiling, counter-recoil, cocking, ammunition feeding, and extraction and ejection of ammunition. Special attention was given to gun malfunctions:

> "*The importance of a thorough knowledge of machine gun malfunctions cannot be stressed too strongly. You gunners must know your guns thoroughly, and have complete knowledge of all the parts and their functions. The machine gun and its care is your responsibility. If you keep your gun in perfect condition by following routine inspections and cleaning procedures, checking to see that there are no burrs, worn plates, bent springs, and so on, the gun will seldom fail you. But if it does you gunners must know the symptom's characteristics of each malfunction, and how to correct it.*"[30]

  During the first three weeks of training, the men also received an hour or two of training each day in aircraft recognition. This was an extremely important course. If Walt and his comrades passed the gunnery-training course and later succeeded in become gunners on a bomber crew, they would have the means in their hands of blowing any aircraft within range of their guns out of the sky. Aircraft recognition training taught them how to distinguish friend from foe, to increase the likelihood that they opened fire only on the enemy. Walt's aircraft recognition instructor explained the importance of the subject:

> "*The ability to distinguish friend form foe is a basic ability expected of every soldier. Nowhere is this more difficult than in air combat, because of high speed aircraft and the short duration of engagements.*
>
> "*The deadly combination of accurate gunfire and faulty recognition has been responsible for unnumbered casualties and incomputable loss to the Allied Air Forces. One B-17 squadron in a raid over France had the inglorious record of shooting down: 2 German Me-109's and 8 British Spitfires. Such a record is tragic.*
>
> "*It is essential to distinguish friend from foe, but this is not enough. The recognition of the exact type and the knowledge of its wing span enables the gunner to set the Target Dimension Dial on Sperry Computing Sights. Recognition of exact types and knowledge of its speed and armament will give you useful information as to the tactics the enemy may employ.*

"As a gunner, you will be primarily interested in fighter aircraft. Therefore, your recognition course will cover current America, British, German, and Japanese fighters. You will be required to memorize wing spans of enemy fighters. Also required is familiarity with certain aeronautical terms which will enable you to understand instructions and questions of Intelligence Officers and to converse intelligently as an air crew member.

"Get acquainted with these famous fighter airplanes, study pictures and silhouettes, and practice identifying them at every chance you get.

"DON'T KILL YOUR FRIENDS!"[31]

By 1943, there were at least 100 different models of aircraft involved in aerial combat on a daily basis across the globe. While fighters and bombers from all air forces had markings to identify the nation of origin, there were many instances in combat situations were these marking would be impossible to see. It could be dusk or dawn, or even nighttime; it could be foggy, cloudy, or too sunny; the other airplane could be traveling too fast or at the wrong angle to see their marking; in all these cases bomber gunnery crew members would have to make a split second decision. If an enemy fighter was not properly identified it could get close enough to shoot down the bomber before the gunners had a chance to defend themselves. If a friendly fighter or bomber was not properly identified, it might be brought down by 'friendly fire.'

Aircraft recognition training developed the ability to recognize combat aircraft of all nations by their shape and flight characteristics. Day after day, the men studied the aircraft in top and bottom view, side profiles, front and back views. Minute differences from one model to another were discussed so that the trainees' ability to correctly identify an airplane model quickly increased day after day. The skills they learned were very much like the skills a birdwatcher develops – the ability to name the species of a bird seen sitting in a tree or at different angles while on the wing, and in different light and background conditions. Like birdwatchers, the men honed their skills by making their own personal sketchbooks of typical combat aircraft.

Walt bought a little 4" by 7" blue covered notebook where he carefully worked on his own sketches over the following three weeks. On the first page he wrote: "Pfc. Walter B. Detroit." On each of the 30 pages of the book he carefully made a light pencil sketch of various aircraft, then when he was satisfied, he finished the sketch in ink. Each page was numbered and carefully labeled to identify the plane: '-1- Lockheed P-38 Lightning,' '-2- Junkers Stuka JU 87D,' '-3- De Havilland Mosquito,' '-7- Consolidated B-24 Liberator,' '-15- Zeke Type OO Zero,' '-24- Boeing B-17 Flying Fortress,' etc.[32] This little notebook fit conveniently in Walt's shirt pocket, so he could study it often or take it aloft later if needed.

Gunnery training began with basic gunnery theory and gun mechanics. This began with learning about the characteristics of handguns, rifles, and shotguns. Later came handheld 'tommy guns' as well as .30 cal fixed machine guns. The sighting course included the effects on firing of various forces, such as temperature, moisture, windage, and wind drift. Bullet drop due to gravity over distance and bullet pattern due to vibration when firing (which spread the bullets apart from the target aimed at) were both covered in theory and in practice. 'Leading' a target based on movement and direction was also covered in the classroom, prior to live firing practice.[33]

The classroom instruction provided hands on familiarity with various weapons. Typically, new guns would be removed from storage in cosmoline, a thick, greasy preservative substance. They would be disassembled and each piece carefully cleaned. Then each gun was reassembled and checked by the instructor to ensure it was ready for use. The crews were trained to name each part, for example the charging handle, bolt, sear, trigger, cover plate, back plate, drive spring, etc., plus how to adjust each piece and ensure the gun was properly assembled. They learned how to 'bore sight' a gun – by using a vice to hold the gun stationary, adjusting the sights, then firing test shots to gauge the accuracy of the sights (and readjusting again if needed).[34]

The big boy of guns they would train (and later fight) with was the Browning M-2 .50 caliber machine gun, which could fire 800 rounds of ammunition per minute to an effective range of 600 yards. Each one weighed 64 pounds and had about 150 parts. The men would be required to strip and reassemble the gun blindfolded while wearing gloves as part of their training.[35] The first two weeks of training also included a three-hour malaria control course and an intermediate first aid course.

The end of the first week of training was Sunday, October 31 – Halloween Day. Walt and his new friends took a bus into Laredo that day on a short pass to attend a special Halloween Party at the

USO that evening, with food, dancing, and a 'Junior Hostess Popularity Contest' to cap off the evening.[36] The USO club in Laredo was a key resource for service men on leave. To reach the club, the men got off the bus running south on San Bernardo Avenue at Montezuma Street. They then walked east along the north side of Montezuma (the Tex-Mex Railroad line ran along the south side of the street) for three blocks till they crossed San Eduardo. The USO was located north of Montezuma between San Eduardo and San Francisco Streets, and bounded on the north by Scott Street. It was a new building which occupied a full landscaped city block, which Walt described as the "…U.S.O. in Laredo…is the nicest I seen in quite a while."[37]

The locals who staffed the USO for the benefit of servicemen provided all types of entertainment and services. In addition to providing overnight rooms, if needed, typical activities included swimming, singing, bingo night on Wednesday, game night on Thursday, and informal dances on Friday and Saturday. The USO included a snack bar, and for service men staying overnight, room service was available. Especially popular were the Junior Hostesses, pleasant young local volunteers who provided dance classes, lessons in beginning and advanced Spanish, and a program called 'talk that letter home' where a service man could dictate a letter to a hostess to send back to his family. This reflected the fact that literacy in the 1940's in the U.S. was far from universal, and no impediment in keeping a man from military service. The atmosphere was wholesome, including a Rosary Hour every night except Saturday for Catholic servicemen, and frequent encouragement for all soldiers to attend religious services somewhere each week.[38] For Walt and his friends, after the Halloween USO party, it was back to base and rest up for the second week of training to begin again early Monday morning.

A week later, Monday, November 8, marked the beginning of the third week of training and a new component was added to the daily routine. This was the beginning of what was known as 'pool period.' This was designed to familiarize the trainees with the high altitude oxygen and breathing equipment they would need to use as aerial gunners, and also to test them physically to ensure that they would be suited to the high altitude gunnery training that would begin in a couple of weeks. The pressure chamber simulated the low-air pressure and low-oxygen conditions the aviators would experience flying at high altitudes for extended periods. Within the chamber, an instructor and several trainees would experience the air pressure and oxygen being gradually reduced as if they were flying at high altitude.

In addition to explaining the use of the oxygen breathing equipment, the instructor would be observing how well each man did under the artificial conditions. As the simulated altitude increased, fingers and feet would swell. For some men, the high altitude low pressure would also cause their stomachs to expand; some experienced the 'bends,' for others heads swelled. Men with such reactions usually washed out of the program at that point. For some, the simulated altitude caused teeth to ache, which was usually a sign of an undiscovered cavity, which earned the man a trip to the base dentist. Eventually, simulated altitudes of as much as 30,000 feet were reached – higher than Mt. Everest, and actually higher than any of them would fly. The men also practiced use of oxygen breathing equipment in the pressure chambers. The instructor would illustrate the importance of using oxygen at high altitudes by having one of the men take off his mask at high altitude, then have him take off his shoes. After they were off, the instructor would tell him to put the shoes back on, but almost always the lack of oxygen would make the task impossible – the man might just stare at his feet, unable to move and unable to concentrate. Then the instructor would walk over, put the oxygen mask back on and almost immediately the man would be able to go on with the task and get his shoes on. This lesson taught the men the importance of oxygen very effectively.[39]

By the end of the third week the weapons familiarization course would be completed and the classroom study of the various bomber gun turrets would begin. By this time the state of the art turrets for both the B-17 and B-24 bombers was the Sperry Upper Local Turret and the Sperry Lower Ball Turret, Model A-2 designed for the B-17 and Model A-13, a retractable turret, designed for the B-24. Walt paid particular attention to the description of the lower ball turret, described as a "spherical metal structure designed for mounting in the lower part of the fuselage….The turret houses the necessary equipment required by the gunner in accurately directing the fire of two caliber .50 machine guns which are mounted in the turret." The turret included a hydraulic system to move the structure in azimuth and elevation, hand controls for directing the movement of the turret, and a Sperry .50 caliber Automatic Computing Sight, K-4, designed to automatically calculate predicted ballistic corrections for the machine gun projectiles.[40]

Out of the classroom and away from the base, life for Laredo's citizens, despite streets more crowded with off duty Army Air Force personnel, continued much the same as before the war. The local newspaper in early November was filled with war news: articles from the front by correspondent Ernie Pyle, news of USAAF bombing raids against Germany, a description of the capture of a German weather base in Greenland, and the death of a key Japanese Army general. However, the local news was much what you would expect from a city at peace: a local youth graduating from Yale, social events for several Laredo 'brides-to-be,' notices of upcoming meetings of the Wednesday Card Club, Four Leaf Clover Club, and the Laredo Chapter of the Order of Eastern Star. The height of the Laredo social scene was a Laredo Civic Music Association concert on November 9 at the High School, by soprano Vivian Della Chiesa.[41]

Walking through Laredo on leave, Pvt. Babinski saw a modern city, but with very strong Mexican influences. The business district was comprised of buildings largely built of white face brick and stone, which reflected the sunlight and shone dazzlingly, giving a sense of cleanliness. Older narrow streets were paved in brick or even creosoted wood blocks. Outside the business district, streets were lined with fine large residences or older sections with adobe and limestone houses. Typically door and window casings on these whitewashed structures were painted in brilliant colors and accented with small flower gardens or flower boxes with roses, geraniums, and bluebonnets. Oleanders and bougainvillea were common shrubs, and many city streets were lined with orange and grapefruit trees. Palmetto and date palm were also common, as well as mesquite, mulberry, and pomegranate.

As Walt and the other gunnery trainees walked through Laredo, its international flavor was readily apparent. At one point Walt would write, "Notice all [the] Mexican girls."[42] Display signs, posters, and advertising were seen more often in Spanish rather than in English, although most commercial establishments went out of their way to try to capitalize on the business potential that young G.I.'s presented. Jarvis Plaza was the commercial center of the city. Bordered on one side by the imposing and modern Hotel Laredo, the plaza was covered by lawn and kept cool by its many trees.[43]

Laredo catered to the tourist and commercial trade with nine other hotels, various lodges and tourist campgrounds, the Casa Blanca Country Club with golf and tennis, as well as five movie theaters.[44] One of the points of interest in Laredo was the Shriner Cactus Nursery, on San Bernardo Avenue, about two miles north of downtown. The bus from the Air Force Base would pass the nursery after making its left hand turn off Saunders Street and eventually most of the young aviators would make the stop and walk through the grounds. The Shriner Nursery had thousands of varieties on its grounds, from the 35-foot high "old man of the mountains" variety, to others so small they had to be handled with tweezers. Other sites they visited included the old "Indian Crossing" on the Rio Grande, out Bruni Street. This is a ledge of limestone extending out into the river, with the ruined earthen walls of the original Fort McIntosh on a bluff above the river.[45]

About five blocks southeast of Jarvis Plaza was Martin Plaza. In 1886 it was the site of a deadly gunfight between two rival gangs. In 1943 it was the site of a more domestic Mexican custom. Here on every Thursday and Sunday evening, along the plaza's circular walks the young men and girls would stroll in opposite directions. Eyeing the maids as they passed, a young man would choose his favorite and nod. If his nod was received with favor, the two would leave the promenade and stroll together. Even though many people gathered to watch the ritual, that did not deter the young suitors from the custom.[46]

An exotic excursion when on pass off the base was a visit to Nuevo Laredo, in Mexico just across the Rio Grande. The city bus, which the off duty troops could ride into Laredo for just five cents, would deposit them on Convent Street, south of Jarvis Plaza. A short walk down Convent Street would take them to the International Foot Bridge across the river. Despite its name, the bridge carried both vehicles and foot traffic. The fare into Mexico was just five cents for pedestrians (and less back into the U.S., depending on the exchange rate).

Once across the bridge, Walt and his friends found themselves in a city of about 25,000. Nuevo Laredo had several small plazas, many casinos, shops, and cafes located down its dusty streets. The main route south through town was the Pan American Highway. For about four blocks south of the bridge, this route was lined with many small souvenir shops and sidewalk stands. Open day and night, and with music blaring from radio loudspeakers in the background, they offered a variety of Mexican curios, including earthenware, baskets, sombreros, serapes, and typical native handiwork.

About three blocks south of the bridge, they would come to the 'Meat Market' a typical Mexican market plaza covering an entire city block. The Meat Market catered both to tourists and the residents of Nuevo Laredo itself. Its vendors sold meats, vegetables, fruits, and flowers for the local inhabitants. Others sold more pottery and baskets, along with perfume and jewelry, some of it quite costly, as well as garish imitation copies of Aztec art. At the center of the plaza they found a fountain fed by an overflowing spring. Chairs and tables nearby provided a place for a cool drink and maybe a bite to eat, while dogs drank from the fountain, little boys played in the water, strolling musicians, dressed in charro costumes with gay sombreros, embroidered serapes, and velvet pantaloons, would serenade the tourists for donations with the folk songs of old Mexico.[47]

Another possible weekend excursion for Walt and his friends was an overnight trip to Monterrey. Located about 135 miles south of Nuevo Laredo, Monterrey was capital of the Mexican state of Nuevo Leon. Like Nuevo Laredo it was a tourist destination for Americans, located on the Pan American Highway to Mexico City. But Monterrey was much more than a tourist town. It was a major diversified industrial center that was prospering in the mid 1940's due to the demands for war production in the U.S., which opened manufacturing markets for consumer products made in Monterrey factories.

Monterrey was an ancient city by North American standards, having been founded in 1579. It had a university, an institute of technology, an ancient cathedral, and the homes of many wealthy Mexicans. Overlooking the city was Cerro de la Silla, or Saddle Back Mountain in English.

Monday November 15 Walt and his fellow trainees began hands-on gunnery training, out of the classroom at last and on to the firing ranges that were Laredo AAF's reason for being. A bit of excitement for Laredo AAF that day was the news in the *Laredo Times* that the previous day two Laredo AAF gunnery instructors had won the big Army-Navy skeet shooting contest held at Brooks AAF in San Antonio. S/Sgt. Dick Shaughnessy had a perfect score of 100, and Capt. D. Lee Braun a 97 for the Laredo AAF Team. Braun and Shaughnessy had recently set a worlds record for the 450 registered target for a two-man team. Their victory in San Antonio represented the seventh consecutive meet won by the Laredo AAF team.[48] Walt and the other trainees were certainly in good hands and sure to be held to high standards of marksmanship.

Gun firing training started with handguns and rifles. Then shotguns were mastered, learning to hit trap fired clay pigeons, or skeet shooting. Walt liked skeet-shooting, writing: "This is about the most fun here we get to shoot about 500 shots a week."[49] In the evenings, the men could practice 'leading' targets on machines that shot BB's with compressed air and provided moving targets. These machines had an unending supply of compressed air, so you could practice until tired out. This became a chief free time activity for the crews...plus there were no more off-base passes until this phase of training was completed successfully.[50]

The next part of gunnery training was not so easy. It consisted of riding in pick-up trucks, traveling at between 25 and 40 mph. The trucks had plywood panels to help the gunners stand up, one on each side, while traveling down a road that curved through a remote part of the base. Spaced along either side of the road were small structures, square and only about three feet tall, with a man inside each one. As the trucks approached, the men in these target shacks would release clay pigeon targets, some straight out, some high, some low, some even heading straight at the gunner. While moving in the bouncing pickup, the gunner would have to figure the angle the pigeon would take to make a successful hit before firing. The distance between each target shack gave each gunner just enough time to reload his shotgun before the next pigeon was released. The course consisted of 25 of these targets. There was lots of practice at this, but on the final day of testing, each gunner had three boxes of 20 shotgun shells, and had to make 16 hits out of each box to pass. A poor showing could get you washed out of gunnery school right at that point.[51]

Once proficiency with the hand held shotgun had been proven, the men moved on to firing from a turret, similar to those mounted in the B-17 and B-24 bombers. Specially equipped pickup trucks would have either a top turret, or a Sperry ball turret mounted. These turrets were armed with only a single 12-ga shotgun mounted upside down for ease of loading, not the dual machine guns installed in combat aircraft turrets. The gunnery students would enter the turret and the truck would drive at about 20 mph over a twisting course lined with fixed targets to the right and left. Later the students would progress to firing from their turret at moving targets, which were mounted on a jeep that would drive a parallel route separated by low hills, allowing the target to appear and disappear randomly. This further developed their skills firing at a moving target.

The challenge of firing at targets with only a single shot helped train the gunners to control their trigger finger for later training with machine guns, to shoot only short bursts of no more than 20 rounds of ammunition. The Browning .50 cal machine gun could theoretically fire 800 rounds per minute, but long bursts of firing created so much heat that the barrel would overheat and melt, or if not, create so much heat that individual rounds in the firing chamber would 'cook-off' and fire unexpectedly. At best this could damage the gun, or at worst it could send a shell into another friendly aircraft. When the men began training on the Browning .50 cal's, Walt wrote: "More fun, shooting the machine guns in the last two weeks."[52]

Out on the firing range, safety was a major concern. Rules were strict:

"1. No one will go beyond the cable in front of the guns for any reason unless ordered to do so by an instructor, at which time the guns will be in a safety position.
2. The guns will be fired only on order from an instructor.
3. All the ammunition on this range is live ammunition. It will not be played with at any time.
4. No student will be on the line working with any gun unless the instructor is present.
5. No student will move a gun from its mount, unless ordered to do so by an instructor.
6. Students will carry out all instructions carefully. Care will be exercised at all times on the line.
7. There will be no smoking on the line nor near ammunition."[53]

Rules were very strict. Walt thought about the similarity between the strict gunnery school instructors and the stern nuns who had educated him for eight years at St. John Cantius. His Catholic school education prepared him well for the rigor of military training.

A lot of time was spent on gun sights and sighting techniques. Walt's instructor explained:

"The Caliber .50 is the aerial gunner's best friend, but you must be able to direct its projectile accurately. This can only be done with the aid of a device designed especially for this purpose. This device, known as a sight, shows the gunner where the gun is pointing in relation to the target. In this respect, the sights used on a machine gun are basically the same as those used on any shotgun or rifle. In addition to showing where the gun is pointing, the machine gun sight must enable the gunner to estimate range and lay off deflection. All phases of the Gunnery School and aerial gunnery training lead to one thing, correct sighting. A poor aerial gunner is more dangerous to a combat crew than all the flak over Berlin. Know your sight and how to use it!"[54]

The students started this aspect of training by handling a single pedestal mounted machine gun, equipped with a rear ring sight and a front post mounted ball sight. They learned to judge the distance they would need to 'lead' a target by mentally calculating the number of rear sight rings, or 'RADS,' so that a target to the side might be 'led' by three RADs ahead and above, to just one RAD above for a target behind.

Range estimation was an important skill for the gunners to develop as well, and it related to their aircraft recognition training:

"In estimating range, the flexible gunner is interested only in knowing if the enemy is in or out of range. The range of a flexible gun in combat is considered to be 600 yards. Although the Caliber .50 is able to inflict severe damage at four miles, man simply is unable to sight accurately over 600 yards. Using a 70 mil sight, it is possible to set up the following range estimation rules:
1. A single engine fighter in range fills ½ rad.
2. A small twin engine fighter in range fills 7/8 rad (53' class).
3. A large twin engine fighter in range fills 1 rad (66' class)."[55]

The trainees also learned the proper technique for 'harmonizing' gun sights with gun bore. This process included adjusting the axis of the bore to actually fire above the line of the gun sight, to allow for the drop of the bullet in flight out to the effective range of 600 yards.

The gunnery instructors explained the nature of the defensive role the men would assume on a bomber crew:

"War has often been likened to sport in that each force has a plan and proceeds to put their plan against the enemy's in deadly competition. In war, as in sport, each side attempts to know what the enemy is expecting to do…intelligence tries to find out what the enemy is attempting to do.

The aerial gunner, being primarily a defensive fighter, must know what to expect from the enemy so that he can be set to shoot. The enemy has been encountered thousands of times by our air men and those of our allies. The enemy's pursuit pilot attacks are very similar to the attacks made by our own pilots. The reason for the similarity is that the fighter pilot has only two alternatives. He can use either the 'Fly Through Attack' or the 'Pursuit Curve Attack.'

The guns of a pursuit fighter are fixed and harmonized to fire directly forward along the flight path of the plane. The pilot must aim his entire plane to accurately fire his guns. He must also make allowance for the movement of the bomber by shooting a certain distance in front of it, because the bomber will move during the time the projectiles take to get from the guns to the bomber. For example, if you were to attempt to hit a moving car from the side with a snow ball you would have to throw in front of the car to hit it. But if you were directly in front or behind the car you could throw the snow ball point blank and hit.'

The Fly Through Attack consists of aiming the fighter at a point somewhere ahead, on the bomber's line of flight, opening up with all guns, and flying straight on through. This puts up a wall of bullets in front of the bomber, so the bomber must fly through it.

The Pursuit Curve Attack starts with the 'Overtake Phase,' where the pilot flies parallel to the bomber and maneuvers into a favorable position to start the attack. The pilot then begins the 'Turn In' by banking and turning towards the bomber so that his guns point directly at the point of aim just in front of the target bomber. Once this point is reached the pilot commences firing while flying the 'Pursuit Curve' which keeps his guns bearing in relation to the moving bomber. The final stage is the 'Brakeaway,' where the pilot gets away from the bomber, usually by peeling off in a dive."[56]

The instructors explained that the gunners' best chance for scoring a hit was during the 'Pursuit Curve', because the movement of the enemy plane would need to be constant. They were encouraged to 'hit first' just as the Roll In ended, keep their guns firing in bursts during the Pursuit Curve, and break off firing during the breakaway.

Because of the massed firepower of bombers flying in formation, fighters would seldom attack without a 3 to 1 superiority. This required that the gunners be able to locate individual enemy planes and distribute the responsibility for defensive fire, much like a zone defense in basketball. All crew members were trained to identify the location of enemy fighters by using the clock system, with 12:00 o'clock being directly in front of the nose, 6:00 behind the tail, and 3:00 and 9:00 the right and left sides. 'High' referred to the sky area above the plane, except that 'Above' referred to the area directly above the plane. 'Low' referred to the sky area below the ship, except that 'Below' referred to the area directly below the plane.

Each gunner had an obvious section of the sky around the plane to defend. "To successfully defend the bomber formation, the approaching fighter must be spotted and the guns brought to bear on it, ready to fire at the proper range. It is difficult to remain on alert on a long mission, but the gunner must realize that in a very few seconds a fighter can dive out of a cloud or the sun, fire, and be gone out of range. The lives of an entire crew are in the hands of each single man in that crew. Each man must be constantly alert." Their instructor summarized the rules for effective defensive firing:

"1. Spot and identify the attacking plane.

"2. Track your target.

"3. Determine the direction of attack from 'Turn In.'

"4. Apply the proper deflection between your target and your own tail.

"5. Open fire at 600 yards.

"6. Fire 2 second bursts, adjusting your deflection as necessary.

"7. Quickly re-determine the position of the fighter and correct your deflection on each succeeding burst.

"Remember: Be alert, do not waste ammunition firing at targets over 600 yards and don't be fooled by a 'decoy.' Study and practice Position Firing and enjoy your medals. The Purple Heart is NOT a desirable medal, especially when awarded posthumously."[57]

After the theory of aerial attack and defense, the men began to also get experience in turrets, using the Sperry compensating sight. This devise allowed the gunner to aim directly at the target, but the Sperry sight included an automatic calculator that sensed the relative direction and elevation of the target (based on the gunner's aim) and automatically calculate for gravity and elevation.[58]

A new aspect of ground firing for the men at Laredo AFB consisted of firing at airborne targets. An AAF AT-6 trainer was used to tow targets at low elevations for practice at leading a target through the pursuit curve. This was done by firing at the target with a shotgun, either from the ground or from a top turret mounted on a tower.[59]

Gunnery and gun turret training also included use of the 'Jam Handy Trainer' in a specially equipped hanger. The Jam Handy system was designed to simulate combat experience for the gunners that would allow them to put the theory of aircraft recognition, range estimation, sighting, pursuit curve tracking, and firing into practice in an environment that could easily be observed by other students as well as instructors. The Jam Handy Trainer consisted of twin movie projectors, screens, and .50 caliber guns in a turret. The guns were modified to project a beam of light onto the screen to indicate where the trajectory of ammunition would strike. The room was soundproof and large enough for 5 to 8 students as well as the instructor. A sound system would reproduce the sound of the planes, as well as the sound of the guns while firing. The movie projection screen in the hanger would project the image of an enemy fighter making a 'pursuit curve' attack, arcing from the side of the defended bomber, to the rear and side again. Because of the variable movement of both the defended bomber and attacking enemy fighter, as well as the differing characteristics of windage, gravity, distance, etc., the gunner had to demonstrate that he could properly compensate for these factors to succeed in hitting his target.

For example, a gunner firing at an enemy directly to the rear would likely only have the effect of gravity on his shots to consider. Firing at a target flying to the side would require him to consider the effects of gravity and windage on the course of his shots as well. The Jam Handy Trainer allowed the instructors to ensure that the students could locate an airplane using the clock system, identify friendly from enemy aircraft, identify specific type to determine when the plane was within the 600 yard effective firing range, properly track through the pursuit curve, and maintain defensive fire in two second maximum bursts.[60]

On Monday, November 22, the men started their first flight training! The whole point of aerial gunnery training was to apply skills learned in the air. The AAF had to determine if the men were physically fit for service as part of an aircraft flight crew. Each of the trainees was put through a couple of test flights to verify that flying duty was the right field for them. These first flights were in an AT-6 training plane. The pilots were flying Staff Sergeants and a few Second Lieutenants. Some of them were very wild and most had very little flying experience.[61]

For the first two flights, the gunners took off in the rear seat, facing backwards. This seat, designed for aerial gunnery training, rotated 180 degrees. It had a ring to attach a hook from the safety belt worn around the waist. The trainee also wore a parachute and had instruction on how to bail out over the side and count to ten before pulling the ripcord. Each man was also given a bag, in case they got sick on these first flights. The first flight usually consisted of dives, spins, rolls, half rolls, stalls, loops, and wingovers. If anyone got sick on the first flight, they had a second chance with a less violent follow-up flight, but if they became ill on this one, they would automatically wash out of the aerial gunnery-training program.

After passing these two flights, the trainees were allowed to rotate their seat 180 degrees and face forward on takeoff and landings. On the next flight in an AT-6, the trainees were instructed to stand up in the rear seat once in the air, and feed an ammunition belt into the machine gun now fitted at the rear. Then they would begin test firing the gun. The guns were limited mechanically in how far they could arc, because some trainees would get excited on their first airborne firing and freeze on the trigger. This could result in shooting off your own airplane rudder, or another nearby plane. A key skill to learn was to anticipate and prepare for the movement of the plane as it maneuvered in the air. When the pilot would dive down toward the tow target, the gunner's feet would leave the floor with the loss of gravity, resulting in a loss of balance. On pull-ups, the gunner's feet would buckle as it seemed like their bodies were being driven into the floor. With time and experience, the gunners learned to anticipate these movements and related forces and to prepare for them. A man had to get used to all these new sensations quickly. If he was scared, the G forces and violent movement were too hard to control. All these tasks and actions quickly separated the men from the boys.[62]

By the middle of that first week of flight training only a very few of the men washed out because they could not cope with the physical and mental demands of flying and operating a gun. Thursday, November 25 was Thanksgiving Day – but it was a day of training for Walt and his pals like any

Laredo AAF, Texas 1943: Scenes around Laredo and the Rio Grande (top), airmen in parade through downtown Laredo (middle), new but cheap buildings at barren Laredo AAF (bottom).
Laredo Army Airfield

Laredo AAF, Texas 1943:
Skeet-shooting (top), classroom instruction (middle), and machinegun practice firing (bottom). *Laredo Army Airfield* booklet as annotated by Pvt. Walter Babinski.

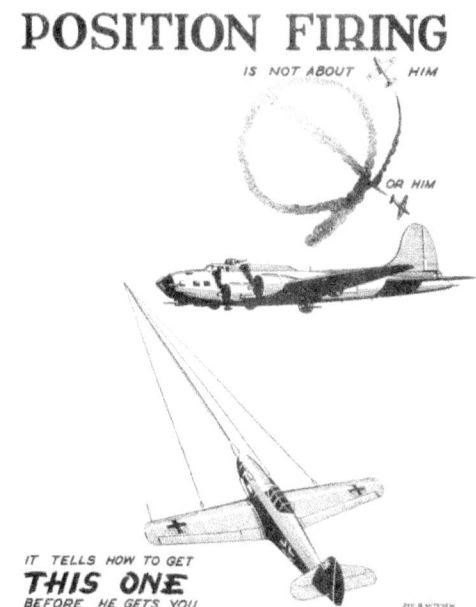

Laredo AAF, Texas 1943:
Skeet-shooting was not just fun, it taught deflection shooting, a critical key skill that every aerial gunner had to understand and master.
Basic Principles of Aerial Gunnery

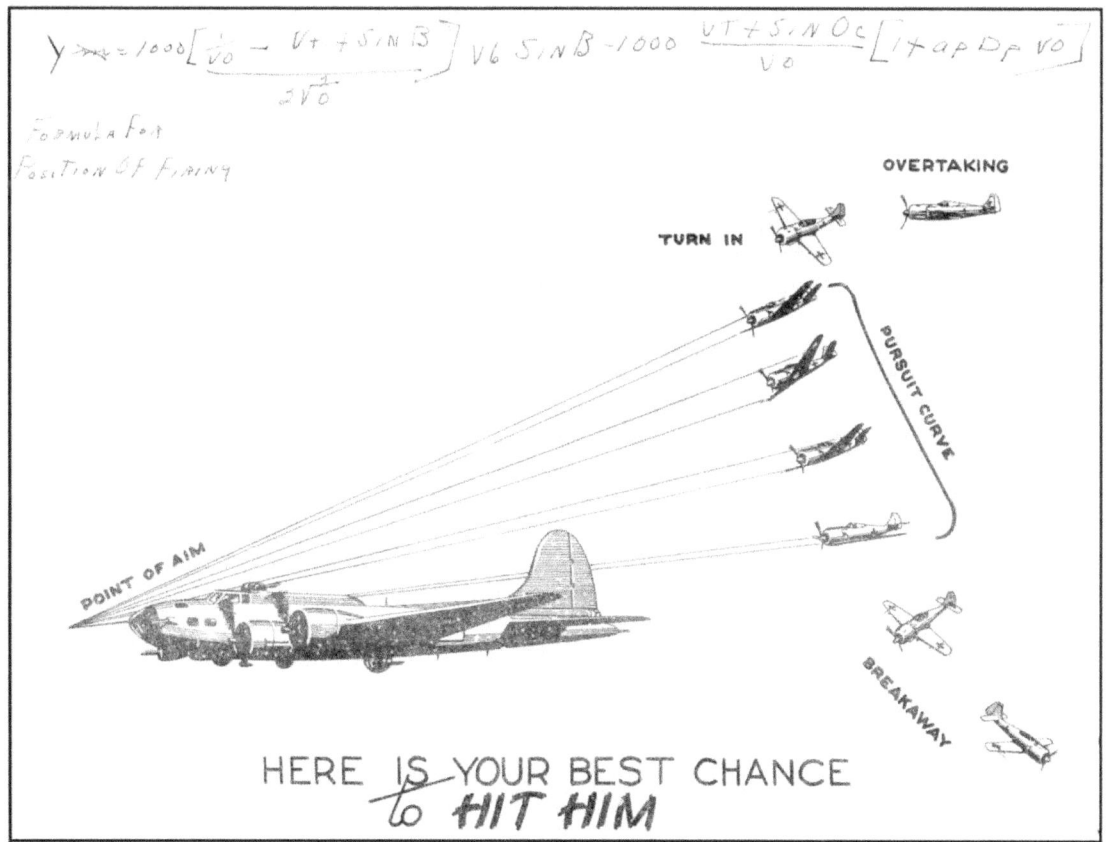

Laredo AAF, Texas 1943:
An aerial gunner had to learn when his best chance was to hit an attacking plane by understanding his opponent's position and likely tactics.
Basic Principles of Aerial Gunnery

To "locate" the enemy, the "clock system" is used by our air force. With this system, the area about the bomber is divided into hours, just as the face of a clock. 1200 o'clock is at the nose, 3:00 straight out the right beam, 6:00 at the tail, and 9:00 straight out the left beam. Planes observed are located by calling out the hour. In addition, the area about the ship has been divided as to altitude. Terms referring to altitude are:

> High -- the sky area above the ship
> Low -- the sky area below the ship
> Above -- the sky area directly above the ship
> Below -- the sky area directly below the ship

For planes approximately level with the formation, only the "clock" term is used. All planes sighted are assumed to be enemies until positively identified as friendly.

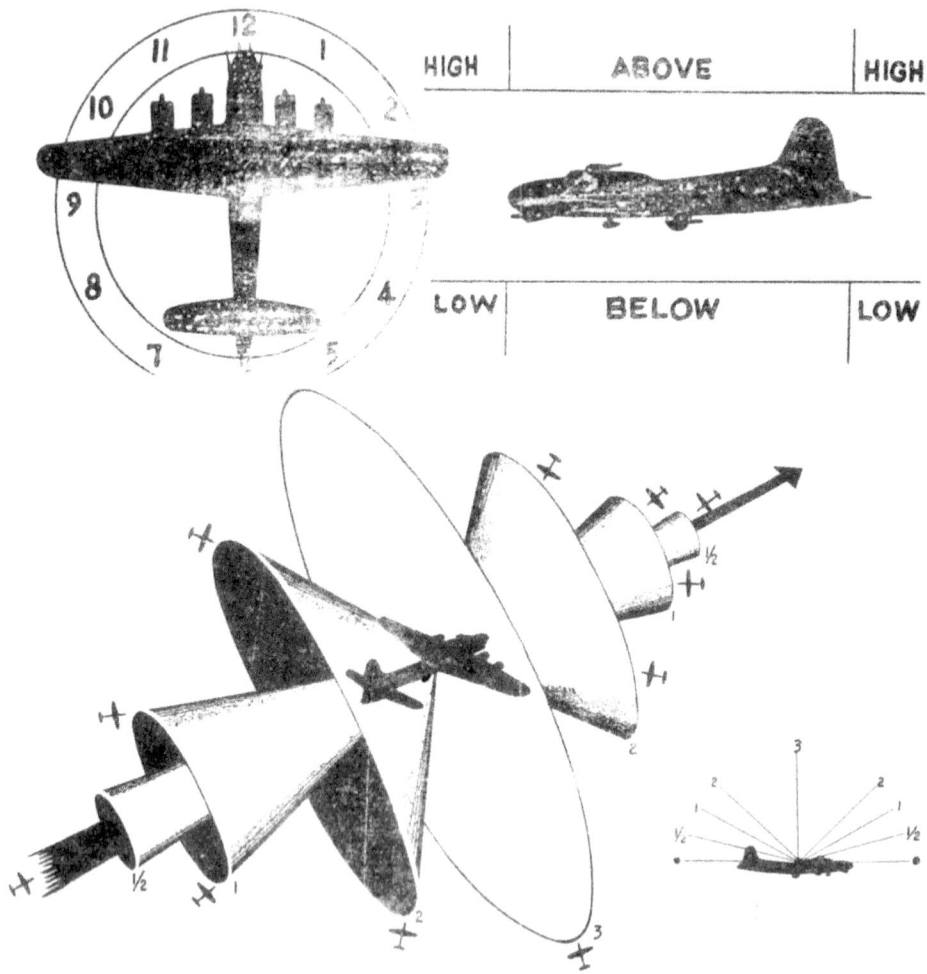

One point to be considered is the fact that the fighter does not stay at any one angular position longer than an instant. Therefore, the gunner realizes that to shoot effectively, he must alter his deflection constantly throughout the attack. If the attack develops at 90° the opening shot is fired with 2 rad deflection, then as the fighter drifts toward the bomber's tail, the gunner must decrease his deflection constantly so that when the fighter is at 45° the deflection is 2 rads. This process must be followed when going from one zone to another.

Laredo AAF, Texas 1943:
An aerial gunner learned that he was a member of a team that needed to work and fight together to defend their ship so that the bomber could get through to its target and return safely to fight again. An effective aerial gunner had to think quickly and act reflexively. The knowledge and skills that Walt first developed at Laredo AAF and Eagle Pass would be refined at Muroc and Guadalcanal and put to the test across the South Pacific. *Basic Principles of Aerial Gunnery*

Laredo AAF, Texas 1943:
Recreation facilities (top), chowtime outside (middle), and students preparing for flight time (bottom). *Laredo Army Airfield* booklet as annotated by Pvt. Walter Babinski.

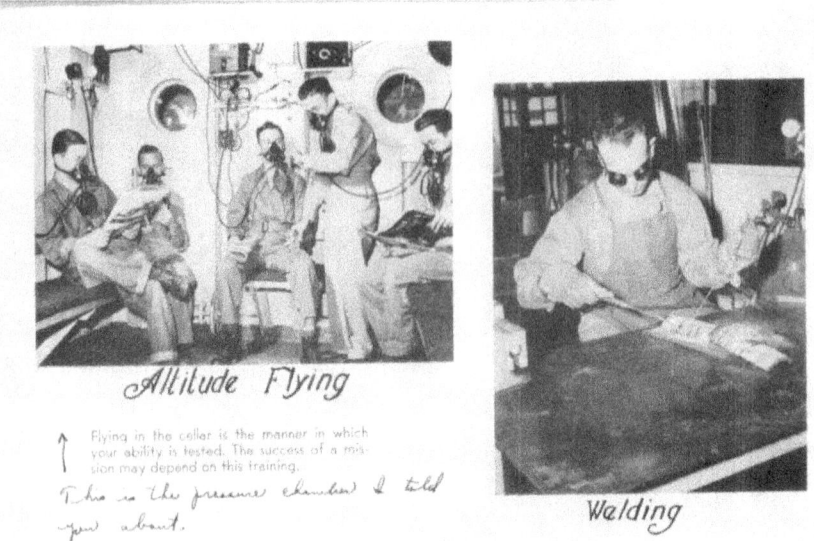

Laredo AAF, Texas 1943:
Students in high-altitude pressure chamber (top), primary gunnery trainer aircraft (middle), and a top turret gunner (bottom).
Walt also trained in B-17s during his week at Eagle Pass AAF.
Laredo Army Airfield booklet as annotated by Pvt. Walter Babinski.

Home on Leave December 1943:
Walt poses with his sister Genevieve in the back yard of the Babinski house on Pulaski Street in Detroit on a mild day in December 1943.

Detroit 1943:
Dorothy Florence Kurzyniec poses with the family dog in the summer of 1943.

Home on Leave December 1943:
Walt with his ma in the back yard of the Babinski house in December 1943.

Keeping in touch with friends in December 1943: Clark and Frank

Keeping in touch with friends in December 1943: Jo Jo Jekielic

Keeping in touch with friends in December 1943: Joe Kurzyniec and Krupka

Keeping in touch: Joe Kurzyniec

other day during the preceding five weeks. Dinner was special though, with turkey, pie and other traditional holiday treats. For the past two weeks there had been no off-base leave while the men were mastering the firing of a variety of weapons under increasingly complex conditions. The pace seemed to have been getting more and more intense, but there was hope for leave the coming weekend. As November drew to a close, local news in the *Laredo Times* had a uniquely border area slant. Mexico had recently declared war on Germany (an event that would have a peculiar consequence for Walt 16 months later). The *Laredo Times* carried reports about Mexico's preparations for war, including the intention of the Minister of Defense, Gen. Lazaro Cardenas, to lead Mexican troops into battle himself. More mundane news included the reopening of the local 'Piggly-Wiggly' store and plans for a new free bridge across the Rio Grande announced by the Laredo Chamber of Commerce.[63] Laredo was also looking forward to the arrival of the Daly Brothers Circus and its star attraction – little eleven-year old Norma Davenport and her baseball-playing elephants, with performances Sunday and Monday, November 28 and 29.[64]

Eagle Pass Army Air Field, Texas

But trips to the new Laredo Piggly-Wiggly or to the Daly Brothers Circus were not to be. An announcement came before the end of Thanksgiving Day that the men were headed to a new temporary base and a week of more intense aerial gunnery training, at Eagle Pass Army Air Field.

Even though Eagle Pass was just 125 miles by road northeast of Laredo on the Rio Grande, the men traveled by train due to the poor condition of the roads and the continuing need to conserve fuel. Friday afternoon, November 26, shortly after lunch, the few dozen men destined for Eagle Pass AAF loaded with their gear onto a bus for the short drive to the MPRR depot in Laredo. They boarded a day coach on a Missouri Pacific passenger train that left Laredo around 3:15 p.m. and arrived in San Antonio about half past eight in the evening. The Missouri Pacific Station in San Antonio was at West Houston and North Medina Street, just off West Commerce Street, which was the main east west thoroughfare in the city. They had to make their way across town, about a mile and a quarter to the Southern Pacific station at 700 East Commerce Street. They could take a city bus for just 10 cents, but it is more likely that groups of three or four men each shared a taxi with a fare of just 35 cents for up to two miles.[65] This got them to the SP station quickly and gave them time to get a little dinner in time to catch the Southern Pacific local train to Eagle Pass that left at 10:30 p.m. This train traveled along the SP mainline route to El Paso, but switched onto the Eagle Pass branch at Spofford, Texas. Walt and his friends likely struggled to sleep as well as they could in the coach seats, for they knew their arrival in Eagle Pass was early in the morning.

In 1940, Eagle Pass had a population of just over 5,000. It was considered a tourist resort for that part of Texas, with narrow streets and a Mexican border atmosphere. Since the days of the Mexican War, there had been a U.S. military presence at nearby Fort Duncan. The town was laid out in 1850 and named El Paso del Aguila (Eagle Pass) for the daily flight of an eagle back and forth across the Rio Grande to nest in a huge cottonwood on the Mexican bank. During the Civil War, Fort Duncan was occupied by the Confederacy. Here on July 4, 1865 the last Confederate flag flew over an unsurrendered force of 500 rebel Missouri cavalry who crossed into Mexico to avoid capture. Federal troops then reoccupied the fort, which was later garrisoned by an unusual force of Seminole Indian-Negro scouts.

By 1940 plans were being made to have the City of Eagle Pass take over Fort Duncan as a recreation area, when plans began to develop an Air Force training facility several miles outside of town, as the nation began to gear up for possible war.[66] The Army Air Force activated Eagle Pass Army Air Field in 1942. It was located ten miles north of Eagle Pass, in dry rangeland, east of the Rio Grande.[67]

On Saturday, November 27, Walt and the other 1027th Flexible Gunnery Technical Training school students arrived at SP's Eagle Pass station at about 6:45 a.m. They then traveled out to Eagle Pass Army Air Field to get settled into their temporary barracks, meet their instructors, rest up and get ready for the next phase of their aerial gunnery training, which would start the following day.[68]

One of the tough and demanding gunnery instructors that Pvt. Babinski met and learned from at Eagle Pass was Sgt. Donald R. Harlow. Sgt. Harlow, originally from Maine, had just entered the service himself in August 1942, completing his own training at the Armament School located at Buckley Field, Colorado. Harlow was quickly recognized as an effective instructor who was also well liked by his students.[69]

A key component of training at Eagle Pass was putting the ground gunnery skills and preliminary flying experience in AT-6's developed at Laredo into practice in the air. The more remote and sparsely populated location of Eagle Pass AAF was ideal for the task.

Sunday, November 28 Pvt. Babinski and his fellow trainees began this phase of training. At first, the aerial gunnery trainees would be loaded into an old worn out B-17 and taken aloft. There they would man a .50 cal machine gun at one of the waist windows, for practice firing in the air. The gunners fired from an assigned ammunition belt. Each belt had both tracer shells and other specially painted shells of a single color. The tracers (every sixth or seventh bullet) were so that the gunners could see the route their fire took to the target to help refine their firing skills.[70]

Later, a gunnery instructor would take a half dozen or so students aloft for target practice, again firing from the waist position. At first, this practice was 'air-to-ground' gunnery, firing at targets on the ground. Sometimes this resulted in shooting a stray cow, which may have wandered off a nearby ranch. In such cases, the rancher would invariably submit a claim for his "prize" bull, or cow, or whatever, and be paid $200-$300. Later the B-17 was used for 'air-to-air' gunnery, firing at a target towed by an AT-6 (usually about 1000 feet behind) and 600 to 800 feet off the side of the wing of the B-17. Each gunner had a belt of ammo with bullets that had been dipped in different color paint. If the gunner hit the towed target, the paint would leave some color on the sleeve, which allowed a rough score to be calculated. A 2% score was required to qualify through this aspect of training. Any hits on the AT-6 itself were highly frowned upon, and did not count towards the final score (and occasionally they were hit).[71]

Aerial gunnery school was voluntary, and once the airmen progressed to training in various types of aircraft, they began to draw flight pay. Walt and his comrades knew that aerial gunnery school graduates were considered specialists and were typically assigned to B-24 Liberators or B-17 Flying Fortresses heavy bombers or B-25 Mitchel medium bombers. These were the aircraft designed to fly into the most distant, most heavily defended enemy airspace, usually beyond the range of American fighter plane escort. The skill and proficiency of the aerial gunner therefore, was a key factor in the ability of these planes to fight their way through to their targets, and then return to base safely in order to fly again. Aerial gunnery school graduates were not the only gunners on B-17's and B-24's though. Typically, the flight engineer (an enlisted man) would man the top turret during aerial combat. The radio operator would also man one of the waist guns (firing out one of two side openings, to protect the aircraft to right or left). These men would have only limited training in gunnery though. The most technically complex and physically demanding gunnery positions in the B-17 and B-24 would be reserved for graduates of the concentrated aerial gunnery course: nose turret, tail turret, and lower ball, or belly, turret.[72]

The week at Eagle Pass AAF was intense and tiring. In just five days they had spent about 40 hours in the air, in addition to time briefing and debriefing with their instructors. Once again, Walt and most of his friends did well and were passed by their instructors. On Friday morning they were ordered to pack up and report for transportation back to Laredo. They arrived in Eagle Pass late in the morning, with just enough time to wander briefly through the border town before boarding the early afternoon train at the SP depot. Reversing their route of less than a week before, they arrived back in San Antonio at around 6:25 p.m., with about five hours free before catching the MP train to Laredo just before midnight. This gave them time for dinner, perhaps time to visit the Alamo less than a half-mile away, and maybe time for a beer or two. Then they boarded the train and settled in to sleep as the train made its way slowly south through the cold Texas December night.

Back to Laredo

At around 6:30 in the morning on Saturday, December 4, Private Babinski and his friends arrived back at the MP depot in Laredo.[73] Later that morning they made their way back to base on the outskirts of town, where they were assigned barracks again. The pace slowed down somewhat for them at this point...a weekend pass into town was welcomed after three or four intense weeks of training. Walking through the base, or up and down the streets of Laredo, Walt and his friends were aware that new trainees continued to pour into the base, and that these new trainees looked at them now as experienced old hands.

Monday the 6th of December, the training continued, but it was clear that their time at Laredo AFB was coming to an end. Ground firing, PT, and more study of turrets continued that week, followed by leave on Friday night. On Sunday, December 12, Pvt. Babinski graduated from Class 43-

50 of the 1027th Aerial Gunnery School. In a brief ceremony, he and other graduates were presented with Aerial Gunner Wings and promoted to the rank of Sergeant.[74]

The Gunner's Creed
"This is my ship. I have faith in the Pilot who flies it. I know the Navigator will direct a true course. I trust the Bombardier to destroy our objective. I believe in the abilities of the crew. Yet the safety of my Ship and the success of our mission depends on me and my guns. I vow solemnly to perform my duty. I am the Ship's Gunner."
-George E. Mathison, 1st Lt. Air Corps[75]

With less than two weeks to Christmas, Laredo was preoccupied with a flurry of social events that drew its attention away from the training and graduation activity at the nearby Army Air Force gunnery school. Sgt. Babinski read in the local newspaper about the many holiday parties, church open houses (which did not forget the local servicemen), and local family reunions, as Laredoans began to return home for the holiday season.[76] Walt's thoughts surely began to focus on his family, which he had not seen for almost 11 months.

Then on Tuesday, December 14, the news that the men had been waiting for was announced. The following day, 91 graduates of Class 43-50 were granted leave to return home, with orders to report to Hammer Air Force Base in Fresno, California after their leave for further duty and assignment. Sergeant Babinski's orders were typical: He was authorized base rations through lunch on the 15th, then to depart the base at 2:00 p.m. He was paid a cash meal allowance in advance for 2-2/3 day while traveling at a rate of $3.00 per day. He was given authorization to travel home by private transportation at government expense, but required to report to Hammer Field on Sunday, January 2, 1944.[77] He was going home!

Private...now Sergeant Babinski was one of 300,000 men who would graduate from gunnery school, the second highest number for any AAF specialty during WW2, after aircraft maintenance.[78]

Back to Detroit on Leave

At 2:00 p.m. sharp, two trucks with Walt and the men of Class 43-50 on board started off, passed through the main gate of Laredo Field one last time, and made their way to the Missouri Pacific Railroad depot in Laredo where the men had arrived two months earlier. Walt and his comrades quickly boarded the waiting train with their duffel bags, and at 3:15 p.m. they were off. The train Walt was on was headed all the way to St. Louis, so he settled in for a good day and a half of travel. Not all the men would remain all the way to St. Louis though. Many would get off earlier to make connections for home in different parts of the country. Days were short in December, so they passed through San Antonio and then Austin in the dark, and shortly after sleep came.

The following morning they awoke traveling through northeast Texas cotton country. After breakfast, the train crossed into Arkansas at Texarkana and sped past Hope, Benton, and other small towns until they reached Little Rock midafternoon. It was dark again before they entered Missouri, and finally at about 11:00 p.m. they pulled into St. Louis Union Station. Walt had checked with the conductor and knew that with luck he could still make a connection to Chicago that night, via the Chicago & Alton Railroad. He made a dash for it and boarded just before midnight. Arriving on this train late, he had a poor choice of seats, but settled in for the overnight trip to Chicago the best he could manage. Arrival in Chicago's Union Station was before 8:00 a.m. Walt took the time to get cleaned up, have breakfast, and then make his way over to Central Station where he caught a New York Central train departing at about 10:30 a.m. bound for Detroit.

Back in Michigan soon, he read, perhaps played cards, and tried to stay calm. At around 2:40 p.m., the train stopped in Battle Creek, and Walt thought of how far he had traveled and how much he had accomplished since arriving at Camp Custer outside Battle Creek the previous January. He was a 'buck private' then. Now he was a sergeant and both an airplane mechanics school and an aerial gunnery school graduate. Night fell shortly after they passed through Jackson, but Walt was alert as the conductor announced cities and towns that sounded more and more familiar. Ann Arbor, Ypsilanti, Wayne, and Dearborn passed by in turn, then the train slowed as it crossed the Rouge River and entered one of Detroit's many heavily industrialized areas. It was now Friday evening, but Walt could see through the car window that men were working late shifts at the mills, foundries, and factories they passed. Detroit, proud of its title as the 'Arsenal of Democracy,' continued to make its contribution to the war around the clock.

Friday, December 17, at about 5:45 p.m., the train pulled into Detroit's Michigan Central Station, and Walt stepped off and onto the platform. It was cold again...no more balmy south Texas or Gulf Coast weather. Walt buttoned his overcoat and walked down the platform and through the concourse. He had a pocketful of money now as a sergeant, so Walt turned left in the concourse and headed to the cabstand where he hailed a waiting cab, got in, and directed it to Pulaski Street, 9050 Pulaski Street. When they arrived, Walt paid the fare, gave the cabbie a generous tip and grabbed his duffle. Walt noticed the fringed victory banner with two stars – one for Johnnie and one for him – hanging in the front porch window. A few steps, then he opened the door to the enclosed front porch, walked up the short flight of stairs and stood at the front door. He knocked and in a few seconds it opened. Whatever they were doing at the time, Walt's ma, pa, Henry, Leona, Helen, Gennie, and Anne were soon there around him, welcoming him home.

Walt would have less than two weeks at home on furlough, but he made the most of his time. Everyone was impressed and proud of his Sergeants stripes and aerial gunner wings.[79] The days ahead, Walt would enjoy no early morning rising and Army discipline. His ma thought he had grown far too thin and she would do her best to fatten him up again with big meals and many extra helpings of the traditional Polish cooking he had missed for almost a year.

Of course, Walt also wanted to see his friends, but most were not home now. Little by little, Walt would piece together the information about who was serving where and what news there was about them. Joe Kazyka, Clark, and Buzz were in the Army, completing their training. Joe Prus (known as Pudgy), was also in the Army and his brother Stanley was in the Air Corps, training at Fort Benning, Georgia. Ed Czopeto, Joe Sklarzek (Fuzzy), and Joe Suchyta (from the family of Delray bakers) were serving in the Army.

Walt's friend Joe Jekielek was in the Air Corps, training to be a fighter pilot at the Southern Aviation School in Camden, S.C. and later at Maxwell AFB, Alabama. Joe was known in the neighborhood as Jo-Jo or sometimes as Tex. He wrote regularly to the Kurzyniec girls throughout the war. Jo Jo's dad was a good friend of Thomas Kurzyniec. Joe Kurzyniec, the Kurzyniec girls' second cousin, was in the Army, along with his friend Joe Krupka. They were in infantry training together in San Luis Obispo, California. Adam Sordyl[80] was in San Diego training with the U.S. Marines, and his brother Theodore was in the Army. Walt's friend Joseph Stoklosa (who lived across Pulaski from the Babinski's) was also in the Marines now. Eddie Chopek was completing training as a paratrooper.[81]

There was news from his brother John in Australia. He had been promoted to Machinist Mate Second Class the previous June, but letters home indicated that he was unhappy being stuck at a repair base, even if Brisbane was an enjoyable city to be quartered in. His latest letters home indicated that he had been transferred to another unit in Brisbane where he would have more of a chance to be assigned to a ship.[82] Walt wondered how he would feel about his assignment when he completed his training.

Walt surely stopped in the Kurzyniec store more than once to say hello to Mr. and Mrs. Kurzyniec, as well as to the three girls. Walt had taken quite a liking by this time to Dorothy Kurzyniec. In addition to helping out at the Kurzyniec Store, Dorothy and her older sister Martha had been working for quite some time at Zokower's Department Store on West Jefferson Avenue in Delray. Mr. Zokower was a nice Jewish man who was pleasant to the girls. He was only 4' – 10" tall, but hard working and fair. He paid Martha and Dorothy each $13 per week. Each Christmas he also gave each girl a $25 war bond as a bonus. Walt took every chance he could to see Dorothy when she was not working at Zokower's or busy at home.[83]

On Thursday the 23rd, Walt saw the Red Wings take on the Rangers to win 5-3. On Christmas Eve, Walt attended midnight mass at St. John Cantius. Christmas Day 1943 was a Saturday, and the family tree would have been trimmed and the candles lit (briefly) as in years gone by. There were presents under the tree, including some that Walt had bought with his Sergeant's pay for his ma and pa, as well as brothers and sisters, including little Henry and Leona.

Perhaps, Walt saw the Red Wings play again on Sunday the 26th versus the Boston Bruins (they tied 4-4 to end the year with a 7-8-4 W-L-T record. Then, almost as quickly as it began, Walt's leave was over. As he did the previous year, he wondered what the coming year would bring and how he would be changed and challenged.

- 7 -
BOMBER CREW PHASE TRAINING 1944: GOODBYE GOLDEN GATE

Hammer Army Air Field, California

Very early on Thursday morning, December 30, Walt, Sergeant Babinski now, dressed in his uniform, finished packing his duffel bag, put on his overcoat, and said his goodbyes to his ma, sisters, and brother Henry. They had just finished a big breakfast with scrambled eggs, kielbasa, and toast. His ma asked Wally if he had the packages of food and candy she had given him for his journey westward. He assured her that he did, giving her a last hug and kiss, then picked up his duffel and went out the front door onto the enclosed porch. He walked down the side steps, opened the porch door and stepped out into the dark and cold of the December morning.

Walt's pa was in the Ford parked in front of the house, keeping the engine warm while he waited for his son. Walt tossed his duffel in the car, got in himself, and off they started down Pulaski Street. They made small talk as they drove the short distance to Michigan Central Station on Michigan Avenue. They talked about Johnny in the Navy, and about relatives and friends, and about Walter senior's job at the Ford Rouge plant, and how busy they all were with war work. As they pulled up to the station, the Ford stopped, they both got out, and Walt grabbed his duffel. His pa told him to be careful, to write his ma, brothers, and sisters often, and to come home safely. Walt thanked his pa for the lift, promised to write and keep safe, and told him not to worry. They shook hands, gave each other a fast hug, and then said last goodbyes. Walt waved as his pa got in the car to drive off to work, then he turned and headed into the warmth of the great station.

Walt checked in at the military travel office to present his orders and get his travel authorization for the trip west. He did not have long to wait before the arrival of westbound train #17 was announced. He boarded and found a seat shortly before its 7:15 a.m. departure. It was just dawn as the train headed west, passing the Ford Rouge Complex to the south, where his pa and several of his friends worked. Both north and south he could see jammed freight yards and smoke belching from many of the factories from the 'Arsenal of Democracy'.

The hours and miles passed quickly and around noon Walt's train pulled into Chicago's Central Station. Walt made his way over to Union Station, where he had time for a bite at a lunch counter, before catching an Alton Railroad train leaving at 3:00 p.m. for St. Louis. It was getting dark by the time they passed Joliet, and Walt dozed or chatted with other aviators or soldiers he met on the train. Shortly after that he managed to get a bite to eat as the train rolled south, then he got as comfortable as the crowded conditions would allow. Just before 10:00 the train arrived in St. Louis Union Station. Walt found his way back to the USO he had visited months before, to wash up and relax for just an hour or so. Then he boarded CB&Q train 25, which left St. Louis at 11:55 p.m., for an overnight journey to Kansas City.

Train 25 arrived in Kansas City at the Union Station at about 7:35 a.m. It was now New Year's Eve, Friday, December 31. Walt had about 14 hours until his next train left, so he found his way to the nearby USO, where he washed up, changed clothes, and had some breakfast. After breakfast, he checked his duffel and decided to explore Kansas City briefly.

Kansas City Union Station was located at 24th and Main Street, about 12 blocks south of the central business district. Walt wanted to stretch his legs though, so on the advice of USO volunteers, he walked south, past the Liberty Memorial. The Memorial was a limestone building completed in 1926 to commemorate the efforts of the allies in World War One. Further south was Penn Valley Park, a 131-acre preserve of wooded hills, small lakes, and picturesque drives. At the western edge of the park, up on a ridge, Walt found a bronze, life size statue titled 'The Scout,' depicting a bronze Indian on his pony, one hand shading his eyes as he stares out over the City's business district. The Scout was modeled after the son of Kicking Bear, an Ogallala Sioux, by sculptor Cyrus Dallin from Utah.[1]

As Walt made his way back north, he passed through Kansas City's Mexican residential district. Working his way along Main Street, he wandered downtown Kansas City around 12th Street, where he likely stopped for lunch. After eating and window-shopping, he wandered across the Missouri to the terminal of Kansas City Municipal Airport. He was tired then so he caught a city bus headed back to Union Station. He went back to the USO as darkness fell, spending the rest of the evening eating supper, talking with other service men, resting, reading, and listening to the radio. At about 9:00 p.m., Walt retrieved his duffel, walked back to Union Station, and boarded his train for California. He noted that this train, AT&SF train No. 1 was named 'The Scout.' He wondered if it was named for the proud Indian he had seen depicted earlier in the day in Penn Valley Park.

Departure of the Scout, Train 1, was at 10:00 p.m., and shortly after, Walt settled in to try to doze off to sleep. However, it was New Year's Eve, and it is very likely that the train was packed with many other young men in the military who were determined to ring in the New Year while wide-awake. It is unlikely that many on that train were asleep, when 65 miles out of Kansas City, just after departing Topeka, Kansas, the clock stuck midnight and the conductor announced it was the New Year. Later that evening as he fell asleep, Walt reflected on all he had experienced during the previous 12 months, and he wondered what the next 12 months would reveal to him.

As Walt finally awoke after dawn on New Year's Day, he discovered himself traveling southwest across central Kansas, having passed Wichita several hours earlier. Soon they entered Oklahoma, passing through Waynoka and Woodward, an area of rolling, broken country, dotted with cattle ranches. Just as darkness fell, the train entered Texas, passing through Canadian and Amarillo. Daylight on Sunday, January 2 came just before they reached Gallup, New Mexico. This was more classic western scenery that Walt associated with the Western movies that he had loved as a youth and still enjoyed then. Throughout the day the train gradually gained altitude, past Winslow, Flagstaff, and Williams, with a branch line to the north to the Grand Canyon. This was a region of deep valleys and canyons, and men became aware of changes in the sound the wheels made on the rails as the route left the ballasted desert or mountain right of way to cross some chasm via a long, high bridge.[2] Early the following morning the train entered California, and then paused at Barstow at 2:15 a.m. Walt may have slept while the Santa Fe switched a number of cars out of train 1, including the one Walt rode in. These cars were quietly moved to a siding then reassembled as ATSF train 9, while train 1 sped off for Los Angeles. Just after 4:00 a.m., train 9 departed Barstow bound for San Francisco. If Walt woke early he got a glimpse of the great Mojave Desert and the new little station at a dusty place called Muroc.

Past the town of Mojave, the train began to climb steadily, curving to the left and then to the right again and again as it gained elevation to reach Tehachapi Pass. More tight turns and steep down grade brought the train into the great Central Valley of California. Across the great flat valley, cotton fields stretched for miles into the distance. The train paused in Bakersfield briefly, and then was off again, past Corcoran, Hanford, and Laton. Later, Walt observed miles and miles of vineyards as the train finally neared Fresno, arriving at 12:15 p.m., just in time for lunch. This was Walt's stop, and he and the few other men destined for Hammer Field got off here. After washing up at the Santa Fe station at Tulare and Q Streets, Walt had lunch in a diner near the station.

Fresno had a population of just over 52,000 by the early 1940's. Known as the center of the world's major raisin producing region, the land was flat in all directions, the only height being attained by a compact downtown core comprised of modern, tall office buildings. In addition to the Spanish-speaking element in the city, the other major ethnic group in the early 1940's was a significant Armenian presence. Walt may have noticed a number of boarded up Japanese restaurants and shops, but their owners had been gone for more than a year and a half now, following the internment of Japanese-Americans after Pearl Harbor.[3]

It was now Monday, January 3, 1944. After lunch, Walt caught the bus that ran from various points in downtown Fresno out to Hammer Army Air Field, about five miles northeast of downtown. The U.S. Government began acquiring land in 1941 to develop an airbase near Fresno. It was named for Lt. Earl M. Hammer, the first California Army aviator to achieve an air victory in World War One, and the first California aviator killed in air combat over Germany.

Eventually covering over 1600 acres, Hammer Army Air Field served during World War Two as an ordnance storage area and night fighter-training center.[4] It was also used as a base for B-25 Mitchell bombers that patrolled the California Pacific Coast from 1942 into 1944, on watch for Japanese submarines and surface ships.[5] By 1944 Hammer Field had also become a placement depot for holding military personnel awaiting assignment to various Army Air Corps bases and as a staging base for B-24's on training flights throughout the Western U.S. This was Walt's reason for being at Hammer – waiting for orders for his next phase of training.

Once on the base, Walt handed in his orders and reported for duty.[6] He was assigned to the 4th Air Force Replacement Depot, Squadron G, Barracks 109.[7] A disturbing rumor going through the base the day he arrived was about a B-24 from nearby Muroc Field whose 10-man crew had been killed on a training flight just that afternoon.[8] There was also talk about four B-24's that had taken off from Hammer Field but that had been lost in training accidents the previous November and December.[9] Walt and the other men arriving at Hammer to wait for orders wondered what they were getting into! Flying bombers seemed to be dangerous, even without Germans or Japanese trying to shoot them down!

Walt's time was spent mostly waiting for his orders to come through, with daily calisthenics and minor base work details. On Wednesday the 5th he visited with the base doctors for a physical exam. He was certified as fit for 'Class 1' flying duty, qualified for flights up to an altitude of 36,000'. He also received a smallpox inoculation and an anti-tetanus booster shot.[10] This was in preparation for his next assignment for bomber crew training.

The following day he found time to write out postcards, including one with an aerial view of downtown Fresno to his sister back in Detroit:

> Hello Genie,
> Well here I am in my base in California two days and haven't seen any sun yet. Haven't much to say but wish you send the pictures as soon as you get them cause I would like to see them. Write more soon.
>
> -Lize[11]

The reality of winter in California's Central Valley is reflected in Walt's card to his sister. He was expecting sunny California winter days as depicted by Hollywood. But Fresno was not Hollywood. The common tule fog and chill air of a Fresno January must have been a disappointment after months spent in Florida and at Biloxi the previous year, but it was still better than a Detroit winter. Temperatures in Fresno in January ranged from highs around 55 during the day, down to 40 or lower at night. There were no clear days at Hammer, instead most days started with fog or drizzle, possibly burning off to hazy sun in the afternoon.

On Friday, January 14, Walt saw orders posted on the order board, instructing him and more than 100 other men to turn in their bedding the following day and report with their duffels for transport by rail to Muroc Army Air Field. Walt reflected as he walked back to his barracks, that it was now exactly one year since he had been inducted back in Detroit the previous January. Where would he be another year from now?

Muroc Army Air Field, California

Walt and his comrades had breakfast early on Saturday the 15th and then they rode by bus from Hammer Field to the Santa Fe depot in Fresno. The men boarded a local AT&SF passenger train before dawn, and then caught up on their sleep as the train made its way south. Around Noon, the train stopped at the tiny platform at Muroc, east of the town of Mojave in the California Desert, and they got off. Muroc station was originally a shipping point for products from mines located out in the surrounding desert, but transiting military personnel was its primary significance for the Santa Fe now in 1944. The men shivered as they left the train, for if anything, Muroc station seemed even colder than Fresno, though less cloudy and no fog. The scene they saw north and south of the tracks was desolate in the extreme. They had been deposited on Muroc Dry Lake, a barren, hard-packed playa, or dry lake, that stretched perfectly flat for as far as their eyes could see, many miles in every direction. The hard-packed lakebed had been used for automobile speed trials over the preceding decades, but since the mid-1930's it had been acquired as the site of a U.S. Army Bombing Range.[12]

The men spied a bus waiting nearby with an NCO with a clipboard standing at the door and realized this was their ride onto the base. They boarded and soon began their journey about seven miles south into the heart of Muroc Army Air Field. They drove through what appeared to be a sandy wasteland, with a sparse covering of sagebrush and Joshua trees, and were soon deposited in front of the base headquarters at Second and 'C' Streets, where the men filed in and presented their orders. Saturday, January 15, 1944, Walt had officially arrived at Muroc Army Air Field.[13] Walt's orders assigned him to the IVth Air Force Bomber Command, 382nd Bombardment Group, 539th Bombardment Squadron.[14]

The 539th Squadron, commanded by Major George H. Felton, was itself new to Muroc. Through November 1943, the 539th had been assigned to bomber phase crew training at Mountain Home Air Base near Pocatello, Idaho. On December 5 the Squadron departed by troop train from Pocatello for Muroc. The squadron spent most of December setting up in their new quarters at Muroc. The first

539th training crews had arrived on December 29. By the end of January 1944, the 539th would have 54 crews organized into five flights under ground school and flight instruction.[15]

One by one the 539th adjutant, Lt. F. J. O'Connor, called the names of more than a hundred new 539th Squadron men. O'Connor's clerk handed each man whose name was called new orders. Walt's orders indicated that he was now assigned to "…duty which involves regular and frequent participation in aerial flights until relieved by competent authority."[16] Walt would begin frequent and regular flights soon, as part of bomber training crew #430.

After initial processing, Walt and the other men were assigned barracks in a separate operations area of the base, about three miles to the east. This was the HBC, or 'Heavy Bomber Command' area of Muroc, on the edge of Rodgers Dry Lake, a portion of the Mojave Desert, located about 100 miles northeast of Los Angeles. One of the other men who had come over from Hammer Field that morning, Jack Riley, had been assigned to crew 430 also, so Walt and Jack agreed to try to find their quarters together. Before boarding the bus with their gear to head over to the HBC area though, Walt had time to walk over to the base post office and send a telegram to his ma and pa back in Detroit:

To Mr and Mrs Babinski
HELLO FOLKS ARRIVED AT NEW BASE MY ADDRESS IS SGT WALTER BABINSKI
ASN 36559500 382 BOMB GP 539 BOMB SQDN MUROC AAF MUROC CALIF
SON= WALTER.[17]

Muroc, which Walt was to call home for the next 10 weeks, had been established as an air base because of Rogers Dry Lake, which covers an area of 44 square miles. In prehistoric times the lake contained water year round, but changing climate patterns dried up the lake hundreds of years ago and left it devoid of vegetation and perfectly flat and hard. The Army Air Corps first came to the lake in September 1933, when then Lt. Col. H. H. 'Hap' Arnold was looking for a remote site for bombing practice for his bomber squadrons from March Army Airfield in nearby Riverside, California. A small detachment of enlisted men from March established a tent camp on the edge of the lake and then laid out bombing and gunnery targets on the dry lakebed.

The facility was initially known as Muroc Bombing and Gunnery Range. One problem was that the base was so remote, that no officer would agree to be based at the site, so for many years an NCO, Sgt. Harley J. Fogleman, commanded the facility. The small detachment of enlisted men would maintain the tents that would be used when bomber detachments from March would land on the lakebed for a week or so of bombing and gunnery practice. The range personnel would score the crews on their performance on the targets they had laid out on the desert, and also provide them with food in mess tents. Living conditions for the visiting crews were austere for their week or so on the base, and the permanent Muroc detachments must have watched with envy when the crews departed after their temporary stay. In 1937, the entire U.S. Army Air Corps fleet of 300 bombers and many fighter squadrons arrived at Muroc for the largest American air maneuvers to date, but then once again the quiet of the desert returned after the last aircraft departed. In June 1940, with war raging in Europe, the U.S. Government completed acquisition of all of Rodgers Lake for Army use, but not much else changed.[18]

In December 1941 however, things at Muroc began to change quickly. Within just a few weeks after Pearl Harbor, Muroc had been renamed Muroc Army Air Field and designated as a key site for final training for fighter and bomber crews. On December 10, 1941, the first groups of Hudson and Liberator bombers arrived at Muroc. They began flying patrols along the California coast by the end of the year. Thousands of men poured onto the base, which had been home to less than a hundred at the beginning of the month. The isolated location, flat lakebed, and usually clear skies above were the key assets of the site for its mission of preparing aircrews for combat.

Muroc started out in 1942 by training entire squadrons of bombers and fighters for deployment overseas. For example, in July 1942, then Lt. Col. Curtis LeMay arrived at Muroc to begin training the 305th Bomber Group of B-17 heavy bombers. Years later LeMay recalled those early days at the base: "Muroc had no hangers and we were forced to do maintenance at night, when the planes had cooled sufficiently to keep one from being burned on contact." Later, squadrons of B-24's, B-25's, B-26's and P-38's were formed up and trained at Muroc. By the end of 1942 the base population reached 6,400, comprised of both permanent base personnel, as well as transient crews of fighter and bomber trainees.[19]

The commanding officer at Muroc was Col. Frank D. Gore. He oversaw the construction of five day-bombing targets, two night-bombing targets, and two fighter ground gunnery ranges. The permanent base organization included base headquarters, air base squadron, range maintenance, medical, guard, finance, signals, quartermaster, band, and aviation units.[20]

By late 1943 and early 1944, the prime mission of Muroc had changed to training individual replacement fighter pilots and bomber crews who would fill out units already in combat areas. By January 1944, more than 50 bomber crews were being trained and processed through Muroc each month.[21]

This was why Walt and Jack had been sent to Muroc. Known as 'bomber phase training,' this was the process of taking individual men who had been trained as pilots, bombardiers, navigators, flight engineers, radio operators, or as Walt and Jack had been, airplane mechanics and aerial gunners, and turning them all into a cohesive crew. The training would consist of three individual one month long 'phases' to assess first how the men performed as individuals, then as members of a B-24 crew, then as members of a larger team comprised of multiple planes.[22]

As Walt and Jack arrived at the HBC area, they found neither a blade of grass nor a tree. The HBC area was remote from the Muroc Base HQ area, and was largely self-sufficient though austere for the needs of the bomber phase training crews. To the south was Rodgers Dry Lakebed, which served as the runways for the squadrons of B-24's assigned to the base. At the edge of the lake was the flight line, with flight control and flight operations buildings, a control tower, and a number of hangers for aircraft maintenance work. Just north of the flight line was Circle Drive, with various buildings for ground school, training, and flight crew briefings. Further north, on either side of 7th Street, were located the PX, the Crew HQ building, the Flying Officer Mess, the HBC Enlisted Men's Mess, Base Theater No. 2 and the NCO Club. Off to the east of 9th Street was the HBC area multi-denominational chapel.[23]

East and west of 7th Street and north of the ground school area were rows and rows of, not barracks, but square tents, referred to as Dallas huts, which were the standard living quarters for bomber aircrews. Walt had not been assigned to Army living quarters as primitive as these Dallas huts since he had been at Harvard Army Air Base in Nebraska the previous April. A Dallas hut was the name for a wooden floored and wood framed structure covered with thick oiled canvas for walls. The huts were about 15 feet square, with a door and a couple windows for both light and ventilation. Each hut provided housing for four to six men in double bunks. They were provided with a central coal stove for warmth and bare electric bulbs for light at night.

Walt and Jack had been provided with directions to find the Dallas hut assigned to the six enlisted men of what was to be bomber phase training crew #430. Walt and Jack found the hut and discovered that a couple of the crew 430 enlisted men had been at Muroc for almost three weeks already. By the end of the day the other two men had arrived and the six sergeants spent time together at the mess hall and back at the hut doing introductions and getting to learn a little about each other.

Clifford J. Llewellyn, from Winona, Minnesota had been at the base for almost three weeks. Soon known as 'Lou' by the rest of the crew, he was the radio operator and also qualified as aerial gunner. Cliff had worked in Milwaukee, Wisconsin before the war, first in a Swiss cheese plant, then in a lamp plant, and finally in a glove factory, before enlisting in October 1942. Cliff had gone through Air Corps radio operator school at Sioux Falls, South Dakota, then aerial gunnery training at Nellis Army Air Field in Las Vegas, Nevada.[24] He had two weeks leave after graduating from Nellis to visit his family in Winona, before reporting to Hammer Field, then Muroc in December 1943. Cliff, or 'Lou', was an expert at Morse Code. This was a critical skill for the radio operator, as voice broadcast by radio could be received no more than 200 miles. However, the B-24 could fly much farther than that. Morse code could be received anywhere around the world. In combat areas the Morse code key was changed every 24 hours to confuse any enemy radio operators trying to listen in.[25]

Edward R. Bretherton was the flight engineer and also qualified as aerial gunner. He was born in 1916 in Portland, Oregon, but his family moved to Waterloo, Iowa. The crew knew Bretherton by the nickname 'Jupe'. Bretherton had also been at Muroc for a couple weeks already. Along with Cliff Llewellyn and three officers, he was part of the minimum five-man 'skeleton' crew needed for B-24 test and training flights.

Jack B. Riley, who Walt had first met at Hammer Field, was from Raleigh, North Carolina. Like Walt, Riley was qualified as an airplane mechanic and aerial gunner. Leonard W. Sherman, another aerial gunner, was from nearby Los Angeles. The crew of enlisted men was rounded out by aerial gunner Francis M. (Frank) Mansir, from Buffalo, Texas, a tiny farming community mid-way between Houston and Dallas.[26] Mansir was the oldest of the crew by about three or four years. He had been regular army before 1943, when he volunteered for the AAF. Because of his longer service he would usually be a little more 'cocky' with officers, especially later in the South Pacific where it was very informal anyway.[27]

It turned out that Cliff and Jack Riley were the only two Protestants on the crew, the rest were all Catholics. Cliff was the only man who did not smoke, but no one gave him a hard time about this...he just never picked up the habit.[28]

As the men introduced themselves to each other, Cliff Llewellyn mentioned that he had an older brother, Lorian, who had volunteered for the Air Corps in early 1942 and was already a B-24 crewman based in the Pacific theater. Assigned to the 7th Air Force, 11th Bombardment Group (Heavy), 98th Bombardment Squadron, Lorian had sent back a number of accounts of life as a flight engineer and aerial gunner on a B-24 to his family that Cliff passed on to his new crewmates over the coming weeks. Cliff also passed around a clipping from his hometown newspaper with a picture of Lorian and a story about how he was credited with shooting down a Jap Zero fighter in a fierce bombing raid over Tarawa Island on September 18-19, 1943. He also recounted other details of life as a member of B-24 combat crew that Cliff had learned from letters that Lorian had sent home to his family. While each man on the crew had relatives or friends in the service, they were all very interested in learning as much as they could from Cliff's stories about Lorian to get a sense of their future life when they went into combat in a few months' time.[29]

Just before lights out that first day at Muroc, Walt read a copy of the camp newspaper, the *Muroc Mirage*. He learned that just a few days before, General 'Hap' Arnold, Army Air Forces Commander in Chief, had visited Muroc on an inspection tour. Arnold had been instrumental in the development of Muroc as a training facility, so his visit was a doubly special event for the base and for the General. Major General William Lynd, Commander of the Fourth Army Air Force, of which Muroc formed a part, also made a separate inspection visit to the base.

Of more immediate interest to Walt and his hut-mates, the *Mirage* reported news of an upcoming USO Camp Show at to the Muroc Post Theater on Monday, January 24, titled 'Come What May.' The *Mirage* reported that the show would feature 'plenty of pulchritude' and would be free of admission. In the athletic news, Walt read that one of two Muroc basketball teams, the 'Whacky Rabbits,' was in first place in the desert League, with upcoming games on base against Yermo Depot on next Thursday and against the Douglas Aircraft team on Sunday the 22nd. All the men discussed another article about U.S. Eight Air Force fliers in England who had begun wearing specially designed steel-plated armor vests (weighing 120 pounds each) to protect them against enemy flak and gunfire.[30] They fall asleep that night wondering about the danger they were getting in to (and the need for 120-pound armor vests)!

Early the following morning, after chapel and breakfast, the newly arrived men were bussed back to the Base HQ area for dental and medical exams and to get their immunizations brought up to date at the dispensary on 'A' Street. Walt checked out OK and only needed a cholera vaccination to bring his immunizations current.[31]

Llewellyn and Bretherton, as part of the 'skeleton crew' had begun flight training in B-17's over the previous weeks. The 'skeleton crew' was comprised of the pilot, copilot, navigator, flight engineer, and radio operator. These five men represented the minimum required to fly a heavy bomber. They had spent the previous weeks training as a unit before the other aerial gunners and the bombardier arrived. Now, on Sunday January 16, the six enlisted men met all the officers and crew 430 was complete.

Lt. Charles W. McRae, the pilot, was from Arcadia, Florida, a town of about 4,000 in an open prairie-like cattle-grazing area in the south-central section of the state. The copilot was Lt. Donald W. Aubrey, from Norwich, Connecticut, a small industrial city on the Thames River upstream from New London. He had started his military career training as a P-38 fighter pilot, but the Air Force later decided that tall pilots were too large for the P-38 cockpit. Being well over six feet tall, Aubrey had been given the choice to learn to fly the B-24, or give up his pilot training altogether.[32] Aubrey was considered by women to be exceptionally good-looking. Lt. Edward T. Dunne, from Brooklyn, N.Y. was the navigator. The last member of the crew was the bombardier, Lt. Jimmie W. Clark, from Welsh, Louisiana, a small town of about 1,500 people in the state's rice belt, mid-way between Lafayette and the Texas border.[33] While all ten men formed part of one crew, there were differences between officers and enlisted men. Both during training and later in combat in the South Pacific, officers' quarters were separate from enlisted men. Officers also had separate mess facilities and their own officers' clubs.

Lt. McRae, the pilot, quickly let the men know that *he* was in charge. They were to be known as his crew, the McRae crew. Their coming days and weeks would be focused on becoming a smoothly functioning combat unit, each man knowing his job and enough of the jobs of many of the other men, so they could take over if an emergency occurred in the air. Lt. McRae wanted his crew to be the best in the 539th Squadron and he would be hard on any man who could not pull his weight. The men would also be having a series of refresher courses over the coming weeks, to ensure that their mechanical, flight operations, and gunnery knowledge and skills met standards for air combat crews.

Lt. McRae told the men that the routine at Muroc for HBC trainees like them was a cycle of seven or eight days of training and flight operations, followed by two days of leave. Each training cycle was supposed to consist of five 4-hour ground school periods, three 6-hour training periods, and six flying periods of six hours each. McRae warned the men that there would be roll-call a half hour before each ground school class and an hour and a half before any scheduled flying missions, and there would be hell to pay for any man from his crew who was late. They were starting training on Sunday, January 16, so their first two-day leave would begin on Sunday, January 23. For men on two-day leave, the 539th squadron had established a bus convoy system running from the base to Santa Monica, near Los Angeles, and return.[34]

Lt. McRae also informed the men that their first training flight as a crew was scheduled for the following day. It was getting late at this point, so after McRae dismissed the crew, the men made their way to the Mess Hall for food, then headed back to their Dallas hut to turn in.

The following day, Monday the 17th, the four new men of the 'McRae' crew were issued gear they would need for high altitude flying. Primarily this consisted of sheepskin jackets and trousers needed to keep the men warm during high altitude flights, as well as leather flying caps and sunglasses.[35] The McRae 'skeleton' crew had been training in B-17's before, but now they were all part of the 539th and they would fly B-24's exclusively in the future. Briefings, equipment checks, checkout of the B-24 itself, operational orientations, all took more time than the men had assumed initially. On their first flight, as on all subsequent training flights, McRae would carefully check each man's equipment, and chew out anyone who forgot even a single required item or who failed to follow the proper pre-flight procedures.[36] There was further delay when it was realized by McRae that there were no 'facilities' out on the flight line, and the men had to walk back several blocks to 'go' before departing.[37]

Today they would be flying in a B-24-D, an early version of the Liberator bomber built by Douglas, Consolidated, and Ford. This variant of the B-24 did not have a power nose turret or a bottom ball turret when manufactured, though many had these added later. The B-24-D they were flying in for that first flight had likely been returned from overseas combat, with bombs painted on the side to indicate the number of missions flown, and swastikas, to indicate the number of enemy fighters destroyed.

Just entering the B-24 was a challenge for the crew. The bombardier, navigator, and nose gunner entered by crawling on the ground on hands and knees and climbing up through the nose wheel opening in the fuselage. The other crewmembers entered through the open bomb bay in the middle of the plane, about three feet off the runway. Inside the plane they would stand upright and then step onto a narrow catwalk then move to their individual position. The pilot, copilot, radio operator and engineer would move to their positions forward, while the other gunners would take positions aft.[38]

Because it was their first flight with a full crew, McRae and Aubrey would also have an instructor pilot along to assist them on this flight. Days were still short in mid-January, so by the time the crew was aboard and had finished preflight check, there was little day light left. They finally lifted off at around 3:00 p.m.

As they gained altitude, the men went about familiarizing themselves with their personal equipment and the plane under flying conditions. This was not classroom learning now but real experience. Unlike a modern commercial passenger plane, the B-24 did not have a pressurized cabin and in fact had open widows for the side gunners, so it was windy and cold in many parts of the plane and noisy from the sound of the engines, propellers, and the wind. Each man observed the restrictions on their movement while wearing their heavy sheepskin flight jacket and trousers, as well as gloves. They were each wearing warm leather flight caps with integral earphones and they also plugged in microphones to the plane's intercom system. They then each practiced voice checks with Lts. McRae and Aubrey on the flight deck. As they continued to gain altitude they were instructed to put on and activate oxygen masks. Air Corps tests had shown that above 12,000 feet, aircrew performance deteriorated without an artificial supply of oxygen. The B-24 had an operational service ceiling of 32,500 feet, so use of oxygen by the men was critical to make use of this high altitude capability. As time permitted, the men looked out at the scene outside the plane. The Sierra Nevada to the north, the scattered lights of small desert communities below, the more concentrated light from the Los Angeles basin to the southwest, and the quickly setting sun more directly west.

Soon dusk came and the men switched on dim interior work lights, as the flight deck illuminated exterior running lights. This was part of their training program also, to become familiar with all the technical details of their airplane in the dark, cold, windy world high above the desert. For that first flight, there was no specific destination, although the pilots did fly to a pre-determined flight plan. Lt. Dunne, the navigator, tried to keep track of their location, taking star sightings in the dark, doing calculations at his desk, and relaying locations to McRae, Aubrey and the instructor pilot on the flight deck. Sgt. Bretherton, Jupe, the flight engineer, spent time on the flight deck too, standing behind the pilots and keeping an extra eye on all the gauges to look for the first sign of trouble or a reading out of normal expected ranges. Sgt. Llewellyn, Lou, the radio operator, kept at his radio station, receiving messages from Muroc ground control and other stations, and sending back pre-arranged messages at set intervals. Lt. Clark, the bombardier, checked out his bombsite equipment in addition to his other personal gear. Walt and the other gunners checked on guns, ammunition supplies, first aid equipment, life rafts, as well as a variety of remote hydraulic and electrical gauges.

Finally, after about four hours in the air, the instructor pilot announced that it was time to return to base and land. He nudged McRae out of the pilot seat and took control for the tricky nighttime landing. The landing went without a hitch on the broad, flat lakebed just before 8:00 p.m. Walt had completed four and a half hours in the air, including three hours of night flying.[39]

The following day the crew had a debriefing about the flight with the instructor. He outlined what had gone well and where the crew's performance and training needed improvement. On Wednesday the 19th, the McRae crew made another training flight, this time in a B-24-J. This was the current combat version of the Liberator bomber. Key differences between the B-24-D, which the men had flown two day before, and the B-24-J, were the addition of power nose and lower ball turrets as standard equipment. These two turrets had been added to give the B-24-J a total of four (nose, tail, top, and lower ball turrets) each armed with two .50 cal. Browning machine guns. In addition, the B-24 was armed with two additional machine guns, which were fired, one each, from the left and right fuselage waist openings. This gave the B-24-J a total defensive firepower of 10 machine guns. One of the things the McRae crew would have to work out was which gunner would be assigned to each of the gun position. That Wednesday, they flew with an instructor pilot again. This was a shorter flight, just two hours, but it allowed the instructor to complete his evaluation of the refresher ground training the crew would need over the coming days before more intensive training flights began.[40]

On Friday, January 21, a rumor spread through the base that the first contingent of WAC's had arrived at Muroc. It turned out to be true – Lt. Jane Musser, a research bacteriologist and chemist in civilian life, along with eight enlisted WAC's had arrived. However, they were assigned to duties in the base headquarters area and the men in the HBC training area had little opportunity to enjoy the female presence. The following day more somber news spread through the base when it was learned that seven aircrew had been killed in the crash of a B-24 out on Rodgers Dry Lakebed, when two of their engines failed on takeoff.[41]

This news had more immediate relevance for crew 430, as they contemplated the many additional training flights they would have to complete before their days at Muroc were finished. More than once they wondered as they observed or listened to McRae and Aubrey, '…how good are they as pilots? Do they know what they're doing?' Jupe had special access to the cockpit as flight engineer, so the other NCOs and perhaps Dunne and Clark as well would have wanted his opinion on their ability. Did they seem to know what they were doing? Did they ever argue or seem confused? The B-24 was considered to be the most difficult plane to fly in the Air Corps inventory. As Lt. McRae sat in his pilot's seat, he faced 27 gauges on the panel, 12 levers for the throttle, turbocharger, and fuel mixture (three levers for each engine), 12 additional switches, plus various brake pedals, rudders, and the wheel, or 'yoke' which was as large as a truck steering wheel.[42] The lives of the men of crew 430 would all depend on the competence of their pilot, as well as on a lot of luck.

Crew 430 had been informed that they would have more than two weeks of ground school and evaluation before their next flight training, although the 'skeleton' crew would continue to fly more frequently. These 'skeleton' crew flights allowed Lts. McRae and Aubrey to gain more familiarity with the flight characteristics of the B-24 and to practice formation flying. They also gave Lt. Dunne more practice honing his navigational skills. Likewise, Sgts. Bretherton and Llewellyn, as flight engineer and radio operator became more familiar with their critical flight duties from the more frequent flying.

Jupe Bretherton and Lou Llewellyn of course were also aerial gunners. Their gun positions on a B-24 crew were determined by their other technical ratings. The flight engineer, Jupe Bretherton for the McRae crew, was always designated the top turret gunner. The reason for this was that the top turret was easy to man quickly and was conveniently located just behind and above the flight engineer's position on the flight deck, aft of the pilot and copilot seats. The radio operator, Lou Llewellyn for the McRae crew, was always designated as one of the side or waist gunners. Again, the waist gun positions were easy to man, so this allowed the radio operator to move from his radio station quickly if the plane came under attack.

For the other four gun positions on the B-24, the four aerial gunner crewmen would have to work out who would be assigned to each. There was the second waist gun position. That was the easiest from a technical standpoint. There were also three other turrets: the nose turret at the very front of B-24-J models and later, the tail turret at the extreme rear of the aircraft between the twin tails, and the lower ball turret, at the bottom of the fuselage. These three turrets were more complex to operate and more difficult positions to man while in the air. Sgts. Mansir, Sherman, Riley, and Babinski would be evaluated for competency at each one of these positions over the coming weeks, but to a large degree, the crew worked out themselves who would be responsible for each on the McRae crew.

Life for Walt and the other HBC crewmembers was different from at their previous bases. While there was still PT – physical training sessions most days, the routine was easier. There was no marching in formation. Each man was just responsible to report for ground school, PT, or flying missions on time. Mess hall was different too. There was no KP for the HBC crews (unless a man committed a serious infraction) and the mess halls themselves were operated on a 24-hour around the clock basis. This was because training flights took place around the clock as well, and when the crews were preparing for a flight or after they had returned from a long flight, they just had to eat.[43]

On Saturday, January 22nd Walt picked up a copy of the *Muroc Mirage*. He quickly skimmed over an article about Gen. Arnold praising the efficiency of the Fourth Air Force. If they were so damned efficient, why didn't they have a latrine out on the flight line here at Muroc? Other articles covered the services provided by the Red Cross staff to airmen, the wonderful job the base surgeons had done saving an injured man's hand, and the new chaplain at Muroc. Walt always enjoyed a good laugh, so he was amused by the story of two G.I prisoners being escorted by an armed guard in New Jersey. The guard fainted and fell to the ground. The prisoners "...jumped on the M.P. and snatched his gun. Then they picked him up and carried him to the dispensary, telephoned the provost marshal and asked to be sent another guard because 'this one isn't any good any more.'" They saved the guard and also had their sentences commuted.

Of more immediate interest to Walt was news of a variety of events on base that he might be able to take advantage of. He read about a big basketball game at the Muroc Base HQ area gym that evening. The Muroc Whacky Rabbits team was playing the top visiting Lockheed team, which included Jerry Gracin, a former all-American star from the University of Southern California. The following two days, Sunday and Monday, the 23 and 24th of January, Walt had leave time. He may very well have stayed on base, because on Sunday, Jack Benny was doing a special radio broadcast, along with his wife, comedienne Mary Livingstone, Rochester, Benny's valet, Phil Harris with his band and singer Dennis Day. Movie star and pin-up girl Alexis Smith would also be appearing. The show was to start at 1600, with lines allowed to form no earlier than 1430 for free admittance to the Base HQ Post Theater, located next to the Gym.

However, there was more for Walt's amusement. The *Mirage* reported that a USO Camp show would be offered at the Post Theater on Monday the 24th at 1830. The bill, nowhere near the same caliber as the Jack Benny show the night before, included: the Wen Hai Troupe, a two man, two woman Chinese juggling act; Marian Burroughs and her violin; Catherine Westfall and her troupe of life size puppets; Carole Dexter, a popular east coast nightclub singer; Russel and Ferrar, two 'clever' girls with song satires; and Hypnotist Howard Klein, who would demonstrate mass-hypnotism. The *Mirage* suggested, "...perhaps this will solve the K.P. situation."[44]

On Tuesday January 25, ground school resumed for the men of crew 430. For the NCOs of the crew, there was an extensive gunnery practice and turret training facility at the base, even if they could not complete the flight line latrines on time. A line of concrete posts mounted .50 cal. machine guns for target firing and practice firing in short, controlled bursts. Special concrete foundations also mounted actual turrets - top, nose, and tail - so the gunners could become familiarized with their operational controls and also practice firing from within the turret. Stout metal frameworks were used to mount ball turrets also, to simulate being suspended below a plane and to allow the men to experience the operational and firing characteristics of this position.[45] Each of the gunners tried each turret. They would try them each again later when they resumed flight training.

Walt's home at Muroc, the HBC training area located at the southwest edge of Rodger Dry Lakebed, was exclusively a heavy bomber phase training facility. However, the men of crew 430 knew that something unusual was going on elsewhere at the base. When they had been bussed south from the tiny Santa Fe station at Muroc on their arrival, those who were awake and alert would have noticed a well-traveled side road leading off to the east across the desert. What they did not know was that this rough road led to a highly secret portion of Muroc, known as North Base or the North Hanger Area. Eventually every man at Muroc would see a small strange aircraft streak by at low level or high across the sky. The noise this aircraft made was strange also, for there was none of the distinctive whine made by a propeller driven plane powered by a reciprocating gas engine.

What the men were seeing was the top-secret XP-59A Airacomet, America's first experimental jet fighter. The plane had been manufactured by Bell at their aircraft factory in New York State, and then shipped by rail under tight security in an automobile boxcar to a siding at Muroc. The XP-59A flew for the first time on October 1, 1942 at Muroc, under the controls of Bell test pilot Bob Stanley. XP-59A test flights continued until February 1944, when Major Everett Leach flew the plane to a new American altitude record of 47,700 feet. However, the Air Force concluded that the XP-59 was underpowered and the plane did not have potential as a combat aircraft.[46]

As the men of crew 430 began arriving in December 1943 and January 1944 another new secret test program had begun at North Base. This was the Lockheed XP-80 Shooting Star. This jet was first trucked over from Lockheed's Los Angeles design center and flown for the first time on January 8, 1944. Frequent tests continued after this first flight, eventually proving that the Shooting Star, with a top speed of 490 m.p.h., was a combat capable fighter jet. By the end of the war in 1945, P-80 fighter squadrons were being formed and trained at Muroc's secret North Base. Crew 430 made frequent sightings of the XP-80 from the ground and while in the air, during their remaining weeks at Muroc.

One story about the XP-59A that McRae and Aubrey heard from the 539th instructor pilots, and that they shared with the other men of crew 430, had to do with one particular test pilot. This was a guy from Bell by the name of Jack Wollems. He had a real sense of humor and on many occasions, while testing the XP-59 in the air, he would put on a gorilla mask and derby hat, fly next to a B-24 on a training flight, then turn and wave at the pilot and crew.[47]

Late in January, Len Sherman invited the NCOs from crew 430 to spend their next two-day leave with his family in Los Angeles.[48] Early on Tuesday, February 1, the six men took the bus headed into L.A. As they left the west gate from Muroc, they noticed a number of buildings at the side of the road just outside the base with a number of cars parked in the dirt either side of the entrance. This was a honky-tonk nightclub known as the Happy Bottom Riding Club. Established in 1935 by a famous aviatrix named Pancho Barnes, her nightclub's business had boomed with the rapid expansion of Muroc before and after Pearl Harbor. The nightclub was always more popular with Muroc officers, because it was easier for them to leave the base than it was for NCOs or enlisted men. Most nights, but especially weekends, the Happy Bottom Riding Club was packed with officers from Muroc having a good time. Most enlisted men and NCOs with a 48-hour pass would head into Los Angeles though, where there were a wider variety of amusements.[49]

Soon Len, Walt, Cliff and the other men arrived at Sherman's house in Los Angeles, where they had a pleasant home cooked meal with the Sherman family. They visited the sites around Len's neighborhood and got a feel for southern California living. The weather was warmer in L.A. and less windy than the high desert winter climate at Muroc. Around midafternoon on Wednesday Mrs. Sherman, Len's mom, drove the men to the nearest stop for the base bus returning to Muroc. The men were always anxious to return to base before their passes expired. Anyone who returned late even a minute lost his pass privilege the next pass period. On the second offense, the man was 'broken,' or reduced in rank back to private.[50]

Three days after their return to Muroc, it was time for the entire McRae crew to resume flight training. Over the preceding couple weeks, the base pilot instructors had completed their checkout of McRae and Aubrey, and now they were cleared by the 539th Squadron instructors to fly on their own. The second phase of the three segments of phase training was now beginning, with less emphasis on individual knowledge and skills and much more focus on the men performing as a cohesive crew.

The 539th had been in operation at Muroc for over a month by the beginning of February. The squadron now had 64 individual 10-man crews under instruction, organized into five flights. Each flight had 13 crews, except for flight E with only 12. The squadrons' instructors had had several weeks to assess the level of fitness and training of the crews and their assessment was not good. Unknown to the McRae crew and the other 63 crews, the official squadron report stated: "The crews assigned to the squadron appeared to have had insufficient preliminary training. The pilots were fairly well advanced, but seemed to lack leadership ability and experience in handling crews. Most of the Engineers were particularly inexperienced."[51]

The 539th was operating under some handicaps. For the 64 crews, they were assigned a total of just 14 aircraft by the end of the month: 12 B-24-D's, one B-24-E, and one B-24-J. This mix of planes was a handicap, because it was the B-24-J model the men would fly into combat. Most of the planes were in need of maintenance, repair, or adjustments, so that factory representatives were needed to get heater and oxygen systems working properly for high-altitude flying. There was also a shortage of flying gear for the men. Nevertheless, the 539th instructors improvised to maximize the progress of the trainees with the equipment and facilities available.[52]

As training flights resumed in February, they took on something of a pattern. Typically McRae and Aubrey, along with Cliff Llewellyn as radio operator, would be informed the day before that they would be flying the following day. Llewellyn was included in the initial notification, because he needed to receive new daily radio code sheets for the following day's flight. Cliff would then inform the other NCOs, to ensure that they were not late, because most flights began early in the morning. Because of this, Llewellyn developed something of a 'mother-hen' relationship with the other NCOs, but they never missed a flight.[53]

The 'skeleton' crewmembers typically had to arrive at the flight line as early as 4:00 a.m. Later the gunners and Lt. Clark, the bombardier, would arrive for final briefing and take off. Roll call for all the men was no less than 1-1/2 hours before scheduled flying missions.

On Saturday, February 5, Walt and crew 430 completed a five-hour training flight in a B-24-D. The following day, Sunday the 6th, they flew again in a B-24-D, taking off in early afternoon and being in the air for six hours, including one hour at night and a nice nighttime landing by Lt. McRae.[54] The men were starting to gain confidence in McRae's flying ability.

Each flight was designed to accomplish various goals. Some flights were with other B-24's to practice formation flying or for airspeed calibration, primarily for McRae and Aubrey. Other flights were for aerial gunnery or for air to ground gunnery practice for the six sergeants on the crew. Bombing was practiced over Muroc or other desert bombing ranges with live bombs, or as long distance camera bombing runs on 'targets' such as airports in Los Angeles, Burbank, San Francisco, or other California targets. These flights helped hone the skills of Lt. Clark and his use of the secret Norden bombsight. On many occasions their flights took them far out to sea to test the navigational ability of Lt. Dunne, without landmarks below.[55]

One feature of Muroc Army Air Field that made it a unique Air Corps training facility was a structure known as the 'Muroc Maru.' Maru is the Japanese term for steamship. The Muroc Maru was a bombing target located out in a remote corner of Rogers Dry Lake. It had been constructed in August 1943 as a full-scale wooden mock-up of a Japanese Mogambi class heavy cruiser, complete with funnels, gun turrets, and even simulated waves sculpted from the lakebed sand.[56]

The Muroc Maru served as a bombing target for the McRae crew and for Lt. Clark. Lt. McRae would begin the high-level approach to the mock cruiser, and then Jimmie Clark would actually take over control of the B-24 via the linkage between the top-secret Norden bombsite and the flight controls of the bomber. The Norden device automatically corrected for aircraft speed and altitude, wind speed and direction, and the target location that Clark kept in view in the bombsite. The Norden bombsite was also connected to the bomb release mechanism in the bomb bay. On most of their training bomb runs the Muroc HBC crews dropped sand bags, not real bombs. Walt, often now in the ball turret, had the best position for observing the accuracy of the bombing mission and would keep the rest of the crew informed via intercom. The Muroc Maru also served as a low level-strafing target for the HBC crews. Crew 430 made a number of practice gunnery runs past the mock cruiser during February and March.

On Monday, February 7, Walt found a postcard in his mailbox from his friend Martha Kurzyniec back on Pulaski Street in Detroit, which she had mailed from Chicago a few day earlier:

Hello Lize,
Well I finally made it and I'm enjoying myself. I saw so much of Chicago, I know more about it than Detroit.
-Martha (Jerk)[57]

Walt wondered how Martha's younger sister Dorothy was doing.

The following morning at breakfast, the men of the 539th were informed that there would be an address to all the crews by Major Felton before ground school classes would begin. When they were assembled, Major Felton described an accident that had occurred the previous day to one of the crews. A squadron B-24 on a training mission had landed at nearby Palmdale Air Field late at night, when one of the crewmembers, 2nd Lt. Jack Burrell walked into a turning propeller on the plane. He suffered severe facial lacerations and a broken nose.[58] Major Felton told the men that there were many aspects of flying B-24s that they did not have control over – the weather, luck, the enemy they were soon to face – but they did have control over their own good safety habits. He told the men that their instructors would again go over safety practices around aircraft with engines running and he expected the men to shape up and be safe. For a touch of humor, he reminded them that their wives, girlfriends, and other girls they would meet on leave would not be impressed with a man who had walked into a turning propeller. He then dismissed them and told them to get back to work.

The following morning crew 430 was scheduled to fly again. The men had learned by that time that Lt. McRae had quite a temper – a 'short-fuse' if you got on his wrong side. He insisted on proper dress while on base so that his crew looked good, but he also insisted that the men were prepared with the proper clothing and gear for a flight. It was his habit to check each man's tote bag, like a big gym bag, when they arrived for roll-call prior to a training flight. He checked these bags to ensure that each man had his high-altitude jacket and pants, as well as adequate warm inner layers. He checked to ensure that each man had his parachute with harness, oxygen mask, gas mask, and many other pieces of required equipment. If any man was missing a piece of equipment, it would delay the start of the flight, and reflect poorly on McRae's own rating by the instructors. But his checking had a practical application also. A man flying without the proper equipment could endanger his own health or well-being, or in some cases present a dilemma if a man forgot his parachute with harness. If the plane developed a problem and the men had to bail out, McRae, as pilot, would be obligated to give his parachute to a man who forgot his. It did not take long for most of the men to develop the responsibility for their own gear, but Jupe and Len Sherman were exceptions, and McRae often spent time chewing them out for things they had forgotten.[59]

Also this morning, McRae reminded the men to watch out for turning propellers! All that having been said, Walt and crew 430 safely completed a 3:25 flying mission in the air on Wednesday, February 9, in one of the squadron's elderly B-24-D's.[60]

It was around this time, before one of their training flights, that the men of the McRae crew had an informal photo taken standing on the right side behind the wing of one of the 539th Squadron B-24's. Len Sherman missed this flight, but the other nine crewmembers are seen wearing heavy sheepskin lined leather flying jackets and pants. Walt and two of the other gunners were also wearing leather flying caps with goggles, to help them stay warm at the open waist gun windows. Several of the men are also wearing parachutes. A few days later, each of the men received a few copies of this photo, and Walt penned the names and hometowns of each of his crewmates on the rear side of his.[61]

The day after this latest flying mission, the men of crew 430 began their third two-day leave. Early Thursday morning Walt and a few of his friends caught the base shuttle bus that headed out to Hollywood and Santa Monica. The bus rattled south through the base then made its way both west and south towards the small desert farming community of Palmdale. Irrigation had made Palmdale a center of alfalfa farms and Bartlett pear orchards. Leaving the flat desert behind, the bus started along US Route 6, climbing up Mint Canyon and summiting the interior range at over 3,400 feet before descending toward Newhall. The bus then headed south into the San Fernando Valley where it left Route 6 and headed south on Cahuenga Highway and crossed one more range into the great Los Angeles basin. They traveled south out of the hills and into Hollywood, stopping at the Hollywood Union Bus Terminal at 1646 North Cahuenga Boulevard, with Hollywood Boulevard one block to the north and Sunset one block to the south. Walt and most of his pals got out before the bus departed again and headed southwest to Santa Monica and the beach.[62]

A lifelong movie fan, Walt wanted to explore Hollywood, the center of the motion picture industry. It was close to noon now, and he had been on the bus for several hours, so he headed south about two blocks, crossed Sunset, and came to a shabby looking clapboard building that used to be a stable. However, this was now the Hollywood Canteen at 1451 North Cahuenga. The Hollywood Canteen had been the idea of two major names in Hollywood, Jules Stein and Bette Davis. It was a USO club extraordinaire. As with other USO clubs, it offered food and a place to relax and rest during the day, but in the evening it was transformed into a nightclub featuring top talent for servicemen only.[63] But for now, Wally only wanted to get a bite to eat before exploring Hollywood.

After some coffee and donuts Walt heading out to explore. Just across Sunset from the Hollywood Canteen Walt had noticed De Valle's photo studio at 1540 North Cahuenga, where a sign advertised that they took photos of service men for family and friends. Walt couldn't resist. He stopped in and after a short wait he was led into a studio where he was given a leather jacket, white scarf, flying hat, and goggles. The photographer took a number of shots of Walt standing in front of a backdrop photo of an Air Corps P-38. Some were in his 'borrowed flight gear' and in one he was wearing his regular uniform with sergeant's stripes and Army Air Corps patches on his sleeves and his aerial gunnery wings on his left chest. He was given a claim check and told to return the following day to pick up his completed photos.[64]

Later in the afternoon he walked north to explore Hollywood. Just a district of the city of Los Angeles, Hollywood had a character that set it aside from LA and from any other city in the world. Walt came to Hollywood Boulevard, the main east-west thoroughfare of the city. Shops, restaurants, movie theaters, and office buildings line its route and luxury abounded, but the stroll down Hollywood still had a small-town feel to it. Even during the war years, many of the men and women who worked in the entertainment business could be seen, casually, but stylishly dressed - polo shirts and sports jackets for the men, women in slacks, and all wearing the ubiquitous sunglasses.

A few blocks to the east Walt came to Vine Street. Hollywood and Vine was the location of the Brown Derby Restaurant, where Walt surely saw dozens of movie fans on the lookout for celebrities at this famous tinsel town eatery. Walt then turned back west down a number of the residential side streets that parallel Hollywood. The typical residence here was very unlike what Walt was familiar with in his industrial neighborhood in Detroit. The prevailing style of house was a one-story stucco bungalow painted a pastel color with a red tile roof. Green lawns and gardens front and rear surrounded each house with a palm tree growing on the lawn and a pepper tree in the grass strip between the sidewalk and the curb.[65]

Making his way back up to Hollywood and continuing west, Walt crossed Highland Avenue and in another half block came to 6925 Hollywood Boulevard. This was the address of Grauman's Chinese Theater. The location of many of the premieres of major Hollywood movies, Walt was familiar with Grauman's from newsreels and movie magazines back in Detroit. Grauman's was designed to appear like an ancient Chinese temple, with four large obelisks embellished with oriental decorations flanking a Chinese gate, which led into the courtyard. At the end of the courtyard, a 30-foot high dragon statue stood guard above the final entrance into the theater. Walt and his buddies may have taken in a show here, but more likely they just walked around the forecourt to the entrance, looking down at the concrete walkway pavement, with the scores of signatures with handprints and footprints of Hollywood's most famous stars.[66]

That evening as darkness fell, Hollywood continued to glow with neon, searchlights, and indoor lighting from restaurants, shops, and nightclubs bathing the streets. They stopped at a bar or two for a beer or perhaps something stronger. Eventually, late that night Walt and his friends made their way back to the Hollywood Canteen where they checked in for a place to sleep. They were directed to a nearby house in a residential neighborhood, where they were assigned bunks to spend the night. Owners of many of the large houses in the area lent them to the USO to provide lodging for visiting servicemen. USO volunteer workers tended the houses.[67] Friday morning they had a big breakfast, perhaps trying to shake off a slight hangover, then wandered off to explore the area more.

Walt took a local bus down Vine to Wilshire Boulevard, where he transferred to a west bound bus. He got off at a place called Hancock Park. North across the 32 acre park was located the La Brea Tar Pits. Walt had heard of this site, a small pool of water covered with oil and with blobs of tar bubbling to the surface. Walt came here because he was intrigued by the story of prehistoric mammals – birds of prey, saber toothed tigers, wooly mammoths, ground sloths, early camels, and many other fossil species had met their demise thousands of years ago when they slipped into the pit trying to get a drink on the dry Los Angeles plain. A small building nearby displayed some of the species recovered from the pit by paleontologists from the Los Angeles Museum of Natural History and the Field Museum in Chicago who worked at the pits much of the year.[68]

After an hour or so at the pits, Walt made his way back to Hollywood and Cahuenga Boulevard. After lunch, he stopped back at De Valle's Photo Shop where he turned in his claim ticket and picked up the photos he had taken the day before. An hour or so later he made his way back to the Hollywood Union Bus Terminal to await the shuttle bus back to Muroc. One by one Walt's crewmates and other men he knew from the base showed up, each with stories about their explorations. Many of the men from another crew had splurged at one of the many Hollywood nightclubs and Walt and his pals decided to try to make a show at a top club the next time they visited L.A. Soon the bus arrived, they boarded and made their way back out into the desert. It was soon dark and many of the men dozed by the time they returned on base, signed in with the duty officer, and returned to their humble Dallas huts for sleep. What a contrast from the center of the entertainment world to their dusty home of sagebrush, jackrabbits, and coyotes.

The following day, Saturday the 12th, Walt mailed a number of the photos he had taken at De Valles back east to family and friends. One, showing him in a fighter pilot's scarf, leather jacket, and flight cap, he sent to his ma and pa with this written on the back: *"To Mother from her son."*[69]

That weekend after their two-day pass, Frank Mansir told the crew some big news. His wife Virgie, with their 14-month-old son Paul, Jupe Bretherton's wife Helen, with their 15-month-old son Dennis, and Jack Riley's wife Nancy were all on their way to Muroc to be near their husbands. The three women had written to each other while their husbands were together in aerial gunnery training in Texas. After the three men were assigned to the same crew at Muroc for their last months of training before being shipped overseas for combat, the three wives decided to drive out to California to spend as much time with their men as possible. One of the women had access to a 1938 Chevy, and they each pooled their gas ration coupons and money and headed west from Texas with the two babies.

A couple days later though, Frank, Jupe, and Jack became worried because the three women were late showing up near the base. Finally, by mid-week, they got a phone call that they had arrived via 'Route 66' and were waiting at Muroc. The three men got a pass approved by Lt. McRae to go out and meet their wives. They were overjoyed to see each other, but there was a problem – where were the women and children to live for the coming weeks while their husbands were at Muroc? Muroc and the small community of Boron further east along Route 66 were tiny communities with no real hotels or rental housing. Plus, Virgie, Helen, and Nancy were not the only wives who had followed their husbands out to this base and others during the course of the war. There were many other wives of both officers and NCOs looking for temporary housing. What to do? The men's passes were just for a few hours, so they decided that for the first night, the women and two babies would have to just sleep in the car. It could get below freezing at night in the desert during February and March, so they would have to do something soon. At least the cramped quarters in the car that first night would help keep them warm from body heat.

The three men returned to base that evening to talk over the problem with Walt, Cliff, and Len. Frank Mansir was older than the other men by three or four years and, unlike the rest of the crew, he was 'regular Army.' That is, he had enlisted before even the war in Europe had broken out, so he tended to be cockier toward officers and authority. The six NCOs were not going to be able to solve this problem, so Frank went to the crew officers' tent to talk to McRae, Aubrey, Clark, and Dunne. Once again, they were sympathetic to the problem of the three men and their wives. One of the officers remarked that during one of his visits to Pancho Barnes' Happy Valley Riding Club, just outside the base, he heard that Barnes had a number of small house trailers on her property that people had rented from time to time. Lt. McRae agreed to arrange a pass for the three men again the following day, so they could have their wives and kids drive around the base from Muroc, to Pancho Barnes place. The following day the men pooled their money and with McRae's help they arranged for the three women with the two kids to rent the trailer. It was still cramped and still cold at night, but it was livable and close to the base.[70]

While the excitement with the arrival of the three women was going on, on Monday, February 14 news came in that a B-24-D on a training flight from Muroc had been forced to make an emergency landing in Owen Valley, to the northeast. One man from the crew had been killed on landing.[71]

A constant challenge for the men of crew 430 and the other crews was keeping their 'Dallas huts' cleaned to the satisfaction of the squadron officers. For most of their weeks at Muroc, the problem was dust, but this week their problem was rain – too much rain that created oceans of mud. 539th

Squadron duty officers and NCOs urged, threatened, and cajoled the men to keep them policed and ready for the frequent inspections that occurred. The rain created another problem. There were limited recreational facilities at the HBC area of Muroc and the rain made the outdoor athletic fields unusable for many days. Fortunately, on Tuesday, February 15, the 539th Squadron's own Day Room opened for use by enlisted men. This provided a large, comfortable lounging area where the men could read, play cards, ping-pong, or pool, or listen to the radio.[72]

Walt and several of the crewmen also took an interest in a mangy mongrel dog that ran semi-wild through the camp area. He was there when they had arrived at Muroc and would be there when they left. No one was quite sure where it slept...perhaps in the crawl space under one of the wooden camp buildings, but it was a frequent visitor to the mess hall and often followed men back to their Dallas hut begging for food. The men of the McRae crew came to look for it and would always report a sighting at the end of the day. When it was reasonably warm the dog would amuse the crews with a game of fetch. He wasn't reliable though and eventually was christened 'Malfunction' by all the crews.

By this time Lt. McRae and crew 430 had decided on the last four gun position assignments for the crew. Leonard Sherman would be the second waist gunner, working opposite Cliff. Frank Mansir would be the nose gunner and Jack Riley would be in the tail turret. Walt would be the ball turret gunner. Walt was an unusual choice for a ball turret gunner in one respect – he was tall. The ball turret was the most confining of all the gun positions on a B-24 or a B-17. Each of the other three turrets was open on one side to permit easy access or exit. The Sperry Ball Turret though was a plexiglass and metal sphere that the gunner would enter from a hatch, which was then closed from behind by a crewmate. The problem was, all the men on crew 430 were as tall as or taller than Walt. Walt did not seem to mind though – he was happy to be an aerial gunner on a good crew, and he had flexible enough joints and he was not claustrophobic. Men who were familiar with the Sperry Ball Turret would always express surprise when Walt told them he was a ball turret gunner, because of his height. Many B-24 crewmen felt that '...the ball turret was the most terrifying place to fly on the plane...You're just suspended there above the earth in a glass compartment.' However, a ball turret gunner with nerves had a great view.[73]

On Wednesday, February 16 the McRae crew was assigned the squadron's lone B-24-J for a training flight. This gave each man, from the flight deck to the gunners, the chance to check out the latest version of the aircraft they would soon be flying into combat. Once airborne, Len Sherman helped Walt into the ball turret. They rotated the guns to point straight down, which moved the hatch to the top of the ball. Walt stepped in with his left foot on the seat, then his right foot into the right foot stirrup. He then moved his left foot onto the range pedal. Walt stooped into the turret and Len closed and latched the door. Settling into the turret, Walt turned on the main power switch, sight switch, gun selector switch, and firing switch.[74]

Walt was too big to wear his parachute while in the ball turret. Being in the most confined space on the B-24, some ball turret gunners were nervous about not having easy access to their chute. Some ball turret gunners would compact their chute with a baseball bat. Not great for the parachute, but it allowed a smaller man to wear it while in the ball turret.[75]

Each man checked and rechecked his controls, switches, and special equipment. Walt tested the limited range of movement he had for his legs, arms, and hands, made worse this February day by the thick warm clothing he needed to wear to protect against the freezing outside air. Walt also snatched time to gaze down through the plexiglass to the California or Nevada landscape slowly passing by thousands of feet below. He had been nervous at first, but gradually he relaxed as everything worked as it had during ground school. As he relaxed, Walt came to appreciate the beauty of the passing scene below. Time passed quickly on this flight. Soon, on Walt's signal over the intercom, Len raised the ball turret back into the plane, unlatched the ball turret hatch, and helped Walt out. Shortly after, they landed, having logged another 4 hours 15 minutes in the air.[76]

Late on Thursday, Cliff let the men know they were scheduled for another flight the following day. On Friday the 18th, as the men of crew 430 were sitting down to breakfast, prior to heading over to the flight line, news came in that five men had been killed the night before when their B-24-D was destroyed shortly after takeoff on a night training flight.[77] The accident had occurred on Runway 7 at 2315 as the B-24-D was on take-off, when one or more of the engines failed. The plane was destroyed, although the pilot, Lt. Robert N. Beeman, and co-pilot, Lt. Paul F. Cooper survived with only minor injuries. The killed included the navigator and bombardier, Lt. Donald F. Kane and Lt. George C. Worthley, plus the flight engineer, radio operator, and one gunner, Sgt. Thomas E. Livingston, Sgt. Leon C. Dethloff, and Sgt. Hugh B. Ellis.[78]

These were men that Walt and the other men of crew 430 knew, maybe not well, but it is likely that they rode the bus into LA, or played cards, or ate lunch with some of these men who were dead now. Walt and the others talked about fear and 'the odds' and just how much they could really trust Lt. McRae and his flying ability. You had to trust the co-pilot too, if for nothing else than to keep an eye on what the pilot did. The navigator could get you killed also. There were many stories of B-24's from Muroc on night flights or flying in fog or clouds that never came back, because they had flown into 'granite clouds' – some peak in the Sierra Nevada or the Rockies, due to navigator error. By this time it had become common knowledge that any man could request to be grounded 'for fear of flying.'[79] Many of the permanent base personnel at Muroc and the 539th Squadron ground staff had been bomber phase training crewmen who requested ground duty. It was just as well. Neither the Air Corps, nor the other flying crewmen wanted to depend on someone who had lost his nerves. The consensus of crew 430 though was that Lt. McRae was a no-nonsense pilot who knew his stuff. The sense was that Don Aubrey was a good pilot too. The only thing that was a little worrying was that Aubrey and McRae did not seem to get along with each as smoothly as the men would have wished. And Ed Dunne the navigator had done fine on their night flights and over water flights also, getting back to Muroc with no problem.[80]

Without much more thought, the crew proceeded with their flight that Friday the 18th. They took a B-24-D up for their first low-level gunnery training. This involved flying low over a remote site on the base where a long row of wrecked autos had been lined up for several miles on the desert floor. The instructor, sitting in the co-pilot seat, told McRae to weave back and forth, about 300 feet above the ground, first to the left of the cars then to the right. Approaching the row, McRae announced over the intercom that the gunners were to open fire on the cars when they came into view.[81] McRae almost jumped out of his seat when all six or seven .50 cal. machine guns opened fire with an effect like rapid sledgehammer blows to the plane. Frank, Walt, and Jack trained their guns from the left to the right as the plane weaved its course, while Cliff and Len took turns firing from the waist gun positions when the cars came into view on their side. The view during a strafing run for the ball turret gunner was sometimes harrowing. One vet ball turret gunner recalled a low level strafing run when sagebrush was getting caught in his twin .50 cal guns. He screamed to the pilot in his intercom '…get more damn elevation on this plane!'[82]This gave the men practice firing in short bursts as they had been trained and also practice clearing jams in their assigned firing positions. After just two passes they had expended all their ammunition and returned to base. Walt and the crew logged in 1 hour and 55 minutes flying time.[83]

The following day, Saturday the 19th, the men of crew 430 began another two-day leave. However, they didn't go in to Los Angeles, partly because they had learned that food, a hotel room, and entertainment there was not cheap and it was now three weeks since their last payday (at the end of the month) and they were likely short of spending money, and partly because Frank, Jack, and Jupe were anxious to spend time with their wives. Walt, Cliff, and Len offered to baby-sit for the two little boys – Dennis Bretherton and Paul Mansir – so that Jupe and Frank could spend some pleasant time with their wives. None of the men other than Jupe and Frank had experience babysitting, but the women coached them and kept an eye on them at first and they quickly figured out what to do and what not to do to keep the babies happy. The wives took pictures of the boys with the men, as they laughed, joked, and played together. Many years later Virgie Mansir would remark 'how cute it was to see Walt and the other men of the crew with Paul and Dennis.'[84]

The arrival of the three wives with their two children and the problems and challenges this presented helped turned crew 430 from a group of young men who had just met into a cohesive team. Lt. McRae, despite his short temper and 'by-the-book' approach, and the other officers had showed that they sympathized and supported the men. The six NCOs, from north, south, east, west, big city, and little town, demonstrated that they would go to bat for each other. Little by little every man grew to both trust and care for each of the other men in the McRae crew. This bonding would be as important as their formal Air Corps training for their success and survival as a crew during the coming year.

As the men of crew 430 were enjoying their weekend leave on and near the base, they may have also noticed an unfamiliar type of aircraft in the air above Muroc. This was the new Convair XB-32 Dominator, a twin tail, heavy bombardment airplane, much larger than a B-17 or a B-24.[85] It was neither the first time nor the last that the men would see unusual and unfamiliar planes in the air above Muroc.

Muroc Army Air Field, California:
The Heavy Bomber Command tent area can be seen in this view, with the flight line shops behind them and Rodgers Dry Lake Bed stretching out into the distance. Walt and the enlisted men of Crew 430 lived in one of these tents during January-March 1944.
Edwards Air Force Base.

Muroc Army Air Field:
The 'Muroc Maru' is seen in profile 'cruising' across Rodgers Dry Lake while a low flying bomber completes a strafing run.
Edwards Air Force Base.

Muroc Army Air Field:
The Oasis Trading Post was the NCO club at Muroc, the place for Walt and the men of crew 430 to play cards, get a haircut, drink beer, and relax.
Edwards Air Force Base.

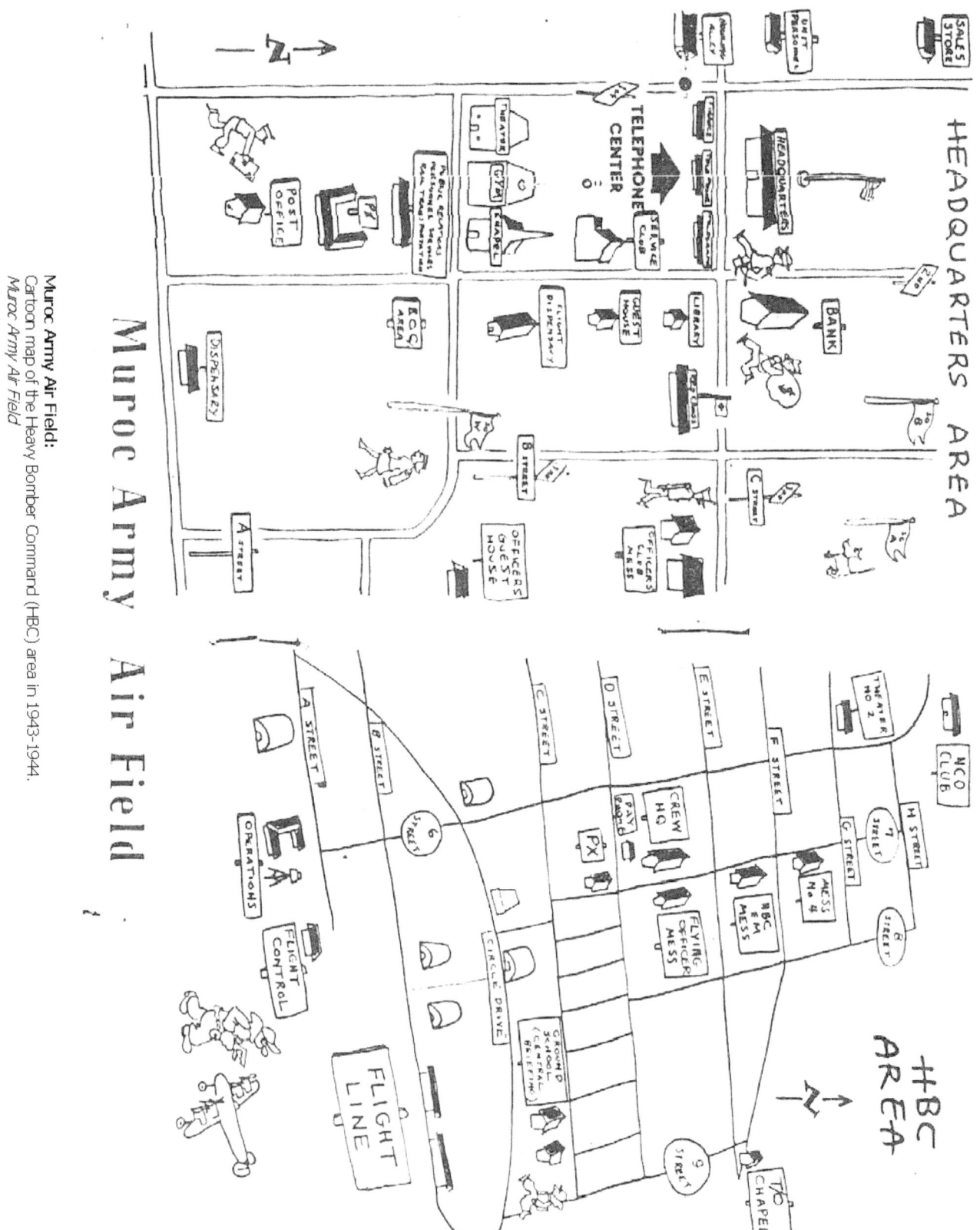

Muroc Army Air Field:
Cartoon map of the Heavy Bomber Command (HBC) area in 1943-1944.
Muroc Army Air Field

Hollywood Canteen USO: Located across Cahuenga Boulevard from De Valle's Photo Shop.
Babinski Family Collection.

Grauman's Chinese Theater: Walt visited here to see the hand prints and signatures in cement of his favorite Hollywood stars.
Babinski Family Collection.

Earl Carroll Theater: Walt and Len Sherman spent a night-on-the-town here in March, 1944.
Babinski Family Collection.

Sgt. Walter A. Babinski, February 1944: Walt had this photo taken at De Valle's Photo Shop (Located on Cahuenga Boulevard in Hollywood) in February 1944. He sent it home to his family with the caption: "To mother from her son."
Babinski Family Collection.

Earl Carroll Theater: Walt (center), Len Sherman (left), and unknown friend enjoy the Earl Carroll show in March, 1944.
Babinski Family Collection.

Muroc AAF, March 1944:
Walt posses with Clifford Llewellyn in the Muroc HBC tent area.
Babinski Family Collection.

Muroc AAF, March 1944:
Walt posses on a cool March day in the Muroc HBC tent area.
Babinski Family Collection.

Muroc AAF, March 1944:
Walt posses with two friends from Detroit, Charles Burden and Russel Cross, in the Muroc HBC tent area.
Babinski Family Collection.

Muroc AAF, March 1944:
L-R: Jack Riley, Walt Babinski, Clif Llewellyn, Ed Bretherton & Frank Mansir pose in front of the 1938 Chevy.
Babinski Family Collection.

The Bretherton Family: Muroc AAF, March 1944:
Helen, young Dennis & Ed Bretherton pose at the 1938 Chevy behind Pancho Barnes place.
Babinski Family Collection.

The Mansir Family: Muroc AAF, March 1944:
Frank, young Paul, & Vergie Mansir pose at the 1938 Chevy behind Pancho Barnes place.
Babinski Family Collection.

The Riley Family: Muroc AAF, March 1944:
Jack & Nancy Riley pose at the 1938 Chevy behind Pancho Barnes place.
Babinski Family Collection.

The Leonard Sherman Family: March 1944:
His sister, Leonard Sherman, & his mother pose somewhere in sunny California. Based on the vegetation, the location could be near Muroc.
Babinski Family Collection.

539th Bombardment Squadron (Heavy) McRae Crew – Ready for a Typical Training Flight:
Muroc AFB, California – March 1944:
Standing (L-R): Sgt. Jack Riley, Sgt. Walter Babinski, Sgt. Frank Mansir, Sgt. Edward Bretherton, Sgt. Clifford Llewellyn
Kneeling (L-R): Lt. Donald Aubrey, Lt. Jimmie Clark, Lt. Charles McRae, Lt. Edward Dunne
Babinski Family Collection.

Sgt. Babinski - Ready for a Training Flight:
Muroc AAF, 1944:
Walt poses in full gear at the waist gun position of a B-24-D before take-off. He has his leather flight jacket opened to reveal the harness for his parachute, buckled over a 'Mae West' life jacket. His intercom connection can be seen for communication with the rest of the crew.
Babinski Family Collection.

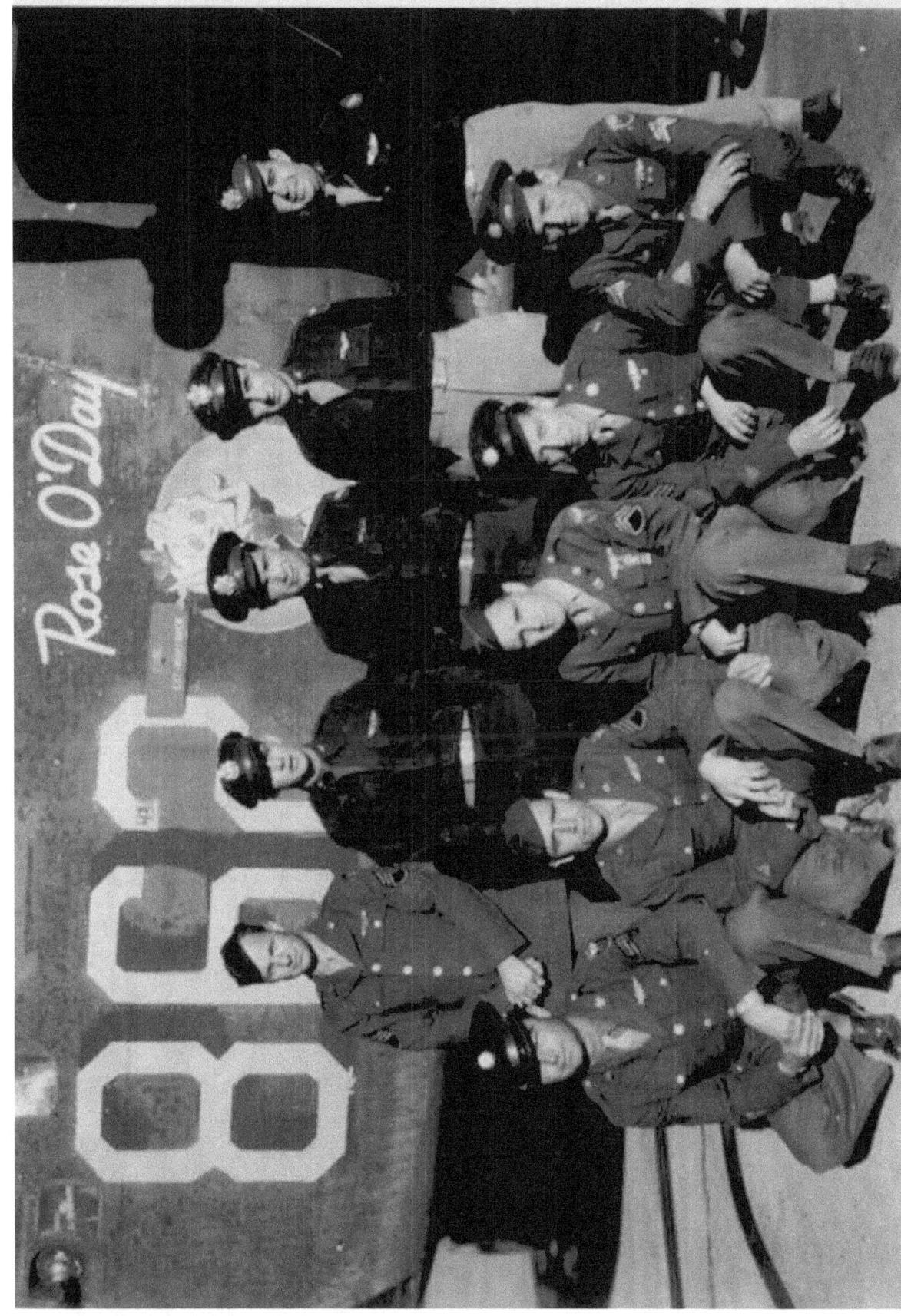

539th Bombardment Squadron (Heavy) Crew 430 – Ready for Overseas Deployment: Muroc AFB, California – March 1944: Standing (L-R): Sgt. Edward Bretherton (Flight Engineer); Lt. Jimmie Clark (Bombardier); Lt. Edward Dunne (Navigator); Lt. Donald Aubrey (CoPilot); Lt. Charles McRea (Pilot); Kneeling (L-R): Sgt. Leonard Sherman (Gunner); Sgt. Frank Mansir (Gunner); Sgt. Clifford Llewellyn (Radio Operator); Sgt. Walter Babinski (Gunner); Sgt. Jack Riley (Gunner) *Babinski Family Collection.*

Muroc AAF, 1944: Flight line in the HBC area.
US Library of Congress, Yale Collection.

Muroc AAF, 1944: Chow line in an enlisted mans mess hall (above). Air crewman in a Dallas hut donning high-altitude gear for a training mission (right).
US Library of Congress, Yale Collection.

Honolulu, Hawaii Territory 1944: Typical street scene that Walt and Jack Riley would have seen on leave waiting for their ATC flight out of Hickam Field.
We'll Say Goodbye

Honolulu, Hawaii Territory 1944: Walt purchased these post cards as mementos of his brief stop in Hawaii in April 1944.

Canton and Fiji Islands: Cartoon *from WWII US Military Newspaper.*

Guadalcanal Airfields 1944: Walt and Jack Riley landed at Henderson Field (foreground) on April 11, 1944)
Photo courtesy 307th Bomb Group Association

Guadalcanal 1944: B-24s on the steel mat on Guadalcanal air fields that startled Walt and Jack on their first landing (left) and 307th BG Dispensary where Charles McRae first had his ear infection treated (above).
Photo courtesy 307th Bomb Group Association

Monday after their leave ended, heavy rains began which closed Muroc Field. The dry lakebed runways had cost the Air Corps nothing to build, but nature could make them unusable. Finally on the Tuesday the 22nd the rain began to decrease and by the following day the field began to dry out.[86] A couple days later the normal dry weather had returned the lakebed to a usable state for takeoffs and landings and the 539th and other training squadrons at Muroc resumed flying operations.

Friday, February 25 the McRae crew flew a B-24-D on a long training flight to practice formation flying. They were to learn that formation flying was not just for show, not just to put on a good impression when viewed from the ground. Proper formation flying was a life or death matter for bombers under attack by enemy fighters. Veteran pilots back from the European Theater briefed the crews that a single B-24 or B-17 was a sitting duck for one or two aggressive enemy fighter pilots. However, a whole squadron of twelve bombers in tight formation presented enemy fighters with not ten .50 cal. machine guns to contend with, but 120. A whole Group of four squadrons could present almost 500 guns to attackers. Like wolves, enemy fighter pilots would concentrate their attacks on slow or damaged planes, or those that could not fly tight formations. One B-24 pilot remembered:

> *"'Get in close! Get in close!' demanded the instructors. 'Don't leave room for fighters to fly through your formation. Keep it tight so your gunners give each other mutual fire support. The enemy always picks on the sloppiest bomber formations first.' One day…one of the student pilots pulled his bomber in so close that the right outboard prop chewed off the lead plane's vertical stabilizer on the left side. Both pilots landed their planes safely…"*[87]

With tales such as this in mind, all the planes of the 539th Squadron practiced formation flying long and carefully that day, while the instructors slowly urged McRae and the other pilots to adjust speed and direction and altitude as a unit with the other planes in the group. The gunners were surely nervous as they saw the other planes get closer and closer, and occasionally lurch dangerously near. They could look at the gunners in the neighboring planes, see the expressions on their faces, and think about what formation flying would be like while under attack. The gunners also thought about advice they had been given by their gunnery instructors, to be cautious about firing widely while in formation. While the power of massed gunnery was a major threat to an attacking enemy, they also did not want to shoot down a neighboring bomber – or be shot down themselves. Finally they landed, Walt and crew 430 having logged 6 hours and 5 minutes in the air.[88]

The frequency of training flights increased during the last week of February and into March. The types of missions increased as well. In addition to formation flying, they went on night flying missions, they practiced takeoff and landing procedures, and they practiced bombing, using sandbags with a small detonator to indicate the location of their hits. Later they practiced with live bombs dropped over remote targets at Muroc or at other bombing ranges in the Nevada desert. Practice with live bombs trained the gunners on the safe method of inspecting the payload for proper stowage in the bomb-bay by the ground crews, and ensuring that safety pins were in place and secured with cotter wires to prevent premature detonation. The gunners also practiced removing the cotter wires and safety pins as their plane approached the bombing run. In combat, dropping bombs that had not been properly armed for detonation on enemy targets was only slightly less of a problem than premature detonation before the bombs were released. Another type of mission the crews practiced was formation bombing from an IP. A squadron or two of planes would fly in loose formation to an IP or the 'initial point' of a bombing mission, where they would then make a sharp, sometimes as much as a 90 degree turn. The IP was an arbitrary location that might be 50 to 100 miles from a target, designed to deceive enemy observers of the actual intended target. Once past the IP, all the planes would then fly in a tight formation to the target, both for protection from fighters, plus to follow the most accurate bombardier in the squadron to increase the likelihood that the group's bombs would all hit the intended target in a tight ground pattern.[89]

To complicate this increased tempo of training activity, in late February aircrews were informed that the Muroc bombing range would be off limits during designated portions of the day due to top secret bombing tests being conducted over the next several weeks. Little did the men realize that these secret tests would have a direct impact on most of them a year and a half in the future.[90]

On Saturday, February 26, the McRae crew was assigned a B-24-D, for a training flight that logged Walt and his crewmates 3 hours of flying time.[91]

One day in late February Cliff Llewellyn got some bad news in a letter from home that shocked everyone on the crew. Cliff's mom and dad had written him to let him know that they had received a telegram from the War Department saying that Lorian, Cliff's older brother, had been reported 'missing in action, January 20, 1944.' His B-24 had been last reported at 0305 over the Marshall Island, with a radio report stating 'Plane badly damaged, wounded on board.' At 0319 a 'key-lockdown' signal had been received at the 11th Bomb Group base on Tarawa Island from the plane until the signal ended. That was the last heard from the crew. Cliff, as radio operator, explained to the other men that 'key-lockdown' was standard procedure for radio operators when a plane was in danger of crashing. The locked-down radio operator's telegraph key sent a constant, clear signal back to home base to help get a 'fix' on where the signal came from to help any rescue attempts.

Despite his calm description of what 'key-lockdown' meant, Cliff was clearly upset by this news. The other members of the crew, including all the NCOs and Lts. McRae and Aubrey told Cliff they would 'go to bat' for him to try to get him deferred from overseas combat duty. This was not unheard of under similar circumstances. Cliff thought about this a little while, but then decided that he would stick with the crew. Things were starting to move fast and they all knew that their training would be completed soon. Cliff thanked his friends for their concern, but let them know he was OK and would be sticking with them. Once again, an incident out of the ordinary and how the men reacted had fostered enhanced crew cohesion that would be important later in combat.[92]

Letters between Cliff and his mom and dad passed back and forth over the coming weeks, expressing both support and the slim hope that Lorian had survived the loss of the plane and might be rescued at some point in the future.

Two last short training flights ended out the month for crew 430. On Monday, February 28, the McRae crew took up one of the squadron B-24-D's for a 2-hour flight. The following day, Tuesday, February 29, they flew in another B-24-D for a 3 hour 30 minute training mission. For the month of February, Walt had flown a total of 35 hours, 10 minutes, including 1 hour of night flying.[93]

At the end of February, the 539th Squadron reported that training was progressing well, with most crews successfully completing the second stage of phase training, except for the bombardiers who were behind on bomb simulator training. The squadron reported that living conditions were still far from optimal, the Dallas huts being very difficult to keep adequately cleaned. "In spite of these conditions, it was felt that Squadron morale was good." Special efforts were made to ensure that all the crews continued 'PT,' or physical training. Basketball, baseball, and volley ball games were organized for every crew. A volleyball court was built by the 539th Squadron near the parachute building and was very popular with all the crews during free time on base.[94]

At around this time, the three wives, Nancy, Virgie, and Helen decided that they would visit Virgie Mansir's cousin who lived in Needles, about 200 miles east on the Colorado River. With the tempo of training flights increasing, it had become more difficult for Frank, Jupe, and Jack to get off base to visit their wives. The trip to Needles would give the women with the two boys a change of pace until the men got their next 48-hour pass. The only problem was that they were short of ration coupons and couldn't buy enough gas to make it to Needles and back. Shortly after that, one night back in their Dallas hut, Walt overheard the three men talking about how they had 'liberated' two 5-gallon cans of gas to ensure that the women could get to Needles and back.[95]

On Wednesday, March 1 the McRae crew started another 48-hour pass. The day before had been payday so the men all had money to spend. However, Frank Mansir, Jack Riley, and Jupe Bretherton were planning to spend the two days with their wives, while Cliff Llewellyn did not feel like leaving the base because of worry about his brother Lorian and his mom and dad. Walt and Leonard Sherman decided to take the base bus to Los Angeles and head back to Hollywood. Three weeks before when Walt had visited Hollywood on an earlier two-day pass, he had gotten the idea of spending an evening at a swank Hollywood nightclub. That evening Walt, Len and a number of airmen from other Muroc crews attended a show at the Earl Carroll Theater and Supper Club. Located at 6230 Sunset Boulevard, the theater was only about a half mile from the North Cahuenga Boulevard Bus Terminal and the Hollywood Canteen USO Club.

New York impresario Earl Carroll, who owned two top nightclubs on Broadway in New York City, had built his new Hollywood theater in 1938, converting an acre of vacant land into a dazzling nightclub in just 75 days. The modernistic façade of the building became a Hollywood landmark, with a 20 foot high neon image of Beryl Wallace, one of Carroll's 'most beautiful girls in the world,' and a 'Wall of Fame' of 150 inscriptions in cement of Hollywood's most glamorous stars. These were enough to draw Walt and Len to the theater. Once there, they learned that for just $3.30 they would have dinner with reserved seats for a view of the show with 'sixty of the most beautiful girls in the world.' They could also dance to two famous bands, watch a CBS radio broadcast, and enjoy a meal cooked by world-renowned chefs. How could they refuse?[96]

Walt and Len bought their tickets early in the afternoon then went back to the USO to relax until later. That evening, looking good in their dress uniforms, they returned to the theater, dazzled by the exterior neon artistry and the luxurious interior. They enjoyed drinks and a good meal, then sat back for the show with the '60 beautiful girls' dancing on a 60-foot wide double turntable on the 80-foot wide main stage. Earl Carroll himself choreographed all his shows and he was a master of mechanical stage wizardry. Other features of the show included three swings that lowered dancers from the auditorium ceiling, an elevator that raised dancers onto the stage, and a rain machine to further awe the audience. During an intermission, while waiting for refills on their drinks, a pretty photographer came by to snap their picture as a souvenir.[97]

Late in the evening, after the last act and last drinks, the men made their way back to a local hotel where they slept long into the following day. After a late breakfast, they wandered the Hollywood streets a little while longer or perhaps just relaxed at the USO, before boarding the bus for the long ride back to Muroc.[98]

At the beginning of March, each of the crews of the 539th Bombardment Squadron began the third and final phase of 'phase training'. This was the last stage of training before crews graduated and were assigned an aircraft to fly to an overseas base. Training flights honed the skills of every member of each crew, but special attention was paid to perfecting the skill of the gun turret operators. The crews also began more frequent bombing practice with real bombs. One aspect of dropping bombs from a B-24 was the special view of the process that the ball turret gunner had. Walt and other ball turret gunners could look forward to make sure that the bomb bay doors were opened, then count the bombs as they were released to make sure they all dropped OK. On rare occasions a live bomb would not release properly, because one side of the B-7 shackle to which the bomb was attached would catch. No one on the crew liked to land a plane with live bombs, so one of the gunners would have to walk out on the narrow catwalk over the open bomb bay doors with a screwdriver, to free the stuck shackle so the bomb would drop away.[99]

More of the flights were at altitudes above 15,000 or even 25,000 feet, as the ground crews gradually solved problems with oxygen equipment on the B-24's. Another special operation the aircrews practiced during phase three was operating special auxiliary fuel tanks. These were designed to fit into a portion of the bomb bay, to extend the range of a B-24. With some work they could be installed or removed in about a day. Later in combat, the McRae crew would use these often. They allowed the B-24 to fly much further than its design range, but they also reduced the capacity of the bomb load that could be carried. The men practiced flying with these tanks and operating the transfer pumps to shift fuel from the auxiliary tanks to the wing tanks as they became emptied in flight. These flights with auxiliary tanks also began to condition the men for the even longer flights that they would encounter within the next few months.[100]

On Friday, March 3, crew 430 was assigned a B-24-D for the first of their third phase training flights. This was likely a long distance flight to an arbitrary 'IP' and then assembly with a number of other aircraft for a mock formation fly-in to a hypothetical target. Each of the crewmembers would have gone through a long checklist to ensure that they performed each of their duties properly. Lt. McRae touched down back at Muroc after 6 hours and 45 minutes in the air, followed by the normal post-flight critique and debriefing.[101]

Even though training would end soon, the instructors of the 539th Squadron felt a great responsibility to ensure that all crews completed all the required ground school and flight training. Additional aircraft were assigned to the 539th during the month, bringing the total to 16 B-24-D and three B-24-J models. This was still not many aircraft for use by 65 crews, even though the planes were each scheduled to fly three or four times per day if the ground crews could keep them airworthy. The 539th Squadron instructors reported a high degree of enthusiasm and cooperation by the 65 Heavy Bombardment crews. This was likely because each man and each crew knew that their survival and success in combat someplace overseas would depend on the skills and abilities they developed in the friendly skies above California. By the end of March, the 65 crews of the 539th would fire 85,000 rounds of training ammunition and drop 1,300 live 100-pound practice bombs.[102]

Walt had somewhat more ground training to do than the other NCOs. As the designated ball turret gunner, he was at a disadvantage, because most of the B-24-D model aircraft did not a have a ball turret fitted. The B-24-J models came with a ball turret as standard equipment, but the 539th started the month with just one J model and only acquired two more by the end of the month. The HBC ground training facility at Muroc did have a number of Sperry ball turrets mounted in a special hangar on a raised framework for instructional purposes. The 65 ball turret gunners from the squadron would spend extra time training on these, which operated the same as a B-24 mounted turret, except that the guns, of course, did not fire. Walt and the other ball turret gunners took their time training in this facility seriously, because without it they would be behind the other gunners in airborne experience with their designated gun position.[103]

This was unfortunate, because the ball turret was the most challenging turrets for a gunner to master. On the few training flight he made in a ball turret equipped B-24 while at Muroc, Walt's routine started after take-off. While on the ground the ball turret on a B-24 had to be stowed almost completed retracted into the fuselage of the plane. The bottom ground clearance under the B-24 was insufficient for the turret to be lowered while the plane was on the ground, unlike the B-17, which had higher ground clearance. The B-24 ball turret was stowed with the twin .50 cal machine guns horizontal and pointed to the rear of the plane so they would fit into special gun wells in the bottom of the craft. Once in the air, the ball turret gunner, assisted by one of the waist gunners – usually Len Sherman on the McRae crew – would activate a hydraulic switch to lower the ball turret onto the azimuth ring, which support the turret and allowed it to rotate a full 360 degrees in a horizontal plane. Then Walt would rotate the turret in the vertical plane to lower the two guns so they were pointing straight down. This was necessary to bring the turret access hatch door up into the aircraft.

Len would then hold the hatch open while Walt entered the turret. Once in the turret, Len would lower and lock the hatch and Walt would be on his own. When the guns were pointed straight down, the gunner's position was almost facing directly down at the ground. When the guns were brought up to the horizontal position, Walt was almost lying flat on his back. Walt controlled movement of the turret with two control handles, mounted in front slightly forward of his body. Walt rocked the control handles forward and back to raise or lower the turret and guns. Tilting the handles to right or left moved the turret horizontally. The triggers for the guns were on top of the control handles. Each trigger controlled one gun, but there was also a switch that allowed both guns to be controlled by one handle, in case a gunner's arm was injured in combat.

Located between the two control handles was the Sperry Automatic Computing Sight. This automated sight performed a number of calculations on distance to target, speed, azimuth of the guns, etc., and made minor adjustments to the aim of the guns. The Sperry sight helped make the ball turret an effective defensive position on the plane, but it required much skill to operate properly.

Exiting the turret also required the help of the waist gunner or another crewman. Walt would lower the guns to the straight down position, and then Len would unlatch the hatch door and help Walt out. In an emergency, if the hydraulic or electrical system failed, Walt had small handles to rotate the turret manually. Once out of the turret, the guns were rotated horizontal and to the rear, and it was retracted hydraulically back up into the plane. It took both teamwork to man the ball turret and self-confidence and a spirit of independence to operate it for hours alone at the lonely bottom of the plane.[104]

On Monday, March 6, the McRae crew reported for an early morning flight, assigned to a B-24-D. Two days previous, the 539th Squadron had started camera gunnery missions for the HBC crews. This involved the bombers flying a previously predetermined route to simulate a bombing mission, then a mock attack by P-38 fighters. The planes were fitted with specially mounted cameras on the guns to allow the instructors to analyze how well the gunners tracked the attacking fighters, after the films were developed. Today the McRae crew was assigned to fly in formation for a simulated camera bombing attack on North Island, the location of Coronado Naval Air Station near San Diego, about 165 miles south of Muroc. After that, they turned north and made another simulated camera-bombing run on the North American Airplane factory at Mines Field in Los Angeles. On the return flight to Muroc, the P-38's made their mock attack and Walt and the other crewmen got their first realistic idea of how quickly an attack could develop and how confusing and difficult it would be to aim their guns at skilled enemy fighters. After this and other camera gunnery or camera bombing missions, each entire crew was debriefed and interrogated to get the crew's view of the mission. This also oriented the crew on the type of actual post-combat mission debriefing they would encounter in just a few weeks' time.[105]

Walt and crew 430 logged 3 hours and 55 minutes on that training flight, but when they returned to Muroc, they were told that they were going up again. After a debriefing and something to eat and then a new briefing for their second flight, they took off again in another B-24-D. However, after just a little while in the air, Lt. McRae decided to land again. It is not known what was wrong with the second plane, but crew 430 logged just 25 minutes in the air on that second flight.[106]

All the men and their instructors knew that they had to maximize training time in the air. But the ground crews were struggling to keep their handful of aircraft operational for all 65 HBC crews.

On Thursday, March 9, the McRae crew flew again in another B-24-D. The 539th Squadron HBC crews were flying more and more long distance training missions and today was no exception for Walt and crew 430. They took off late in the afternoon and likely flew about 600 miles to a simulated target like San Francisco, Salt Lake City, or El Paso. Completing their practice 'bomb run' at dusk, they then headed home in the dark to test Ed Dunne's ability to navigate successfully at night. This was critical training, because many combat missions would either begin or end at night so that the bombers could arrive over the intended target at the correct time of day for either tactical or lighting considerations. No one rushed Lt. Dunne on that return flight in the dark, and Lt. McRae slowed down and kept plenty of altitude, until Ed was confident that his star sightings had been checked and double-checked and his computed location corresponded with their dead-reckoning position.[107] They then made a slow descent until the runway lights were sighted and they could make a safe landing. Walt and crew 430 logged six hours and 30 minutes in the air, including three hours and 30 minutes of night flying time.[108]

Successful missions like this enhanced crew 430's self-confidence. They all had an object lesson on what could go wrong a couple days later when crew 460, piloted by Lt. Frank Gengler, ran out of gas while returning to Muroc. Gengler did an excellent job of crash landing safely out on the dry lakebed 12 miles from base and no one was injured, but a crew like that would always question the reliability of the pilot, co-pilot, and flight engineer.[109]

Crew 430 got another two-day leave on Saturday and Sunday, March 11 and 12. The officers headed into Los Angeles, but Len and Walt were short on cash following their last trip to Hollywood and the Earl Carroll Theater. Walt spent a quiet weekend on base. He read in the camp newspaper, the *Muroc Mirage,* that on that Saturday evening he could attend a USO show at the Muroc Post Theater. M.C.'d by Bill Brown, the show included the Musical Johnstons, a xylophone act, Emmet Oldfield & Company, a comedy acrobatic act, the dancer 'Peanuts' Bohn, and the Three Dixon Sisters, Pat, Mary, and Joan, singing 'Night and Day.'[110]

That Saturday, before the show, Walt and Cliff Llewellyn wandered the HBC barracks area with Cliff's small camera and had their pictures taken with some of their crewmates. Walt ran into two acquaintance from Detroit, Charles Burden and Russel Cross, who were on different crews. Charles, Russel, and Walt had their picture taken together, then spent a long time talking about the training they had been through, the various bases they had been stationed at, and what their crewmates and officers were like. They promised to look each other up back in Detroit after the war. Walt had his picture taken alone, and also with Cliff Llewellyn.[111]

Sunday morning at 0900, Walt and Jack Riley (another Catholic on the McRae crew) attended mass services held at the T-O Chapel on East E Street, performed by Father Shannon, Muroc's Catholic Chaplain. Sunday morning, after mass and breakfast, Jack invited Walt to join him and the other crewmen out at the wives' trailer. The six NCO's spent a pleasant Sunday afternoon with the three ladies and the two little boys, Dennis and Paul. Virgie, Nancy, and Helen were always grateful for the visits from their husbands, and they would talk about how cute it was to see Walt, Cliff, and Len amuse the two little boys.[112] Using Cliff's camera, Len took a picture of Frank, Walt, Jack, Jupe, and Cliff standing in front of the women's '38 Chevy outside of Pancho Barnes nightclub. Later that afternoon, the men retired to the Happy Bottom Riding Club for a few drinks, before heading back onto base in time for evening mess and the end of their 48-hour pass.

When the McRae crew returned to barracks late on the 12th, they began to hear rumors that a secret study had begun at Muroc on the effects of parachuting from aircraft at high altitudes and high airspeeds. This must have made the men wonder what the Air Corps had in mind for them and what they thought about their probability of returning to base unharmed after the combat missions they would be expected to fly.[113] They also heard that the 539th Squadron had begun taking graduation photos of each crew, posed in front of one of the squadron B-24's. These were intended to be part of the official squadron history, and also as a souvenir for each crewman. This was a good indication that their training was nearing completion.[114]

On Monday, March 13, crew 430 was scheduled to have their photo taken. The men were instructed to report to the flight line in their dress uniform by mid-morning. However, after breakfast, as the men headed back to their Dallas hut to change, they noticed that the wind was picking up quite briskly. With the wind, the dust began to blow and soon the base was assaulted by an intense dust storm. Soon word came that that day's photo session was cancelled. This dust storm was so intense, that the entire field was closed to flying operations and 539th planes that were already in the air were diverted to alternate fields. The men were instructed to change into their fatigues and report to the flight line to help cover B-24 engines with canvas sheets to try to keep the dust out. The dust storm was very intense and continued into the night. The following day, as the wind gradually died down, all the men of the HBC squadrons were instructed to attend hastily organized two-hour lectures on escape procedures. These were designed to give the men some useful instruction while the ground crews checked over the planes for dust damage and performed any needed cleaning or repairs so the squadron could resume training flights later in the week. The 'escape lectures' did provide useful ideas from senior 539th Squadron officers who had returned from European combat duty on how to avoid capture if they were shot down over enemy territory and various escape methods.[115]

A couple days later, early in the morning, the men were told that they would have their graduation crew picture taken right after breakfast. Again attired in their best dress uniforms, the men were lined up by the photographer in two rows to the left of a parked B-24, just forward of the left wing. They were facing east into the rising sun, which was still low in the morning sky. The plane was one of the 539th Squadron B-24's named 'Rose O'Day,' A/C 862[116]. In the back row, standing, were the four officers, plus Jupe Bretherton. In the front row, squatting, were the five other NCO's, including Walt. Lt. McRae smiled broadly as the photo was taken, looking proud of his crew. Walt and Len Sherman were smiling also, but Cliff, Jupe, Jack, and Frank looked more serious. Did Cliff still worry about his brother Lorian, and were the other three men worried about their wives and the two little boys they would soon be leaving behind? Quickly the photo session was over and the men headed back to barracks, while another waiting crew lined up in front of 'Rose O'Day' for their photo.[117]

On Thursday, March 16, crew 430 was assigned another camera-bombing mission in a B-24-D. This was to be a very long-range mission to camera bomb Portland, Oregon and Seattle, Washington. However, after they were in the air, the mission was called back because of reports of very inclement weather over the target cities. With 14,000-foot peaks in northern California and Washington to contend with, IV Bomber Command did not want to take chances with the crews.[118] Walt and the rest of the McRae crew completed 4 hours and 40 minutes flying time in the air on the cancelled training mission.[119]

The following weekend the crew had another two-day leave. It had been apparent from their instructors and what Cliff Llewellyn, Lt. McRae, and the other officers had heard that their training was nearly completed. Walt and some of the other crew may have gone back in to Los Angeles that weekend, but it is just as likely that they remained on or near base. Frank, Jupe, and Jack spent most of the time with their wives. When the crew shipped out of Muroc, the three women, along with little Paul and Dennis, would head back east and to their various homes. The men made sure the women had all the gas ration coupons they had saved (and perhaps additional cans of 'liberated' gasoline) plus plenty of money for the return trip.

At that point, none of the men knew what their next base would be or where their crew would be assigned. From what the men had learned, HBC crews that had graduated from Muroc in the past had been assigned to both the European Theater and the Pacific Theater. They could be going to England, or North Africa, or Italy, or Alaska, or some island in the Pacific, or even to India or China. 650 men, the men of the 65 or so crews that the 539th Squadron had been training for the past nine weeks, were all wondering the same thing that weekend at Muroc.

The following Monday, March 20, the McRae crew was assigned another B-24-D for an early morning training mission. This was a long flight – actually both the last and the longest training flight of Walt's time at Muroc – 6 hours and 55 minutes. This was probably another camera bombing mission, but not a formation flight. Walt's total training flying time for the month was 29 hours and 10 minutes.[120] The McRae crew had been honed (they thought) to a high level of skill and efficiency. The men trusted each other and liked each other as well.

Later that week, instead of flying and ground school, the men went through various military processing and administrative procedures. The men were required to return their flying gear, including leather jackets and pants, goggles, and so on.[121] There were more physical exams and dental exams for each of the men. Physical training continued. As part of their overseas processing at Muroc, the men were briefed on their ability to start or increase allotment payments. These allowed a service man to voluntarily allot a portion of his monthly pay to his family. The HBC crews were told that they would be getting increased combat pay soon, and would also likely find few opportunities for spending their money at a remote combat airfield. Walt had contributed money while he was working to help his ma and pa while he lived at home in Detroit, so he signed up for a $50 per month allotment, to be sent to his mother beginning with his May, 1944 pay.[122]

Jupe, Frank, and Jack made a fast trip out to the little trailer at Pancho Barnes' place to say their last good-byes to their wives and sons. The men explained the increased pay allotments they made out to their wives. They made sure again that the women had plenty of cash for the trip back east in the '38 Chevy. There were many promises to write each other and the wives told the men to write often, be careful, and take no unnecessary risks. Last hugs and kisses were followed by tears as the men boarded the bus for the trip back onto base. While they watched them wave as the bus pulled away, the three men wondered if they would ever see their wives and sons again. Then they sat quietly as the bus bumped across the desert, through the gate and back to the HBC camp area.[123]

Finally, on Friday, March 24 the 539th began issuing clearance orders for the graduating crews. Saturday the ground school was formally closed. Crew 430 was among the first 20 of the 539th HBC crews to be ordered out.[124] Walt and the other men of the McRae crew received identical orders on Sunday, March 26, transferring them from Muroc and the 539th Squadron to the IV Bomber Command Processing Unit located at Hamilton Field, California, north of San Francisco. The 539th Squadron Operations Officer, Capt. Herbert E. Lindhe, signed Sgt. Babinski's orders.[125] This was also the end of the 539th Bombardment Squadron, for on March 31, 1944, it was deactivated and all its instructors and permanent personnel transferred to other units.[126]

How did Walt and the rest of the crew feel, what did they think, now that their training was ending? Sgt. Babinski had been training now for 14 months. He and his newfound friends would soon be putting their lives on the line. The Army Air Force had taught the men – overtly and sometimes subtly – that they were the cream of the crop of the fighting men the U. S. was preparing for the next wave to take on the Axis powers and win the war for the Allies. They had graduated from the largest educational organization ever implemented before or since. Army Air Corps pilots and crewmen did regard themselves as highly skilled technicians and professionals. Many of them had seen friends wash out of basic training, or fail to make the cut when tested for aircraft mechanic school, or fail the aerial gunnery course, or lose their nerve during bomber crew phase training. Those who made it as far as Walt and his buddies had, now walked with a swagger, proud to be masters of "...this super-toy, this powerful, snorting, impatient but submissive machine." Flying a heavy bomber gave them a "...feeling of aggressive potency bordering on the unchallenged strength of a superman and an overwhelming sense of the vastness of the universe." And each crew thought of itself as the best.[127]

Hamilton Army Air Field and Fairfield-Suisun Army Air Field, California

Very early on Monday, March 27, Walt and the other men from the first 20 crews boarded buses for the ride out to the Southern Pacific Railroad depot at Rosamond, just west of the base gate near Pancho Barnes' place. There a special troop train was waiting and the men boarded with their duffle bags in hand and orders in their pockets. They left shortly after the last of the men boarded, then headed north over Tehachapi Pass and down into the Central Valley of California. They soon passed Fresno, where Walt and the other men had arrived three months prior at Hammer Field. Their train traveled at high speed for an hour or so, then be switched into a siding as two or three long Southern Pacific freights passed. They made their way north, past Merced, Stockton, and Lodi, with the Sierra Nevada crest visible far to the east every now and then. There was a long wait in Sacramento, and then their train headed to the southwest via the SP mainline through the small farming communities of Davis and Fairfield. At Fairfield, the train was switched onto an SP branch line that traversed southern Napa County. In Sonoma County, the train switched onto a branch line of the Northwestern Pacific Railroad, then headed into Marin County. At Ignacio, the train switched once more onto the NWP mainline, for a few short miles to a small platform adjacent to Hamilton Air Field. Walt had arrived at his next base.[128]

Hamilton Field was located in Marin County, 25 miles north of San Francisco along U.S. Highway 101. Located in a broad, open valley, the base covered 928 acres of drained marshland that sloped very gently down to the northwest fringes of San Pablo Bay. The base had been constructed beginning in 1932 at a cost of $7 million. It was named for Lt. Lloyd Andrews Hamilton, a Marin County native who was shot down in France in August 1918 while serving with the 17th Aero Squadron. Built originally as a bomber base, by 1939 it was home to 900 men, 125 officers, and 60 planes. In 1940 it was converted into a pursuit plane base, but by 1943 had reverted to its original purpose.[129]

Architecturally, Hamilton Field was a much different environment from the other bases Walt had been stationed at before. It was not only relatively new, but it had also been built before the construction boom that followed the outbreak of war. Hamilton had been carefully planned and much of it had the look and feel of a pleasant California residential suburban community. Driving through the stuccoed portal gate of the base, the men traveled down a palm-tree lined road that led to the gleaming white California mission-style headquarters building. Off to one side the officers club resembled a colonial hacienda, while hangars, the hospital, mess halls, and barracks all were predominately white stucco with red tile roofs. Landscaping of green lawns with tropical shrubs and flowers completed the picture.[130]

After Walt and the rest of the crew checked in at headquarters they were assigned to barracks. Lt. McRae told the crew that they had been put 'on alert' and would be getting orders within a few days for assignment overseas. Whether it would be to Europe, the Pacific, or some other destination, he did not know yet. He told them that when the time came, he would notify Sgt. Llewellyn, who would then notify the rest of the crew. They were to check in on a day-by-day basis in the morning and might be able to get a short leave if orders to depart were not issued for that day. Then they went off to rooms assigned in the enlisted men's barracks, which were in a beautiful three-story white stucco building. HBC crews in transit found Hamilton to have a top-rate enlisted men's day room, with typewriters for the men, radios, record players with a good collection of records and a piano. The food in the mess halls was very good too.[131]

A day or two after arriving at Hamilton, shipping orders for the crew had not yet been received, and Walt and a few of the crew decided to head into nearby San Francisco, after clearing with Cliff that they would not be getting orders that day. After securing the necessary passes, Walt and his friends boarded one of the buses that made the shuttle between the base and the City. Heading south on U.S. Highway 101, the bus left Hamilton and climbed up San Rafael Hill to about 700 feet above sea level, then dropped down into San Rafael, the seat of Marin County with over 8,500 residents. Highway 101 through San Rafael passed old-fashioned brick commercial buildings and large homes set back behind green lawns and shade trees. South of San Rafael the route passed a marshy area that stretched to Greenbrae at the mouth of Corte Madera Creek. Corte Madera had just a hundred residents or so who lived on a colony of houseboats in the mouth of the creek. A little further south the highway went over another low hill then descended again to skirt Richardson Bay, an arm of San Francisco Bay, on a half-mile long redwood bridge.

Once again the highway ascended a low hill, this time crossing the summit via deep cuts and a long tunnel to emerge above the picturesque town of Sausalito. As the bus descended toward Sausalito, Walt saw the hulks of a large number of abandoned sailing ships: square-riggers, brigantines, and schooners, slowly rotting in the mud flats north of Sausalito. A community of just 3,500 residents, with a single commercial street that paralleled the Bay and residential neighborhoods occupying the steep hills that rose to the west above the bay, by the 1940's Sausalito depended on a little commercial fishing, a Coast Guard Station, and a small artist colony.[132]

One last time the bus headed back up into the Marin Hills, finally rounding a bend to bring it into view of the north tower of the Golden Gate Bridge. Walt had traveled big bridges before. Detroit's Ambassador Bridge, not many blocks from his Delray home, was the longest international suspension bridge in the world at the time. But the Golden Gate Bridge, opened less than seven years earlier, was already an icon of America. It was not just because of the engineering and construction superlatives associated with the bridge that this was so. At 4,200 feet, it was the world's longest between two suspension towers and its roadway height of 220 feet above the water the world's highest. It was not just because the bridge had been a symbol of America's ability to get itself out of the great depression while it was under construction more than a decade ago. And it was not just because of the beauty of the bridge and its ability to blend in with and compliment the beauty of San Francisco, gleaming white to the south, and the beauty of the Marin hills to the north, at that time of year mostly various shades of green.

There was something in the very location of the bridge, clearly tying northern and southern California together, where 'the palm tree meets the pine'. The bridge also afforded a view to the east and the great harbor of San Francisco, crowded with ships of war and ships of peace of all kinds, with branches of the Bay disappearing out of sight to the left and right, and Oakland and the East Bay beyond, all beckoning to the great continent beyond. Crossing the Golden Gate Bridge, Walt glanced to the right also, to the west and out into the Pacific Ocean. This was another feature of the bridge that made it a powerful American icon – it also led to a wider world beyond America's shores, to islands and lands of mystery that were much different than even the foreign lands that most American immigrants had come from. Walt's eyes fixed on that western horizon and briefly wondered what those distant lands might hold for his future not many days from then.

Then the bus jerked to a stop at the tollbooth at the north end, the driver paid the toll, and they continued into the city. Thoughts of the future and what might await beyond the bridge vanished quickly as the bus turned east into the Presidio, the old Army Base in San Francisco that looked more like a park. The bus continued down Lombard Street then turned south on Van Ness. At some point along its route, after determining where and when to catch the last bus for the return trip to Hamilton, Walt got off to begin exploring the City.

There were many things for him to see and experience...Fisherman's Wharf, working home to 2,000 men and 350 vessels that landed 300 million pounds of seafood per year. Along the wharf side, Walt found the sidewalk lined with huge cauldrons simmering over open fires, where live crabs were boiled after buyers made their selections. He saw other diners sitting at restaurant windows looking out over the fishermen's basin as the men unloaded their catch or cleaned their gear and repaired their nets. North Beach, further east and inland (no longer 'on the beach') was home to the City's vibrant Italian community, with distinctive restaurants, delis, and shops. South of North Beach, along Grant Street, Walt wandered into Chinatown. Home to 16,000 people of Chinese descent – half immigrants and half second generation or later, Walt found congested sidewalks with souvenir shops next to oriental markets with roast ducks glazed in salty wax on display, along with whole roast hogs, eels, octopi, and shark. Walt surely winced when he saw this display, as well as those Chinese produce markets with strange 12-inch long string beans, peas with sweet edible pods, bamboo shoots, bean sprouts, and lotus roots. Later Walt had a late lunch in a nearby Chinese restaurant, where he dined with relish on his 'chopsuey' made with many of the meats and vegetables he had earlier looked on with disfavor.

One thing Walt noticed about 'Frisco was that it was crawling with Navy men, because of the City's status as the major West Coast port. Soldiers or Airmen on leave felt like a minority. Looking up or down the crowded streets, it seemed like all he could see was the white caps of sailors.[133]

Perhaps a beer or two at a local bar and a cable car ride back to the Fishermen's Wharf area, and then it was time for Walt to catch the last bus back to Hamilton. It was dark when the bus crossed north back over the bridge and crept through Sausalito, past Greenbrae, and San Rafael before pulling up to the gate at Hamilton.

On Thursday, March 30, Walt and the rest of the McRae crew were ordered to gather their gear and report for transfer to nearby Fairfield-Suisun Field.[134] Fairfield-Suisun Army Airfield was located near the two small rural California communities of Fairfield and Suisun City, midway between San Francisco and Sacramento, along U.S. Highway 40 and the main line of the Southern Pacific Railroad. Planning for the base began in 1941 after the Pearl Harbor attack and initial construction was completed in 1942. Located on low, flat, marshy land near the delta of the Sacramento River, the base had been intended originally as a medium attack bomber base. It was also used by the Navy to train fighter and attack plane pilots in aircraft carrier deck landing. The outline of a carrier deck was painted on one of the runways for this purpose and the strong, steady maritime winds in the region aided in simulated carrier deck landings. Consolidated Aircraft had moved their major B-24 maintenance facility from San Diego to Fairfield-Suisun in December 1943, and many new Liberators received their final factory inspections and adjustments at the base.[135]

However, by mid-1943 Fairfield-Suisun was also beginning to specialize as a base for the Air Transportation Command, the arm of the Air Force responsible for overseas air transportation of personnel and vital military cargo, while nearby Hamilton Field continued to specialize mostly in the movement of Air Force bombers headed overseas, both to the European and the Pacific Theaters. Throughout the war, facilities at Fairfield-Suisun were crude compared to the pre-war construction at Hamilton Field.[136]

The following day there was a buzz of excitement amongst the crew, as Lt. McRae announced that they were being assigned to the Pacific Theater. When they received this news, HBC crews at Hamilton or Fairfield-Suisun would be instructed to not make phone calls nor send telegrams out of the base, and in letters home they were not to say where they were going or when they would leave. They were instructed to send their winter uniforms back home – they would not need them for a long time. They spent time to bring their wills, insurance, immunizations, and pay allotments up-to-date.

Then each man was issued his own combat flight equipment: fatigues, heavy shoes, leggings, a steel helmet, a .45 pistol and belt with ammunition, a sheath knife, first aid kit, musette bag, poncho, mess kit, and canteen. They also were issued a lightweight tropical flight suit, aviator helmet, headset and throat microphone, oxygen mask, flight jacket and trousers (it would still be cold at 20,000 feet in the tropics), life preserver, and jungle survival kit.[137]

Walt also had his medical cards (forms 206 and 8-117) checked to ensure that his inoculations were all up to date. He then reported to a special room in the quartermasters department where he had new dog tags made that reflected his completed inoculations.[138]

The following day there was more news. They had been assigned a plane that had just been released to the AAF by Consolidated. The men went down to the flight line where they saw a brand new Ford built B-24-J, unpainted and gleaming silver in the California sunshine with serial number ending in '075' painted on the side. It looked good to the men and they all hoped it would stay with them as they went into combat in the coming weeks. Walt checked out the ball turret, relieved that it was not an older D-model, like those they had done so much of their training in back at Muroc.

Because they had flown mostly B-24-D models at Muroc, the skeleton crew had to become qualified to fly overseas in the B-24-J. This meant that the five men of the skeleton crew, McRae, Aubrey, Dunn, Llewellyn, Bretherton would be doing some extra flying. The skeleton crew was also expected to check out the fuel consumption of the new B-24, by flying long routes north and south over calibrated courses. Careful readings were taken of fuel tank levels before and after these flights, which were then compared to fuel gauge levels. This information provided each plane with fuel gauge consumption correction tables, which were critical on very long flights to enable the crew to know exactly how much fuel remained throughout the flight. In coming months, the men would appreciate how critical this was for many of the long-range missions they would be assigned. Lt. Jimmie Clark, the bombardier was not needed for this qualification and fuel gauge testing process, nor were the other four gunners, including Walt.[139]

The crew remained 'on alert' at Fairfield-Suisun Field for shipment overseas, to where exactly they still did not know. There were no passes available to men 'on-alert' at this base, although crews were always buzzing about various schemes to get out and into town. After a few days, the men got the news from Lt. McRae that they would be shipping out soon in the new B-24-J, but that they would be expected to carry some combat cargo. To save weight on the plane and increase the amount of combat cargo they could carry, they were looking for volunteers to ship out early – likely just by a few days. The men who volunteered would travel by Air Transport Command (ATC) to their overseas base, to be joined in a few days by the rest of the crew. Walt and Jack Riley talked about this for a while and then both agreed that they were fed up with waiting and were willing to volunteer to go on ahead. On Monday April 3, Walt was issued a new 'Soldiers Pay Book' in preparation for shipment overseas.[140] Later that day Walt and Jack both received orders to transfer back to Hamilton Field where they would catch the first available ATC flight to their overseas base.

On Tuesday, April 4, Walt and Jack said goodbye to Cliff, Jupe, Frank, and Len, and caught a bus going back to Hamilton.[141] Two days later, on Thursday April 6, Walt and Jack received further orders – they were to pack their gear and report to the Hamilton air departure area that afternoon for a flight to Hickam Field in Hawaii, enroute to the 13th Air Force Bomber Command on Guadalcanal in the South Pacific. There they would be assigned as a complete crew to a specific squadron. To make it more official, their pay records were marked to indicate that would start drawing an additional 20% foreign duty pay starting that day.[142]

Walt and Jack boarded the waiting ATC C-87 transport. The C-87 was a transport version of the B-24-D, 13 of which operated out of the main ATC base back at Fairfield-Suisun. The C-87 could carry 12,000 pounds of combat cargo, or it could be fitted with removable seats for up to 25 passengers. The ATC flights to Hawaii originated at Fairfield-Suisun, because of that base's superior maintenance facility, but most passengers were loaded at Hamilton Field, because it was closer to San Francisco and the major base facilities around the Bay that funneled replacement troops onto the overseas flights. The ATC flights to Hawaii and beyond were actually operated by employees of a company called Consairway, a division of the Consolidated Vultee Aircraft Company. Consairway employees wore Army Air Force uniforms, but with a special insignia, so Walt and Jack may not have even realized that their pilot and crew were civilians.[143]

In the late afternoon, with all her passengers on board, the C-87 taxied away from the Hamilton terminal, headed to the end of the runway, turned, and paused. The sounds were familiar to Walt and Jack as the pilot revved the engines and then released the brakes on the plane to start down the runway. The C-87 slowly lifted off and gained altitude steadily over the Marin County pastures and then out over San Pablo Bay. The plane continued just east of south, gaining altitude all the while, until it reached about even with Angel Island, then banked to the right and headed out to the west. Walt and Jack got a glimpse of San Francisco in the early evening light, then the Golden Gate Bridge below. They heard other passengers murmur, 'there it is – I hope to see that bridge again soon.' Perhaps they saw the Farralon Islands with its lighthouse about 30 miles off the mainland, then that was it. They were headed overseas and out of the U.S. as it then was.

The flight from Hamilton Field to Hickam Field in Hawaii was well within the range of the C-87. Yet this would be the longest flight that Walt and Jack would have experienced to that point, and it was over water the entire way[144]. Many thoughts must have raced through their minds as they tried to sleep – would the pilot and plane find Hawaii OK? Would they connect again with the rest of the McRae crew at the end of their journey to Guadalcanal? What would life in the south Pacific be like and how bad would combat be?

- 8 -
SOUTHWEST PACIFIC 1941-1944: HOW LONG AND WHAT COST?

Grand Strategy and the Nature of War

On February 8, 1904, Japan launched a sneak attack against the Pacific Fleet of the Russian Navy, while it lay at anchor in its main base of Port Arthur in Manchuria. The attack came completely without warning or a declaration of war. A squadron of destroyers led the Japanese assault that day, launching a torpedo attack that sank two Russian battleships and a cruiser and killed or wounded hundreds of Russian sailors.

The very battle flag that was flown by the flagship of the Japanese destroyer squadron that made the sneak attack at Port Arthur was flown again 37 years, nine months, and 29 days later by the aircraft carrier Akagi, flagship of the Japanese Fleet that launched the sneak attack against the U.S. at Pearl Harbor. Again, Japan made an attack completely without warning or a declaration of war, resulting in the sinking of major US Navy battleships and other fleet units, the destruction of scores of USAAF aircraft on the ground, and the loss of thousands of lives.

No American was unaware of or unchanged by the attack on Pearl Harbor. Before December 7, 1941 the war was something across the ocean that, while big news, most Americans could and many did ignore. The war had been dragging on for more than two years in Europe, while in Asia, Japan had been at war for 10 years against China. The vast majority of Americans sympathized with the British and the Chinese, and even with the Soviet Union after Germany and its Axis allies had struck it in June 1941. While many American's supported President Roosevelt's growing financial, material, and moral support to the Brit's and the Chinese, very few indeed would have supported American entry into the war prior to December 7, 1941.

On Monday, December 8, 1941 that attitude had changed forever. Roosevelt's call to Congress and the nation to avenge the 'day that would live in infamy' was met with firm resolve by all Americans. War was declared against Japan, and a few days later Germany and Italy, in solidarity with their Axis ally, declared war against the U.S.

During the 13 months after Pearl Harbor until Walt was drafted in January 1943, a lot had happened in the war. It was likely very difficult for most citizens to have an understanding of the broad strategy for defeating the Axis, much less to have a clear idea of the specific plans and priorities of the Allies. Certainly the news had been bad when the U.S. entered the war and it stayed that way for quite some time. In the winter months of early 1942 the Germans were on the outskirts of Moscow and the defeat of the Soviet Union seemed quite likely. The German blitz of England continued and in North Africa, the Italians had been joined by German Panzers divisions, which continued to threaten British control of the Suez Canal and the shipping route through the Mediterranean Sea.

In the Pacific and in Asia, December 1941 and the early months of 1942 brought one major disaster after another. Not only was Pearl Harbor attacked on December 7 (December 8 in the Far East), but so also were the Philippine Islands (which of course was an American Commonwealth before the attack, being then prepared for early independence) and the British colony of Malaya. Before the end of 1941, two American island possessions, Guam and Wake Island, fell to the Japanese. The British surrendered Hong Kong on Christmas Day, Borneo was invaded, and Japanese Navy bombers sank two Royal Navy battleships trying to stop the invasion of Malaya.

In January 1942, Manila fell to the Japanese Army and the remnants of the American and Philippines Armies were cut off on the Bataan Peninsula under command of Gen. Douglas MacArthur. Malaya, with its rich tin mines and rubber plantations, had been conquered by Japan and the British were under siege in Singapore. The Celebes in the Dutch East Indies were invaded by Japan, and further east the Japanese took the Bismarck Archipelago with the important harbor and base of Rabaul from Australia.

Across thousands of miles of ocean and islands, the Allies were trying to organize their defenses. In February, a small combined British, Dutch, Australian, and American fleet under a Dutch admiral sailed out to try to stop the Japanese invasion of Java and Sumatra. The allied fleet was utterly crushed with little damage to the Japanese fleet. Nothing could seem to stop the advance of Japan.

Before the end of the month, both Singapore and Borneo, with rich oil fields and refineries, had fallen.

March 1942 brought more bad news. The British colony of Burma was invaded and its capital of Rangoon surrendered. In the Dutch East Indies, both Sumatra and Java were attacked and fell to Japan in less than a month with more valuable oil fields. New Guinea was invaded, both the western Dutch half and the eastern Australian half of the great island. Things were not going well with the defense of the Philippines. Gen. MacArthur was ordered by Washington to leave the remaining defenders in Bataan and Corregidor and make his way to Australia to head the allied defense there. That task was made more urgent by a Japanese submarine attack in Sydney harbor and the first of many Japanese bombing raids against Darwin, in Northern Australia, which would continue into 1944.

How could things get worse? In April, Mandalay in Burma fell to Japan, giving her more oil fields and allowing the Japanese to close the Burma Road, which had been funneling supplies into China to help keep the Nationalist Army in the field fighting against Japan. From Mandalay the Japanese Army could launch an attack into British India. The Japanese Navy also sailed deep into the Indian Ocean, sinking a British Royal Navy aircraft carrier and a number of cruisers, as well as merchant shipping all up and down the east coast of India and Ceylon. The Royal Navy withdrew the remainder of its Indian Ocean fleet to Kenya in East Africa to keep it out of the reach of Japan until reinforcements could arrive. There was real fear that India would rise in revolt against Britain.

In the Philippines, Bataan surrendered, with tens of thousands of U.S. and Philippino soldiers entering into years of Japanese captivity, if they were lucky. Thousands who were not lucky would die on the Bataan Death March, on their way to hellish Japanese POW camps. The remaining defenders in the Philippines were now concentrated on the small fortress island of Corregidor, in Manila Bay.

The first boost to allied morale in the Pacific occurred on April 18, when Col. Jimmy Doolittle led 18 USAAF bomber crews in B-25's off the deck of the U.S.S Hornet. Medium bombers like the B-25 had never been launched off a carrier before, which aided the surprise attack against Tokyo and several other Japanese cities. Like the Germans in 1940, the Japanese had been told by their leaders that their Islands were impregnable to attack. The bombs of the Doolittle raid, while causing little material damage, were a blow of the greatest importance to Japanese morale, and likewise a boost to American morale and allied resolve to do whatever it would take to defeat Germany and Japan.

Early in May, the last American and Philippino defenders on Corregidor, the U.S. Army fortress in Manila Bay, surrendered to the Japanese. At the same time far to the southeast, an American Fleet had been sent to stop the Japanese Navy's drive to cut Australia off from the U.S. In the Coral Sea, south of the Solomon Islands, a great naval battle took place between the U.S. and Japanese fleets. Both sides lost one aircraft carrier sunk. The U.S. Navy could afford the loss less than Japan, and requested their British allies to send a Royal Navy aircraft carrier to Pearl Harbor, to reinforce the U.S. fleet temporarily. However, the loss of the one carrier also caused Japan to pause for the first time in its drive to expand across the Pacific.

The U.S., Britain, Australia, and New Zealand knew that Japan would soon continue their attacks, but where and how to defend against them was a key questions. The grand strategy the allies adopted shortly after Pearl Harbor complicated their problem. The U.S., Britain, and the U.S.S.R. agreed that Germany, not Japan, was the major enemy and that the vast majority of resources would go to defeat the Nazis and Italy first. Then the Allies would turn their full attention against Japan. However, Japan could not be allowed to run wild across the Pacific and East Asia unopposed. Nevertheless, through 1944 the battle against Japan would get just a small proportion of the Army, Air Force, and Navy resources that the United States, Australia, New Zealand, and Britain could deliver. In Europe and North Africa, American forces built up to dozens of well-equipped divisions, thousands and thousands of bombers and fighters, and the majority of US Navy resources. The U.S. in the Pacific never had more than a few Army and Marine divisions to work with, and hundreds, not thousands of planes to try to control the skies over millions of square miles of ocean.

In June 1942, Japan launched a new grand strategy to try to knock the U.S. and Australia so hard that they could never threaten Japan again. This strategy had three parts. To the far north, they sent an invasion fleet to seize a number of the Aleutian Islands in Alaska. Their real goal in doing this was to try to draw off a major portion of the US fleet from Pearl Harbor to the north. The central part of the Japanese strategy was to seize a U.S. Pacific Island that they could use to bomb Hawaii in preparation for an invasion. In June 1942, Japan launched this attack against Midway

Island, but a scratch fleet of three US aircraft carriers was able to find the Japanese fleet and sink all four enemy carriers, to the loss of just one U.S. carrier. The result of this battle was that the situation in the central Pacific would remain static for many, many months.

Far to the south, the third prong of the Japanese strategy was an attempt to seize Port Moresby, the last Australian stronghold on the southeastern tip of New Guinea, as well as the last of the Solomon Islands. They then intended to follow up by seizing New Caledonia, Fiji, the New Hebrides and other islands that could be used as air bases to stop American reinforcements reaching Australia and New Zealand. The New Guinea attack landed on the north coast, and then over several months, the Japanese fought their way 120 miles towards Port Moresby. Finally, within 30 miles of their goal, the Australians stopped them. Then began many, many months of the Australian army pushing them back the way they came across mountains covered by steaming dense jungles.

The other part of the Japanese southern strategy began with the seizure of the last of the Solomon Islands, Guadalcanal. Soon after, they began constructing an airfield on the island, which was identified by American photo reconnaissance flights. If the airfield went operational, Japan would be well on its way to threatening the shipping routes from the West Coast to Australia. With Australia challenging the Japanese advance across New Guinea, on Guadalcanal the U.S. took the lead to stem the Japanese tide, landing the U.S. First Marine Division on August 7, 1942. However, the Marines were not strong enough to push the Japanese off the island quickly. Japan was able to land reinforcements on the Island, and the U.S. had to follow suit. For six months, the battle for Guadalcanal raged back and forth, with both sides using soldiers, ships, fighters, and bombers to try to gain the advantage. Guadalcanal was really one of three great pivotal battles that were raging across the globe at the end of 1942. In Russia at the same time the Nazi and Red Armies were locked in a life-or-death struggle at Stalingrad, while in North Africa, the British Army under General Montgomery had launched their attack against the Germans and Italians under Field Marshall Rommel at El Alamain.

When Walt was drafted in January 1943 the battle for Guadalcanal was still in doubt, but in February news came that Japan had finally evacuated most of their troops from the Island. However, during the 14 months since the war in the Pacific began, Japan had been able to ship hundreds of thousands of troops, plus thousands of planes to dozens and dozens of its island strongholds across the South Pacific. It had taken Japan five months to clear US troops off Corregidor and the battle for Guadalcanal had raged for six months. Japan hoped to win the war by making the Allies fight for each island in the Pacific, one after another like pearls on a string, to be able to get at Japan. It was their hope that the Americans and Australians would find the cost in men and machines too great as they fought their way westward.

Indeed, after Guadalcanal the bitter fighting continued island by island up the Solomons. The major Japanese stronghold in the area was the great harbor of Rabaul, on New Britain Island. Rabaul could accommodate multiple fleets of ships and Japan had ringed the base with almost 100,000 troops and many hundreds of planes scattered over half a dozen airbases protected by powerful anti-aircraft gun batteries.

As 1943 wore on and Walt continued his stateside training, the U.S. Navy, Army, and Air Corps began to organize itself in the Pacific into two separate but coordinated commands. The Central command would be primarily a Navy show under Admiral Chester Nimitz, with the goal of pushing the Japanese back through the Central Pacific, across the hundreds of small atolls and volcanic islands that the enemy had fortified. The Southern command was initially under Navy control also, with General Douglass MacArthur as Army commander. MacArthur pushed for overall control of the Southern Command. However, as he got it rivalry for resources developed between the Navy in the Central Pacific and the Army in the Southern Pacific. MacArthur's goal for the Southern Command was to push the enemy back out of Rabaul, New Guinea, and finally out of the Philippines, which would become a base for the eventual invasion of Japan.

Nimitz in the Central Pacific and Macarthur in the South Pacific both lacked the resources to win the war as the Japanese intended, island by island. Somewhat in parallel, both commands developed and began to refine the strategy that would give them the chance to succeed with the limited resources they did have. This strategy was to use the limited ground troop resources available, not to seize each and every Japanese held island, but to capture and fortify widely separated bases hundreds of miles behind the most advanced Japanese bases. Once the Americans took an island, it would be defended from counter attack and an airfield would be constructed quickly. These airfields would allow the Air Corps to base fighters along with medium and heavy bombers that could then fight the Japanese on adjacent islands for control of the air. Once the air battle in an area was won, the Air Corps could prevent the Japanese Navy from reinforcing their other island bases over many thousands of square miles. Then the process could be repeated by advancing to still more distant bases. The by-passed Japanese held island could then surrender or starve, in either case with a fraction of the cost to the U.S.

Through the second half of 1943 and into 1944, this strategic process slowly advanced. In the Central Pacific, the Americans took the islands of Tarawa and then Kwajalein. The Japanese, who tended to fight to the last man, fiercely defended both islands, costing thousands of casualties. But they provided the first advanced bases to put the new strategy in place. Tarawa was where Lorian Llewellyn had been based when his B-24 was lost. In the South Pacific, Munda in the Solomons was invaded by the U.S. in August 1943, followed by Choiseul in October and Bougainville in November.

Walt and the other men of the McRae crew heard the names of many of the islands and battles that had figured in the war through 1943 and into early 1944, but with very little concept of the overall strategy. That is the nature of war for most of the men who do the real fighting.

What was the organization of the Air Force units that would fight the war in the Pacific? The Central Command was a Navy-led operation, but it was supported by the Air Force. Nimitz and the Navy controlled the U.S. 7th Air Force, with fighter, medium bomber, heavy bomber, and transport groups. Farther south, MacArthur would come to have two U.S. Air Forces under his control, as well as various Royal Australian Air Force, Royal New Zealand Air Force, and U.S. Marine air units.

The two U.S. Air Forces at MacArthur's disposal were the 5th Air Force and the 13th Air Force. The 5th Air Force was much the larger and would tend to protect MacArthur's left flank on the south as he made his push towards the Philippines. The smaller of the two was the 13th Air Force, which generally operated on MacArthur's right flank to the north and out into the Central Pacific.

The 13th Air Force came into existence on January 13th, 1943 (the thirteenth day of the thirteenth month of the war for the U.S.). On that date, a number of fighter, bomber, and support units already operating in the South Pacific were placed under the command of Maj. General Nathan F. Twining in Noumea, New Caledonia. Eight days later Twining moved his headquarters to Espirtu Santo in the New Hebrides Islands, where it remained for a year. Also activated on January 13 were the XIII Bomber Command and the XIII Fighter Command. Units under Twining's command in January 1943 included the 5th and 11th Bomb Groups, each with four squadrons of heavy B-17 bombers, the 42nd Bomb Group with two squadrons of medium B-26 bombers, the 347th Fighter Group with three squadrons of P-39 and one of P-38 fighters, the 18th Fighter Group with one squadron of P-39 and one of P-40 fighters, plus one photo-reconnaissance and one C-47 transport squadron.

Operational control of the 13th Air Force was under COMAIRSOLS (Command Air Solomons), which in turn was subordinate to COMAIRSOPAC (Command Air South Pacific) and COMSOPAC (Command South Pacific). COMAIRSOLS controlled all Army Air Force, Navy, Marine, and New Zealand air units in the Solomon Islands. Either a Navy Admiral or Army or Marine Corps General would command COMAIRSOLS. In April 1944, it was commanded by Maj. General Hubert R. Harmon, USAAF, who also commanded the 13th Air Force.

By 1944, the 13th Air Force was operating from airfields on Guadalcanal. During the previous year, as Walt was going through his 14 months of training back in the U.S., the composition of the 13th Air Force had changed as well. The 11th Bomb Group and its B-17s had departed back to the U.S., to be replaced by the 307th Bombardment Group, equipped with longer range B-24s. The 5th Bomb Group had also been re-equipped with B-24s. The 42nd Bomb Group had been re-equipped with B-25 medium bombers and expanded to five squadrons. The 347th Fighter Group was reduced to three squadrons, but now all operating twin-engine P-38s, while the 18th fighter group was completing re-equipping its three squadrons with P-38s as well. In support, the 403 Transport Group had been expanded to three C-47 squadrons.[1]

In January 1944, 13th Air Force Headquarters was moved to Guadalcanal. The heavy and medium bombers of the 13th began a campaign to destroy Japanese air, navy, and army forces at Rabaul, on New Britain Island far to the north. The 13th Air Force also provided air protection as U.S. Army and Marine units, as well as the New Zealand Army, continued taking small islands in the Solomons further to the north. At about this time, General MacArthur put the 13th Air Force into play in the leap-frogging strategy to bypass most Japanese held island and move quickly across the Pacific. On February 29, 1944, the Army landed the 5th Cavalry Regiment of the First Cavalry Division on Los Negros Island in the Admiralties chain, far to the northwest of Rabaul. Their mission was to drive out the Japanese forces around Momote airstrip. The Japanese had built Momote on Los Negros in

1942 to help protect their base at Rabaul. It took the 5th Cavalry Regiment just one hour and 35 minutes to drive the Japanese out of Momote and secure the perimeter. Momote airstrip was not up to USAAF standards and the U.S. Seabees began work to expand and improve the field. They also started work on a second airstrip at nearby Mokerang. Enemy troops on Los Negros were reinforced from nearby Manus Island and counter-attacked, but the U.S. First Army Division, which had arrived to reinforce the American hold on the island, drove them off.[2]

While 13th Air Force Headquarters and many of its rear echelon support units remained at Guadalcanal, by April when the McRae crew left the U.S., most 13th Air Force operational units were based at Los Negros or en-route.[3]

So this was the military situation in the South Pacific in early 1944. President Roosevelt, the U.S. Chiefs of Staff in Washington, Nimitz, MacArthur, and their chief military deputy commanders knew most of the situation as described here. Newly arriving heavy bomber crews knew some of the situation, but not much. One thing that no one knew, from the President right on down to the private on KP duty at a remote base was the question that most of them really cared most about. That question was how long would it take to win the war and at what cost? In early 1944 no one knew. Could the war be over in a year, two years, four years or more? Could it take 10 years? Any service man or woman leaving the U.S. in 1944 must have pondered that question, and also the question, would I come back?

- 9 -
OVER THE ISLES WE FLY[1] WITH THE 13TH AIR FORCE

Hickam Field, Territory of Hawaii

Friday, April 7, 1944 was Good Friday. The day before, Sgt. Walter Babinski and Sgt. Jack Riley had left Hamilton Field in California aboard a C-87 transport plane of the U.S. Air Transport Command. Now, after almost 2,400 miles and more than a dozen hours in the air, the four-engine transport made a slight jerk to one side that let the dozing passengers know that they were turning to head towards the Hawaiian Island of Oahu and their destination of Hickam Field.[2]

Off to the right they could see Honolulu and the white sand of Waikiki gleaming in the dawn sun. Ahead if they had a view, or to the left as the plane came in for its landing, they could see the great U.S. Navy base of Pearl Harbor, with the hulk of the U.S.S. Arizona visible above the water and beyond it Ford Island and more Navy installations.

Finally, the plane touched down and taxied slowly. Along one side of the runway, they could see the flight line crowded with many different types of aircraft. Beyond the flight line, there were massive concrete hangars.[3] After they stopped at the ATC Air Terminal Building and their plane was secured, Walt and Jack walked down the ramp with their duffle bags and into the tropical morning air. Once in the adjacent Terminal Building, they were met by a passenger-handling officer who instructed them to fill out customs forms and had them clear through customs (they were not in the United States now). They then made their way to the Passenger Traffic Office, where they presented their travel orders. They were told that ATC normally ran a single flight each week from Hickam to Guadalcanal, which was scheduled to leave on Mondays. A file card indicating their final destination was prepared for both Walt and Jack and they were told that they should check the Passenger Notification Board in the enlisted men's billeting area each morning before 10:00 a.m., to see if they were scheduled on the next flight. Passengers were selected based on available space and priority status. All things being equal, passengers who had arrived at Hickam first were given space on an outbound flight first.

The Passenger Traffic officer told the two that there was a good chance that they would be on the flight the following Monday. He then directed them to the registration desk for billeting, in nearby building T-561. They walked down whitewashed pebble paths past large stucco houses with landscaped yards, flowering shrubs, and tall palm trees. Tiny doves pecked at the pebbles and in the grass. Jack said 'Hey Walt, I thought that the Army only built tarpaper shacks in the mud or the dirt!' Walt laughed and shook his head. These were quarters like none they had ever seen.[4] At T-561 Jack and Walt were issued beds in one of about 20 nearby barracks buildings, along with an issue of bedding. Part of the registration process included filling out an individual locator card which would remain on file until they departed Hickam, then it would be sent to an Air Force processing center to help trace their movement and location. At Hickam, transients like Jack and Walt were told that they had to pay 25 cents each per day for an orderly who helped keep the barracks room clean and for laundering their bedding. Walt and Jack were still responsible for making their own beds and keeping their area of the barracks swept and clean. They also checked the location of the alert list in the billeting registration building. They were told that they could return after 10:00 a.m. for a pass to leave base that day. They were hungry though from their flight, so they cleaned up, and then headed over to the nearby Transient Mess in Building T-120 at F Street and Vickers Avenue.[5]

As they walked over to the mess, they could still see signs of the damage remaining from the Pearl Harbor attack in 1941. Construction of Hickam Field had begun in 1935 with the clearance of 300 acres of land. By 1938, the base was ready for operation with the necessary runways, hangars, barracks, and other service buildings. The field had been names for Col. Horace M. Hickam who had been killed in 1934 in an airplane crash at Fort Crockett, Texas. When the Japanese attacked on December 7, 1941, there were 51 planes on the ground at Hickam, and another 12 B-17's were in the air, en-route from the West Coast of the U.S. While most of the first wave of Japanese Navy fighters and bombers concentrated on U.S. Navy ships in the harbor, they strafed and bombed Hickam as well, to prevent the Air Corps fighters from interfering with their attack. The Japanese attack, with Zero fighters and Val bombers, concentrated on U.S. aircraft on the ground. During the attack, the 12 B-17's, by then low on fuel and unarmed, landed at Hickam, but many were subsequently destroyed after they had taxied to a stop. At 8:40 a.m., a second wave attack began, with the Japanese now also concentrating against the barracks area at Hickam and the mess hall was hit as well. By the time the attack ended, 121 men had been killed, 274 wounded, and 37 were missing and half the planes were destroyed.[6]

It was Good Friday, so Walt and Jack went over to the Catholic Chapel at Sixth and Singer Boulevard for Good Friday services. Afterwards, the NCO Club at Porter and Wilson Streets started serving beer at 3:00 p.m., and there was also an informal dance held every Friday beginning at 8:00 p.m. They learned that there was a weekly USO show every Friday at 7:30 p.m. at the Starlight Bowl in Manselman Circle. Other recreational facilities available for Walt and Jack at Hickam included the Base Theater at Vickers Avenue and G Street, with shows at 6:15 and 8:15 p.m. daily, as well as Saturday and Sunday matinees at 2:00 p.m. Newspapers and magazines were provided for enlisted men at the Base Service Club adjoining the Theater, as well as ping-pong, card tables, and a collection of both classical and swing records.

Nearby Memorial Gymnasium was the consolidated mess hall prior to the Pearl Harbor attack. With the later growth of the base it was converted into the largest gym in the Hawaiian Islands. In addition to basketball and volleyball courts, a boxing ring, and exercise room, the gym also included a ten lane bowling alley (cost 15 cents per line) six pool tables and a snooker table. There were also nearby tennis courts, five softball fields, a swimming pool and many other athletic courts.[7]

Saturday after breakfast, Walt and Jack verified at 10:00 a.m. that they were not on the alert list for a flight that day.[8] They did not expect to be, since the scheduled plane for Guadalcanal was not due to leave until Monday. However, they were now free to leave the base and explore Honolulu. Jack and Walt, wearing the regulation cotton khaki uniform with khaki garrison cap and necktie, reported for their passes, then caught the base shuttle bus to the main gate. Past the gate, they caught one of the commercial city buses that ran every 10 or 20 minutes into Honolulu.

Once in the city, they noticed that many of the shops were different from in the U.S. They were all open in front, with no real door, wall, or windows...just a roll down shutter for at night.[9] There were not many normal tourists. However, the influx of military personnel, mostly Navy, but many Army and Air Corps as well, made up for some, but not all the lost business. Walking through Honolulu, servicemen always commented on the many beautiful dark-skinned Hawaiian women. Some were native Hawaiian, some from other Polynesian Islands, some Chinese, and some Filipino.[10]

Walt and Jack made their way out to Waikiki Beach and looked across the bay to Diamond Head. Both of them commented on how the winds had been calm the previous day at dusk and again at dawn, but now the Northeast trade winds blew strong during the day at 20 to 30 miles per hour. The wind was not unpleasant though, with palms swaying in the breeze and spray blowing off wave crests. The sky had been pure blue at dawn, now clouds had rolled in with the breeze during the day, to clear again at night. Other images included geckos clinging to shady walls on the base and around town, with birds calling 'caa-caa-coo,' the sound of wind in palms and farther off, of waves breaking on beach, and intense flower-scented air. They browsed the tourist shops after lunch, perhaps bought leis and postcards, likely swam for a while at Waikiki Beach, and toured the grand Royal Hawaiian Hotel, a Honolulu landmark.[11] Later they had dinner at a fine Honolulu restaurant, perhaps had a few more beers at a nightclub, and then caught the bus back to Hickam Field sometime before midnight.

Sunday, April 9th was Easter Day. Jack and Walt dressed in their class A uniforms and walked over to the Catholic Chapel located at Sixth Street and Signer Boulevard for Easter services at 8:00 a.m. Having fasted to take communion, they were hungry after church, but the enlisted mess served breakfast only until 7:30 a.m. They walked over to the Hickam Air Terminal, where the Terminal Restaurant was open continuously from 6:30 in the morning until past midnight. After Easter breakfast there, they went back to the Transient Billeting desk to check the alert list. Sure enough, both Walt and Jack were listed as passengers on the next morning's southbound flight out of Hickam. They were still cleared for a pass until 11:00 p.m. to leave the base, but warned that they had to be ready to leave and report to the Air Terminal early the following morning.

Jack and Walt likely spent at least part of the day heading back in to Honolulu. After all, they had no idea when they would again be in a city with modern conveniences and attractions. They stopped in at the USO in Honolulu. This USO had developed a reputation for the banana splits it would serve to GI's on demand. One GI recalled 'those splits were what made life worth living.'[12]

Walt and Jack also spent time at shops in Honolulu or at the PX at Hickam, stocking up on magazines, cigarettes, candy, and a bottle or two of whiskey to stuff into their duffle bags. They both tried to imagine what items they would find difficult to come by in the South Pacific as they made these purchases. They stopped in the base barbershop for a last haircut. They also spent time writing letters to their families back home. Jack wrote to his wife Nancy and to his father in Tennessee. Walt wrote to his ma and pa in Detroit, although what he could say about where he was headed was limited and he knew the censors would be checking his letters. He included a number of postcards he bought in Hawaii. These cards gave an idea of the sights and experiences that made the greatest impression on Walt. They included scenes of lei sellers, pineapple fields, fern forests, broad sandy beaches, and palm trees silhouetted in a tropical sunset.[13]

The following morning Walt and Jack arose early, had breakfast, and then packed up their gear. They were still on alert and required to remain either in barracks or at the mess hall. Shortly after, the Passenger Services Office contacted the Billeting Office with the final list of passengers for the day's flight. The Billeting clerk told Sgts. Babinski and Riley that they would be departing later that morning. He informed them that they had about three hours before flight departure. They policed up around their bunk then took their bedding to the registration desk where they received a clearance slip. Then they checked out, the clerk removed their locator cards and wished them luck. They didn't have long to wait for a Passenger Services bus, which took them from the barracks back to the Air Terminal that they had arrived at just a few days before.

Once in the Terminal the Passenger Handling Officer checked their names and travel orders against the flight list. They had to be weighed in and cleared through customs at least an hour and a half before departure time. Finally at some point in the morning their flight was announced and the Passenger Handling Officer met the men at the loading gate and led them to the waiting C-87. They went up the loading steps by order of rank. Once the plane was loaded, crewmen moved the steps away. The Passenger Handling Officer stationed himself at the left wingtip, in full view of the pilot. When the engines had been run up and the wheel chocks pulled, the Officer released the plane for departure by giving the pilot a military salute.[14]

Across the Pacific: Canton, Nandi, and Mysterious Ndeni Island

The sound of the four engine C-87 was familiar to Walt and Jack, but they knew that they were leaving most of their familiar life behind. How many of the men aboard wondered if they would return? The plane climbed after takeoff, then turned to a course south by southwest. Their destination was Canton Island, 1,912 miles and about nine or ten hours in the air. Midway to Canton, remote Palmyra Island had a radio station and emergency landing strip, in case of problems. Finally late in the afternoon on Monday, April 10, 1944, after having crossed the equator into the southern hemisphere, Walt and the other men aboard the C-87 could see a low horseshoe shaped island off in the distance.[15]

They came in for a landing on a long coral runway aligned along one leg of the atoll that formed Canton Island. The total circumference of the atoll was about 15 miles, but in no spot was it wider than 600 yards and in many spots it was only 50 yards wide. Maximum height above sea level was 20 feet, but most of the island was only five feet in elevation. The island was covered by low scrub and contained only one single tree, which was one of the chief sights.

American whaling ships had visited the island in the 19th Century. One of these, a ship named the Canton from New Bedford had run aground at the south end of the island in 1854. The Canton's captain and crew made an epic journey in the ship's open boats, reaching Guam and rescue after 49 days. An American expedition in 1873 named the island Canton in honor of the ship's captain and crew. The island was a British possession, but had been turned into an American base early in the war. It was used primarily for refueling aircraft on their way across the ocean and as a base for patrol aircraft looking for Japanese ships or aircraft. Several times during the war, Japanese submarines were sighted off shore, which created alarm on the island.[16]

Walt watched as the plane dropped down towards the water. Soon he could make out individual waves as they descended lower and lower, and then he saw them cross a short stretch of white sand beach, then a patch of scrub, then they were down on the ground. At the end of the runway, a jeep was waiting for the C-87. It led them slowly to where they came to a stop near a group of tents that served as the terminal area. They would refuel and spend the night here, before continuing on the following day. As the men stepped off the plane, they saw thousands of empty 55-gallon gasoline drums and nearby a big freighter was docked unloading supplies...and more fuel drums. Walt and Jack got their gear and checked into a tent for the night. Strolling around for an hour or so before dinner, they gazed out at the breakers on the ocean side of the island, and then quickly crossed over to the lagoon side, where there was a soft sandy beach and calm water. They looked at the lone palm tree and shook their heads over the remoteness and desolation of this spot.[17]

They took saltwater showers and shaved, although the saltwater turned their soap into something more like slime, not suds. Soon it was time for supper and afterwards an outdoor movie was shown. Walt and Jack spoke to some of the men stationed on the island. They had all seen this movie many times before. They had all seen all the movies available many times. There was nothing else to do here and the standard tour of duty was at least a year. Walt and Jack both wondered how these men kept from going crazy here and concluded that maybe many of them did go crazy.

The following morning, Walt, Jack, and the other passengers were instructed to assemble for re-boarding the aircraft shortly after breakfast for the next leg of their journey: Fiji.[18] Once again the passenger list was checked as they boarded and shortly after they took off down the long runway. The C-87 was just a few dozen feet off the ground when they were out over water again. They continued to gain altitude steadily and soon the pilot set a course to the southwest. After just under six hours and 1,270 miles of gazing down at nothing but empty ocean, the plane descended to just about water level, they passed over breaking waves, a white coral beach, and then they landed at USAAF Nandi, in Fiji.[19]

Guadalcanal: British Solomon Islands

The passengers and crew had an hour for lunch at Nandi as the C-87 was refueled and then they re-boarded. Their next leg was another 1,300 miles to the northwest, to Guadalcanal. Late on the afternoon of Tuesday, April 11, 1944, Walt and Jack arrived at Henderson Field on Guadalcanal.[20] As the plane touched down, Jack and Walt were startled by a loud unfamiliar noise the wheels made on landing. They later learned that the runway surface and taxiways were covered with perforated steel mats. These were needed on Guadalcanal because of the wet, soft soil. A heavy bomber would sink into the mud without steel mats spreading the weight of a plane landing or taxiing.

Climbing down the boarding ladder and onto the steel mat, the first impression of new arrivals to Guadalcanal was the oppressive humidity and the vast number and variety of insects.[21] The daytime temperature was somewhere around 82 to 85 degrees. The nighttime temperature would be a consistent 80 to 84 degrees. It rains five or six days per month on Guadalcanal during April and May, creating an ideal environment for mosquitoes and other insects. All across the South Pacific, Walt, Jack and the other allied soldiers, sailors and airmen would pretty much find the same monotonous temperatures.

Walt and Jack rode a covered trailer with open sides behind a truck to the personnel receiving office at Henderson Field. They reported in with their orders. Walt and Jack were assigned to the 13th Air Force, XIII Bomber Command, Combat Training Center (CTC).[22] Walt and Jack explained that the rest of their crew would be following in a day or two from Hamilton Field with their new B-24. The Personnel Officer explained that when the rest of their crew arrived, they would begin a month or so of final combat training. He directed them to Koli Point, where they would find tents for training crews, informing them that they were on their own to find a tent that had been vacated by a crew that had already completed their combat training. They were told to leave their location at the CTC headquarters and also check in each day for the arrival of the rest of the McRae crew.

Koli Point was near Iron Bottom Bay, the landing area for supplies coming in by Navy ships. Most 13th Air Force personnel still on Guadalcanal were quartered there, along with 5th Air Force, Navy, and other military personnel. Walt and Jack found a suitable tent and stowed their gear. The tent was typical of those they would call home for almost a year. A tall wooden pole in the center and others along the sides supported a large canvas tent roof. Most tents at Guadalcanal had been improved by previous occupants with raised wooden floors to keep off the wet ground. They were typically open on all four sides for ventilation, so the cots were usually grouped near the center of the tent to help keep dry during rain. Tents often had crude tables and chairs, candles or kerosene lamps for light, and various storage racks or boxes for the men's gear and belongings. An important consideration was mosquito netting – always in short supply, some tents were completely walled in by netting, while others used a strategy of separate netting for each cot. Each tent was large enough for the six enlisted crewmen of a single plane. Officers had their own four-man tents in a separate area of the base. By the end of their first day on Guadalcanal, Walt and Jack had secured a place to sleep and figured out where the mess hall and other facilities were located.

The following day, Wednesday, Walt and Jack checked in at the XIII Bomber Command CTC at Koli Point as required and verified that the rest of their crew had not yet arrived. They spent the rest of the day on their own, getting to know their neighbors, figuring out where the PX was, sports facilities, the camp theater, and so on. They asked about the sporadic machine gun fire they heard off in the distance and were surprised to learn that though the Japanese Army had ended its struggle for Guadalcanal more than a year ago, there were still many pockets of Japanese troops who had not yet surrendered in the rugged hills of the island. Covering over 2,000 square miles, with mountains over 7,000 feet high, Guadalcanal was extremely rugged. While most of the enemy survivors remained far away, occasionally they would get within a mile or two of the fields, mostly hoping to steal food or supplies.

By 1944, there were three major airstrips on Guadalcanal used by the 13th Air Force. In addition to Henderson, there was also Carney Field and Koli Field. Carney Field was used primarily by the B-25's of the 42nd Bombardment Group (M) as well as for B-24's assigned to the CTC. Koli Field was used by the B-24's of the 370th and 424th Squadrons of the 307th Bombardment Group. Walt and Jack asked about the Combat Training Center. They had been 'training' now each for 15 months. Why did their crew need additional training? Why wouldn't they be sent into combat as soon as the rest of the crew arrived? Men they met told them to be patient – the CTC program would increase their chances of survival. Of course there was nothing Jack and Walt could do but wait.

The days passed by and each morning they checked in at the CTC office, but still no word of crew 430.[23] Walt and Jack likely began writing letters to their families back home to let them know they had arrived safely. These first letters and others they would write later throughout the South Pacific would be censored heavily to remove references to where they were located, the unit they were assigned to, and any details about combat missions. While none of Walt's letters from the South Pacific to family and friends appear to have survived, letters written by other 13th Army Air Force crewmen can give an idea of what they could tell their families of their initial impressions and also what they might like to have sent to them from home. One young B-24 gunner wrote during his first weeks in the South Pacific with the 13th AAF:

> *"Dearest Mom, Dad, and Sis,*
> *"We Arrived here in good condition...I Haven't the slightest idea where we are....don't you worry about me because I am getting along just fine. However I can't say this is the best place that I have ever been, but neither can I say it is the worst. If you can find any sweets, peanuts (in cans), etc., anything that is candy and will not perish, I would appreciate it a million."*
> *"So far everything is going along just swell, but I suppose it will take quite a while to get used to not having 'cokes' and other soft drinks. Those are about the only things I really miss."*
> *"Do hope that you are all well and that everything is going along swell at home. As yet I still haven't heard from home, but I suppose that we will be getting some soon, I hope. I Hate to say it, I believe I'm getting used to this place...We usually hit the hay about 9:30 P.M. and get up at 6:30 A.M., eat breakfast and start out the day....we had a pay day yesterday. About ten pounds. This screwy money system really has me mixed up. Sixpence, eightpence, pounds. I don't have the slightest idea how much to pay...it's really a mess. Chow has been improving quite a bit. Hope it keeps up. By the way, I still have about four weeks laundry to do....Most of the fellows have theirs done, but me-no!"*
> *"As usual, there is very little news here. Things that we aren't allowed to talk about are naturally the things that would make the best news, etc., and so on....I'm afraid that I'll have to send my watch back home. There seems to be moisture under the case and it is forming on the crystal. It must be the moisture and heat, etc. We've been playing quite a bit of football lately. Now we're trying to schedule a game with our officers."*
> *"...You asked me if there was anything that I needed. Well this time I believe there is. Do you happen to remember those real thin sleeveless shirts that we wore in the summer? They're called 'Tee' shirts. They're very similar to undershirts, made of the same material practically, only they have quarter sleeves. I really would appreciate if you could send a dozen of them."*[24]

Each evening Walt and Jack made their way to one of the outdoor theaters for the nightly movie. Each group had their own open-air theater, run by a Special Services Officer. The typical theater had a wooden framework from which an improvised screen was suspended. A small projection booth faced the screen and an outdoor sound system was rigged. Seats were boards supported by empty fuel barrels or shipping crates. In the tropics, dusk was short and it was dark by seven o'clock. There was never a question of sitting in the balcony or downstairs, as there was no balcony. There were plenty of exits and smoking was permitted anywhere. Admission was always free. Mosquitoes were a problem, but smoking helped keep them at bay. It might rain too, but most men would just bring a poncho, sit, and watch through the rain.[25]

Late one night Walt, Jack and others were awakened by a fellow screaming and saying, 'Shoot it! Shoot it!' After a few minutes, with men searching in the darkness and hoping no one was going to shoot, everyone figured out what was happening. One of the crewmen had too much to drink and fallen asleep and his arm had falling asleep too. He was lying on his back and when he woke up; he felt this foreign object, like a snake, on his stomach. So he grabbed himself by the wrist and then it became a cobra with a hood staring him in the face. Soon he was screaming to shoot him in the hand. By the time they got the flashlights on, they found out what it was. That poor guy was really scared thinking that he had a cobra across his stomach! After a bit of cursing and laughing, everyone went back to his cot and back to sleep.[26]

By the end of the week, Jack and Walt were becoming bored because of inactivity and worried about the delay of the rest of their crew arriving. They probably attended a baseball game or two of the 'Guadalcanal Coast League'. By 1944, the 307th Bomb Group all-star baseball team was beating all comers, including teams from the Navy, Marines, Seabees, and other Army units. Pick-up volleyball and basketball were popular sports on the island. They may have had their first contact with the Guadalcanal natives, being surprised to learn that they were all Christians, with highly developed culture and artistic ability.

Sunday April 16, Walt and Jack attended Catholic mass at a beautiful thatched chapel that had been built by Guadalcanal natives near Henderson Field. Both Catholics and Jews shared this chapel for their services, while Protestants used a separate chapel at Koli Point.[27]

Finally the next day, Monday, April 17, Walt and Jack got news that the rest of their crew had arrived on Guadalcanal. They had landed their new silver B-24J earlier that morning at Henderson Field. All ten men were glad to be reunited as a crew again. While Lts. McRae, Aubrey, Dunne, and Clark went off to arrange for their own tent in the officers' quarters area, Jack and Walt showed Cliff, Jupe, Frank, and Len the tent they had secured for the six of them. Jack and Walt described their journey to Guadalcanal, and then the other four men explained their delay. They had left California nine days earlier, on Saturday April 8, arriving late the same day at Hickam. They described how they were all dead tired the following day, due to the lack of sleep resulting from an all-nighter in Sacramento the day before their departure. Apparently they weren't the only ones tired on the flight, for Jupe reported that Lt. McRae slept most of the time in the air, letting Lt. Aubrey do most of the flying all the way to Guadalcanal.[28]

They were delayed at Hickam due to a faulty carburetor in the number four engine, which took three days to repair. They left Hickam for Canton on the 11th, arriving in the afternoon, a few hours after Jack and Walt left Canton the same morning. However, into the flight, the number four engine torched again, so they shut it down and grounded the plane on Canton. They radioed back to Hawaii for yet another carburetor, which stranded them for five days on the island. While on the island, Len Sherman became badly sunburned and had to spend most of his time in the tent. Finally, the replacement carburetor arrived and they were ready to depart on Sunday, April 16. They reached Noumea in New Caledonia after a five and a half hour flight. They explored this French colonial city, and then the following morning made another five and a half hour flight to Guadalcanal.[29]

After the four newly arrived men had settled in, Walt and Jack told them it was time for chow. The typical routine at Guadalcanal, as it would be for later jungle island bases, was for each man to retrieve his own metal mess kit from his tent, then get in line at the closest mess hall. Soon one of the mess attendants would open up a door to the large screened-in tent and the line would start moving inside. Each man opened the two halves of his mess kit to form a double plate, then moved down a serving line where mess crew served out breaded spam, dehydrated potatoes, and canned corn with canned peaches for dessert. A slice of fresh coarse bread with a slab of canned butter topped off the meal. The meal was washed down with heavily chlorinated water. Each man was also expected to take an atrabrine tablet, a bitter tasting pill that suppressed the symptoms of malaria.[30] They continued to talk through dinner at one of the long wooden mess hall tables. After dinner, each man washed his own mess kit at a line of 55-gallon drums. First, they scrapped any scraps left into the first drum that served as a garbage can. The next drum was filled with boiling soapy water for washing the mess kit. The last drum had clear boiling water for rinsing. Then they carried the mess kits by the long handle and returned to their tent for sleep.[31]

XIII Bomber Command Combat Training Center

At some point late the following day, Lt. McRae called all the men together and told them what he had learned about the CTC and the additional training they would go through. The CTC had been moved to Guadalcanal in January. It was under the command of Major Henry E. Jones, who was from the 307th Bomb Group's 424th Squadron. Its mission was to orient newly arrived combat crews to the combat conditions, tactics, and survival techniques that were unique to this theater of the war. CTC instructors were veteran combat crewmen who had volunteered to stay in the South Pacific to help with training. At the completion of the course period, the CTC would also assign whole replacement crews or individual crewmen to one of the XIII Bomber Command's combat squadrons.[32]

The following morning, their CTC training began with a briefing by Major Jones. He began by outlining the history, organization, and role of the 13th Air Force. Using a large wall map of the area, Major Jones explained that their current objectives included destruction of enemy airfields and other support facilities on New Britain Island. They were also tasked with destroying Japanese supply lines on and around New Britain and the other northern Solomon Islands. They also had to help support ground forces mopping up Japanese resistance on Bougainville Island and support Army and Marine amphibious assaults on other islands around Rabaul. Major Jones read from a recent Japanese news report that indicated the success of the American tactics:

> "The...enemy has advanced since Guadalcanal into Rendova, New Georgia and Bougainville. The means of their advances are always identical...they land under the cover of bombing and naval bombardments. Then they build airfields, thus extending the sphere of air influence on toward the north...As the enemies themselves say, the advance is very difficult...building up airbases there, their operations are to intercept our supply lines. Whenever the enemies build an airbase on an island, our communications and supply with the other islands becomes difficult. Thus, even though we have got great naval victories, they thread their way through the lines to land and gain footholds. This is the way the Americans operate."

Major Jones told the men that the Japanese news report got it right and that they would continue these tactics to the north and west.[33]

He then told the men of the McRae crew that he knew they were disappointed that they needed additional training, but the 13th Air Force had developed statistics after many months of war that 75-80 percent of combat losses occurred with crews during their first three months of combat. The 30-day course would give the men refresher training in an environment that could not be duplicated back in the United States. However, to give the crew a fair chance to see if they might be ready for combat, Major Jones explained that they would be given a variety of flying and written tests.

The flying tests were conducted by just the skeleton crew. One test was designed to see if Ed Dunne, the navigator, could successfully find a remote island after a nighttime takeoff and then return to base successfully, as they might have to do on a long mission without city lights below to guide them. Lt. McRae was assessed on his ability to land on a rough and unfamiliar island landing strip, as they might have to do on a long mission or if their plane was damaged or low on fuel. Another test assessed if Jimmie Clark, the bombardier, could get his bombs away quickly under conditions similar to real combat. Another tested Jupe Bretherton's ability to correctly transfer fuel on an extremely long flight to minimize fuel consumption and to maximize range. The crew failed all these and the written tests as well. After the bad news, Major Jones explained that no crew had ever passed the tests, but that their performance was no worse than any other newly arrived crews.

After this testing the month long training began. All men would receive about 120 hours of ground training in a variety of subjects, along with 90 hours of flight training for the skeleton crew (though less time in the air for the other gunners). The training facilities included a full operational mockup of a B-24, including each of the types of gun turrets. A number of well-lighted and ventilated classrooms were used. There was also an intelligence war room, which included a collection of technical pamphlets and reference books. The intelligence room also included large-scale tactical maps, which were marked up each day to show the previous day's strikes and other tactical actions. The men also received briefing reports from their instructors on recent missions, which were used to help illustrate various important aspects of their combat training curriculum.[34]

On Wednesday April 19, their CTC training began. Their first four days were spent on about 15 hours of intelligence related training, partly classroom work and partly out in the field. They spent time in the war room, viewing maps and aerial photos of current 13th Air Force bombing targets, to help familiarize them with what they would see on their own combat missions. Intelligence officers and experienced combat crewmen described the variety of means that the Japanese would use to defend their bases, including fighter attack techniques and anti-aircraft flak defenses. They emphasized the importance of tight formation flying both to defend against the fighter attacks and to concentrate bombing patterns over their targets. Common bomber crew combat accidents were described to educate the men of the types of problems they would encounter under the stress and excitement of their first combat missions.

They spent some time in the nearby jungle, learning how to live in such an environment if forced to bail out or crash land behind enemy lines. They practiced deploying and getting into an inflatable life raft and the best ways to survive on the water. This survival training also included learning the use of the items in the survival kit each of them would take into combat. Each man would carry a notebook and pencil; web belt with a first aid kit, jungle knife in its sheath, and a canteen; a .45 cal. pistol carried in a pocket with three clips of ammunition; two packages of cigarettes with matches and a candle; a head net and bottle of mosquito repellent; a folding machete; a pair of canvas gloves; and another knife strapped to one leg. They also carried a separate plastic emergency kit that contained a compass, two blocks of jungle chocolate, a box of caramels, four sticks of gum, adhesive tape, anti-sleep tablets, atrabrine tablets, aspirin, halazone tablets, ointment, and a waterproof match safe. They would also wear a parachute harness to attach a backpack parachute, a Mae West inflatable life jacket with dye sea marker and shark repellent, and a flak helmet.[35]

One day they were driven into a small jungle clearing for training in intense survival techniques provided by a group of Australian soldiers in their distinctive wide-brimmed digger-hats. They told the men that their qualifications came from months as coast watchers and radio operators on Japanese held islands. They were not going to teach the American fliers how to fight the Japanese, but how to stay alive in enemy territory. The Aussie told them it was easier to bring home live men as opposed to dead bodies. They learned how to use tree fronds and branches to build and camouflage a jungle shelter. They were shown how to strip the inner bark of certain trees to weave into cords and ropes. One Aussie showed them how to determine the directions of the compass by observing the shadow of a stick stuck in the ground. The men split up and followed an instructor into the dense jungle to identify fruit and palm cabbage, and even distinctive mounds of earth that indicated where bird eggs might be buried. Walt learned that it was better to drink rainwater trapped in bamboo stalks, rather than from streams that might be polluted or infested with parasites. They were told that natives might help them, but maybe they would not. Natives lived with the Japanese presence, and even though they did not like the Japs, they would not put themselves or their families in jeopardy. Each man would have to make his own judgement before approaching natives for help.[36]

The final aspect of intelligence training was to teach them about pre-mission briefings and the types of information that the pilots, bombardiers, navigators, radio operators, flight engineers, and gunners would each receive. The instructors explained that the briefing information was intended to improve the chances that the mission would be successful and that they would return safely. Lastly, they were told about how intelligence officers would question each one of them during post mission briefings and the types of information and details they would be expected to bring back. [37]

On Sunday April 23, after chapel and breakfast, the men started two days of aircraft recognition training based on the Renshaw system. They had all trained on how to recognize both allied and axis aircraft back in the U.S., but this refresher course emphasized the Japanese Air Force and Navy types that they would encounter in combat, as well as how to safely distinguish friendly aircraft flown by the U.S. Army Air Force, Navy, and Marines, as well as the Royal New Zealand Air Force and the Royal Australian Air Force. The Renshaw system had been developed by Samuel Renshaw, a psychologist from Ohio State University, who created a series of projected images that each showed a different aircraft type in various views. Some views showed the plane from the bottom, some from the side, some from the front, back, or at an angle. These were flashed to men quickly, eventually for as little as 1/100th of a second.[38] This training developed aircraft recognition skills designed to simulate the type of brief view and quick response that would be needed in a combat situation. The CTC instructors explained that in combat the gunners would often have to make a split-second decision about an airplane they sighted coming out of a cloud or diving down from above. An incorrect identification could be deadly. Failure to identify an enemy plane quickly could allow it to get in close and make a kill. Failure to identify a friendly plane quickly and accurately could result in friendly fire being directed at one of their own planes.

One feature of the Sunday night movie at Guadalcanal was a current events talk given by Lt. Pestell, 307th Bomb Group S-2 officer before the films began. This helped keep the men informed of world war news and what was happening back in the states.[39]

Around this time, the McRae crew learned that most of the remaining 13th Air Force units were starting to move north from Guadalcanal. The destination would be new bases in the Admiralty Islands far to the north. Of the heavy bomber squadrons, two (371st and 372nd) had already been flying out of bases on Munda Island for several months. Munda put them closer to Rabaul, which was the major enemy target during the first months of 1944. On April 20, the last two B-24 squadrons (370th and 424th) had been ordered to move to the Admiralties as well. Walt and the other McRae crewmembers could observe the heightened activity as the 370th and 424th crewmen packed up and began moving out.[40]

Later that week all the men had a short three hour range estimation course – learning how to estimate the distance of a plane sighted in the air or to estimate their height above the ground or the ocean. The following three days the crewmembers completed 15 hours of turret gunnery training. This training was in each of the various turrets the B-24 was equipped with. These turrets had been removed from older planes and were mounted on stationary frameworks on the ground. The training permitted experienced gunners to check out the newly arrived crews on latest equipment and actual combat conditions and techniques in the theater. All the men checked out on each turret, in case it became necessary for someone to replace an injured gunner during a combat mission.

Saturday started with an hour of instruction on basic life raft navigation for all the gunners, if they had to ditch or bail out over water. The B-24 was equipped with an emergency life raft in a hatch on the top of the fuselage, which could be quickly released after a water landing. Later in the day, they began a two-day ordnance course. This focused on safety around bombs and fuses. Only rarely would the crew load bombs on their own planes, but it would be their job to check the load at the beginning of a mission to ensure that safety pins and safety ties were in place to prevent accidental detonation. They would also need to know the proper techniques for removing the safety pins while in the air before a bomb run, and if necessary, replacing the pins if a plane had to return to base without dropping its load.[41]

By the end of April, all 13th Air Force B-24 squadrons and their attached personnel had departed Guadalcanal for the Admiralties.[42] By the end of the month, it occurred to Walt that he and the other crewmen had not done any practice flying yet.[43] CTC training was supposed to include up to 90 hours of flight training. However, a few weeks after arriving on Guadalcanal, Lt. McRae had developed an ear infection. This proved to be very painful for him during high altitude flights with the skeleton crew after they had arrived. The base doctors diagnosed the problem as a rare tropical fungus infection. Consequently, none of the required crew flight training had begun. Rumors began to move through the crew that without a pilot, they would be broken up and assigned as individual replacements to other crews.[44]

They had only been together as a crew for about four months, but this prospect was distressing to them all. Lt. McRae was under full time medical treatment by this point and there was no imminent prospect of improvement for his ear infection. However, shortly afterwards a solution to the problem was found. In April and early May 1944, a large number of replacement 'Mitchell' B-25 medium bomber crews had begun arriving on Guadalcanal for their own XIII Bomber Command CTC training. These crews were destined to be assigned to the 42nd Bomb Group (Medium) of the 13th Air Force. The 42nd Bomb Group was known as the 'Crusaders' and provided medium range bombing with their twin-engine aircraft. The B-25 was similar in appearance and in flight characteristics to the B-24, but of course with shorter range and less bomb capacity.[45]

Along with the newly arrived replacement B-25 crews there were a number of extra pilots. At some point during Lt. McRae's ear fungus problem, it was arranged by XIII Bomber Command to ask a B-25 pilot to volunteer to take over the McRae crew. A young Texan, Lt. Carl Appling, volunteered for the job. It is not known if Appling was one of the newly arrived replacement crews or if he was already serving as a 42nd Bomb Group (M) pilot. In either case, he must have been an unusually skilled pilot to make this transition. During the first few days in May, Appling met the other officers and crewmen. They learned that he came from Kingsbury, Texas, a small ranching community about 75 miles northeast of San Antonio. Right from the start, it was apparent that Appling's personality was very much different than Lt. McRae's. Where Charles McRae was strict, gruff, and 'by the book,' Carl

Appling was more of an easy going leader, willing to take risks and fun-loving too.[46] His personality and leadership qualities must have been exceptional, because there was no apparent conflict between Appling and Don Aubrey, who probably thought that he should have been given the chance to take over the crew after McRae's ear fungus grounded him.

It was agreed with XIII Bomber Command that the crew would have to curtail the total number of flying training hours, in order to give Lt. Appling time to concentrate on the transition to flying the B-24. On Monday May 1, the crew began eight hours of aviation medicine training. The focus of this course was tropical first aid and also a refresher on the use of oxygen equipment. This was followed by a day of chemical warfare training. This was mostly a refresher on the use of gas masks if needed during a Japanese bombing raid, as well as training on the use of smoke and incendiary bombs. Thursday May 4, a four-day armament course began, which provided more gunnery and turret refresher training for all the crewmen, but double work for the gunners themselves.[47]

Late on Thursday, news came that they would fly their first training mission as a complete crew the following day. That afternoon they had a pre-flight briefing with Lt. Appling and the other officers. Early on Friday, May 5 the crew was awakened, ate breakfast and then loaded in to a 6x6 truck for the ride out to Carney Field. Carney was used both by the XIII CTC and by the 42nd Bomb Group (M), so it is possible that Lt. Appling caught sight of some of his old B-25 crew comrades. Walt, Jack, Frank and the other crewmen may have felt some initial misgivings about flying with a pilot who was relatively new to B-24's, but that feeling quickly faded away as Appling proved his competence flying the Liberator. Their mission that day was to practice day formation assembly in a B-24J. While one instructor observed pilot, copilot, navigator, flight engineer, and radio operator, another instructor observed the gunners. Ed Dunne solved a navigational problem to get them a few hundred miles from base, where they formed up with a few other B-24's for a short period of tight formation flying. They then returned to Carney Field where they landed safely after four hours in the air.[48]

Their instructors critiqued the crew after this first flight. That was their job. No matter how well the crew did, the instructors would be working hard to find fault and thereby transfer as much of their combat experience and South Pacific know-how to these ten green men. Late on Friday, the crew was debriefed. The skeleton crew would continue to work to get Lt. Appling fully proficient with the B-24, while the rest of the crew completed the CTC ground curriculum.

They went back to the armaments training they had started on Thursday. Special attention was given to the importance of cleaning guns carefully and thoroughly after each mission, due to the dirty, moist tropical environment. On Tuesday the 20th they started six days of flight engineering training. The aim of this instruction was to provide a refresher course on all aspects of in-flight aircraft operation and emergency repair. Particular attention went to fuel transfer procedures and hydraulics systems.[49]

By mid-May, all 13th Air Force bomber squadrons were now up north at new bases in the Admiralty Islands. Word was filtering back to Guadalcanal that the move of the squadrons to the Admiralties had not gone smoothly. Supply ships carrying heavy equipment and supplies had been loaded quickly and not carefully, resulting in delays in unloading and sorting through gear and equipment. Also, the Navy was shorthanded for the unloading, so bomber crews had to work as stevedores and then later as construction crews to help build mess halls, repair sheds, headquarters buildings, and the dozens of other facilities need for an efficient air base. Nevertheless, the XIII Bomber Command was starting to fly missions again. The targets were new though – to places called Truk and Biak.[50]

By around Tuesday, May 16, the men of what was now the Appling crew were finishing their last CTC ground course. This was code blinker training – to ensure that each man could send five words per minute in Morse code. The code blinker light was used to send messages from plane to plane while flying in formation during radio silence, or to send or receive messages from downed fliers in a life raft or on the ground.[51]

On Wednesday the 17th Lt. Appling alerted the men for a pre-flight briefing that evening. Before dawn the following day the crew took a B-24D on their second CTC training flight. The early departure before daybreak was to test their nighttime takeoff and navigation ability. Appling was instructed to take the plane up to 20,000 feet, for high altitude formation flying. This gave the men a chance to test their oxygen equipment procedures again and also drive home the need for warm gear, even in the tropics, when flying at extreme elevations. Walt and the rest of the crew logged four hours and thirty minutes on this flight. The following day they were up again for a third training flight, which lasted four hours.[52]

That afternoon, during the post-flight debriefing for this third training flight, their CTC instructor made a major surprise announcement. Regular CTC training for the Appling crew was over. Tomorrow they would be flying their first combat mission! This would still be part of their combat crew transition and they would still have a CTC instructor on board who would critique their performance, but they would be flying against a real enemy target. The men were told that there would likely be at least one other combat mission after this one, but that after these were completed, they would be transferred from the XIII Bomber Command CTC and assigned to one of the 13th Air Force heavy bomber squadrons.

The debriefing transitioned into a briefing for this combat mission. Two other CTC crews joined them. Their target tomorrow would be Rabaul, the major Japanese army, air, and naval base about 600 miles north on New Britain Island. The briefing officer explained that Rabaul had been a major target for both the Fifth and 13th Air Forces for the past three months. The US Navy and Marines had been targeting Rabaul as well. The briefing officer explained that the Japanese had ceased major air defense of the base. At the beginning of 1944, they had as many as 400 aircraft available at various airfields around Rabaul, but most had been shot down, destroyed on the ground, or removed to remote and hidden jungle airstrips. The Japanese still had perhaps a hundred thousand troops on the island, but as long as the Allies controlled the sea and the air in the region, they could neither easily be reinforced or supplied, nor pose a threat to other US bases in the area. To keep Rabaul neutralized, the 13th Air Force was tasked with photoreconnaissance over the base, as well as frequent bombing missions to keep the many Japanese held airfields damaged and unusable. Their specific target tomorrow would be Tobera Air Field, located in an old plantation about a dozen miles south of Rabaul town itself. Their goal was to drop their bombs along the runway, to keep it unusable for any Japanese planes that might try to sneak in. The briefing officer told the men that this should be an easy mission – a 'milk run' – that would still give them a feel for combat conditions. He warned them that the Japanese still had good anti-aircraft (A/A) gunners around Rabaul, so they could get some enemy fire while over the target. That night Walt and the other men likely talked about nothing but their upcoming initiation into combat. Did they write letters home, or pray, or drink if they could? They all knew that a good night's sleep would be important, but could they sleep easily as their minds focused on the following day's new experience?

The following morning after breakfast, the men assembled early for the ride out to Carney Field. As with all South Pacific bases, planes were dispersed some distance from the main runway in individual revetments. These were spaced out so that an enemy attack would find targets spread out and more difficult to destroy. Once at their assigned plane, an older B-24D that had been fitted with a ball turret, each man went right to his job. Lt. Appling and Jupe checked first with the maintenance crew chief to ensure that everything was OK with this plane. The crew chief assured Appling and Bretherton that everything was in good shape and he also told them to bring the plane back to him the same way. Appling checked in with the instructor who would accompany them as an observer and advisor on the flight. Ed Bretherton went through his pre-flight engineering checks while Cliff Llewellyn checked out the radio equipment. Each man checked his guns carefully, then they also checked the load of 500-pound bombs along with Jimmie Clark, to ensure that the fuses were installed properly and each one was secured properly with a safety wire, so they would not go off prematurely or if they had to be jettisoned before they reached the target. Finally, the time came for Appling and Aubrey to start each engine and check the various engine settings. Each man went to his takeoff position, Lt. Aubrey in the seat next to Appling, Ed Dunne at his navigators station and Jimmie Clark at his bombardier position forward of the cockpit, with Jupe Bretherton on the flight deck behind Appling to help monitor gauges and call out settings. Cliff Llewellyn was at the radio operator's seat while the other gunners sat near the middle of the plane to help keep the center of gravity balanced with the heavy bomb load aboard.

Finally, the time came for Lt. Appling to rev up the engines and begin taxiing from the revetment out onto the taxiway. Along the way they were aware of the other two B-24s doing the same thing. Finally, they reached the main runway and after Lt. Appling got clearance for takeoff, they accelerated down the strip and into the air. Each plane was instructed to navigate independently to a designated IP near Rabaul, where the three planes would assemble for their bombing run. After they were in the air, each of the gunners moved into his assigned position, Frank Mansir to the front turret, Jack Riley to the tail turret, and Jupe to the top turret, while Len Sherman helped Walt into the ball turret and then hit the hydraulic switch that lowered him below the fuselage. Len went to the left waist gun position, with Cliff at the right window. They were going up to 12,000 feet for this mission, so each man checked his oxygen, in addition to connecting his headphones and throat mike. Lt. Appling went on the intercom and checked in with each man to ensure he had communications and that there were no problems. After each man settled in to his position, they also checked their guns by firing a few short bursts.

Old crew 430 had never flown a training mission of more than six and a half hours. Today they would break that record. Rabaul is about 660 miles from Guadalcanal. While the B-24 could fly at more than 300 miles per hour, on bombing missions the speed was limited to about 175 mph, to conserve fuel and allow them to carry a heavier bomb load. As Walt settled into his turret he had lots of time to observe the route below. They were flying northwest, up the water channel known as The Slot, with the individual Solomon Islands to the right and left. The Slot was the sight of many furious and famous battles between the US and Japanese navies, while the islands they passed below were equally famous for the fighting between the Japanese Army and the US Army, Marines, and New Zealand Army. Walt observed New Georgia to the left and Choiseul to the right, and then they passed directly over Bougainville past mountain peaks over 10,000 feet high. Later came a stretch of open water and then Lt. Appling announced that they had finally reached their IP.

The other two B-24s were sighted and the trio formed up and they started their final short approach to the target. The safeties on the bombs were removed. The men watched in all directions for possible enemy fighter opposition as Lt. Appling announced that they were on their final approach to the target. At this point on a bombing mission, the bombardier took over control of the plane via his Norden Bomb sight, which made slight automated adjustments to the plane controls to put it in the proper position to release the bombs. The bomb bay doors were opened hydraulically by Jimmie Clark. Walt began to notice little red flashes on the ground below, then a few seconds later there were a number of explosions in the air around them, each with a puff of black smoke and a concussion that rocked the plane gently. This was anti-aircraft fire that the Japanese were directing at them. It occurred to the crew that there were men down below who were trying to kill them! Walt thought 'why are they shooting at us? We never did anything to them.' A few of the A/A bursts were close enough so that the flak fragments pelted their plane, with a sound like gravel hitting the thin aluminum skin. Later Cliff would say that with each burst he would try to pull his head down into his shoulder blades, but that did not stop him from keeping watch out his waist window.

Soon enough they were directly over target, Jimmie Clark hit the switch, the bomb load released, and the plane lurched up slightly with the sudden reduction in weight. Walt could see the bomb load fall away and he continued to watch as they approached the ground. As soon as the bombs were away, each pilot resumed control of the plane from the bombardier, and turned quickly to get away from the A/A fire. Walt kept his eyes on the bombs though and saw that they hit the Tobera runway with a good tight pattern. The ball turret gunner had the best view of the bombing, so Walt went on the intercom to describe the results for the rest of the crew. He likely also sighted Rabaul itself and its magnificent harbor – a sunken volcanic crater, further off to the north as Appling turned the plane around to head back south. Ed Dunne set their course back to Guadalcanal and then wrote out a preliminary bomb run report, which Cliff radioed back to Guadalcanal.[53]

In somewhat less than four hours, they were approaching Carney Field again. Carl Appling put their plane down safely and they taxied back to the revetment. May 20, 1944, after a mission lasting eight hours, they were now a combat crew with one mission to their credit.[54]

After the engines cut off, they each gathered up their gear and assembled for the ride in a 6x6 back to the CTC debriefing room. As they rode, they talked about every aspect of the mission. Realizing that enemy fighters could have been at them at any moment created a new tension as they approached Rabaul. The A/A fire was also a new experience. They got through it OK, but they knew that if it had been closer, it could have hurt them. All in all, they agreed that they had been more excited than scared on this mission. Back at the debriefing room, they gave their reports to the intelligence officers, who were interested in how close to target the bombs fell, the intensity of the A/A fire, and where it seemed to be coming from on the ground. Walt and the other two ball turret gunners were questioned particularly closely on these aspects of the mission, because of their unrivaled view. At the end of the briefing, the instructor who had gone along with them as an observer complimented the crewmen on a good first mission.[55]

By this time the men were all hungry, so they headed back to their tent to drop off their gear, then off to the mess tent for chow. Many bomber crewmen after their first combat mission thought about the men on the receiving end of their attack. Their target was a jungle runway, but there were young men down there too. Did Walt and the other men on his crew wonder if they had killed anyone that day? Whether they did or not, their patriotism, combined with the facts of the war, the propaganda that both sides used, and the urge to support their crewmates and their new found home in the 13th Air Force ensured that any such thoughts would not stop them when they got the call for their next mission.

The following day, Sunday the 21st, after church and breakfast, the gunners had to head out to Carney Field to clean the guns on the B-24 they had used the previous day. Other than this routine, the pace slowed down for the Appling crew. There was no more CTC training – they had completed that. They were basically waiting for transfer orders now. At the end of the week, news came that they had been assigned another combat mission. They were assembled for a briefing with four other CTC crews and learned that early the following day (Saturday, May 27) the five B-24's were to return to Rabaul and bomb Tobera Field again.[56]

Early the following morning Walt, Cliff and the rest of the crew followed the same routine as before. First breakfast, then back to the tent for their gear in time to catch the 6x6 out to the revetment and their assigned B-24D. As they assembled and began to load their gear and check over the plane, Lt. Appling told the men that he had just learned from the CTC mission CO that one of the five crews was not ready and would miss the mission. Whether this was due to pilot illness or some other reason was not clear. The men continued to prepare for the mission and soon they were all aboard with the engines warming up.

Lt. Appling got the signal to begin taxiing down the steel mat toward the runway behind the other three planes. After just a couple minutes of slowly taxiing along though, the B-24 lurched to one side and slowed down. Len Sherman looked out one of the waist gun windows and yelled in his throat mike 'we've got a flat tire!' Appling cursed while Don Aubrey came back to confirm for himself. Aubrey cursed about how he did not like the steel runway mats. Appling was quickly on the radio to report the problem. There was no possibility of flying this plane now. Within a minute or so, Appling was instructed by the Carney radio operator to pull the plane into the nearest empty revetment, stop the plane and get the crew out with their gear.

As the crew was getting their gear off the plane, a 6x6 truck and jeep came speeding up. The CTC group Captain was in the jeep, yelling at the men to get their gear and get into the 6x6. He got Lts. Appling and Aubrey in the jeep and they sped off with the 6x6 behind. Soon they stopped at another B-24 – the one intended for the fifth crew that did not fly that day. This plane was fueled, armed, and ready. The ground crew already had the engines warming up. The CO and Lt. Appling told the men to get their gear on board quickly. They were going to try to catch the other three planes! Within half an hour of the flat tire, they were in the air but still 30 minutes behind schedule. Try as they might they were not going to catch the other three planes. Lt. Appling announced that if they could not catch up with the rest of the squadron, they were instructed to bomb the target on their own.

They arrived at the IP still too late to meet the other three planes, so Lt. Appling again announced that they would bomb on their own. The gunners were at their stations after the bomb fuses had been checked and the safeties released. They knew that if there were any Japanese fighters in the area they would be on their own to fight them off, so all the men were tense. They got to the target area OK. Jimmie Clark took over control of the plane from Appling with the Norden Bomb Sight controls as they began their bomb run. The A/A fire was very light, so they had no problem releasing their bomb load right over the target, then they began to turn to get away and head back south. As Walt and Jimmie Clark watched the bombs fall away toward the target, all of a sudden they both noticed a group of US Marine Corsair fighters far below strafing the Tobera airfield! Walt and Jimmie held their breath, not believing what they were seeing as the bombs fell towards the low-level fighters. Fortunately, the bombs and the bomb blasts missed the fighters and Walt and Jimmie breathed a sigh of relief. They went on the intercom to describe what had happened to Appling and the crew.

On the way back to Guadalcanal everyone on the crew was distressed by what had almost happened. The gunners especially were aware of the danger of 'friendly fire' injuring or killing other Americans while firing their guns in a tight formation of bombers. This danger had been drilled into them during aerial gunnery training, then again during bomber phase training, and also by the XIII Bomber Command CTC instructors. Lt. Appling told the men that there was no mention of the Marine fighters during their pre-flight briefing – why were they there? After they landed at Carney Field, they filed into the briefing office and explained to the intelligence officers what had happened. By the time they had each finished their complete mission report, one of the intelligence officers came back with the explanation – the Marines had timed their strafing run on Tobera to occur shortly after the original B-24 attack. However, no one notified the Marine air commander that one of the B-24's –

the Appling crew – took-off late and would bomb late. Also, no one told Appling to be on the lookout for the Marines below. The S-2 officers told the men they had done nothing wrong – Walt and Jimmie had proven their mastery of aircraft recognition skills to correctly identify the planes far below as US Marines Corsairs. Command and crewmen that day all learned the lesson that things could easily go deadly wrong because of human error. It was a dangerous business they were in. Just as they were leaving the briefing room another intelligence officer came in and said, "…hey you guys know who that was with the Marines at Tobera? That was Pappy Boyington!"[57]

Walt's second mission had lasted eight hours and 15 minutes. The men were all tense, hungry and tired. After supper though, Cliff told the men that they were to report to XIII Bomber Command CTC HQ. There Lt. Appling told them that he had learned that they had all been transferred from XIII Bomber Command, to the 307th Bombardment Group (Heavy), 424th Bombardment Squadron of the 13th Air Force.[58] They were to be prepared in a few days to pack up their gear for a flight to Los Negros in the Admiralty Islands, where the 424th Squadron was based. Once there they would go into rotation with the other combat crews of the squadron. Carl Appling also told the men he had learned that there was a general movement of combat units and support personnel to the north and west. It appeared that major new operations were beginning in the South Pacific and they were going to be part of it. Even the XIII Bomber Command CTC on Guadalcanal was closing in a few days. In the future newly arrived bomber crews would go through final combat training at a 5th Air Force facility at Nadzab in New Guinea.[59]

- 10 -
Los Negros: The Long Rangers

The 307th Bombardment Group (Heavy)

The 307th Bombardment Group came into existence on January 28, 1942, just seven weeks after the attack on Pearl Harbor. Initially just a 'paper' formation, on April 15, 1942 the first five enlisted men were assigned to the 307th at Geiger Field in Spokane, Washington. Lt. Col. William A. Matheny was assigned as first commanding officer on May 22, 1942. The Group was organized into four squadrons: the 370th, 371st, and 372nd Bombardment Squadrons and the 35th Reconnaissance Squadron. Shortly afterwards the designation of the 35th Squadron was changed to the 424th Bombardment Squadron. By May, the Group personnel expanded to more than 400 enlisted men and training began on B-17 maintenance. On May 26, 1942, the 307th was transferred to Ephrata Air Base, located on a dry lakebed in central Washington. 1,800 enlisted men, 360 officers, and 16 B-17's were assigned to make the group operational. The 307th was assigned to Ephrata to protect Seattle and Spokane in case of a Japanese attack on the Pacific Northwest. This was before the battle of Midway and there was great uncertainty as to where the enemy might strike next.[1]

In September 1942, the 307th went to Sioux City, Iowa, where it became one of the first Army Air Force groups equipped with the new long-range B-24 bomber. By November 1942, it had been transferred to the Seventh Air Force at Hickam Field in the Territory of Hawaii, where it not only continued to train but also flew long-range patrol missions. In December 1942 and January 1943, the 307th initiated its first long-range bombing missions against Wake Island by staging through Midway. In May 1943, the 307th was transferred again to Guadalcanal, where it was assigned to the 13th Air Force. In April 1944, when the 307th moved from Guadalcanal to Los Negros, the Group was commanded by Col. Robert F. Burnham.[2] By May 1944, the 307th was one of the most experienced units in the Air Force, with over 16 months of almost continuous combat.

Each heavy bomber squadron of the 307th Bomb Group (and those of the Fifth Bomb Group as well) was authorized 24 combat crews. Each combat crew was authorized one 1st Lt. (pilot), three 2nd Lts. (copilot, bombardier, and navigator), and six sergeants (flight engineer, radio operator, and four gunners). Every fourth combat crew was authorized a Captain in place of the 1st Lt. as pilot, to act as a flight leader. Including ground support personnel, each squadron comprised a nominal total of 115 officers and 432 enlisted men.[3] In May 1944, when the Appling crew was transferred to the 424th the squadron was commanded by Major Laurence F. Krebs.[4]

Los Negros Island, in the Admiralty Islands, is northwest of Rabaul. The U.S. First Cavalry Division had landed on Los Negros on February 29, capturing Momote Airfield from the Japanese defenders. Over the next two weeks the First Cavalry as well as elements of the Marines completed securing Los Negros, as well as nearby Manus Island.[5]

The Seabees (Navy construction battalions) quickly followed to bring Momote Field into operation and to begin constructing various base facilities needed for Los Negros to be an effective and efficient bomber base. Momote Field was lengthened and revetments added nearby for parking the bombers and fighters that would be stationed there. The first combat air elements to arrive at Los Negros was a RAAF (Royal Australian Air Force) wing equipped with Spitfire fighters, which would provide air defense of the base. By April, the 5th Bombardment Group (sister to the 307th Bombardment Group) of the 13th Air Force began arriving at Momote. As the 5th Bomb Group began flying operations, a serious defect of Momote Field was discovered. The runway was not level – it sloped up gently from either end to a crest at the center that was several feet higher. This caused many accidents on takeoff, as inexperienced pilots believed that they had taken off after they passed the crest in the middle of the runway, when in fact they did not have sufficient airspeed yet. This defect was never corrected, because the airstrip could not be taken out of operation.[6]

As an alternative to reconstructing Momote, the 13th Air Force's 821st Engineering Aviation Battalion began constructing an all-new airfield at a coconut plantation about 20 miles north called Mokerang. They worked around the clock. At night they used anti-aircraft searchlights for illumination. The 821st had an 8,000 runway at Mokerang ready for use by May 26.[7] Well before the airstrip was completed, advanced echelons of the 307th Bomb Group began arriving by C-47 cargo plane to start planning and laying out the camp area at Mokerang for the rest of the unit. The

Seabees had cleared an area of trees and leveled the ground using bulldozers to provide sites for headquarters buildings, mess tents, latrines, and crew camp tents. Because this was a virgin building site, exotic pests were encountered, including eight-foot long jungle snakes and various other coral snakes and adders. Mosquitoes were found to carry elephantiasis, so men were instructed to wear long sleeves during the day as protection, until breeding areas could be eradicated.

Between May 4 and May 12, the other air echelons and ground personnel of the 307th arrived at Mokerang. This was the first time in many months that all four squadrons of the 307th were located at the same base. Prior to this, the 370th and 424th squadrons had been at Guadalcanal, while the 371st and 372nd squadrons were based at Munda on New Georgia Island. At first, there was not even electricity at Mokerang, and fresh water had to be trucked 20 miles from Momote. Clothes had to be washed in the sea to conserve fresh water and the only bathing was in salt water.

The Navy controlled Mokerang. It had two parallel runways, each 7000 feet by 100 feet wide, with 50-foot shoulders and 500 foot cleared and graded overruns at each end. The surface was compacted coral: no more perforated steel mats! The runways were aligned NW/SE. The SE approach was over flat ground, cleared for 1000 feet. The NW approach was over open water.[8]

There were approximately 13,000 native Melanesians on Los Negros. They provided a ready source of willing and able labor to assist with a variety of construction duties. They were particularly adept at thatching roofs for the mess halls and headquarters buildings with woven palm fronds, as was normal in their own villages. Most of the natives near Mokerang lived in clean villages built at the edge of the sea, which they kept spotlessly clean each morning.[9]

Los Negros marked a change to the Allied axis of advance. Before, the thrust had been north. Now the airfields at Los Negros allowed the advance to pivot west, towards Biak, Noemfoor, and the Philippines, with protective air strikes north to neutralize Japanese bases in the central Pacific. Momote and Mokerang brought many new targets into range for the B-24's of the 5th and 307th Bomb Groups. Truk, Yap, Palau, and Biak would be the new targets for the aircraft and crews of the 307th.

During May, prior to the arrival of the Appling crew on Los Negros, the 424th Bomb Squadron (BS) had completed 20 individual bombing missions. 14 of these missions were to Biak Island, far to the west, off the north coast of Dutch New Guinea. The Biak missions were in advance of the U.S. invasion of the island, intended to prepare it as an air base for the allied drive to the west towards the Philippines. Three bombing missions were to Woleai Island and one to Alet Island, key supply points between Japan and Truk. Two 424th BS missions were to Truk, the huge Japanese naval and air base. Destruction of Truk as an effective enemy base would be critical to allow the U.S. Navy to proceed with its advance towards Japan through the Central Pacific, north of MacArthur's more southerly line of advance. During May, the 424th put six planes in the air for the typical daily mission. They dropped 240.5 tons of bombs, of which 214 tons were judged to have hit their target. 424th gunners fired more than 24,000 rounds of .50 caliber ammunition during May, resulting in the destruction of one Japanese plane. The squadron lost no men during the month.[10]

The 307th Bomb Group did suffer casualties during the month. One plane crashed on takeoff with loss of life. Another Group B-24 had a Jap Zero crash into it in the air above Truk, shearing off the left rudder and stabilizer. The plane fought its way to within 200 miles of Los Negros, before finally hitting the water and exploding.[11]

Early on the morning of Tuesday, May 30 Walt, Cliff, and the rest of the crew made a six hour and 15 minute flight in a C-47 from Guadalcanal to Mokerang field on Los Negros.[12] They hopped into a truck for the drive to the camp, noticing that Seabee bulldozers were digging long trenches to bury the Japanese dead. They unloaded their gear and reported for orders. They were shocked to learn that the commander of the 424th Squadron, Major Laurence F. Krebs, had decided that he was going to break up the crew and assign the men as replacements for other existing squadrons. The men were told to find quarters, but wait for follow up orders. Dejected, the ten officers and men puzzled over this decision. Except for Lt. Appling, they had been together as a unit for about five months since Muroc. They knew and liked each other, and Lt. Appling was likeable too and seemed a top-notch pilot the men came to trust after their two combat missions from Guadalcanal.

The men went to the squadron quartermaster's depot to get their tents, cots, and tools, and commenced to erect their living quarters – a six-man tent for the sergeants, and a four-man tent for the officers, each in their own designated area of the base between the runway and the beach to the northeast. Mosquito netting was supposed to be draped over each cot to keep the bugs out at night, and a single light bulb in the center of the tent was provided for light until the electricity was cut off at ten each night.[13] The campsite was pleasant enough, within the trees of the old plantation, and work on their tent took their minds of their situation for a day or two. But no one seemed to have any mosquito netting to provide them though, which resulted in a plague of insects dusk to dawn.

Then nothing happened. Other crews were busy flying every other day and the base was buzzing with activity, but they just seemed to sit. They checked in at squadron HQ each day and were told that Dunne, Clark, and two of the gunners would get orders soon to transfer to another crew. Finally, they started hearing rumors that explained their predicament. They were still listed as the McRae crew, and it turned out that Major Krebs and Lt. McRae had served together years ago as flying cadets. The two were big rivals as cadets and McRae had been picked over Krebs to be acting leader of the training class. Krebs carried a grudge and saw the opportunity to break up McRae's crew while he was out with his ear infection. This would make it very difficult for McRae to get a crew again later, if and when he did return to flying duty.

Unknown to the crew though, on May 30, 1944, the day that Walt and the Appling crew arrived on Los Negros, Major Krebs was replaced as commanding officer by Captain Clifton L. Fowler. Krebs had been transferred to 307th Bomb Group Headquarters.[14] Finally, after Krebs departed and Fowler settled into his new job, he realized that he had a crew that was doing nothing. Krebs blew his stack, calling the Appling crew early in the morning on Friday, June 2 for a talk. All transfer orders were cancelled. They would remain together as a crew and they were to prepare for their next combat mission the following day![15] Formal orders were cut assigning the Appling Crew to the 424th.[16]

Truk

The crew was happy – they were going to stay together and they would finally get into action with the rest of their new squadron. However, later in the day, news started coming in from the 424th crews that had been in combat earlier on June 2. Five 424th B-24's, in addition to bombers from the other 307th Bomb Group squadrons, had taken part in the strike against Truk earlier in the day, arriving over the target at 1106 at an elevation of about 20,000 feet. The bomb drop against Dublon Town, the Japanese base area on Truk, was judged to be successful, with two huge explosions observed on the ground, as well as many smaller fires. Then the B-24's were attacked by 30 to 40 enemy fighters. Most of the enemy attacks came from above and from 10 to 2 o'clock. Some of the enemy passes came within 100 yards and the Japanese pilots seemed very determined. The aerial gunners managed to fight off the attackers, destroying six of the enemy planes with two additional 'probables.' However, at about 1130, a Zeke made a determined vertical pass from 11 o'clock at Lt. Townsend's plane. Townsend's copilot, Lt. Harvey R. Vanderslice, was hit by a 20 mm shell that entered through the right windshield. The shell passed through Vanderslice's wrist and entered his right leg, where it exploded. Three 7.7 bullets also entered his chest. Enroute back to Los Negros, Vanderslice died at 1345.[17] In addition, a pilot and navigator from a 372nd squadron plane had also been killed during this attack.[18] The news of Vanderslice's death, the death of the two 372nd squadron men, and the reports of the determined defense at Truk gave Walt and the rest of the crew something to talk about that evening, and to think about as they went to sleep in preparation for their combat mission tomorrow.

Before sleep though, the men assembled on Friday evening with four other 424th crews for their pre-mission briefing. The 307th and Fifth Bomb Groups would each put four six-plane squadrons in the air to attack Truk and Eten Island. At about 1830 the briefing tent was packed with several hundred babbling men when a major strode to the front of the tent and a master sergeant bellowed out 'Atten-hut!' Two-dozen crews snapped to attention in silence, and the sergeant announced 'At Ease!' and they sat down. The major strode to a large map with a pointer in his hand and announced 'The 307th BG target will be the aircraft and repair facilities on Eten Island in the Truk group.'

The Group was instructed to fly at 9,000' at an indicated airspeed of 155 mph to the IP, designated as 07°16'N-161°45'E. Once formed up at the IP, the Group would make a 20 degree turn to the right and come into the target at a heading of 40 degrees magnetic. The 372nd Squadron would lead the attack at an elevation of 20,600' with the 370th Squadron following on their left, but at 21,000'. A second section would be led by the 371st Squadron at 20,200', with the 424th Squadron following on their left flying at 19,800'. The bombing run speed was designated as 160 mph. The lead bombardier of each squadron was instructed to sight for range and deflection, with all other bombardiers sighting for range only. The 371st and 424th Squadrons were instructed to aim their bombs 200 feet to the right of the angle formed by the Eten Island apron and runway and paralleling the runway. After bombs away, the squadrons were instructed to make a right turn at 165 mph. Each of the planes was armed with nine 500-pound general-purpose bombs, instantaneously fused.[19]

The major told them that Truk was one of the Marshall Islands that Japan, as one of the Allied nations, had seized from Germany during World War One. The League of Nations had confirmed Japan as the mandate holder for the Marshall's in 1920. Despite the League of Nations prohibition against fortifying the Islands, Japan had spent 20 years turning Truk into the 'Gibraltar of the Pacific' with naval facilities to support hundreds of ships, airfields, barracks, anti-aircraft gun emplacements, and thousands of troops for defense.[20] As recently as February, it was home to four Japanese battleships, four carriers, 11 cruisers, dozens of destroyers, and hundreds of land based fighters and bombers.[21] Lt. Vidmar would lead the mission for the 424th, with the other four planes under his command. Lt. Appling, as a new pilot was warned to stay in a tight formation over the target, both to keep their bombing pattern tight and also to concentrate the defensive firepower of the gunners against attacking fighters.

The men were up early, before dawn. After breakfast, they were out at the airstrip where they found their B-24, number 144, nicknamed 'Whadahell', ready and waiting for them. Checking and rechecking their gear, they nervously awaited the order to proceed. Lt. Vidmar made a point of checking with Appling and the rest of the crew, to ensure that these green men were not forgetting anything critical. Finally, shortly after dawn, they took off down the Mokerang runway, gaining steady altitude over the sea as they made their way in loose formation. Their target was to the northeast, but they took a circuitous route to try to avoid being jumped by a large concentrated mass of fighters like the mission the previous day.

Once in the air and at elevation, Walt descended into the ball turret, hooked up his oxygen and intercom, test fired his guns, and settled in for a long flight. The weather to and from the target was good – with scattered cumulus from 3,000 to 7,000 feet. Over the target the cloud cover was 3/10 with some towering cumulus to 23,000 feet. At around 1100 as they approached Truk, the five planes formed up into their tight bombing formation at 21,000 feet and prepared for the attack. They approached the target at Eten Island at 1115 and began their bombing run. Heavy, accurate anti-aircraft fire began to pepper the planes as they pressed on. As they did so, Walt heard a single urgent word over the intercom: 'Bandits!' 15 Jap fighters approached high above them. Someone else confirmed 'There they are.' Before beginning their attack, a few of the Jap fighters dropped phosphorous bombs above and in front of the approaching bombers. This was a danger the Walt and the Appling crew had heard about, but not yet seen. Phosphorous bombs were fused to explode shortly after being dropped by the attacking fighter. They were packed with a jellied phosphorous substance that would burst into dozens of individual burning globs that were intended to fall onto the wings and fuselage of the bombers. The globs were sticky and burned intensely from the same chemical used to ignite a match – they could burn through a wing into the fuel tank or engine, or through the fuselage and into the plane itself.[22] Walt could not see the planes dropping phosphorus bombs, but he heard Carl and Don chatter over the intercom as they burst above. Then he noticed some of the smoking streams of burning phosphorus as they dropped below the level of the bombers.

After the phosphorous bombs were dropped, the fighters attacked the 307th with their guns. The quintet of 424th Squadron B-24's bore down on their target, dropped their bombs, then by 1118 they turned away while still under attack. While alert for attacking fighters, Walt did notice that most of the bombs fell in the water and not on target. As they turned away, the A/A fire decreased and then stopped, but the 15 attacking Japs pressed their attack. Many of the planes attacked from behind the formation and low, due to thick cloud cover above. They did not seem to come in as close as was reported the previous day though. But now this was the first time that Walt saw a fighter with the red sun on its wings and side – a Jap Zero. This was it, this was real. But Walt did not fire. All he could do was look at it sped by. The next time he had a Zero in his sight he did not hesitate though.[23]

Walt and the other ball turret and rear turret gunners got their bursts of shots away as best they could. However, the enemy seemed to stay outside the range of their guns, a tribute to the skill of the American aerial gunners. Carl Appling spoke calmly over the intercom: 'Short bursts gunners, short bursts, don't overheat your guns!' S/Sgt. Chester Vaughn, left waist gunner on Lt. Lewis' crew, hit one Zeke that got in close. This enemy came in level at 9 o'clock and got to within 600 yards, when it turned away in the face of Vaughn's fire, then exploded in mid-air, with the pilot seen to bail out just before the explosion. Lewis' top turret gunner, T/Sgt. Byron Benoit, poured fire into a Zeke coming in at 11 o'clock high. At about 600 yards, it broke away and a large part of its tail was seen to break off, after which it went into a dive, although it was not observed crashing. Shortly after the attacking planes all departed as the B-24's made their way to the southwest and back to Los Negros. Radio checks soon confirmed that all planes were accounted for and no one had been injured.[24]

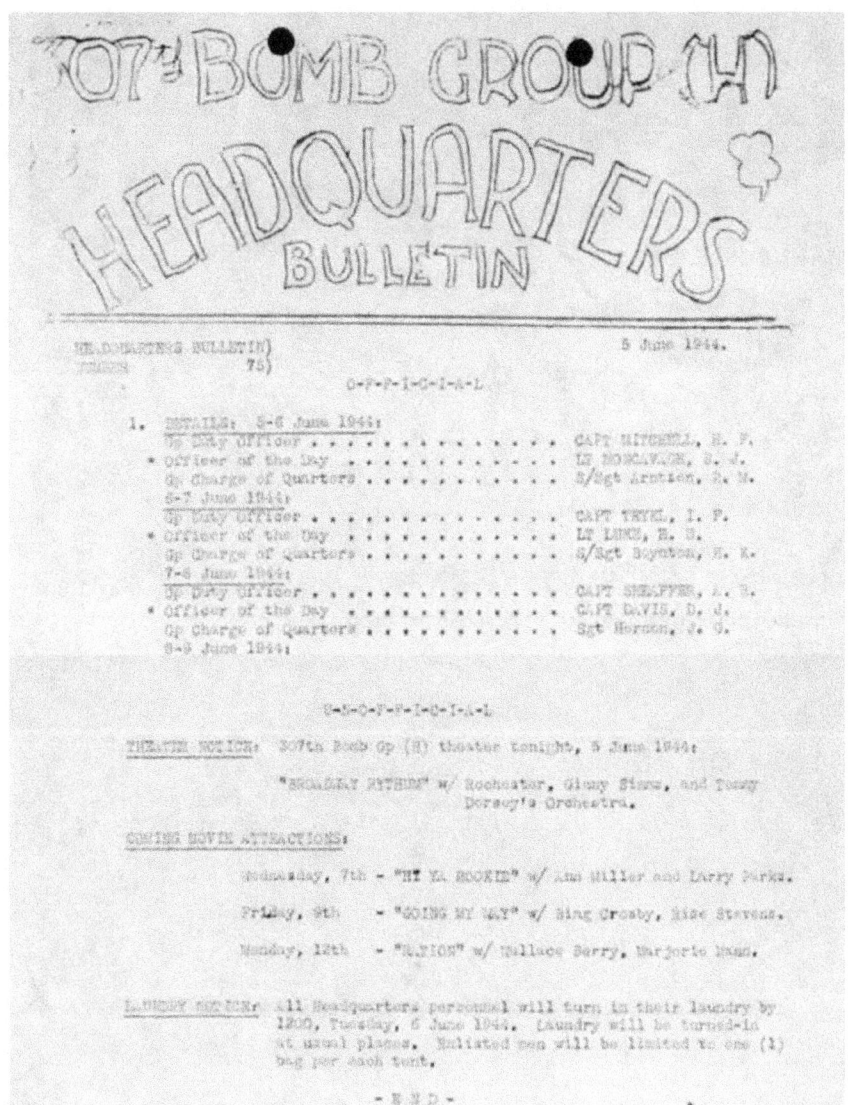

307th Bomb Group Bulletin: – Los Negros, June 5, 1944 with movie schedule and laundry note. This was published during Walt's first week on Los Negros after assigned to the 307th Bomb Group.

Los Negros: A/A guns outside Los Negros HQ area

Los Negros: Invasion of Los Negros to secure new bomber base. *USAAF Information Poster*

Los Negros: Mokerang Airdrome. Walt's home from May 30-August 19, 1944. *Airdromes Guide Southwest Pacific Area, p. 27*

Los Negros: Funeral ceremony at USAAF chapel.

Los Negros: Walt crouches at his ball turret while injured in late July 1944.

Los Negros: Crewmen with A/C 540: Standing (L-R): Ed Bretherton; Walter Babinski; Leonard Sherman; Jack Riley; Frank Mansir; George Volchko. Kneeling (L-R): Ed Dunne; Jimmie Clark; Carl Appling; Don Aubrey. Photo likely taken on June 11 or July 3, 1944. Note that the censor has cut out a portion of the photo where the secret radar unit would be located.

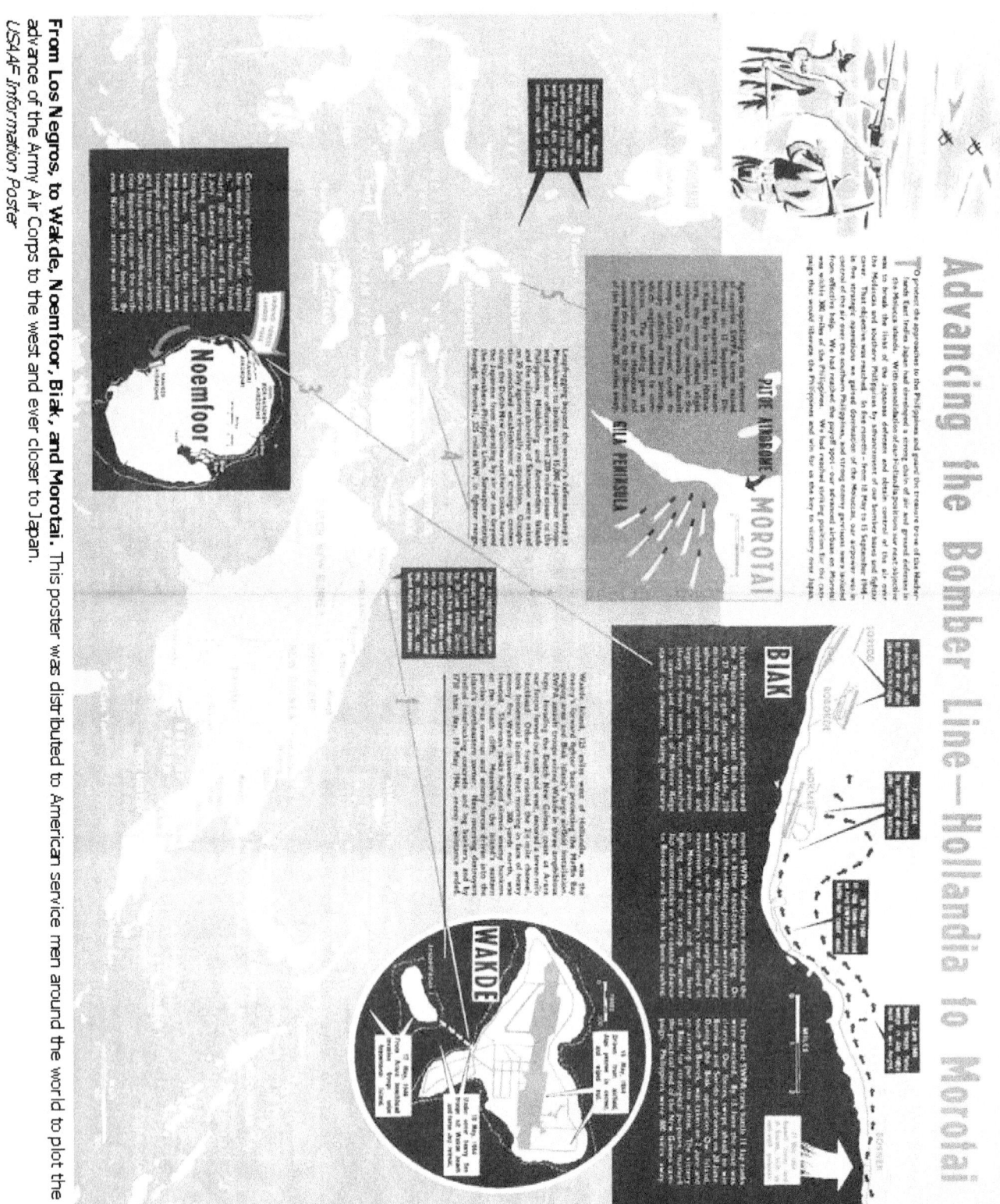

From Los Negros, to Wakde, Noemfoor, Biak, and Morotai. This poster was distributed to American service men around the world to plot the advance of the Army Air Corps to the west and ever closer to Japan. *USAAF Information Poster*

Wakde: 307th Bomb Group tent area. *Missing Aircrew Project.*

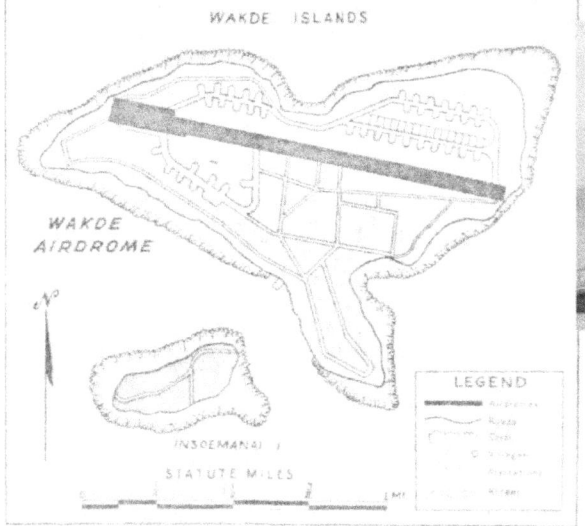

Wakde Airdrome: Walt's home from August 19-September 26, 1944. *Airdromes Guide*

Wakde: 307th Bomb Group shower with lister bags for hot water supply.

Wakde: View of tent area with the New Guinea mainland in the distance (above) and Japanese religious monument on Wakde used as a telephone pole (left—*Missing Aircrew Project*).

Wakde: Bob Hope USO Show in August 1944 finally begins.

Wakde: Bob Hope USO Show with dancer Patty Thomas. 'Hey—sit down in front!'

Wakde: Bob Hope USO Show singer Frances Langford.

Wakde: Bob Hope USO Show. Bob talks with 31 year old singer Frances Langford.

Wakde: Bob Hope USO Show with dancer Patty Thomas.

Wakde Airdrome: C-47 Skytrain landing on Wakde, hopefully with fresh meat, fruit, and beer from Australia. And maybe a lift for Walt for a week-long leave in Australia.

Sydney Australia: - A Bobby patiently stops for yet another picture for an American GI.

Sydney Australia: -Luna Park Amusement Park, just north of the Harbour Bridge.

Sydney Australia: Civic Theatre where Walt and Johnnie saw a double feature.

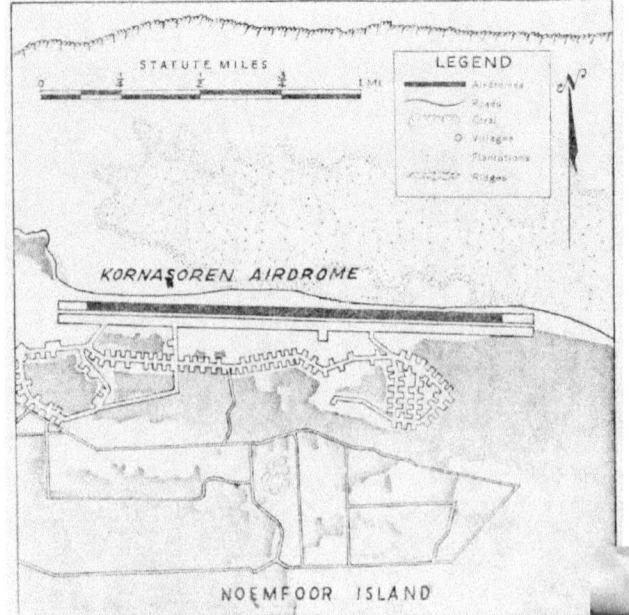

Noemfoor: Walt looking skinny but happy.

Noemfoor Island and Kornasoren Airdrome: Walt's home from September 26-November 10, 1944. *Airdromes Guide Southwest Pacific Area, p. 83.*

Noemfoor: The airfield was littered with wrecked Japanese airplanes.

Noemfoor Island: 307th Bomb Group camp area.

Noemfoor: Walt (second from right) and his pals relax on a bomb dolly on Noemfoor in October or November 1944.

Noemfoor: Walt (back row left) and his crewmates at the 424th Squadron tent area on Noemfoor. Not a pleasant location.

Noemfoor Esprit-de-corps: Leather bomber jackets with the new 424th Bomb Squadron patch were purchased, photoed, then put away for safe-keeping.

Noemfoor Bravado: Four friends pose on A/C 547 in October, 1944 on Noemfoor. Frank Mansir on his nose turret (above), Jupe Bretherton on his top turret (above right), Walt at the left waist gun (below left), and Jack Riley on the tail turret (below right).

Noemfoor: A B-24 burns at the end of Kornasoren airstrip.

Clouds Over the Molucca Sea: A .3 layer of alto-cumulus with a lower .2 layer of strato-cumulus clouds. Photo by the author over the Molucca Sea in October 2013. Walt passed over these same seas in October 1944, to and from Balikpapan.

Balikpapan Fires burn in the Balikpapan refinery and storage tank areas as Long Ranger B-24s start on the long and dangerous return flight to Noemfoor or Morotai. *307th Bomb Group.*

Capt. Zbigniew Babinski of the Polish Air Force. His whereabouts unknown since 1939.

Lt. Julian Gebolys (center) with Lt. Maurice Newnham and Polish Paratroops at Ringway, England. *Fronczak Collection, Buffalo State University.*

V-Mails: While the fate of relatives in Europe was a mystery, American GI's kept in contact with family and friends back home via V-Mails.

Above left, Pvt. Joseph Prus wrote to 'Mr. & Mrs. Kelly." from New York during Easter '43 and again from a field hospital in England at Christmas 1943.

Below, left, Cpl. Joseph Stoklosa, USMC, V-mail to Mr. & Mrs. Kurzyniec.

Below, right, Walt wrote to Dorothy Kurzyniec from Los Negros in June 1944.

Southwestern HS Grads:
Friends Dorothy Kurzyniec and Genevieve Babinski, Southwestern High School graduation in 1945.

More V-Mails: Walt sent this one Martha Kurzyniec: 'The Japs have surrendered, but the natives are giving trouble.'

V-Mail: Walt sent this one to Dorothy Kurzyniec. 'Be merciful, Oh God, I pray, Take care of her while I am away.'

Dorothy Kurzyniec:
Southwestern High School Graduation Yearbook photo.

Walt and the other ball turret gunners noted many enemy ships maneuvering around Eten and Truk, apparently taking evasive action. A formation of 16 ships included cruisers and destroyers. Six other scattered ships were sighted, as well as six to eight barges. All of the 307th BG aircraft returned safely to Mokerang between 1512 and 1537.[25]

During their debriefing, Walt and the other ball turret gunners were questioned on the effectiveness of the bombing. The results for the 424th were disappointing. The poor bombing was blamed on faulty aiming by the lead bombardier in Vidmar's plane and also some mechanical failure. The rest of the group did better though, with about 50 percent of the bombs considered to be on target and many fires observed in the hangar and repair facility area. The mission had taken exactly nine hours and each man had flown over 1400 miles. Capt. Fowler was not happy with the results, but he was glad that there were no casualties or aircraft damage. He told the Appling crewmembers that they had done OK and to expect to be called to fly missions about every other day.

The following day, Sunday June 4, Walt and Jack Riley attended mass at the 307th Bomb Group chapel, located on the shore of the Bismarck Sea. At this point the chapel was just a large open tent with a rough alter at one end and wooden planks for the men to sit on.[26] The only decoration was a lattice of palm fronds behind the altar. The 307th Bomb Group had three chaplains, Lt. Frank D. Dennis, presided over services for Protestants, 1st Lt. Max Lidz held Jewish services, and Chaplain Patrick C. Coyle led services for Catholics.

There were no missions scheduled for the 424th on June 4 or June 5. On June 6, the men of the 307th Bomb Group received a windfall – free beer and cigarettes! The 307th PX originated in Hawaii and it was one of the first Post Exchanges (PX) in the southwest Pacific. For a period the 307th PX was the only one available on Guadalcanal. However, on Los Negros it was decided that unit PX's would be consolidated into a single base PX. However, the Los Negros base PX would not buy all of the stock from the 307th, so it was decided to ration out the cigarettes, tobacco, and beer to all the men of the Group.[27]

That same day, Tuesday June 6, a storm blew through the Admiralties and all missions were cancelled. That afternoon the Appling crew was notified that they were to fly a mission the following day. The pre-flight briefing officer told them that they would be returning to bomb the airstrip on Eten Island, near Truk, the following morning. Six 424th Squadron B-24's would be part of a 48-plane strike by the 13th Air Force, which would include planes from each squadron in both the 307th and 5th Bomb Groups. The Appling crew was assigned to aircraft (A/C) 215, nicknamed 'Bunny-Boo.' Wednesday morning came and after the preflight routine, they took off from Mokerang between 0627 and 0653, gained altitude and began to make their way towards Truk. Walt descended into the ball turret, checked his guns and other gear, and settled in for the flight. The weather seemed fine as the group took off, but the farther north they flew, the worse the weather became. The 48 B-24's were to fly independently to a designated point near Truk, then form up for the bombing run. Heavy wind and driving rain made the flight very rough, especially for Walt in the ball turret. Appling and Aubrey both agreed that they had never seen weather this bad before. They checked with Ed Dunne who reported he was having a difficult time navigating, due to the bad weather and the buffeting the plane was taking. Above 14,000', moderate icing was encountered, along with turbulent winds and even sleet![28] Appling, Aubrey, and Dunne were worried. They were still a green crew, and they did not want to turn back if there was still a likelihood that they could make the rendezvous with the rest of the mission, but the weather continued to deteriorate. Finally, after about three hours in the air, Appling and Aubrey agreed that they could not proceed and needed to return to base. They radioed their decision back to Los Negros, were they touched down after 6 hours and 35 minutes in the air. They were disappointed that they had returned, but later learned that they were not alone. Of the 48 planes, only 10 made it to the target, the rest returning to base due to the weather. All the 424th planes aborted the mission.[29]

Two days later the Appling crew was called for another mission to Truk – this one to attack Dublon Town. They were back in Whadahell, A/C 144, armed with six 1000-pound GP bombs.. Once more, Walt settled into his ball turret after they were airborne shortly after 0708. Again the weather eventually turned bad as they flew north. The flight of 424th Squadron bombers pressed on to the northeast though, hoping for a break. About two and a half hours into the flight though, Ed Bretherton, the flight engineer, noticed that one of their engines was sputtering and then caught fire. Appling quickly killed the engine and fortunately the fire blew itself out. They could not proceed with just three engines though, so at 1000 Cliff radioed back to Los Negros that they were turning back on three engines. They were instructed to drop their bombs over the ocean. They would not want to try landing on three engines with a couple tons of live bombs in their hold. The crew also jettisoned the .50 cal ammunition for the tail and waist guns to lighten their load.[30] Now everything depended on Appling and Aubrey, along with Ed Bretherton. This trio kept an eye on every gauge and the condition of the remaining three engines. As they approached Mokerang, the base radioed for their condition and told them that a fire and rescue crew would be waiting as they landed. Appling put the plane down with no problem though and brought it to a halt after exactly six hours in the air. Their confidence and respect for Appling, Aubrey, and Bretherton grew after that flight. Eventually they learned that a blown cylinder had caused the fire in the #3 engine. They also learned later that weather had once more forced all the 424th Bomb Squad planes to return without reaching Truk, although some of the other 307th BG aircraft made it to the target that day.[31]

Soon the weather then turned more settled. In fact, Walt would find the weather on Los Negros monotonous. The typical daytime high temperatures would be about 85 degrees; while at night it would get down to about 76. Each month at Los Negros would see 10 inches of rain, or more. The rain would come in thundershowers and a day without rain was the exception, not the rule. The temperature was made somewhat pleasant by mostly gentle winds up to about 15 mph that would blow mostly from the southwest, east, and southeast. Rumors began to circulate about the invasion of Europe by the US, Britain, and Canada that had started a few days before. Detailed news was hard to come by though and the men felt isolated from what was going on in the rest of the world.

As June progressed, life started to settle into a routine for the men of the 424th at Mokerang. Food was plentiful, though monotonous. Spam was the staple meat, along with all kinds of stew, potatoes, bread, and canned peaches. Dehydrated eggs were available for breakfast.[32] A policy was established that every man stationed in a combat zone like the South Pacific would receive free tobacco, compliments of the American tobacco companies back home. Whether he smoked or not, every week each man was given one carton of cigarettes, six cigars, and a plug of chewing tobacco. The men would discover that in the South Pacific there would be long periods of monotony between missions and within a couple months almost every man smoked. The cigarettes were good brands too, while people back home were lucky if they could buy third-rate smokes.[33] Of course Walt had been smoking for many years, so he did not need convincing.

Sports were organized, with baseball (naturally), volleyball, and badminton being the most popular with the men. War news was difficult for the men to come by, so on June 23, Capt. Standart, from the 307th headquarters, began weekly Friday talks for the men on the progress of the war. The officers had the added advantage of an Officer's Club, which was steadily enlarged.[34]

Swimming was an obvious pass time, but the Los Negros beaches were bounded by coral reefs. Determined swimmers though could wear old shoes and walk across the sharp coral out to deep emerald pools at the far edge of the reef. There the men could explore the colorful coral formations and the exotic fish that inhabited the pools.[35]

The crews were in constant contact with the natives who lived in the area. Most natives lived in family groups of 10 to 15 people in lean-to's and huts near the beach on the road from the 424th camp area out to the airfield. Near each hut was a large black kettle that the family would cook in, each family member dipping his or her bowl in for poi, made from the plentiful coconuts from the pre-war plantation. Most of the natives seemed to have had some education from a Christian missionary who lived on the plantation before the Japanese came in 1942.

Walt and the rest of the crew noticed that there were not many native men around and those they did see were mostly old. The belief was that the Japanese had killed the younger men or worked them to death. The children were good swimmers, and soldiers and airmen often amused themselves by throwing small coins in the water and watching the kids dive down to come up with the money. The men and women wore loincloths or short skirts. The story was told how the 307th Command tried to get the women to cover up their breasts when the AAF first arrived at Mokerang. The native chief replied, 'OK, our women will cover up, but you'll have to supply the material.' The 307th supplied rolls of burlap that had been used to cover machinery on the way to Los Negros. Soon the island native women were wearing the scratchy burlap over their bodies, but they cut large holes in the front for their breasts. After that, 307th HQ gave up and the women went back to just wearing skirts. Most servicemen did not care what the natives wore or did not wear.[36]

Those able-bodied native men that could be found were paid by the Australians to work for the American military. Two dollars per month, one pound of rice per day, and a weekly supply of canned meat, sugar, fish, tobacco, tomato juice, and soap was the standard compensation. The natives constructed the headquarters buildings for the 307th and began work on a chapel to replace the tent used for religious services.[37]

On Sunday June 11, the Appling Crew was assigned another mission. Capt. Fowler himself led six 424th Squadron B-24's to bomb Dublon Town on Truk. The 424th was accompanied by other 307th Bomb Group squadrons, while the Fifth Bomb Group attacked Peliliu Island. Late in the morning, after about four hours in the air, the Appling crew in A/C 540 formed up with the other 424th Squadron planes at the I.P. As they approached Dublon Town, there were thick clouds that obscured the target, so Capt. Fowler decided to try one of the alternate targets, choosing the airdrome on nearby Moen Island. A break in the clouds allowed the bombardiers to get a good sight on the target as they made their bomb run, but as they turned after bombs away, the clouds obscured the target so they could not see the effect of the bombing. There was no flak, but four Jap fighters came up to attack the planes as they made their escape. There was no damage to either side during this attack, and soon the enemy broke off as Appling and the other crews pressed on south towards Mokerang. They all returned safely after a total of 8 hours 25 minutes in the air.[38]

After this mission, Walt went almost a week without flying. Around this time too, individual crewmembers began to fly missions with other crews on occasion. With an assigned strength of 12 B-24 bombers and 24 crews, the 424th was expected to provide as many aircraft available for as many missions as possible. It was the ground crews' responsibility to inspect, maintain, and repair the planes as soon as they returned from their missions. The goal was to have at least half the strength available for each mission. The extra crews ensured that the men could remain rested and refreshed between missions. However, it was inevitable that individual crewmembers would be sick, injured, or otherwise unavailable to fly with the rest of their crew on occasion. When that happened, volunteers would be requested from other crews. Cliff Llewellyn was the first to start flying as radio operator with other crews. He would not return to the Appling crew for many weeks. In Cliff's place, Sgt. George Volchko would often fly with the Appling crew as radio operator.[39]

There was another rumor going through the base on Wednesday that a huge Japanese Navy task force had been sighted between Los Negros and Rabaul – it was described as two aircraft carriers, two battleships, and six cruisers. Crews were put on alert, but nothing ever came of it.[40]

The Appling Crew was finally called for its next mission on Saturday, June 17. The 307th along with the Fifth Bomb Group would be hitting Truk again – with the fuel storage tank farm at Dublon Town the target. The plan was to start the attack at 1120 flying at 160 mph with the 372nd Squadron first and the 371st echeloned to the right and 300 feet above. The 370th was instructed to lead the second wave with the 424th echeloned to the right and 300 feet below at 18,500'. Once again, lead bombardiers in each squadron would sight for range and deflection, while all other bombardiers would sight for range only. The 424th was instructed to bomb an area to the right of the troop concentration area on Dublon. Each plane was armed with at least nine 500 pound GP bombs, fused instantaneously (eight carried 12 bombs and one carried 10 bombs). Withdrawal after bombs away was to be a turn to the left at 165 mph.[41]

Lt. Vidmar led the six 424th BS bombers assigned. The group took off from Mokerang between 0623 and 0643. Once in the air, Walt settled into the ball turret and checked his guns. The weather was good with low broken cumulus clouds with tops at 6,000 feet from base to target. The four squadrons rendezvoused at the IP. The group formed up and began their approach to the target on a 50-60 degree magnetic heading. All the gunners were alert and tense. Soon Appling announced to the crew via intercom that Jap fighters were in view ahead. An estimated 20 to 25 Zeke's, one Tony, and one Betty bomber were sighted. The Japs began their attack about five minutes before the bomber formation reached the target. First the Betty bomber and a few of the fighters dropped phosphorous bombs over the formation. Then the fighters pressed their attacks.

At this point, the 307th BS B-24's were on their final bomb run and the bombers had to fly straight and steady. Accurate A/A fire began to burst among the bombers and Walt wished he had more in front of him than just plexiglass. He could see phosphorous globs dropping down from above and A/A bursts from below, all the while trying to spot any attacking fighters from below. Each burst of A/A would rock the plane and often flak fragments would strike, sounding like pebbles being tossed against the thin aluminum. The Japs began to press their attacks aggressively. After what seemed like an eternity though, they were over the target and Jimmie Clark announced 'bombs away.' The plane lurched as the two tons of bombs left the B-24. While still trying to remain alert for attackers, Walt watched the bombs as they made their long drop to the target. The results were spectacular. Walt saw a huge fuel explosion and fire just north of the tank farm, along with many smaller fires within the tank farm itself and in the adjacent base area. Five large orange explosions were seen, followed by dense, black, billowing smoke. The black and grey smoke would be visible to the crews 30 miles from the target as they made their retreat.[42]

However, they were not yet clear of the Jap fighters. Five 424th BS B-24s poured machine gun fire into a black Zeke, which had attacked from 1 o'clock high. The pilot of this plane was seen to bail out. S/Sgt. Hoyt A. Shanks, nose gunner on Lt. Hamilton's crew exploded another Zeke coming in at 10 o'clock high. The 371st BS gunners downed another five enemy fighters. The Japs kept up their attack for 25 minutes after bombs away, but after losing so many of their planes, their eagerness soon faded. Miraculously, no one was injured, although two of the 424th BS planes had A/A damage and one was damaged by fire from one of the Zekes. All planes returned safely to Mokerang. Walt and the Appling crew logged 8 hours and 45 minutes in the air with a 1,434 mile round trip.[43]

At about this time Walt began keeping a hand-written personal log of each combat mission that he was assigned to. This log listed each mission by sequential number, date, the name of the pilot of his airplane, the place (target), flying time, and total time in the air. There were also columns for 'fighters' and 'flack'.[44] As of June 17, Walt had flown seven combat missions, including the two with the XIII Bomber Command from Guadalcanal to Rabaul, and logged exactly 55 hours combat flying time. Many other men kept similar logs. The 424th BS Operations Section kept the official record[45] for each man of course, but most men kept a personal record of their own; to ensure that there were no clerical errors.

Two days later, on Monday, June 19, the Appling crew was called for another mission to Truk. Led by Lt. Rinder, six 424th Squadron B-24s were assigned to attack Dublon Town again. Shortly after takeoff, Walt settled into the ball turret and tested his guns. Within a couple minutes though, Lt. Appling came onto the intercom to announce that they were having some mechanical problems and would have to return to Mokerang. Walt made his way out of the turret as Appling brought their plane back to Mokerang safely after just an hour and a half in the air. One other 424th crew had to turn back - also due to mechanical difficulties – only four of the squadron's bombers made it to Truk that day. Unfortunately, the Appling crew would not get credit for a combat mission, because of the turnback.[46] It was clear that the ground crew mechanics had a lot of work to do, to keep the planes flying to and from target without mechanical problems.

The following day, Tuesday, June 20, Walt was summoned to 307th Bomb Group HQ. He found about a dozen other aerial gunners assembled there. What could this be about? To their relief, they were informed that they had each been promoted. Walt now had seven combat missions to his credit and this promotion was recognition. He was promoted from Sergeant to Staff Sergeant, with a slight increase in pay and another stripe to add to his sleeves. His pay and allotments were also brought up to date, not that there was much to spend it on in the island PX or local Los Negros villages.[47]

Yap

Three days later, Friday June 23, Appling's crew was assigned to fly with the 424th to attack Yap Island. Truk lay to the northeast and was about an eight hour round trip from Los Negros. The Appling crew learned during their briefings that Yap was northwest of Los Negros, midway between Truk and the Philippines. Yap was an important Japanese airbase and supply point that would need to be neutralized to support the westward advance of Nimitz to the north and MacArthur to the south. The mission to Yap would be their longest to date, about a 12 hour round trip, all over the ocean with no friendly bases below. The 307th had hit Yap for the first time just the day before – surprising the defenders and destroying over half the 48 planes visible on the runway while they were still on the ground. It was unlikely that the Japanese would be surprised again though.[48]

The Fifth Bomb Group would take the lead, supported by three six-plane squadrons of the Long Rangers. For the 307th, the 372nd BS would lead the attack, followed by the 424th, and then the 370th. The planes were each armed with three 1000-pound GP bombs, fused with 0.1 delay in the nose and 0.01 delay in the tail. The designated target was shipping and the runway at Yap Island.[49]

Walt and the crew had to wake up in the middle of the night that Friday. Near the equator, day and night were each approximately 12 hours throughout the year. For a 12-hour round trip mission, they would have to take off before dawn in order to return to Mokerang while there was still light. Lt. Hamilton led four 424th BS bombers. The group got into the air between 0552 and 0609 and Walt again settled into his turret. The weather started out good. A couple hours out though, there was a problem again with their plane. The #2 engine developed low oil pressure. Appling, Aubrey, and Jupe worked on it for a while, before the engine conked out altogether and they concluded it was unsafe to proceed. Appling radioed the problem to Lt. Hamilton and then turned back for Mokerang at 1048. Jimmie Clark jettisoned their bombs out over the ocean, Walt exited the turret, and they made their way back, landing safely after 5 hours and 15 minutes in the air. Once again, they would get no credit for a combat mission. One of the other 424th bombers also later developed a mechanical problem, though it was able to bomb an alternate target on Sorol Island.[50]

On Sunday June 25, shortly after mass, Walt and Frank Mansir learned that the Appling crew was alerted to bomb Yap again tomorrow, along with the 424th and the rest of the 307th. That evening bad news came in from the returning bombers that had been sent to strike Yap earlier in the day. Between 25 and 30 enemy planes had attacked the bombers, in addition to the typical phosphorous bombs and A/A. The gunners managed to shoot down three of the attackers and another was destroyed on the ground, but these Jap pilots seemed to be much more aggressive and skilled than earlier defenders. Three of the bombers took considerable damage. The B-24 piloted by Lt. Gerald D. Coleman of the 372nd BS received direct hits from 20mm fire – the two inboard engines were hit as well as the cockpit. Flames were seen coming out of the cockpit and bombardier's windows by the tail gunner of the lead plane in the formation. Then the nose of Coleman's B-24 was seen to point straight up and then the craft did a loop, fell off to the right in a spin, and crashed into the water. Jap planes machine gunned it on the way down and after it hit the water. No parachutes were seen – just a bloody foaming spot in the water.[51]

Walt and the other members of the Appling crew listened to this description intently that evening during their pre-mission briefing. There was speculation that the 307th had encountered fresh, experienced pilots recently flown into Yap to stiffen the defense. They were facing a tough mission tomorrow. Capt. Fowler himself would lead five 424th Squadron B-24's along with a like number from the 371st Squadron. Their target was the airstrip and nearby buildings on Yap. The briefing included slides projected on a screen showing detailed target maps, photos, and the weather report for the following day. The designated location for each plane in the bombing formation was indicated and the detailed route out and back described.

Preparation for any mission affected not just the ten crewmembers assigned to each plane, but also most of the other squadron ground echelons. HQ of course both processed paperwork and directed the squadron staff to implement the operational orders that came down from 13th AAF headquarters. Intelligence staff analyzed flight crew reports and photos from the previous missions and prepared briefings for the crews assigned to combat the following day. The aircraft mechanics worked feverishly to repair and service the bombers that would be used the following day. They checked off repair items from the previous crew, patched flak and bullet holes, and also performed any scheduled routine maintenance based on the manufacturer's guidelines. This could have been Walt, if he had not gone on to aerial gunnery school, after Keesler.

By midafternoon, crews of armorers began to spread out to the dozens of bomb dumps dispersed throughout the base to load up the correct type and number of bombs designated for each plane for the following day's mission. The bombs were loaded onto trailers using mobile cranes, and then pulled to each bomber, where fins and fuses were attached. It then took about 20 minutes for the crew to load the bombs into the bomb bay, making sure that the safety on each fuse was wired to prevent detonation. Other armorers moved from plane to plane, unloading belts of .50 caliber ammunition that were loaded into the turret guns or placed adjacent to the two waist window guns.

As dusk fell and the last bombers from that day's mission began returning, ground crews began gassing the other bombers for the following day. This was done from 55-gallon drums, which were also stored in dispersed fuel dumps throughout the base. Each plane required 3200 gallons (almost 60 drums) of gas in main, wing, and bomb bay tanks for the 2000 mile round trip. Fueling would usually end just as the flight crews arrived at their assigned plane in the early morning darkness.[52]

The men woke up around 0400. After breakfast, Walt, Frank, Jupe and the other crewmen got greasy spam sandwiches and cans of peaches from the mess, which they put in cardboard boxes. This was their lunch during the mission. They rode trucks out to their assigned plane in the dark. Despite the early morning tropical warmth, each man wore a summer flying suit and a leather jacket. At bombing elevations of 10,000 to 20,000 feet, it would be cold for much of the flight. Each man also had his .45 pistol with a couple ammunition clips, knife, Mae West life jacket, parachute harness and parachute, oxygen mask, flying helmet with built in intercom headset, and goggles. Some of the gunners also wore flak suits, but Walt did not because there was no room in the ball turret.[53]

Walt did a ground check on his guns and the turret so he could respond that he was ready to go, as Appling and Aubrey checked with each man as their departure time approached. By 0600, engines were running and the planes jerked into motion and began to taxi towards the Mokerang airstrip. Once again, after they were in the air, Walt was helped into the ball turret, where he connected his oxygen and intercom. He tested the operation of the turret and the guns by firing short bursts. Because of the 12 hour round trip to Yap and the desirability of taking off and landing at Mokerang while it was light, the attacking 307th BG bombers would be over Yap right around noon. The crews knew that the Japs would be waiting for them. After the ten planes of the two squadrons formed up at the IP and started their bomb run, they noted that there was fairly thick 9/10 undercast – or broken clouds below the elevation of the planes. Fortunately, this did not hinder the bombardiers aiming, but as they passed over the target, the clouds closed in and Walt and the other ball turret gunners could not see below to judge the effectiveness of the bombing. As they had approached, the A/A was meager and mostly inaccurate. Then an estimated 25 to 30 fighters attacked. A large number of phosphorous bombs were dropped, mostly over the 371st Squadron and four planes had to fly through streamers, but 20 – 25 were dropped over the 424th as well.[54]

The attacking Zekes employed tactics the squadron had not seen before. They attacks came from all around the clock, and from low, level, and slightly high, versus the typical high frontal attacks they had seen over Truk and during earlier visits to Yap. The attacks were coordinated as well, with planes in two and four ship formations. Sometimes a pair of Zekes would fake an attack from one quarter, and then a pair would press the attack from another side. S/Sgt. Jack Cowell, left waist gunner on Lt. Randolph's crew damaged a Zeke that came in low at 7 o'clock to within 500 yards. A large piece was seen to fly off its fuselage before it disappeared into a cloud. Sgt. Elliot Lott, left waist gunner on Lt. Guild's crew likewise damaged another Zeke that came in low at 7 o'clock, getting within 200 yards before it peeled away trailing a stream of fragments. All of a sudden, the guns of Appling's plane began to blaze away above Walt – A Zeke had come in high from 11 o'clock, getting within 100 feet before both Jupe in the top turret and Frank in the nose turret hit it. Pieces of the cowling were seen to fly off as it dove away and into the clouds. Through this all, Walt and the other ball turret gunners blazed away at attackers that came within range low. In addition to the three planes damaged by the 424th, the 371st had four kills. After about 20 minutes, the defenders broke off their attacks as the planes of the 307th BG pressed on for home. As the tension eased after the attack ended, the crewmen agreed that these new defenders on Yap were tougher than anything they had encountered so far. This was also their longest mission to date. Walt and the Appling crew touched down after 12 hours and 35 minutes in the air – more than a 2000 mile round trip.[55]

After Appling brought their plane down and taxied to a stop, the men were surprised to find women – Red Cross girls who had been driven out to their plane in a jeep with a canister of cold lemonade and fresh donuts for the men. This was a welcome sight, as most of the crews remained on the airstrip until all of the other B-24's had returned safely. They were lucky – for all of the skill and aggressiveness of the Japs that day, there was only minor damage to a couple planes and no injuries. The Red Cross refreshments would also sustain the men until dinner, because they had to be debriefed first by 424th Squadron intelligence. The information the crews provided about the bombing run, any damage noted, the ground and air defenders and their tactics would help plan the next day's mission. Other crews would depend on information they brought back, as they had and would again depend on those same crews for their future missions.[56]

A couple days later, the Appling crew was alerted that they would fly a mission on Friday, June 30. The day before, during their mission briefing, they were surprised to learn that they would not be going to Yap, but instead to Noemfoor Island, almost 900 miles due west off the north coast of Dutch New Guinea. MacArthur had ordered the Army to seize Noemfoor and its airbase to support his advance toward the Philippines. The Army invasion was scheduled for Sunday, July 2. The 307th BG mission was to attack the Japanese airfields at Namber and Kamiri, to neutralize any air opposition to the invasion.[57] The 307th would be supported by two squadrons from the Fifth Bomb Group. The attack plan called for the group to fly from Mokerang via Cape D'Urville to the IP at 00°54'N-134°53'E, and from there to the north shore of Noemfoor. The bomb run would be made with three squadrons (370, 372, 371) in a "V" with the 424th following behind the 371st Squadron. On the right, the 370th would attack at 9,300', with the 372nd in the center of the "V" at 9,600', the 371st at 9,000', and the 424th at 8,700'. Once again – lead squadron bombardiers would aim for range and deflection, with the others aiming for range only. The 424th bomb aiming point was a deflection 5,000' south of the shoreline at the east end of the runway, with the range on a line running north and south through a point 2,500' east of the east end of the runway. The planes were each armed with four 1000-pound GP bombs instantaneously fused.[58]

Lt. Lewis would lead the six 424th BS bombers. The Group bombers all left Mokerang between 0651 and 0716. The weather on Friday was good as the bombers took to the air to make their way to Noemfoor. From base to around Wuvulu Island, they found 8/10 to 10/10 low cumulus with tops at 9000 feet, and 9/10 overcast cumulus with tops to 30,000 feet. One of the 424th BS planes had to turn back due to a failure of the fuel transfer system. Approaching Noemfoor at about 1258, the 371st planes were tasked to bomb Kamiri, with the 424th targeting Namber. Kamiri was covered in thick clouds and the 371st could not see the target on their first run, so they turned around for a second try. The airfield was still covered in thick clouds, but they were able to identify and bomb a portion of the base facilities south of Kamiri. Meanwhile, after about a half hour searching for a clear target, Lewis, Appling, and the other three 424th planes had better luck at Namber. There was no A/A and Walt had a clear view of the bombing, which hit the airstrip and surrounding area. Many small fires were started and one large fuel fire. One large building in the center of Namber village received a direct hit and large fires were observed in the center of the village. Three strings of bombs were observed to 'walk' through the target area and another string across the taxi loop.[59]

As they turned to head back to Los Negros, a small force of 10 to 15 fighters attacked the 424th and 371st bombers. Walt and the other gunners blazed away to protect the tight formation of bombers. S/Sgt. Conlon and Sgt. Wilson on Lt. Kaestner's crew each scored a kill for the 424th. The 424th Squadron plane flown by Lt. Donald Balovitch was hit in the right wing, knocking out the #4 engine, but they made it back to Los Negros – almost 1,000 miles, on just three engines. All the 307th BG planes returned to Mokerang between 1810 and 1829, just before dusk. Walt and the Appling crew logged 11 hours 40 minutes of flying time.[60]

During June, the 424th Bomb Squadron had flown 24 missions. Of these, the Appling crew had participated in nine, although they got credit for only seven, due to turning back because of mechanical problems on two occasions. In total, the 424th had carried 245.6 tons of bombs and succeeded in dropping 193.1 tons on target. The squadron lost one man killed during the month. It destroyed 10 enemy planes, assisted in the destruction of one other plane, and damaged a further 11 planes.[61] The 307th Bomb Group as a whole flew twice as many missions during June as compared to March, April, and May. However, the monthly tonnage of bombs dropped did not increase, because of the reduced bomb loads on the much longer missions flown.[62]

At the end of June, Walt and the other Appling crewmen reported to the squadron armament tent. Every month each man was required to report there, clean his .45, and then present his pistol to the armament officer for inspection. Walt disassembled his gun and cleaned each part with an oily rag. After he reassembled it, he handed his .45 to the armament officer who slid open the action and pulled the trigger to snap the hammer down. Satisfied with Walt's work, he handed the gun back and put a check next to the name 'Babinski.'[63]

As July rolled around, Walt and the rest of the crew must have formed a strong identity with the 13th Air Force, 307th Bomb Group, and the 424th Bomb Squadron. These organizations each had their own unique character that helped mold the sense of identity and belonging of the men who served with them. The 13th Air Force was known as the 'Jungle Air Force.' From the first day that he had been assigned to the 13th on Guadalcanal, Walt had seen jungle every day – indeed he lived in the jungle – hot, steaming, mysterious, and alive with strange sights by day and the noise of insects, birds, snakes, reptiles, and mammals by night.

At about this time the 307th Bomb Group began to be known as the 'Long Rangers.' After the missions to Truk had started in May, the group was temporarily nicknamed the 'Trukers.'[64] A popular radio show that almost all Americans were familiar with during the late 1930's and early 1940's was 'The Lone Ranger.' Originating from radio station WXYZ in Detroit, The Lone Ranger featured a mysterious masked lawman and his Indian companion Tonto, who would ride out of the wilderness to bring law and justice to the old west. It was because of the extraordinary capability of the B-24 to fly great distances with a powerful bomb load that the 307th Bomb Group came to adopt the nickname 'The Long Rangers.' This name would characterize their ability to appear across great expanses of ocean to deliver their deadly payload to Japanese held bases that the defenders had considered to be safe from attack.

The 'Long Rangers' was an immensely more popular name than the 'Trukers.' Being from Detroit – and a fan of the western movies, Walt was familiar with the 'Lone Ranger' and identified with the image of the cowboy. With a number of long 12-hour missions under their belts, Walt and the crew understood the significance of replacing 'Lone' with 'Long' in their name – and there would be more and even longer missions in their future. The men began sporting patches with the 307th BG insignia

– a falling bomb being ridden by a cat, wearing a cowboy hat. The 424th BS did not have a nickname, but its insignia represented their mission: a bomb falling straight down with a sun half below the horizon in the background. Not the 'rising sun' of the Japanese imagination, but the 'setting sun' that American warplanes were promising for Japan's future. On July 1, the 424th BG had a strength of 101 officers and 451 enlisted men.[65]

On Monday, July 3, the Appling crew was assigned another mission back to Yap. During the briefing they learned that the job of the 307th during the coming days would be to keep Japanese airpower on Yap neutralized to help protect the flank of the impending American invasion of Tinian Island in the Marianas and of Guam. Capt. John Vanderpoel would lead six 424th Squadron B-24's. The other three squadrons would participate in the attack as well. The attack plan was for the 372nd Squadron to bomb West Yap Town, the 424th to bomb North Yap Town, the 370th to bomb South Yap Town, and the 371st to bomb the runway. The designated squadron bombing elevations ranged from 12,000' to 12,900' with an attack airspeed of 165 mph. After bombs away, the plan was to turn to the left at 170 mph. Each aircraft was armed with three 1000-pound GP bombs, 18 of them set with 0.1-second delay nose fuses and 0.01-second tail fuses, and the rest fused instantaneously.[66]

At some point between 0550 and 0623 on July 3, the Appling crew took off in B-24 A/C 540. Walt settled into the ball turret as usual, while Jack Riley and Frank Mansir manned their turrets. Jupe stood behind Appling and Aubrey, assisting them as flight engineer, while ready to man his nearby top turret if needed. Jimmie Clark and Ed Dunne were at their bombardier and navigator stations, with Len Sherman and George Volchko at the waist guns. The squadron flew at about 8,000' elevation to Sorel Island, then to the IP at 09°30'N-138°14'W at a speed of 157 mph. From base to 100 miles south of Sorol, the weather was 4/10 cumulus with bases at 2000' and tops at 8000'. A mild weather front was then penetrated for about 70 miles, and then the weather cleared again. Once at the IP the three squadrons formed up individually to each begin separate bomb runs. The 424th found fairly heavy cloud cover below making the target difficult to find, although it also meant that there was no A/A fire to speak of. A slight break in the clouds gave them a glimpse below and an airstrip was sighted. Led by Capt. Vanderpoel, the 424th planes dropped their bombs on this target, but unfortunately it turned out to be not Yap proper, but adjacent Gagil-Tamil Island and an airstrip that was once under construction, but never completed. After this mistake, there was nothing to do but head back home along with the other two squadrons.

As the big bombers were turning for home, they were attacked by about 16 Japanese fighters. The attack on the 424th started with 10 to 12 phosphorous bombs dropped ahead of the formation. Then the Zekes started aggressive attacks from all around the clock and mostly from high. There were occasional low attacks too, which kept Walt and the other ball turret gunners busy. Cpl. Witham, ball turret gunner on A/C 264, and Sgt. Dick Reis, nose gunner on A/C 101 collaborated by pouring fire into a Zeke coming in from 4 o'clock low, passing under the formation, then breaking off toward 8 o'clock. The plane was then seen to turn over on its back, the pilot bailed out, and the plane was seen to crash into the sea below. Suddenly Walt could sense Jack Riley yelling into the intercom and firing away toward the rear of their plane. A Zeke approached at 7 o'clock high, but as Riley fired away it broke off and went in to a long shallow dive. Sgt. Lott, waist gunner on the Guild crew, finally saw it crash into the sea about 3 or 4 miles behind the formation. This was another kill for the Appling crew. Meanwhile, another Zeke came in low from 2 o'clock then turned toward 7 o'clock. As Walt and the other ball turret gunners brought their fire to bear, Sgt. Rees, ball turret gunner on the Guild crew hit it. Fragments started flying off the fuselage as it disappeared into a cloud. After about 15 minutes, the intensity of the Jap attack decreased, but the enemy continued to circle the formation and get in occasional shots at the bombers for another 30 minutes.[67]

Meanwhile, the 371st and 372nd Squadrons were also having luck, destroying four more Zekes. Of note was a kill by S/Sgt. Marshall, ball turret gunner on the Adair crew (371st), flying in A/C 323, nicknamed 'Frenisi.' Frenisi was one of the most famous B-24's in the 13th Air Force and this was its 100th mission! The 372nd BS did catch some damage. The planes flown by Lt. Dorries and by Lt. Steffy each took A/A fire. Lt. Dryer's plane was hit near the wing fuel tank between the #1 and #2 engines, while Lt. Baldwin's plane was hit by 20 mm cannon fire, cutting hydraulic lines and caving in the bomb bay door. Most seriously, Lt. Bowles plane got hit by flak from beneath which put a hole in the fuselage near the wing just in front of the bomb bay and hit the #1 and #2 engines, the nearby flaps and horizontal stabilizer, and the left wheel. The oil line to engine #1 was cut, so it had to be shut down and its prop feathered.

This damage slowed down Bowles plane, which meant that the rest of the group returned to Mokerang to await the return of the injured plane. Walt and the others watched and held their breath as this last plane approached, with rescue trucks on the standby along the airstrip. When Lt. Bowles landed (with just three engines) the left tire blew, but he managed to control the landing and bring the plane to a stop without further damage.[68]

Walt and the Appling celebrated their 2,000 mile, 12 hour and 15 minute's mission. For Walt, this completed 10 combat missions. Although the bombing results were disappointing, Jack Riley and the crew got credit for another enemy plane destroyed. The following day, the ten men of the crew lined up in front of A/C 540 for a crew photo. With the four officers crouching in front and the six sergeants standing in back, this was the classic bomber crew photo. They all looked confident. Walt and most of the others looked thin and it is obviously hot – several of the men have an extra shirt button or two undone. We can only imagine that they were anxious to get out of the sun and back into the shade of the trees seen behind the plane on the other side of the airstrip.[69]

Two days later, on Wednesday the Fifth, Walt turned 22 years old. We do not know if there was any celebration or even recognition by the other crewmen of a man's birthday. The crew had no mission scheduled that day, so at the very least, Walt had time to be thankful that he was still alive, even though he was a third the way around the world from his family, his home town, and his Detroit friends. However, he was surrounded by other new friends on Los Negros. These were men he liked and trusted. They ate, slept, and hung out together, they shared their occasional beer and whiskey rations together, and they did their work together. He also trusted the men who were his friends to fly combat with and he took satisfaction that they trusted him likewise to carry his weight on the long missions over the Pacific that they still faced. Perhaps he had some letters or even a package from home to help brighten his birthday. We can imagine Walt at some point late in the day on his birthday walking through the plantation down to the water's edge and gazing out across the Pacific as the sun set behind him to the west. As he gazed at the horizon, did he think about how it led back to the Golden Gate and the rails beyond that could take him back home? Then he likely hurried back towards the camp to be in time for chow and the evening outdoor movie.

On Friday, July 7, the Appling Crew was called for another mission. Capt. Vanderpoel would lead five 424th Squadron B-24's, joining similar numbers from the 370th and 372nd Squadrons in another attack on the Yap Island airfields. During the briefing, the men were told that this would mark the 300th combat mission of the 307th Bomb Group – the Long Rangers. About four hours after takeoff as he sat in his ball turret, gazing at the ocean below, Walt heard Lt. Appling come in on the intercom. They were having engine trouble – Appling, Aubrey, and Bretherton were worried that they would not be able to complete the mission and return safely to Los Negros. After a few minutes, Appling came on the intercom again to alert the men that they would abort the mission to Yap. However, they would bomb a secondary target, Sorol Island to the southeast of Yap. Sorol was the site of a Japanese radio station that was used to alert Yap of approaching American bombers, so it was an important target in its own right. Appling brought the plane in over Sorol, handing over control to Jimmie Clark, the bombardier. There was no A/A or fighter interception. After bombs away, Walt watched the target then reported that the bombing was on target. Appling brought them back to Los Negros successfully, despite their mechanical problems, after exactly ten hours in the air.[70]

Once back on the ground and after their debriefing, the Appling Crew waited for the rest of the Group to return. They soon learned that Sorol had been bombed a second time that day by a 372nd Squadron plane that had also developed hydraulic problems. After the rest of the group returned, they learned that the bombing had been successful and two Jap fighters had been shot down. One 372nd Squadron plane had a 20mm shell explode in an ammunition box, stunning the two waist gunners. Walt was particularly interested in the experience of a 370th Squadron plane piloted by Lt. Kelly. They had a flak shell explode just behind the loaded bomb bay near the ball turret, blowing a two-foot wide hole on the right side of the plane. It killed one of the waist gunners, S/Sgt. Donald Miller, and mangled the ball turret. The other waist gunner and tail gunner were also injured, but they managed to get Pvt. Donovan, the ball turret gunner out of his ruined turret. Donovan was wounded in the head, eye, calf, upper thigh, and had three fingers blown off. While this was happening, S/Sgt. Gundy, the top turret gunner shot down an attacking Zeke that made a pass from above. As the plane turned away from Gundy's fire, flames broke out in front of the cockpit then engulfed the whole ship. Donovan had made his way on the catwalk through the open bomb bay to the flight deck as Kelley completed the successful bomb run. The copilot and navigator stopped the flow of blood in Donovan's wounds, and then administered a plasma transfusion. The copilot and bombardier then went to the rear of the plane where they found Miller dead and the other three gunners in need of first aid. The rudder cables were found to have been damaged by the flak explosion, so the crew had to splice cables salvaged from the wrecked ball turret to make the repairs. Through all this drama, Kelley kept the plane flying home alone on the flight deck, landing successfully where ambulances were waiting to take off the wounded crewmen.[71]

On Wednesday, July 12, Capt. Fowler was succeeded as 424th Bomb Squadron Commander by Capt. John A. Vanderpoel.[72] Fowler was later transferred back to the U.S. for reassignment. Vanderpoel was a tall, well-liked New Englander from Litchfield, Connecticut. He had proven himself as a competent and careful mission leader on several of the strikes that Walt and the Appling crew had been assigned to, so they felt good about this change. He was also open to hearing the gripes of the men under his command.

One of the problems many 424th enlisted crews faced on Los Negros was a shortage of mosquito netting for their tents. Much of the netting they had was torn or tattered, providing poor protection against flying pests as the men slept. Despite frequent requisitions for many weeks past, adequate supplies were not reaching the men. Shortly after taking command, late one night Vanderpoel enlisted the aid of a friend to correct the situation. They removed the insignia from their uniforms, then drove a jeep slowly over to 13th Bomber Command HQ, were there were tennis courts reserved for the brass, each surrounded by triple layers of brand new nylon mosquito netting. They paused in the dark to check if anyone was nearby. Once they were sure the area was clear, Vanderpoel drove near one corner of one of the courts where they tied the ends of three layers of netting to the back of the jeep. Vanderpoel then quickly drove around the outside of the court, ripping off the three layers of netting. They drove back to the 424th area where Vanderpoel hid the netting under the floor of his tent for a week. After the 'heat was off' over at 13th Bomber Command HQ, Vanderpoel distributed the netting to the neediest crews in his squadron.[73] Acts like this served to strengthen the sense of pride and cohesion within the Squadron and secured Vanderpoel the respect of his men.

The day after Capt. Vanderpoel took over command of the 424th, the Appling crew was assigned to bomb Yap Island again. On the morning of Thursday, July 13, Major Pennington led six 424th Squadron B-24's into the air towards Yap. Planes from the 370th and 372nd Squadrons accompanied them as well. Once in the air Walt settled in for the long flight. Before reaching the IP, two of the 424th planes developed mechanical problems and were forced to turn back, although they each managed to bomb the Sorol Island radio station again. After the three squadrons had formed up and were still five minutes away from the target, they were met by about 15 Japanese fighters. Phosphorous bombs were dropped above the massed bombers and several of the planes were forced to fly through the streaming hot chemicals to stay on course. One 372nd Squadron plane had burning phosphorous drop onto the nose and pilot's window, but it was only slightly damaged. 370th and 372nd Squadron gunners each brought down a Zero before bombs away. A/A fire on the bomb run was heavy, slight, and inaccurate[74]. Although Walt's plane did suffer minor flak damage, Lt. Clark kept on target for the bomb run with Lt. Appling ready to take over if serious damage developed. With one eye out for possible attacking planes from below, Walt watched the 424th Squadron bombs fall, spreading an effective bomb pattern from the shore road near South Yap Town to the wharf area in Tomil Harbor. He saw one very large explosion and damage to many of the buildings in the area. As the 'Long Rangers' made their turn for home after the bomb run, they continued to be attacked by the remaining Jap fighters for 40 more minutes.[75]

Just a few minutes after the bombs were away; the 370th Squadron plane piloted by Lt. Ball received a hit in #4 engine, then another in #1 engine. As this plane began to fall behind, five of the Jap fighters concentrated on it, flying in lazy eight patterns and making approaches from front, rear, and high. Soon another shell hit the flight deck, knocking out the radio and a gas leak started, pouring fuel into the bomb bay, which soon ignited. The plane descended rapidly with the gunners still fighting off the Jap attackers, but Lt. Ball still managed to set it down in the water. The plane broke in two on impact and the men tried to scramble out as best they could. Lt. Ball was never seen again. Lt. Vaughn, the copilot and Lt. Merrill, the navigator got to a raft, where they were soon joined by S/Sgt. Morrison, the tail gunner and S/Sgt. Hawes, the right waist gunner. Cpl. Clendennen, the ball turret gunner had been cut badly about the head. He was seen briefly on the surface calling for help when he grabbed the tail of the plane, which then sank and he was gone too. None of the other crewmen was ever seen again. Vaughn, Merrill, Morrison, and Hawes spent the rest of the day, the following night, and most of the following day in the raft. A US Navy submarine rescued them late on the following day.[76]

There had been nothing the rest of the group could have done to protect the doomed plane. Despite the success of the bombing run, the crews were all in a somber mood when they returned to Los Negros that evening. Walt and the Appling crew had completed another mission – this time their longest yet – 13 hours in the air.[77]

The following day the Appling crew enjoyed their rest after their longest mission up to that point. Responsibilities were few the day after a mission. If the plane they had flown in was not called for another mission the following day (which would have been rare after a 13 hour flight), the gunners would be expected to return in the morning after breakfast to disassemble and clean their guns. They would usually go as a group out to the revetments off the main runway where the planes were stored and serviced by the ground crews. Cleaning and servicing their guns was something they took seriously – their lives and the lives of their crewmates or the crews who lived in neighboring tents could depend on them doing their job right. Later in the day, after their work was done, they might spend time on the beach, read a magazine, write a letter home, play volleyball or softball, or wander off to the native village. After mess hall in the late afternoon, the men would look forward to a movie or two at the outdoors theater once the sun went down in the evening. Around dusk, they would listen for the sounds of the returning bombers that had been sent out on the mission for the day.

The following day, on Saturday the 15th, as their evening movie was ending, a buzz started to spread through the theater. Soon everyone was talking about the news that had come back with the crews that had returned from that day's mission. All four of the Long Ranger Squadrons had sent up planes to strike at Yap again. Walt learned of what had happened from crewmen who had returned and completed their debriefings. After the bomb run on Yap, the Jap fighters made their usual determined attacks at the departing bombers. Several Zeros were brought down, but still they came on, concentrating their attack on the 370th Squadron formation. Suddenly one of the bombers was hit by the Zeros and started to swerve to the left, as the pilot lost control. They were flying the normal tight defensive formation and the left wing of the damaged plane slowly overlapped then settled down on the wing of a neighboring bomber. The props on the two planes chewed up the other planes wings, which quickly crumpled. The two planes were then seen by many observers in the group to burst into flames and drop quickly away. Some parachutes were seen to open but many of the crewmen saw the Jap fighters go down to strafe the men in their parachutes and in the water.[78] Walt and every other crewman felt the horror of what had happened to their comrades. Sleep that night, when it came, was uneasy with the thought of the day's loss.

Sometime shortly after this, Walt and a number of other crewmen were on their way out to the planes to do some maintenance on their guns. The men often traveled from the camp area to HQ, the airstrip, or elsewhere via flatbed 6x6 trucks operated by the Group motor pool. On this particular day, the men had just left from the mess hall after lunch and Walt was finishing a can of peaches. He had eaten the peaches and climbed aboard one of the 6x6 trucks with the can in one hand. He sat on the edge of the flatbed, legs dangling down, drinking the peach juice on the way to the airstrip, laughing and talking with Jack, Frank, and the other gunners. The suspension on the trucks was 'stiff' to begin with and the jungle roads were rutted and rough. At some point before the airstrip, the truck hit a big bump. Because Walt was holding on to the peach can with one hand he lost hold and went flying off the truck. He hit the road hard and as he came down his left hand hit the jagged open end of the peach can, cutting it badly. As he lay in the dirt, the other men called for the truck to stop. Frank and Jack ran back to Walt to find him conscious but dazed and bleeding. He didn't seem to have any broken bones but the wound on his hand was bad. There was a small first aid kit in the truck and the wound was cleaned and a bandage quickly applied. Jack flagged down a truck headed back to camp and Walt was helped in for a ride to the Group aid station.

Within a couple of hours Walt's wound had been cleaned, treated, and re-bandaged by a Group medic and a doctor had also checked him over. The cut was bad and the doctor told Walt that it would be at least a couple weeks before he could fly again. Within a few more hours, Walt was back on his feet, with a couple of weeks of easy duty to look forward too. A photo taken a week or so later shows Walt crouched next to the ball turret of one of the Squadron B-24's. His left hand is bandaged from the injury from the base of his fingers to well below his wrist. Despite the wound, Walt looks fit and happy, with a big grin on his face and his hat characteristically cocked to one side.[79]

The morale of the men of the 424th was described during July as high; despite the extremely long unescorted missions the men were assigned. Maintenance of the planes was considered good. There were few disciplinary problems during the month. Morale was helped by the female Red Cross workers who now met each returning combat mission 'with a smile and a cool and very palatable drink' which made the men more amenable to the post mission interrogation they needed to complete while the details were fresh in their minds.

Another occasional source of entertainment with the chance to hear a female voice was Radio Tokyo and 'Tokyo Rose.' A few men had radios that could pick up these transmissions to get American popular music, along with Japanese propaganda dispensed nightly by the sweet, sultry voice of Tokyo Rose. Tokyo Rose would try to rile the Americans by saying things like 'Hey American flying men, those boys back home are stealing your girls and sleeping with your wives.' However, most men just laughed and enjoyed the music.[80]

It was typical for the crews to have at least one and sometimes three or four days between missions, with little to do. The seaside and beach looked inviting, but sharp coral and snakes made swimming a challenge. A squadron baseball team was formed of the best players, with the 424th beating the 104th Seabee's, but losing to the 168th A/A Battery. A volleyball league was also started with a tournament that was still in progress by the end of July. Many men started various types of hobbies. Butterfly, shell, and photo collections were common, and Walt started using a small brownie camera to take snap shots around the base. Reading and card playing were very popular and Walt often indulged in pleasant games of pinochle or poker.

Wally received the occasional letter or package from home – most men did. Wally learned a little about conditions back home – though the censors would likely remove any blatant complaints that could affect the men's morale. Walt learned a little about how his brother Johnnie was getting on in Australia. Walt wrote home on occasion too, though almost none of this correspondence remains. The men were encouraged to send V-mails. These were standardized letter formats that the men could address and write a message, perhaps even include a picture. Once passed by the base censors, these V-mails were then photographed in a group, and then the film was flown back to the States, where it was reprinted on photographic paper, placed in envelopes and mailed to the addressee. Walt wrote at least once to Dorothy Kurzyniec from Los Negros. His V-mail to her depicted a bare breasted beauty wearing a grass skirt, with a flower in her hair, standing on a tropical beach at the water's edge. Wally wrote:

> *Admiralty Islands*
> *"Why would I leave the balmy South Seas*
> *"When I'm surrounded by things like these?*
> *"Because my heart's across the Sea*
> *"And you are the one who appeals to me."*
> *Love, Wally*[81]

The 307th held a ship and aircraft recognition contest, open to men from all four squadrons. Walt participated, studying the notebook in which he had sketched various allied and enemy planes. But the 424th placed a disappointing third in the contest. The camp area was very pleasant and clean, and the mess hall and food was considered by 307th veterans to be better than at previous bases.[82]

While Walt was grounded, the 307th Group Chapel was completed by contracted native labor. Named *Chapel by the Sea*, it had open sides with a thatched roof made from sago palm fronds. It had an altar set back into the wall at the front, with a sanctuary and office for the chaplains on the landward side. Walt and Jack attended the dedication on Sunday, July 23. Lt. Col. Robert Burnham, 307th Bomb Group C.O., presided, followed by remarks by each of the group's three chaplains.[83]

Two days later, on Tuesday, July 25, Walt and many of the other crewmen were called to 307th HQ to be read the Articles of War as required by Article 110.[84] Walt had first been read the Articles of War back at Camp Custer in January 1943 by old Lt. Brower. Now out in the South Pacific, they were likely glossed over much more quickly – discipline at the base was lax, although there were still limits and this was a formality that had to be completed.

By late July, quite a bit of intermixing had occurred among the 424th Squadrons. Walt, Jack Riley, and Frank Mansir had managed to stick together on all missions, and Lt. Appling was still the only pilot they had flown with. Cliff Llewellyn however had been flying regularly with Lt. Randolph's crew and George Volchko had flown several times with Lt. Townsend's crew. Toward the end July of a number of new crews arrived, which tended to add to the intermixing of men. Walt, Cliff, Frank, and Jack were excited to learn that Lt. McRae had finally returned, cured of his persistent ear infection, with a new crew of his own. Walt and the other men from old crew 430 from back at Muroc each stopped by to wish McRae well. McRae, or Mack as he came to be known by the men, had been helped by the replacement of Capt. Fowler by John Vanderpoel. Mack let it be known that he got along well with the new CO and that he was glad to be back. He introduced Walt, Jack, Frank, and Cliff to the other enlisted men on his crew: Sgts. Pierce, Wachter, Raysor, Johnson, Klein, Riddle, and Gilchrist. Walt was still on the mend from his injured hand, so he had time to explain the ropes to the new crewmen. He learned that they had gone through their Combat Training Center, not at Guadalcanal (which had been closed shortly before Walt left there), but at Nadzab, high in the central mountains on New Guinea. This was a joint training center for both the 13th and 5th Air Forces. Mack and his new crew had still flown the standard bombing missions to Rabaul, to ensure that their training was adequate and to give the crew realistic combat experience over a relatively safe target.

As July ended, the 424th Squadron had completed 13 combat missions, of which Walt had participated in three. The 424th also completed an additional four search missions. They had carried aloft 100.75 tons of bombs, of which 86.3 tons were dropped on the primary intended targets, and 7.95 tons dropped on alternate targets. There had been no personnel losses in the 424th (unlike the rest of the Bomb Group) and six enemy planes were destroyed, with an additional probable, plus two enemy planes damaged.[85] As August began, the 424th had an aircrew strength of 92 officers and 147 enlisted men, plus 15 ground crew officers and 343 enlisted men. The squadron was assigned ten B-24J's and four older B-24D's, as well as one C-47A cargo plane.[86]

After the medics judged his wrist sufficiently healed, Walt was ready to fly again. However, there was a problem – Lt. Appling had gone on rest leave and with the mixing of crews, there was not really a designated crew that Walt belonged to at that time. Walt, Jack, and Frank would have been happy to join Lt. McRae's crew, but that was not an option because all his crewmen were each anxious to start getting combat missions of their own under their belts. The trio missed missions on August 2 and 4 to Yap, but then the following day they were called to fly with Lt. Jack M. Arnett.

Arnett was a thin and particularly young looking pilot from West Virginia. But there were rumors about Arnett. Apparently, in the middle of July most of the men who had gone through bomber crew phase training in Tonopah, Nevada, had balked at flying with him. Arnett has a reputation as very domineering, not only over enlisted men and NCOs, but also over the other officers on his crew. Unlike Charles McCrea, Don Aubrey, Jimmie Clark, and Ed Dunne, who had befriended the NCO's back at Muroc, Arnett was never friendly to the NCOs. Worse, there were rumors that Arnett had a major drinking problem. There were even stories that he had been seen with a bottle in his hands while at the controls on combat missions. Relations between Arnett and his crew got so bad, that Captain Vanderpoel decided to shake things up. Charles McCrea was assigned to take over the enlisted men from the Arnett crew and Arnett was assigned most of the men from the McCrea crew.[87]

Arnett's copilot was Lt. Carlton 'Carl' Haglund. Haglund normally flew with the MacMillan Crew. Lt. John Hepfer would be navigator – he had been ill for a time with stomach problems, so he had been looking to get back onto a crew, just like Walt, Jack, and Frank.[88] Their old crewmate Jimmie Clark would be bombardier. Tall and lanky Sgt. Hugh Charboneau would be flight engineer, Sgt. Boris Hidalgo, from Brownsville, Texas was the radio operator, and Sgt. Raymond 'Gene' Schreiner, from Fort Collins, Colorado, was waist gunner.[89]

Hidalgo had an interesting story to tell the others. During a mission over Truk back on April 6, in a B-24 piloted by Lt. Fred M. Lewis, ack-ack knocked out their #2 engine. On the way back to Los Negros, they ran low on fuel and had to ditch. After nine days in the water, they were picked up by the destroyer USS Anthony. After returning to base the Lewis crew was offered two weeks of R&R leave, but Boris declined, fearing that if he did not start flying again, he might lose his nerve.[90]

Schreiner was from the outskirts of Fort Collins, Colorado. Because of his German sounding name, he was soon dubbed 'Schultz' by the other men in the crew, and he took to calling Walt 'Pollock' – all in good nature.[91] Walt, Jack, and Frank had met the other enlisted men on Arnett's crew before, but this was to be their first mission together.

The mission for Sunday, August 6 was Yaptown, on Yap Island. This mission was assigned to six planes from the 424th and similar numbers from each of the other three Group Squadrons, for a total of 23 planes.[92] Capt. Vanderpoel led the attack for the 424th Squadron in A/C 792. Lt. Arnett and his patched together crew would fly in A/C 605, and Lt. McRae would pilot his new crew in A/C 547. The planes were each loaded with 20 or 30 - 100 pound GP bombs, fused to explode instantaneously on impact. Three of the planes also carried cameramen, to get photos of the base to see if the Japanese were making any efforts to repair the damage that had been inflicted by the 13th Air Force the previous month. After take-off, Walt, Frank, and Jack Riley manned their turrets. During the first third of the way to target, the weather was good, with 4-6/10 cumulus cloud cover, with the clouds towering from 3,000 to 15,000 feet high. Farther north, they encountered a storm front with turbulence and heavy rain. The clouds had given way to cirrus coverage at 20,000 feet. F/O Teague, pilot of A/C 101 found it difficult to keep up with the other bombers in the heavy weather, so he salvoed 10 of his 100 pound bombs at about 0815, to lighten his load. As they reached the target, they encountered 4-5/10 alto-cumulus cloud coverage, which fortunately left them a view of their designated target.

Capt. Vanderpoel formed up the small six-plane formation with the remainder of the Bomb Group at Sorol Island, the designated rendezvous point, at about 8,400 feet and led them over the target at 1222-1224. As they approached the target, A/A fire opened up from below, mostly heavy and medium, slight to moderate and inaccurate. As the bombs were dropped over the target, Lt. McRae's plane had two sets of bombs hang up in the bomb bay. It took a minute or so for the problem to be fixed, so when the bombs fell they hit well west of the intended target. There was no fighter interception, so Walt and the other ball turret gunners were able to concentrate on watching the bomb results. Even though the bombs burst right through the target area, there appeared to be only one small fire started. There was another fire seen some distance from the target. About ten minutes after the planes had left the target area, Walt and other crewmen saw this fire explode, followed by a heavy column of white smoke. Walt and the other ball turret gunners made a number of other significant observations of note while over Yap. The precise location of four A/A guns was noted – these could be the target for a future raid. A small vessel was seen five miles northwest of Yap and more distantly – about 35 miles away the men sighted three ships – one very large and the other two medium sized. In addition, the Squadron came back with 40 aerial photos taken while over the target. These would be analyzed carefully by Bomb Group Intelligence officers back at Los Negros.[93]

Returning to base, the six planes flew back into the frontal system they had passed through earlier in the day. The last third of the way back to Mokerang, the clouds were heavy cumulus with 8/10 coverage and tops at 9000 feet and bases at 1500 feet. The weather at Los Negros was so bad that when they approached the six planes were told to circle, to await a break in the weather. Finally, after a half hour or so, it cleared and the six planes were OK'd to land.[94] After 12 hours and 30 minutes in the air, Lt. Arnett brought their plane back to ground safely. Walt had completed his twelfth combat mission and his first with someone other than Carl Appling as pilot.[95]

Later, the 40 photos taken the 424th were analyzed by 307th Intelligence Officers. The 23 planes of the 307th that made it over the target dropped a total of 570 100-pound and six 500-pound bombs. The photos showed that the island seemed to be very well prepared to oppose an amphibious attack, with intensive beach defenses, landing craft obstructions in the water, and extensive barbed wire. Extensive base facilities were identified that seemed to be not associated with the Japanese airfield, but rather to support the troops intended to defend the island in case of an invasion. The photos showed 267 individual bomb bursts, out of the 576 bombs dropped. The report detailed the quantity and type of buildings destroyed or damaged by the bombs. This information would be forwarded quickly to 13th Air Force Headquarters, for dissemination up the chain of command.[96]

Four days later, on Thursday, August 10, the Arnett crew, with Walt, Jack, and Frank, along with Lt. Clark from the old Appling crew were assigned another mission. It was still somewhat of a scratch crew. Ed Dunne would join them this time as navigator, and Lt. Kenneth Rider would serve as Arnett's copilot. Sgts. Charboneau, Hidalgo, and Schreiner would be the other gunners, and Sgt. Mario Campora would join them as photographer.[97] Boris Hidalgo had possibly flown on other crews before as ball turret gunner – in any case, he and Walt agreed that Boris would man the ball turret on this mission and Walt would man one of the waist gunner positions. The target again was Yaptown, specifically, the A/A gun positions they had noted on the mission four days ago, with six 424th Squadron planes assigned, again led by Capt. Vanderpoel. They carried 250-pound GP bombs, with ten in three of the bombers and 14 loaded into the other three. The 424th would be accompanied by a handful of planes from the other Squadrons as well.

Shortly after dawn on Thursday morning with the crew and bomb load aboard, Arnett coaxed A/C 453 into the air. The weather during takeoff was good, with 3/10 stratocumulus cloud cover with tops to 8,000 feet and 9/10 altostratus overcast at 20,000 feet. About an hour out, they passed through a weak front then into more clear weather beyond. Walt helped Boris Hidalgo into the ball turret, and then tested his single gun at the waist window. Once they approached the rendezvous at Sorol Island a second, much heavier front was encountered with limited visibility. They were still about 30 minutes from the target and it was nerve wracking flying, with the planes needing to keep in sight of each other while maintaining radio silence.

Walt's hair stood on end as he watched the other planes in the formation appear, then disappear again into the thick clouds. The wind picked up, reaching 35 knots out of 190°. Soon one of the six planes disappeared altogether, but the other five pressed on. Once they reached Yap, at about 1215 at an altitude of 12,000 feet, they found limited visibility below. A/A fire opened up as they approached their target – heavy, slight to moderate and inaccurate. Suddenly – Boris Hidalgo emerged from the ball turret, seeming very agitated. As Walt and Sgt. Campora the photographer helped him out, Hidalgo said – "I heard a transmission on the interphone – 'Am on fire and going down over target' – are we OK?" Walt and Campora assured him their plane was OK. Bombs away was at 1217 and as they turned away, only one of the ball turret gunners in another plane was able to glimpse the bomb bursts below, due to the heavy weather. At least they had not dropped the bombs in the ocean. Hidalgo returned to the ball turret, switched his interphone to VHF, and then again heard a message "...am on fire and going down over target." He switched back over to intercom to report this message to Arnett and the rest of the crew, but no one else said they heard it.[98]

South of Yap the 307th bombers emerged again from the front and its heavy clouds. More than one plane seemed to be unaccounted for, including one of the six 424th Squadron planes – A/C 540 flown by Lt. Sasser. No one had seen any Japanese fighters, only the A/A fire, so it was unknown what might have happened to the missing planes. Half way back to Los Negros, another front was encountered with moderate rain and heavy turbulence, but the weather improved further south. Jack Arnett brought A/C 453 down onto the runway at Mokerang after 12 hours and 25 minutes in the air. Walt had completed his 14th combat mission.[99] One by one the other planes of the group returned...the men from the 424th were relieved to hear that Lt. Sasser and his crew was OK. After they became separated in the thick clouds, Lt. Sasser decided to bomb Utagal Island, one of the designated alternate targets. The Squadron also brought back 31 photos of the bombing run, although Sgt. Campora was not able to take any photos, due to the commotion over the target when Boris Hidalgo emerged momentarily from the ball turret.[100]

Relief to see Lt. Sasser's crew back safe at Los Negros soon faded though as reports spread through the base that one plane did not return. Lt. Don Anthony and his crew from the 371st was last seen entering the front about 30 minutes before reaching Yap. The 371st Squadron leader reported that Anthony had tried to contact him over Yap, but the radio message was garbled. None of the crew who returned that day reported seeing any enemy fighters. The following day, a 371st bomber flown by Lt. Adair and another from the 424th flown by Lt. Hughes returned to the vicinity around Yap to search for any survivors from the Anthony crew. Before reaching Yap, Adair was attacked by three Japanese fighters, so he turned back towards Los Negros. Neither Adair nor Hughes had seen anything. Later that evening, as men on Los Negros tuned in to Radio Tokyo to listen to Tokyo Rose, they heard her report: "The Japanese garrison on Yap Island shot down one enemy plane from an enemy formation of nineteen heavy bombers which raided the island yesterday." The following day another 424th bomber flown by Lt. Hamilton returned to the Yap vicinity to search for survivors, but with no luck. All subsequent searches for any survivors from Lt. Anthony's crew were negative. It was believed that they had dropped below the clouds and were hit by A/A fire or jumped by the few Japanese planes that remained on Yap. It was Lt. Anthony's plane that Boris had heard, as it was going down alone and for the last time.[101]

Then all missions by the Long Ranger bomber squadrons ceased. The 307th was on the move again. All bombing raids against Truk, Yap, and other targets in the Caroline Islands were terminated. The 307th was moving to Wakde Island, off the coast of New Guinea. Where was Wakde? Walt learned it was about 600 miles due west from Los Negros, just a couple miles north of the Dutch New Guinea coast. The 307th had started the move on August 11. All the men were put to work packing up, loading planes and ships, and helping with a hundred other little details associated with the move. On Monday, August 14, Walt and the rest of the crew had their pay brought up to date, as the administrative clerks began to close up shop on Los Negros for the move.[102]

- 11 -
WAKDE 1944: THANKS FOR THE MEMORIES?

The Southwest Pacific: 1944

Walt and most of the other members of crew 430 from Muroc had flown their first two combat missions from Guadalcanal just three months ago. The target had been Rabaul, then an isolated base that still held 100,000 Japanese troops. The Allied strategy was to bypass bases like Rabaul, using American sea and air power to leapfrog across the Pacific to the heart of the Japanese empire. Walt and the other men of the Long Rangers of the Jungle Air Force were a part of that strategy.

For the past two and a half months, Walt had been based on Los Negros in the Admiralty Island with the 307th Bomb Group. The 13th Army Air Force, including the 307th and the 5th Bomb Groups was tasked during that period with providing direct support to the US Navy, Marines, and Army as they proceeded to seize island bases in the Central Pacific. In June, the 307th had concentrated on bombing Truk, to neutralize this Japanese base and leave it impotent during the American invasions of Saipan and Tinian Islands, 650 miles to the northwest in the Marianas. The 307th succeeded in its mission and Japan withdrew its ships and fighters from Truk (or simply did not replace those destroyed on the ground or by the aerial gunners. First Saipan and then Tinian were invaded in June. It would be many weeks before they were cleared of Japanese defenders, and more weeks of construction work would be needed, but by late November, they would provide air bases for the 20th Army Air Force to begin regular bombing of Japan itself.

By late June and into mid-August the 307th turned its attention to Yap. Like Truk, this was a major Japanese base that could threaten the Allied advance across the Central Pacific. The next planned target for the US Navy and Marines was Guam, American territory that had been seized by Japan in early 1942. Guam is just 500 miles northeast of Yap. Once again the 13th Air Force and the Long Rangers of the 307th Bomb Group succeeded in neutralizing Yap to eliminate it as a threat to the invasion of Guam in July.

One tactical interlude for the 307th was to support the invasion of Noemfoor, a small island off the north coast of Dutch New Guinea. The 307th flew a few long-distance bombing missions to Noemfoor in late June, in preparation for the early July invasion.

The next target for the 307th would be Palau – another highly fortified Japanese base in the southern Caroline Islands. Palau was important because it threatened both the American advance across the Central Pacific, to its north, and the planned invasion route to the Philippines, to its south. Supporting the invasion of the Philippines would become a major focus for the Jungle Air Force and the Long Rangers of the 307th for many coming months.

The Philippines were a major target for three reasons. First was a sentimental reason, or perhaps a matter of pride. The Philippines, like Guam, was American territory. Gen. MacArthur had commanded the U.S. Forces in the Philippines when Japan attacked in 1941 and he was determined to wrest the islands back from them. Philippino guerillas still fought the Japanese in the hills, and there were many American civilians and military personnel held in prison camps on the Philippines. Second, the Philippines was a major Japanese base which held hundreds of thousands of enemy troops, a major portion of Japan's Air Force, and many, many safe bases for the last remnants of Japan's Navy. These Japanese military capabilities were just too important to ignore as the Allies approached Japan itself. Third, the Philippines also provided a relatively safe route to ship the oil, rubber, tin, and rice from the East Indies and Malaya that Japan would need to defend its home islands.

It is about 1,100 miles from the allied air bases on Los Negros to Palau. By August though there was a better way to get at Palau – from an airfield on Wakde Island, just 700 miles away. The U.S. Army's 41st Division had seized Wakde, a small island to the north of Dutch New Guinea, in May 1944, after a fierce three-day battle. Of the 763 Japanese defenders on Wakde, 759 were killed and only four surrendered to the Americans. The U.S. Army lost 40 men killed and 107 wounded in the fight.

By the end of May, the Fifth Air Force had moved two squadrons of P-47 fighters and one squadron of B-25 medium bombers to Wakde, to help support the fighting against the Japanese forces that was still in progress on the New Guinea mainland just a few miles to the south. And the Japanese could still strike back. On June 6, the Japanese launched a bombing raid of their own against Wakde, destroying six U.S. planes on the ground and damaging 80 more. Now in August, the 13th Air Force was moving its bombers from the Fifth and 307th Bomb Groups to Wakde, to get at Palau and other targets north and west.[1]

Wakde

Shortly after breakfast, early on Saturday, August 19, Walt packed up his duffle bag with his things and made his way from the pleasant Los Negros camp area, sheltered by the plantation palm trees, down to a waiting truck for the short ride to Mokerang airstrip. There he boarded a waiting C-47 transport plane, along with 20 or so other enlisted men. As they took off, Walt glanced down at the beach, the Chapel by the Sea, the camp theater, native village, and other landmarks that had been his home for the proceeding eleven and a half weeks. Their route took them almost due west out over the Pacific. After 3 hours and 50 minutes, about 580 miles in the air, they landed at Wakde.[2]

Those men who were awake and alert as they were coming in for their landing would have been amazed at what they saw below. Wakde was a small coral island, just a mile and a half long by three quarters of a mile wide. So small indeed that the single 7,000 foot long by 200 foot wide runway extended from one end of the island to the other. It had a bituminous surface over compacted coral, aligned WNW/ESE with no approach obstructions at either end.[3]

On either side of the runway were revetments for parking the dozens of fighters and bombers that crowded the island. Beyond the revetments, the arriving men could see roads laid out through the trees leading to various campsites, maintenance and stores tents, and landing ships unloading supplies at the beach. But there was a curious thing about the trees on Wakde. None of them appeared to have tops and many were just shattered stumps. Once the C-47 came to a halt after landing, Walt and the others unloaded their gear and then were met by trucks to take them to HQ to report and be assigned to a campsite. Men from the Fifth Bomb Group were directed to a truck headed north of the runway where their camp area was located. South of the runway, most of the island's base service personnel, maintenance and stores tents, and fuel depots were located. At the very south of the island, on a little peninsula, was the Long Rangers HQ and campsite.

As Walt loaded his duffle onto the truck headed south, he asked the driver 'What happened to the tops of all the trees here?' The driver explained that back in May, when the Army stormed the island, the Japanese defenders put up a fierce resistance. 'The Japs had over 100 pillboxes dug into the coral defending the beaches. The Navy destroyers had to pound the island with their guns and the Army landed Sherman tanks to protect the troops as they pushed across the island. This island used to be a coconut plantation before the war, but after all the shelling by the Navy and tanks, you'd never know it.'[4]

Walt reported in at HQ and was told where he could find a tent. The 307th had not been able to ship the wooden floors they had built for their tents on Los Negros, so they were pitched on the bare coral. Much of the ground was uneven, because of the old Japanese foxholes and bomb craters. It soon became apparent that there was a peculiar odor in the air – Walt soon learned that this was the smell of rotting Japanese flesh, defenders' bodies that had been partially buried by the bombardment and not yet buried by the Americans. The first days at Wakde, flies were horrible. There were no screens yet on the tents and mess halls, so a man's fork would be clustered with flies in the time that he raised it from his plate to his mouth. Fresh water was in short supply – there was none for bathing until new wells were completed.[5] The men learned to make do with limited water though. In the morning, a man would fill his helmet with fresh water which he would use to shave with. Then he would use the same water for a sponge bath, then wash his socks and underwear, and then and only then throw it out.[6]

Walt soon found Frank Mansir, Jack Riley, Boris Hidalgo, Gene Schreiner, and Hugh Charbeneau, and found a spot in their tent. He learned that they would likely be assigned to fly again with Jack Arnett, as well as Don Aubrey, Ed Dunne, and Jimmie Clark. So much, though not all of old crew 430 would be together. Walt heard that Cliff Llewellyn was being kept busy as radio operator on C-47's flying supply runs back and forth between Los Negros and Wakde, to complete the move of the last of the 307th.[7] That night, as darkness came, Walt heard rumbling off in the distance. It was not thunder, the others explained, but artillery firing on the New Guinea mainland, just two and a half miles across from Wakde. The Japanese front line was just a few miles down the beach and the artillery would rumble off and on for as long as they were on Wakde. In the coming days, many men would find Japanese 'souvenirs' – discarded equipment, flags, even guns and bayonets, on Wakde.

Some would actually arrange a trip via boat to the mainland and then hitch a ride from the Army to see what the front lines were like.

On August 24th and 25th, the heavy equipment of the 424th arrived from Los Negros aboard two ships, the USS Stanley Matthews and the USS Luis Arguillo. This was good news, but it made more work for the men to assist with the unloading and setting up of equipment and moving of supplies. It was hard work, but kept the men busy and out of trouble.[8]

With the tops of the trees blown away, there was no shade to speak of on Wakde. The 307th campsite on the southern peninsula provided sea breezes through the tents most of the time and the temperature was quite pleasant at night. The beaches on either side of the peninsula were strewn with wrecked equipment though, so they did not provide a pleasant vista and the ragged coral in any case would have made swimming difficult. The tents were pitched on a low ridge that ran the length of the peninsula. A main road from the airstrip through the camp ran past the Group HQ area on the left, then past the crew tents on either side of the road. The road then split off to the left and right. To the left it ended in the Group transportation area where the motor pool was located and where men could pick up rides out to the airstrip or other sites on the island. To the right, the road headed along the beach to an area set aside for unloading landing ships that brought in heavy equipment, fuel for the planes, bombs, other ordnance, and heavy supplies. Just beyond the junction where the road split, was the 307th theater area, with a stage and movie screen just behind the beach, rows of improvised seats, and a projection booth for the nightly movies. If you were seated in the theater on a moonlit night, you could see the coast of New Guinea behind the screen, just a couple miles distant – at least until the movie started and lit up the screen. When it rained during a movie, men would sit on their helmets wearing their ponchos, with the opening in front of their eyes.[9]

Sanitation was basic. There was an enlisted men's latrine built on a little dock extending out over the water near the tip of the peninsula. Nearby was the enlisted men's shower, close to the salt water used for bathing until more fresh water wells could be blasted into the coral. P-47 fighter belly fuel tanks were used to store water for the showers, mounted on a wooden frame above the open-air shower floor. The HQ area, back up the road, was laid out in a rough quadrangle. At one corner was the kitchen, which adjoined the officers' mess on one side and the EM mess on the other. Other tents laid out around the quadrangle included the orderly room, dispensary, operations room, and intelligence room. There was also a supply tent, personnel supply tent, small arms tent, generator, barbershop, and mailroom.[10]

Within the combat squadrons like the 424th in the South Pacific, the concept of rank lost its strict meaning and protocol was lax. Officers had some little privileges, like four men to a tent, not six, their own latrines, and sometimes a little officers' club. Officers wore their bars and stars on their shirt collars, but most enlisted men and NCO's like Walt did not. It was just too much bother and anyway, everyone on a crew knew his job and who was in charge. There was very little saluting of officers, except maybe when a man had to report to the Squadron HQ. One day a new officer reported to the 424th HQ and complained to the Officer of the Day that the enlisted men and NCO's were not saluting him. Captain Vanderpoel overheard the man's complaint and told him to cool-off. He told the new officer *'When you have been in combat as long as these men and have gone through as much as they have, then maybe you can come back and complain to me that they are not saluting you!'* He never came back.[11]

Space was at a premium on Wakde, with very little room between the tents. The only recreation was a little swimming, movies at night, and softball played in the motor pool lot. One bright spot though was a steady improvement in the quality of the food. The 13th Air Force had started rotating men back to Australia for rest leave with priority being given to men with the most combat missions completed. Transport was by the workhorse C-47s making the journey to Sydney via intermediate bases in New Guinea and northern Australia. Everyone on Wakde benefited by these regular flights, because the planes coming back from Australia could carry fresh fruit, vegetables, and meat. Spam, reconstituted eggs, and powdered mash potatoes did not disappear from the diet, but every few days the men would find fresh apples and oranges with breakfast along with real eggs, or fresh beef and salad with real butter for their bread at dinner.[12]

Other problems associated with setting up a new base persisted. There was neither an officer's club nor a day room for the enlisted men. The 307th PX had not reopened. The electrical generators could not handle all the demand, so there was little lighting available after dark, although the movie theater was given priority.[13]

Wakde must have seemed like as desolate and isolated a spot as possible to the men stationed there. There was no native village. The only curiosities were the remains of the whitewashed plantation house, much damaged by the earlier fighting, and a Japanese religious tower – now used as a pole for electrical and telephone wires. You could walk around the perimeter of the island in just a few hours. Some men hitched rides on their days off to the mainland – via one of the many Navy operated patrol or supply boats that were then running up and down the New Guinea coast. Some airmen went out to the front line up the coast to hunt for Japanese souvenirs, until they were chewed out by the Army veterans who pointed out that the Japs often booby trapped guns and equipment they left behind.

A day or two later, as Walt and the other men were sitting down at breakfast, a buzz went through the mess hall. There was going to be a USO show that very afternoon, here on Wakde, and it was going to be first rate. Bob Hope was coming to Wakde, along with a bevy of other stars. The war was still going on of course, and unfortunately some crews were assigned missions that day, but Walt, Jack, Frank, Boris, and Gene were lucky – they were not yet scheduled. It was clear that most everyone on the island who could get away would be there. Well before the announced time, the little 307th theater was packed and most had to stand. MP's were there to keep order and to help escort the stars to the stage when they arrived. The minutes went by then an hour under the hot sun, past the announced time for the show to start. Walt and the others baked in the sun, wondering if the show was really going to happen. A second hour went by, and then a third, with the temperature seeming like at least a hundred. Finally a buzz went up as a band assembled on the stage and tuned up briefly. Then Judith Anderson, movie star and 'Mistress of Ceremonies' appeared on stage and the crowd went wild.

Judith Anderson was born in Australia, where she trained as a classical actress. She had been a Hollywood film star for 10 years, with an Academy Award nomination to her credit. Anderson stepped up to the microphone, greeted the assembled airmen, and then quieted the crowd by quoting a few lines from Shakespeare. She then said, 'I know what you are all really waiting for. Here's Bob Hope.' Hope hopped on stage from the right wearing khakis, a pith helmet, and carrying his signature golf club. Hope stepped up to the mike and as the cheering died down he said, 'I love this beautiful island, with its magnificent palm trees, two of them with tops!' With that, all of the discomfort from the heat and waiting was forgotten. With his casual style, Hope poked fun at the men, he poked fun at their island base, and he poked fun, most of all, at himself. His style was magical and for the time being Walt and all the other men forgot where they were and what they were there for.

The crowd went wild when Hope introduced Frances Langford. Langford was one of the most popular actresses and singers in Hollywood during the 1940's. Her first big hit had been the song 'I'm in the Mood for Love,' recorded in 1935. She had been featured in a couple movies every year since then, and most of the men on Wakde had seen her in recent films like 'Yankee Doodle Dandy' with Jimmy Cagney and 'This is the Army.' At one point, as Langford was about to sing another song, a trio of P-47 fighters taking off for a strike mission up the New Guinea coast buzzed the stage at tree top level. Langford looked terrified as the fighters passed overhead, then Hope put an arm around her and waved to the fighters with the other arm and quipped into the mike, 'Now I know what happened to the tops of all your trees.'[14]

Hope and Anderson introduced other entertainers. Helen McLure, a semi-classical singer entertained the men, followed by Ann Triolo, accordionist and singer with a more popular selection of songs. Jerry Colona, film star, singer, and comedian came on stage. He was a big baritone with a distinctive bushy mustache, known for his campy renditions of popular and not so popular songs. Many men said that Colona almost blew the audience off the island with his powerful voice. Later, Bob Hope joined him along with guitarist Tony Romano as they sang a couple of trios. Patty Thomas, a young attractive dancer entertained the audience as well. A well-known tenor from the 13th Air Force, Capt. Lanny Ross came on stage and sang a song accompanied by Tony Romano.

Once again, toward the end of the show, Hope was telling more jokes, when a P-38 flew low over the stage, disrupting his punch line. Hope spun around, shaking his fist at the plane, then said, 'Doesn't he know there's a good show going on right here?' One of the airmen in the crowd yelled back, 'He went to Hollandia to see a Burns and Allen show.' Hope replied, 'There's always a comedian in the crowd!' The crowd roared and roared with laughter.[15]

During the show, Walt and every man with a camera took pictures of his favorite stars. After a couple hours tough, it neared its end. Hope called each of the troupe members on stage for a final bow, acknowledging the orchestra as well. Then joining hands, they all sang Hope's theme song,

'Thanks for the Memories.' And that was it. The crowd cheered and applauded, as the entertainers stepped from the stage. The MP's were there to help escort them back to the landing ship that would carry them up the coast to their next show. Many men were fortunate to get autographs as Hope, Langford, Colona, Thomas, and the others made their exit.[16]

The show had been a real morale booster for the men. Spirits picked up and everyone talked about the smallest detail regarding the show for days afterwards. The squadron photographers were flooded with requests to develop and print the pictures the men had taken, and as the prints were finished, they spurred a new round of talk about the show.

Palau

Shortly after arriving on Wakde, Walt and the other gunners were required to attend a briefing provided by a Marine Captain on the island of Palau and what to expect if shot down over the Island. He explained that it was possible to find friendly natives on many islands. Palau had been under Japanese rule for 30 years and the natives were believed to be friendly to the Americans. However, like most of the Japanese held islands, there were enemy troops all over the place, so eluding capture would be difficult. A question was asked as to what to expect from the Japanese if captured. The Captain explained that they would most likely be beaten, insulted, and executed if information was not provided during interrogation. He said they would have three choices if interrogated. They could tell the Japs everything they knew, they could just give them name, rank, and serial number – but likely be killed, or they could give information of no value – everything but the most vital details of strength at bases, radar, future targets, and so on. The Marine told them the third option was what he recommended. He told the men also that the Japanese liked to be able to provide a large amount of paperwork to their superiors, so you could go into great detail about your life history, where you went through training in the U.S., with dates, locations, and so on. He concluded by saying that if you were shot down over Palau, hiding would be difficult, but you would not have to wait long for rescue, implying that the Marines would invade the islands soon.[17] He also told the men that there was a Geisha house on Palau, which they should try not to hit if possible…which broke the tension with a little laughter.[18]

On August 23, the last of the 307th personnel had completed the move from Los Negros to Wakde. That day, Capt. Vanderpoel took a single B-24 on a reconnaissance to Palau, bringing back valuable photos showing the island targets and defenses. Two days later the first bombing mission against Palau was sent out, with B-24s from both the Fifth and 307th Bomb Groups. The news Walt and the others heard about this mission at the end of the day was not good. After the bomb run, about 35 miles from the target, two 372nd Squadron planes collided and went down. Lt. Walter Davis and Lt. Kenneth Kienzle piloted these planes. They had been flying in formation and had just entered a heavy weather font when the accident occurred. An explosion was seen and one of the planes was seen turning over on its back then going into a dive. Another observer saw a burning parachute. The 372nd Squadron radioed the position to a U.S. Navy submarine that was reported to be patrolling in the area, but no survivors were ever found.[19]

In addition to the 307th loss, one of the Fifth Bomb Group planes, piloted by Lt. Grant Rea was lost over the target as well. One of the Fifth Bomb Group gunners reported:

> *"Our first raid from our new base at Wakde took place to-day, and it was on Palau. They told us at briefing last night we would probably surprise the Japs. May be my nerves are shot but today I saw one of our own planes shot down. All I can say is please God never let me see another one. 25 Zekes jumped us and A.A. was heavy and accurate. This is a tough target."* [20]

On August 26 and 28, the 424th returned to Palau. On the August 28 mission, the plane piloted by Capt. William Dixon, CO of the 372nd Squadron, was lost. Approaching the target, someone broke radio silence and asked 'How far below base altitude are you going to bomb?' Dixon replied 'We're going in at 11,500 feet at an indicated air speed of 150 MPH.' Someone else radioed 'I hope the Japs didn't hear that.' Shortly after that, Dixon's plane was hit by A/A fire in the left wing root, near the pilot's seat. The plane burst into flames but was seen to salvo its bomb load while continuing to fly steady. Then the damaged wing broke off and the plane fell straight down in flames. One body was seen to be thrown clear but no parachutes opened as the plane crashed into the ground below.[21]

A reporter from a newspaper in the U.S. made his way out to Wakde around this time. Detailed operational information was obviously off limits, but he was able to include the following statement from Col. Robert Totten, 13th Air Force Bomber Command Headquarters:

> *"Palau has been unusual in one respect. Usually at a new target the ack-ack is punk at first but gets better. As you eliminate the Zeroes you find flak getting as sharp as a whip. But at Palau, where the Zeroes gave up early, the flak was sharp right from the start. They can't keep us off the target but they've managed to make it pretty tough."* [22]

That evening, after dinner in the noisy mess tent, many of the men on Wakde wandered down to the southern tip of the island. A major firefight was developing on the New Guinea mainland, where American tanks and ships offshore were firing round after round into the Japanese positions. As the flashes of light from the guns and then the explosions lit up the sky red, yellow, and orange, the men of the 307th cheered. It was good to see someone else doing the fighting, if even just for a few hours. [23]

The Arnett crew was assigned to fly the next mission to Palau. Capt. Vanderpoel would lead six 424th Squadron B-24's. Jack Arnett would be flying A/C 540, with Don Aubrey, Ed Dunne, and Jimmie Clark, along with Hugh Charbeneau, Boris Hidalgo, Gene Schreiner, Frank Mansir, Jack Riley, and Walt in the ball turret. This was at least the third time that Walt, Jack, and Frank were assigned a mission in A/C 540, so perhaps they felt it was a good luck plane. The night before, five of the planes were loaded with 15 250-pound bombs and one was loaded with 40 100-pound bombs, all instantaneously fused. During their preflight briefing, the men learned that the primary target was a causeway linking the base at Koror Town on Koror Island with a seaplane base and other Japanese installations on nearby Arakabesen Island.

Early on Tuesday morning, August 29, Walt, Frank, Jack and the other Arnett crewmen were awakened at about 3:30 am by George the CQ (charge of quarters), banging on the mess kits hanging at the side of their tent. He told them 'get up – take-off time is 5:40.' Then he moved on to wake up all the other 424th crews on duty that day. Some men chose not to get up then so that they could grab an extra half-hour sleep at the expense of their breakfast. The men went to the latrine, washed up and then went to the mess tent. Usually the mess crews prepared the food late the night before and just kept it warm. It was usually Spam, steamy pancakes, dehydrated scrambled eggs, toast, and marmalade. After several hours, it did not taste good, but it was filling and nutritious.

One of the crewmen would pick up a cardboard box with food for the crew on the flight – usually two loaves of bread, two large cans of orange or grapefruit juice, and a five-pound can of Spam.

After breakfast, each man went back to his tent, made up his bed a little, and gathered the gear needed for the mission: flight jacket, maybe a sweater, flak vest, parachute with harness, .45 cal pistol, a pouch with chocolate bars and extra ammunition, maybe an extra shirt. By 5:15 am they were on the GI truck for the ride out to their waiting plane. The ride was short on Wakde, and each man quickly got to work, loading his gear and checking his guns, ammunition, and other equipment. The flight engineers started the little put-put engine so they had lights in the plane before the engines were started. As each man completed his preparations, they gathered and sat under the plane waiting for orders. No one smoked. Smoking was forbidden within 300 feet of a plane.

Soon a jeep drove up with the Operations officer who called out: 'Everything OK, Arnett? OK – start your engines in five minutes.' Arnett ordered the men aboard – except for the flight engineer, who stood with a fire extinguisher in front of each engine as Arnett started them one by one. Once they were all running smoothly, the FE climbed aboard. Soon the 'clear to taxi' signal came on the radio and the flight engineer released the brakes.

Shortly after dawn, Lt. Arnett got the signal to fall in line with the rest of the little group of B-24's, and then one by one they took off. Arnett had been told that there was a slight problem with the 7,000-foot long runway on Wakde. The Japanese had built it quickly and had not leveled it properly, so that there was a big dip in the surface near the east end.

As Capt. Vanderpoel's plane took off, the other pilots and crewmen watched it roll down the runway, picking up speed. When it hit the dip though, it seemed to disappear from sight, only to reappear as it rolled up the far end of the dip. This was a tricky takeoff for the pilots and certainly very unsettling for anyone watching. Nevertheless, all six planes made it into the air with their heavy loads of bombs, fuel, ammunition, and crewmen. Walt and the other men were relieved that Jack Arnett seemed as good a pilot as either Charles McRae or Carl Appling. Usually the 424th planes would form up at about 1500 feet, turn and fly in formation over the runway as a sign of pride and appreciation for the ground crews who had worked on the planes through the night before. [24]

The weather was not good. They flew through thick frontal conditions the whole way, with many periods of rain showers. The men were bounced around a lot, unless they braced themselves for the turbulence. For the first half of their flight north, they encountered 10/10 coverage of altostratus clouds up to 14,000 feet. Closer to Palau the clouds opened up a bit to 7/10 coverage. Ed Dunne soon came on the radio: '10 minutes to IP. Are you ready to bomb?' Jimmie Clark replied 'roger' and the flight engineer went to the bomb bay to remove the cotter pins that were the safety to keep the bombs from detonating. They were still held in safety by a wire that ran through each bomb housing. As they were dropped, the wire would be pulled out, fully arming the bombs.

The six planes were unable to locate the IP, because as they approached the target, the clouds changed to 6/10 coverage of towering cumulus up to 18,000. The squadron climbed to 19,900 feet to get well above the clouds. Lt. Roderick, flying the lead plane with Capt. Vanderpoel, had to lead the other planes in a circle until they got a glimpse of the target area through the clouds below. Arnett asked Jimmie Clark if he could see the target. 'Not clearly' he replied. Then Walt in the ball turret reported that he could see it, giving Clark and Arnett the bearing.[25] One of the six planes, piloted by Lt. Townsend, developed low fuel pressure and could not climb with the rest, so decided to bomb at a lower elevation separately. Lt. Roderick brought the remaining five planes in toward the target at a magnetic heading of 355°. As the target was approached at 1156, the bombardiers took control, and then released the bombs. Wally watched the slow mesmerizing fall of the bombs. Two of the strings of bombs fell in the harbor near Koror Town, then walked across the causeway. Walt and the other ball turret gunners were sure that two or three of the bombs hit the causeway to Arakabesen Island, as intended. When Lt. Townsend came in later at 1204 for his lower bomb run, his crew noted two towering columns of black smoke arising from Koror Town, indicating that some of the bombs hit targets on land.[26]

As the planes turned away and headed back south, Walt heard someone say over the intercom 'Hey did anyone see any Jap flak or fighters?' It soon dawned on them that there had been absolutely no opposition. Returning to Wakde, they passed through the same thick frontal conditions, but they were all able to make a safe landing back at base. Walt completed his 15th combat mission after 9 hours and 5 minutes in the air.[27]

Around this time, Walt found out that the 424th program of rotating combat crews back to Australia for R&R (rest and relaxation) leave might provide him with an opportunity. The most senior combat crewmen were given priority, with 15 completed combat missions being the minimum needed to qualify. R&R leave applied only to combat crews. Ground crews, many of whom had been in the South Pacific with the 307th for a full year or even longer, were not provided with R&R leave. The final decision was made by the Squadron CO, Capt. Vanderpoel, and he was authorized to take into consideration the operational needs of the unit. Still, Walt knew that his chance could come up, and he was looking forward to a change of scene after almost three and a half months of combat.

There had been quite a bit of mixing of crews during the previous month. Walt, Jack, and Frank had now flown three missions with Jack Arnett and they had come to like and trust the young pilot from West Virginia, even though he was sometimes caustic in his comments. It was comforting that Don Aubrey, Ed Dunne, and Jimmie Clark from old crew 430 were flying with them also. They had come to like the new gunners, especially Gene Schreiner and Boris Hidalgo. Carl Appling was on an extended leave and it was not clear when he would return, but they were glad to see that Charles McRae was starting to chalk up successful missions with a crew of his own. Cliff was flying with another crew and Walt, Jack, and Frank missed him. George Volchko had started flying regularly with the Townsend crew and seemed to fit in well there.

At the beginning of September, the 424th Bomb Squadron had 604 men assigned. There were 92 officers and 138 enlisted men among the flying personnel, and 15 officers with 365 enlisted men with the ground personnel.[28]

Walt, Jack, Frank, Ed, Don, and Jimmie were surprised to find that on September 1, Lt. Arnett was assigned to fly a second bombing mission to Palau, but this time with a different crew and not with them. Arnett would be filling in for a crew normally commanded by Lt. Norman Coorssen. Coorssen's crew was scheduled to fly on the First of September, but apparently he took ill the day before and Arnett was asked to volunteer in his place. Walt, Frank, Jack and the other gunners knew most of this crew. Sgt. Robert J. Stinson from California was flight engineer, Sgt. Jimmie Doyle from Texas was radio operator, and Sgt. Charles T. Goulding from New York was ball turret gunner. The other gunners were Sgt. John Moore from Arkansas, Sgt. Leland J. Price and Sgt. Earl E. Yoh, both from Ohio, and Sgt. Alexander Vick from California. The officers, who Ed Dunne, Don Aubrey, and

Jimmie Clark knew well, were the copilot, F/O William B. Simpson from North Carolina, navigator Lt. Frank J. Arhar from Pennsylvania, and bombardier Lt. Arthur J. Schumacher from Minnesota.[29]

Walt would probably have been aware of the early morning noise as the other crews assigned to fly that day were roused from their sleep. Men who were not flying usually just stayed in the sack during the relative cool before dawn, despite the unavoidable clatter of men getting their flying gear ready and then heading out to the airstrip. To a man lying in his bunk, the sound of the trucks carrying the crews out to the airstrip would then die away, to be followed by the return of the sounds made by the insects, reptiles, and small mammals of the jungle – or the little of what was left of the jungle on Wakde. At the first glimmer of dawn, the sounds of tropical birds would then join the South Seas symphony, to be followed by the distant sputtering and revving of B-24 engines as the bombers would begin to assemble for take-off. At the appointed time, one by one, each bomber's engines would roar into full volume in response to the control tower's signal to begin the takeoff. This could go on for a half an hour as one by one the big bombers took their turn to get airborne for the day's mission. Walt was familiar now with the sounds and the routine that the sounds signified, both as a crewman on a plane flying that day and as a crewman staying behind, waiting to fly another day. Today, as he drifted in and out of sleep, Walt listened to the sounds from afar, then finally awoke, washed, ate breakfast, and got on with his day.

Later that day, as Boris, Gene, Frank, Jack, and Walt were wiling away the hot afternoon, they got news that the bombing of Palau had gone well – the Japanese base at Koror Town was pounded, there were large explosions followed by flames and smoke. However, one of the planes was missing and it looked like it was a 424th Squadron plane. Walt and the other crewmen headed out to the airstrip to await the return of the planes. It soon became apparent that the Arnett plane was missing. As the surviving planes landed, the returning crews were questioned by the other men of the squadron, so little bits of what had happened began to be known. The crews that had flown that day were obviously agitated though, and tired, and they had to go through their post-mission debriefing. Before the end of the evening the details were widely known. Flying A/C 453, Arnett was right over Koror at 1106 when the plane got two direct hits of A/A fire in the left wing. The wing folded up then broke off and the plane went into a steep dive, the fuselage then broke in two with the parts hitting the water between Koror and Babelthaup Islands. Lt. Lewis, 424th commander, and another pilot, Lt. Sasser, each saw two men bail out and two parachutes open. Another man from the 424th reported seeing three men bail out.[30] A plane from the 371st thought that five parachutes were seen to open, landing near the Krakabesan Causeway. Lt. Lewis and two other planes circled the area for almost an hour seeing no survivors, although a small launch was observed headed back to the Japanese base from where one of the men seemed to have landed in the water.[31]

Walt, Jack, and Frank made their way quietly back to base. The Arnett crew was lost – but just yesterday, *they* were the Arnett crew. If they had flown today with Arnett, would they be dead now? Each man must have thought about fate, and luck, and the hand of God, and how little really they could control their own destiny out here at war. One flier in a similar situation would write:

> *It is funny in a way, after going on quite a few missions a guy gets so that he isn't afraid to die; he just wants to get home again awful bad, that' all. When a friend goes down and is lost, we feel pretty bad for a while, but we try to forget it soon.*[32]

Arnett had seemed a brave but careful pilot and the men would have all flown with him again. So much was just random – a small crew of Japanese Army A/A gunners had woken up earlier that day on Palau, had manned their gun when the bombing alert was sounded, had been in the right position at the right time to put a couple shells at the right altitude, aiming where they thought the bombers would be, and had brought down the American bomber. Not for the first time nor for the last during his career in the Air Corps, sleep was long in coming for Walt that night.

The following day, a 371st Squadron plane was hit by A/A fire over Koror. The Pilot, Lt. Parenti, was able to complete the bombing, but headed back south toward Wakde, his engines began to falter and die one by one. Realizing he could not make it back, he decided to ditch. Parenti, Lt. Hill his copilot, and two of the gunners, Cpl. Perlowitz and Cpl. McGaffey, managed to get in a life raft, but the rest of the crew apparently went down as the plane sank. A couple days later, the survivors were rescued by a PBY Catalina patrol plane and taken to hospital.[33]

Australian Interlude

Walt, Frank, and Jack now began a period of inactivity. With no pilot available, the enlisted men had time on their hands while they waited for something to change. On September 4th and 5th, the 424th participated in follow up bombing raids to Palau, while Walt and his crewmates watched and waited.

The organized amenities on Wakde were lacking, compared to Los Negros. A few crewmen made their way to the mainland a couple miles to the north and hiked out to the front lines, to see how the Army lived and fought in the jungle. This practice was halted eventually, because of the malaria and scrub typhus that was common on the mainland. Most men were content to remain on the island to do their job, while the army did its job against the Japanese in the jungle.

To amuse themselves, Walt and his mates searched the island for souvenirs. Old ammunition, uniform parts, stamps, coins, even discarded guns could be found from the Japanese forces that had occupied the island. Surrounding the island, the coral reefs were covered with many varied and interesting shells, and many men began to amass collections of shells that they would classify and compare. A regular trade in souvenirs and shells developed as collectors bartered back and forth.

In addition to the debris from the Japanese forces scattered across the island, the men found many discarded P-47 belly tanks. These metal tanks were designed to fit under the fuselage of a P-47 fighter in place of a bomb, to permit it to fly much farther than its normal range. For some reason the supply of belly tanks greatly exceeded the need on the island and ingenious crewmen began to put the surplus to use. By cutting out a portion of the top of a belly tank and adding ballast, one or two men could use it as a boat. Frank, Jack, and Walt went to work on a tank and soon they had a suitable boat of their own. A seat was fashioned in the bottom and paddles carved from discarded lumber. Walt, Frank, Jack, and many other crewmen spent many hours cruising in these boats around the island. Some men rigged sails and one even found a small motor to fit in his boat. Belly tanks were also lashed together and covered with planking to create offshore swimming platforms. Wakde was primitive, but the men made the best of it and moral remained OK, despite the heat and discomfort.[34]

After about a week and a half of inactivity since his last mission on August 29 with the Arnett crew, Walt received notice from the 424th HQ that he would be granted 10 days of R&R leave in Sydney, Australia. Walt had completed 15 combat missions now, which was the minimum for R&R. On a Monday, he was notified that the following day he would be on his way to Australia! Walt knew some men who stockpiled their ration of free cigarettes, and he bargained for half a dozen cartons to take to Australia. He had heard that in Australia, you could sell a carton of good American cigarettes for 10 Australian Pounds – about $10 US.[35]

Early on the morning of Tuesday, September 12, Walt woke around 6:00 am, showered, and dressed. After breakfast, he reported to 424th HQ to pick up his orders authorizing his R&R.[36] S/Sgt. Babinski was told to get the duffel he had packed the night before and report to the transport tent adjacent to the runway. Walt said goodbye to his pals in the crew tent then made the short walk to the transport tent where he checked in.

He had a little while to wait, as a portion of the 424th Squadron planes were taking off for a mission led by Lt. Rodwick to bomb the Japanese runway and installations at Lolobata on Halmahera Island in the Dutch East Indies.[37] Finally, still early in the morning, the B-24's had all cleared the runway and the men in the transport tent were told that they could board the waiting 424th Squadron C-47[38] as their names were called. Walt shouldered his duffel and boarded the waiting transport plane. The co-pilot met him at the door, checked his orders, and told him to take a seat anywhere.

Soon the plane was headed down the runway, into the air, and steadily gained altitude. Glancing out a small window, Walt could see that they had crossed the short stretch of water that separated Wakde from the mainland of New Guinea and were headed into the interior. Their destination that day was Rockhampton, Australia, via Nadzab, New Guinea. Walt likely recognized a few of the other men on the plane, also headed to Australia for their 10-day leave. The little C-47 was important to all the men of the 424th, because in addition to ferrying them to and from the relative pleasures and civilization of Australia, it was also the source of fresh fruit and vegetables, meat, eggs, beer, and liquor that the crew would buy in Rockhampton and then carry back to Wakde. All of this was purchased with money that the officers, NCO's, and enlisted men would pool, to stock their respective mess halls and clubs with products that were not standard fare transported by the Navy and provided by their Air Force Quartermasters.

Nadzab was about 650 miles southeast of Wakde, located about 30 miles from the coast up the Markham River Valley. Nadzab had developed into a major allied air base, with five runways. As the little C-47 came in for a landing, Walt may have seen the high New Guinea mountain ranges that bracketed the Markham River to the north and south. Below he would see hundreds of American and Australian fighters, medium and heavy bombers, and transport planes. As his plane came to a stop, the door was opened and the men stepped out into the mid-afternoon heat of the New Guinea interior. The heat and stifling humidity was something that the men from the 424th were not used to. Here the lack of a cooling ocean breeze quickly convinced the men that there were advantages to the island bases that they inhabited.

Walt learned that they would not leave for Australia until the following day. The men were directed to an army 6x6 truck to take them to temporary quarters. They tossed their bags into the truck and took seats along the side. The driver picked up a few other men waiting for transport, then headed up a steep dirt road away from the airstrips. The road followed a stream that came out of the mountains, and soon they reached a level valley that had been cleared of vegetation where a tent city had been established. They stopped at an orderly tent where they were given instructions on where to sleep, location of the transient mess hall, and when and where to report the following day to continue their journey. Beyond the tents, Walt could see that the dense jungle continued again up to the mountain ridges that ringed the valley.[39]

Walt and the others quickly got themselves out of the sun and made themselves as comfortable as possible. They learned that the intense heat and humidity made sanitation and personal hygiene especially important. Men who did not carefully wash their own mess kits were prone to come down with diarrhea here, and for those who did, the latrines were particularly unpleasant, crawling with flies and maggots. They learned that sometime the maggots got so bad that the men tried burning them out with oil, but sometimes they burned down the latrine![40]

Walt learned that Nadzab was now the location of the Far Eastern Air Forces Combat Replacement and Crew Training Center. This was the successor to the XIII Bomber Command Combat Training Center on Guadalcanal that Walt and the McRae/Appling crew had graduated from earlier in May. Shortly after they were transferred to the 307th Bomb Group, the XIII Bomber Command Combat Training Center was closed and the Fifth Air Force took over the final combat training duties for both of the U.S. Air Forces in the Southwest Pacific area here in Nadzab.

Walt's dinner that evening was standard Nadzab fare. At the appointed time, the screened-in mess tent door would be unlocked and the line of waiting men would file in, mess kits in hand. Food included breaded spam, fried dehydrated potatoes, canned corn, and canned peaches, along with fresh but coarse bread and a hunk of canned butter.[41] Soon after chow, night came and the men made their way to check out the evening movie or perhaps just back to the tents. Here and there, men played cards or dice, or talked or read.[42] Soon all was quiet except for the jungle sounds and men snoring behind mosquito netting.

The following morning the procedure for Walt in Nadzab was reversed. After a visit to the latrine and perhaps a fast shower, Walt ate his breakfast, then packed for the ride back out to the airstrip to continue south on the little 424th Squadron C-47. Soon they were in the air again, gaining altitude along the Markham Valley, then turning south by southwest over the lofty Owen Stanley Mountain Range. Soon they were over the Gulf of Papua and the Coral Sea. All the men grew more and more relaxed as they realized that over these seas and in these skies, there was no longer a danger of Japanese fighters to worry about.

It was about a seven-hour flight to cover the 1173 miles from Nadzab to Rockhampton on Australia's East Coast. Rockhampton was a major air transport hub for the Allied War effort, operated by the joint U.S.-Australian Directorate of Air Transport for the South West Pacific Area.[43] The operation at Rockhampton, located about 25 miles from the Pacific Ocean in the Queensland outback, was very efficient. After landing and saying goodbye to the crew of the 424th Squadron C-47, Walt and the others checked in and were told that they would depart shortly for Sydney aboard a Directorate of Air Transport passenger plane. Rockhampton had something of the look of the 'wild west' to most Americans, with dirt streets, wooden sidewalks, overhanging stores, and many saloons.[44] After a fast bite to eat, the men were airborne again, making their way south on the last leg of their journey – 730 miles from Rockhampton to Sydney. Finally, early in the evening on Wednesday, September 13, Walt arrived in Sydney, with 10 days free.[45]

After arriving in Sydney, Walt and the other men beginning their 10 days of leave were transported by bus into the center of the city, to an area known as King's Cross. Walt and the others were dropped off and led to a small transportation office in King's Cross operated by the Air Force.[46] The men were told that they were at liberty for ten days – until Saturday, September 23, when they were to report back bright and early in the morning at the Transportation Office, for the bus back to the airfield and their flight back north. An officer at the Transportation Office gave them directions to a nearby Red Cross facility that provided various services for the men and arranged quarters for them. Walt quickly made his way over the Red Cross, where he learned that his cheapest option for lodging was in a nearby King's Cross dormitory that charged U.S. service men six pence per night for a bunk and shower.[47]

The Red Cross staff let the men know that they could help them with any information they might need about attractions, entertainment, or other assistance. Walt had an idea – it was a long-shot he knew, but he explained to the Red Cross staffer that his older brother John was stationed with the U.S. Navy in Brisbane. Walt knew from letters he had received from Johnnie and also from his family back home that John was a Mechanics Mate Second Class, stationed at Brisbane Navy Base 134.[48] Would it be possible for Walt to get in touch with John and possibly arrange leave for him so that John and Walt could spend some time together? The Red Cross officer told Walt that it might be possible. He took the particulars from Walt and told him to check back each day.

Walt then made his way over to the dormitory and arranged for his lodging. After shaving and showering, he changed into a clean uniform and set out to begin exploring Sydney. Then it hit him. He was back in civilization. Looking up and down the streets of King's Cross, Walt saw things he had longed for but had not seen for five months since he left Hawaii – restaurants, department stores, tobacco shops, ice cream parlors, florists, little kids selling newspapers. King's Cross was one of the oldest parts of Sydney and its streets were laid out in a haphazard, old-world fashion. This helped to enhance the sense of surprise as Walt would turn a corner and come across something new – a delicatessen, a florist shop, a movie theater. Walt likely found other Air Force NCO's wandering King's Cross who he spent the rest of the evening with – certainly a good meal with a thick steak and eggs, Australian style, with lots of good Australian beer. One of the strongest memories Walt would have of Sydney was the wonderful food, especially the beef.[49]

In the days to come, Walt experienced more of Sydney's attractions. Streetcars provided convenient and inexpensive transportation for just a 'hay penny' a ride. Walt visited the Sydney Zoo to see kangaroos, wallabies, and koala bears in eucalyptus trees. The Sydney Harbor Bridge was a prominent landmark and likely reminded Walt of other bridges – the Golden Gate Bridge in San Francisco, or better yet – the Ambassador Bridge in Detroit, near his home. Walt probably spent time in Mark Foy's Department Store, shopping for things to take back with him when his leave was up.[50]

One day, Walt took a streetcar across the Sydney Harbour Bridge to spend a day at the Luna Amusement Park. Luna Park greeted visitors with an illuminated smiling face over the ticket booth and entry turnstiles. The park had been first opened in 1937 and operated about 10 months every year. Throughout the war, Luna Park attracted servicemen on leave. Many Australian girls waited outside the park, hoping for a serviceman to be their date for a day and evening in the park. Luna Park had about 10 rides and a big roller-coaster, as well as places to eat, drink, play carney-style games, and other amusements. In the evening, the park's external lights were 'browned out' in case of a Japanese sneak attack on Sydney.[51]

A popular daylong excursion was a street car ride out to Bondi Beach, about 10 miles from King's Cross. Walt had not seen a beach like this since his time in St. Petersburg, Clearwater, and Biloxi the year before. September was springtime in Australia, with temperatures usually getting to near 70 degrees almost every day, and pleasant nights with temperatures around 50°. Bondi Beach is surrounded by low hills on two sides, with pleasant cottages, shops, amusement halls, and cafes ringing the beach. The steady breakers brought out Australian surfers and even a few of the American service men brave enough to try the sport. There was a Red Cross at Bondi Beach that provided facilities for service men and pointed out nearby dance halls and nightclubs. After a full day exploring Bondi, Walt caught a late streetcar back to his King's Cross dormitory.[52]

Walt spent part of each day catching up on the news – Sydney papers provided the most recent world news, but the Red Cross had a few older American papers that were of interest also. Walt learned that in Europe, the British Army was pushing into Belgium and approaching southern Holland. The U.S. Army had landed in the south of France and was racing north. Walt read about U.S. Navy air raids that were hitting Palau and also the Philippine Islands. The news also reported that President Roosevelt had just begun meetings with Prime Minister Churchill in Quebec, Canada.

Then, one morning at the Red Cross office in King's Cross, Walt got the news that his brother Johnnie had gotten the OK to come to Sydney for a short leave. Early the next morning, there he was – John in has Navy uniform and Walt in his Air Corps uniform, together for the first time since early in 1943. Walt and Johnnie hurried off for a big steak dinner, a few glasses of beer, and a long talk to bring each other up to date. In March 1943, John had shipped out from San Francisco for Brisbane, where he was attached to the U.S. Seventh Fleet. For the past 17 months, he had worked in a Navy machine shop, responsible for repairs to refrigeration and cooling units on small navy vessels. John had been promoted, from Seaman first class, to Mariners Mate second class, and then in June to Mariners Mate first class, with two men assigned to work under him.[53]

John told Walt what life in Brisbane was like. Smaller than Sydney, but booming as a U.S. Navy base, Brisbane was a pleasant place to be stationed. It was relatively easy to get from the Navy base into the city of a third of a million people when on leave. Queen Street, the main commercial center of Brisbane, was a pleasant palm-tree lined thoroughfare. The American Red Cross on nearby Adelaide Street was similar to the Red Cross in Sydney, with meals for service men in a large dining hall, a game and recreation room, and library.[54]

As they finished their meals and started on dessert, Walt described training, his crewmates, and what the combat missions he had been on to date were like. At this point John said that he was unhappy as a mechanic in a backwater area like Brisbane – he wanted to be assigned to a ship for combat duty. John announced that just the day before, he had received orders to travel back to the U.S., to the '...nearest West Coast port for assignment to a new ship, after 30 days annual leave in the U.S....'[55] They talked about home and family and all the friends left behind. After dinner, Walt and Johnnie walked through King's Cross and talked late into the night.

During the rest of Johnnie's short leave in Sydney, Walt showed him around the city. They traveled to Bondi Beach, saw the movie 'Mission to Moscow' at Sydney's Civic Theater, and enjoyed more hearty meals with good Australian steak and beer. Finally the time came for John to leave, make his way back to Brisbane, and board his transport ship for the return journey to the U.S. Johnnie promised to say hello for Walt to all of his friends back in Detroit, and of course give hugs and kisses to their ma and pa, and their sisters and brothers. Finally, the bus pulled up to the King's Cross transport office. John said his last good-byes, boarded, and waved as the bus pulled away and out of sight. Walt sighed as he left and wondered when it would be his time to return home.

For a couple more days, Walt enjoyed Sydney. He wondered what was going on with his crewmates back on Wakde. He read in the papers that the conference between FDR and Churchill in Quebec was continuing. An uprising of the Polish underground 'Home Army' was reported in Poland and the Red Army had reached the suburbs of the Polish capital. In Holland, British, U.S., and Polish airborne troops were reported to have seized bridges over most of the branches of the Rhine River, with British and American armored forces racing north to try to outflank the German Army.[56] Perhaps the end of the war in Europe would come soon. In the Pacific, Walt read of the invasion of Palau, which was meeting fierce Japanese resistance, and also the invasion of Morotai, a small island in the Dutch East Indies.

Rumors circulated among the U.S. service men in Sydney that it was not uncommon for a man's ten-day rest leave to be extended, sometimes for a week or more, because transport was not available to return to base. After 10 days of leave, on Saturday, September 23, Walt checked in at the Transport Office and learned that there was not a place for him on the plane traveling back north to Rockhampton. He had another day in Sydney, but he was told not to expect many more free days – something was up back at Wakde and the 13th Air Force was anxious to get all men on leave back as quickly as possible. The following morning, sure enough, Walt's name was on the list.

Walt departed Sydney in the afternoon of Sunday, September 24, flying north to Rockhampton. In Rockhampton he learned that he would have to wait till early the following morning for the 424th Squadron's C-47 return to base. The allied Directorate of Air Transport facility in Rockhampton provided a place for Walt and the other returning airmen to eat and sleep. Walt began hearing rumors that the 13th Air Force was preparing for some huge operation and that many of the planes and crews on Wakde were preparing to leave for a new base. The following morning, Walt boarded the little C-47 for a 984-mile flight to Port Moresby on the southeast corner of New Guinea. The plane spent only a couple hours at Port Moresby – only enough time for the plane to be refueled and the crew and passengers to be fed. Soon they were in the air again, climbing high above the New Guinea mountains, before turning to the northwest and Wakde, another 765 miles away.[57]

- 12 -
NOEMFOOR: THE DAMNED 13TH AIR FORCE STRIKES AGAIN

Back to Wakde - September 1944

As the little C-47 approached Wakde[1] on Monday, September 25, Walt could see that things had changed dramatically on the island in less than two weeks since he left on leave. After they landed, Walt retrieved his duffel, reported back in to 424th Headquarters and made his way over to his tent. Along the way he learned that the entire 307th Bomb Group had wrapped up operations on Wakde and was preparing to move to a new base and a new major bombing offensive.

During the two weeks Walt had been away, the 424th had conducted seven bombing missions – one the day he left on September 12 to bomb Japanese installations at Lolobata on Halmahera Island in the Dutch East Indies. The 424th returned to bomb Lolobata each of the three following days. The next two days and again on the 20th, the 424th was sent to bomb Haroeko on Ambon Island.[2]

Near his tent, Walt ran into Cliff Llewellyn. Walt learned that Cliff had participated in three of the missions to Lolobata and one to Haroeko. On one of the missions, Cliff's plane had lost an engine, but made it back to Wakde OK. Just a couple days before, on the 21st, Cliff had flown on a mission to ferry 424th Squadron supplies to their new base – an Island called Noemfoor. Now Cliff was in a hurry, preparing to leave again with men and supplies via a 424th Squadron B-24 to the new base.[3]

Walt wished Cliff good luck and good-bye, and then continued on to his tent. He wondered how much time he had left on Wakde. Soon he found three of his old crewmates – Jack Riley, Frank Mansir, and Boris Hidalgo. They spent the rest of that Monday swapping stories – Walt telling the men about his furlough in Sydney and seeing his brother John, the other three men about what they knew of the new base and the possible major operations coming out of the rumor mill.

The Situation in the Southwest Pacific - September 1944

The pulse of the war across the Central and Southwest Pacific was of critical importance to the men of the 307th Bomb Group, but its details were known only in vague, general terms. Various landings, battles, and base movements were reported and many men charted the progress on maps they kept in their tents. There were maps and information published by the AAF that were posted sometimes in the squadron HQ, to help inform the crews as much as possible. What lay in the future of course was not well known at all and their part in the drama only unfolded day by day.

By late September, it appeared that the war in Europe was going well. The Allies had broken out of their Normandy beachheads and were racing across France, with the British and Canadians on the left and the Americans on the right. The Allied armies had broken into Belgium and had linked up with the Americans and French who had landed in Southern France just the previous month. In Italy, the Allies were well north of Rome now, and had taken Pisa and Florence. On the Eastern Front, the Red Army was racing across Romania and Bulgaria, but inexplicitly was held up on the outskirts of Warsaw, where the Polish Home Army had risen up against the German garrison. Optimists thought the war in Europe could be over before the end of 1944.

In the Central Pacific, U.S. troops had conducted amphibious landings on Saipan and Tinian Islands in June, followed by a landing on Guam in July. That September came news of further landings on Ulithi and Palau. The 307th crews knew of Palau because of their bombing missions there just the previous month. No doubt, news about Palau caused the men of the 424th to recollect the loss of the Arnett crew near the island. They would have wondered what became of the men who were seen to parachute out of A/C 453 at it plunged towards the water. What happened to those two, three, or more men who might have survived the crash and made it to Palau?

In the Southwest Pacific, U.S. amphibious forces had captured Noemfoor in July and Morotai, farther to the west in September. Noemfoor and Morotai were strategically important because they protected the left flank of the U.S. advance in the Pacific towards both the Philippines and Japan itself. The two islands, once they had been developed into suitable bases, would allow the 13th Air Force to operate further to the west, bringing more Japanese controlled territory within range of the big long-range B-24 bombers.

Unknown to the men of the 307th Bomb Group, the next big strategic target in the Southwest Pacific Theater was the Philippine Islands. The Philippines were a key component of the Japanese Empire, linking the rich oil, rubber, tin, and rice resources from the Dutch East Indies and Malaya, with the Japanese home islands. The Japanese defended all their possessions, but the Philippines were especially well protected. Hundreds of thousands of Japanese troops, plus a substantial portion of Japan's air power was concentrated there. The bulk of the Japanese Navy was also based in the Philippines, where it could strike to the northeast, east, or southeast, wherever the next attack from the American Navy came from. The Japanese also knew that the Philippines held an emotional appeal to the Americans – the Philippines were American territory. Tens of thousands of Americans and Philippinos had died in its defense or had been imprisoned by the Japanese. Americans knew of the Bataan Death March and heard reports of how bad the conditions were for the military and civilian prisoners still held by Japan in the Philippines.

When he left the Philippines in 1942, General Douglas MacArthur had promised 'I shall return.' The Japanese knew that MacArthur was approaching closer and closer from the east and they were determined that he would fail to keep his promise.

The critical geo-strategic priority for the Japanese was to defeat the Americans in the Philippines. They had moved a substantial portion of their army, navy, and air force to the Philippines to stop the Americans. However, the Japanese Navy and Air Force, and to a lesser degree their Army, relied on oil and refined petroleum products from the Dutch East Indies, particularly from the great Island of Borneo. There were just no sources of oil in the Philippines (or China, Indochina, and Japan itself). To succeed in stopping MacArthur, the Japanese needed oil from Borneo to fuel their Navy and they needed gasoline refined in Borneo to fuel their airplanes, trucks, and tanks.

Put simply, the job of the 13th Air Force and the Fifth Air Force was now to destroy the Japanese oil fields and refineries on Borneo, along with nearby Japanese shipping, to cut off the supplies that were needed in the Philippines and in the Japanese home islands. Of course, the Japanese realized the danger and were busy strengthening their defenses around the Borneo refineries.

Staff officers from the 13th and Fifth Air Forces shuttled back and forth throughout September making plans. They knew (though most officers and enlisted men did not) that the invasion of the Philippines was scheduled for late October. Even with new bases, what they were being called upon to do would be almost impossible. While the individual B-24 crews and the support staff went about their job of striking camp on Wakde and preparing to move to Noemfoor, the HQ staffs were planning bombing missions the likes of which no B-24 units anywhere had ever been asked to perform before.

If the 13th Air Force failed to destroy the Borneo oil fields, the Philippines invasion could be in jeopardy. What lay ahead would stretch both the B-24's and their crews to the very limit.

Noemfoor – September 1944

When Walt arrived back on Wakde, on September 25, he was immediately ordered to prepare to depart again the following day for a new base. The 307th BG area on Wakde was in feverous activity as the move to their new base was in progress. Walt had seen this apparent chaos just five weeks before during the move from Los Negros to Wakde, and now it had begun again. Walt, Frank, Boris, and Gene knew that they were just little cogs in the big wheels of the war, but still they wondered what lay ahead. Later Walt saw his old friend Cliff Llewellyn, who was busy as radio engineer on a crew flying back and forth between Wakde and the new base: Noemfoor.[4] Cliff told Walt that something very big was up. Cliff had heard rumors that B-24 factory engineers from the states had been seen on Wakde, apparently advising the 307th BG headquarters staff on how to increase the range of the Group's bombers.

The following morning, Tuesday, September 26, Walt and others boarded a B-24 being used to ferry men of the 424th forward to their new base.[5] Their duffels were packed full and they brought on board as much of their camp possessions as they could. The move was being conducted on a very rushed basis. Apparently something very big was planned for the 307th. Even necessary but slightly bulky possessions (tent pieces, cots, lanterns, etc.) that would normally be entrusted to the Navy for transport by ship were being ferried to the new base by air.

There were few regrets to be leaving Wakde, with its cramped space, barren appearance, and painful memories of comrades lost. One last time they experienced the unique take off from Wakde airstrip, with its dip in the middle before lifting into the air at the far end. Walt's route was just a little north of due west for 294 miles. For about half the flight they skirted the Dutch New Guinea mainland to the south, then passed the south end of Biak Island before landing on Noemfoor Island. Walt had been to Noemfoor before, on June 30 during a 1745 mile, 11 hour and 40 minute mission from Los Negros to bomb the Japanese airfields in preparation for the U.S. landings two days later.

Noemfoor was bigger than Wakde, at 129 square miles. Located in Geelvink Bay, midway between Biak Island and the part of the Dutch New Guinea mainland known as the 'Voegelkopf' (meaning 'birds head' in Dutch, because of its shape on the map). The island was timber covered on low limestone/coral hills with a maximum elevation of 670 feet. The island, with three Japanese built airfields, had been invaded on July 2. By July 7, it was declared secure, although there would be pockets of Japanese holdouts in the mountainous center of the island for a few more months. MacArthur himself had ordered that the airfields be put into operation as quickly as possible. In mid-July, one of the airfields, Kornasoren Drome, was assigned to the 307th BG and MacArthur ordered construction to be accelerated to allow the Long Ranger bombers to conduct operations. By July 25, the strip extended 6,000 feet and the following day it was extended another 1,000 feet. By September 2, a second parallel 7,000-foot strip had been completed at Kornasoren, although trees still had to be cleared to allow it to be used by fully loaded bombers.[6]

Kornasoren is on the north end of the island, aligned E/W within a couple hundred feet of the beach. The runway was compacted coral, 7000 feet long, with 50-foot shoulders and 500 foot overruns at each end. The east end approach was clear from over the water and the west approach had been cleared through the trees to allow a 1:60 glide angle.[7]

As Walt and the others approached Kornasoren, they looked down on a jungle-covered island circled by an offshore reef as far as they could see. The airstrip paralleled the coast with the base and camp area inland from that. After landing Walt learned that he was to be assigned to a crew comprised of a few of his old friends from Muroc: Frank Mansir and Jack Riley, as well as Boris Hidalgo, who Walt had come to value as a friend since their three missions flown together earlier in August. Two new faces joined the crew in the enlisted men's tent: Sgt. Richard Cosgrove and Sgt. Raynor. Cosgrove, from Lakeland, Florida, had more combat experience than the others did, so he was a good addition to the crew, although he was somewhat shy and quiet.

Later that day they learned that the officers of the crew would include three old trusted friends: Don Aubrey as co-pilot, Ed Dunne as navigator, and Jimmie Clark as bombardier. Their pilot would be someone new to them: Lt. Norman Coorssen, a big, blond-haired, blue-eyed man from a well-to-do Massachusetts family.[8] His first mission had been as an observer pilot, from Los Negros to Yap, back on June 23. Walt had flown that day with Lt. Carl Appling, but their plane had to turn back due to mechanical problems. The 5th Bomb Group had led the attack on Yap that day, and Coorssen watched with horror as one of their planes up ahead was hit with flak. The plane started to drop off and the pilot radioed out that he would have to ditch the plane. Norman watched as two Jap Zeros followed the plane down machine gunning it along the way. The wing of the crippled plane then folded and snapped off. Soon the plane was in the water, with five more zeros machine-gunning the water. After it dropped its bombs, Coorssen's plane and others returned over to the crash area and searched for about an hour at very low level. Coorssen and the others saw just a single survivor thrashing in the water trying to reach a raft that had been dropped to him, but there was nothing else that could be done. The following day Coorssen would be in another plane on his first mission in the pilot seat. However, the experience from that first mission stuck with him.[9]

Walt and the others remembered that back on September 1 Coorssen's crew, but with Jack Arnett, not Coorssen, had been shot down and the entire crew lost over Palau. What did Walt and his friends know of the reason for the change? What, really, did they know of Norman Coorssen?[10]

Noemfoor Island was home to about 5,000 natives who were happy to see the American's with their rations and money. They were readily available to help the HQ, supply, maintenance, and other departments of the 307th get their building and other facilities set up as quickly as possible. The 307th camp area on Noemfoor was in an area mostly clear of trees, with tents along either side of about 10 rows of paths that ran back at right angle from the main dirt road that ran out to the airfield. The weather was rainy and unfortunately the camp area was low and poorly drained, so it was a sea of mud. The men, assisted by natives when possible, quickly constructed little plank bridges and coral walks. They also dug a large ditch along the main road to drain away as much of the water as possible from the camp area.[11] The 424th Squadron camp area was near the 307th mess tent, which had a wooden floor, tables, and benches.[12]

At the end of September, the 424th Bomb Squadron personnel had increased slightly to 102 officers and 150 enlisted men in the flying group and 15 officers and 372 enlisted men in the ground staff, for a total of 639.[13] Even though Noemfoor was farther from Japanese held areas than Wakde had been, it was still within range of enemy air bases and it was bombed at night a few times.[14] Over the next few days of frenzied activities, the men were also given briefings of what was planned for the 307th, the rest of the 13th AAF and the 5th Air Force as well.

Balikpapan

The invasion of the Philippines was first set for late December 1944. But in mid-September, while Walt was in Australia, MacArthur had decided to push the date up to mid-October. To support the invasion it was decided to try to knock out Japan's major source of oil and aviation fuel: the refineries and oil production facilities at Balikpapan on the east coast of Borneo. The Dutch had first discovered oil there in the 1890's and the oil field was put into production by Royal Dutch Shell. By 1942 when Japan invaded, Balikpapan was one of the largest producing oil fields in the world. Indeed, access to oil was one of the reasons that Japan attacked Britain and the Dutch in Southeast Asia. The Imperial Japanese Navy and Air forces depended on oil from Balikpapan for their war effort. If the oil field and refineries could be put out of commission, Japan's war effort in general, and their ability to resist the American invasion of the Philippines, would be greatly hampered.

Balikpapan had been bombed before, by B-24's from the 380th Bomb Group flying out of Darwin in Northern Australia. However, these were only small raids and because of the strong Japanese defenses, they were only conducted at night. With the need to put Balikpapan out of service, MacArthur's air commander, Gen. George Kenney, requested some squadrons of the new Boeing B-29 Superfortress bombers. However, the B-29's were just coming into operation and were all earmarked to begin the long-range bombing campaign against Japan itself, so none could be spared for Balikpapan.

The 13th Air Force, under General St. Clair Streett, was given the lead to conduct the bombing campaign against Balikpapan, with the 5th Air Force B-24 squadrons assisting. For effectiveness, it was agreed that the bombing would have to occur during the day. From Noemfoor to Balikpapan was 1,080 nautical miles or over 1,250 statute miles. Because of the distances involved, the planes would have to take off at night, to be able to reach the target and then return and land while it was still light. To inflict serious damage, each plane would have to carry a 2,500-pound bomb load. This is what the B-24 bomber factory rep's had been working on...how to carry the required load a distance beyond the normal design range of the bomber. To allow extra fuel to be carried in special bomb bay tanks, the ammunition load for each gun on the planes was reduced by 60%. It was felt that Japanese air defense around the refineries was weak enough that this step could be taken. There were some P-47 fighters that could escort the bombers to the target, if fitted with belly tanks, but it was felt that this protection was not necessary.[15]

To make the long distance round trip flights of up to 2,600 miles possible, the XIII Bomber Command Operations Office, under Lt. Col. Guy Hudson and working with the Willow Run B-24 factory technical representatives, had developed very specific 'Standard Operating Procedures for Long Ranges' which were issued to all the crews on Thursday, September 28.[16] These procedures were organized to instruct the pilot and crewmembers, as well as the supporting ground crews, on dozens of little steps necessary to conserve fuel so that the long missions could be flown to the maximum possible limits of a B-24.

Ground crews were instructed to warm up each plane and then top off the fuel tanks no more than one hour before take-off. Flight crews were instructed to start their engines only at the specific time in their mission plan. Moveable weight on the plane was carefully positioned: pilot, copilot and flight engineer in the pilot's compartment; bombardier, navigator, radioman, and assistant engineer on the flight deck; and nose, tail, and ball turret gunners at the ball turret. Parachutes, flak suits, ration boxes, cameras, and life rafts were positioned at designated locations forward. This precise placement was meant to counter the extra weight of fuel and bombs in the bomb bays.

Taxiing was to be done slowly, and even slower in turns because of the excess weight. Any crippled plane was to taxi or be pushed off the runway so not to delay other planes behind. Take-off however was to be at full power to get the plane into the air before the end of the 7,000-foot strip. The pilot was to raise the nose as quickly as possible to eliminate friction from the one front wheel and to generate lift as quickly as possible. Because it would be dark, Flight Engineers were instructed to stand behind the pilot to call out the air speed and altitude because the pilot '...should be concentrating on flying the airplane.' Full power was to be maintained until all the landing gear was up or the plane was at 1,000 feet. Once in the air the crew was shifted around, with three extra men shifted to the nose immediately after take-off.

Climbing to altitude was to be done at precisely 155mph and as soon as possible the ship was to be put on AFCE (Automatic Flying Control Equipment) because, despite what he might think '...no pilot can get as much out of the ship manually as can be done with a good automatic pilot.' But the pilot was instructed to ensure that the air speed did not drop below 150 mph. Cruising speed and altitude towards the target was to be 155 mph at 8,000 feet altitude, precisely. Only when approaching Balikpapan where the men to shift to their combat positions, then the aircraft were to climb slowly to bombing altitude and form up into their attack formation. All power adjustments were to be made very gradually to allow the planes to conserve fuel so that they could make it back to Noemfoor. One hour after leaving the target the tail gunner and one waist gunner were to return to the flight deck. Theoretically, according to the technicians from Willow Run, these procedures would leave each plane with about 600 gallons of fuel when it returned to Noemfoor. Theoretically! As further preparation, each pilot made three practice take-offs and landings at night.[17]

In just a few days, the 307th Bomb Group and the entire 13th Airforce Bomber Command had completed the move from Wakde to Noemfoor. Everyone was hard at work preparing for the strike against Balikpapan. Walt was in a bit of a daze. On Sunday he had left Sydney. By Monday, he was back on Wakde and the following day he was settling in with his friends on Noemfoor. By Wednesday all the men of the 307th were receiving briefings and ordered to start making preparations for Balikpapan. On Thursday the 28th, Walt, Jack Riley and Frank Mansir ran into their old pal Cliff again. He had just gotten back from a ferry run from Noemfoor to Wakde and back to pick up more crewmembers.[18] Cliff had it from his pilot that they would be on the first wave attack against Balikpapan, in just a day or two. Walt told him that the rumor was that the Coorssen crew was scheduled for the second wave. Walt and Cliff wished each other good luck.

The next day everyone knew that the first attack would be tomorrow. The crews learned that their mission would be at least 16 hours, 2,500 miles, farther and longer than they had ever flown before. They all knew this mission would be dangerous, and many went to church that evening. Everyone was aware of what was happening. Just after midnight on Saturday, September 30, trucks picked up the assigned crews and drove them in the dark to the flight line, where they found their assigned planes. At about 0100, one by one in quick succession the planes started taxiing then making the long, long run down the strip and then slowly into the air. 48 bombers from the 307th and 5th Bomb groups of the 13th AAF, and a similar number from the 5th Airforce climbed into the dark night sky and slowly headed west. Each plane with 10 or 11 nervous men...about 1,000 men total in about 100 planes each flying in isolation towards their rendezvous IP point. Amazingly all of the bombers made it into the air. At about 1030, most reached the rendezvous point, Cape Karang in the western Celebes, just across Makassar Strait from Balikpapan. The planes formed up into two tight squadron groups and climber to 18,000 feet. Approaching the target from the south through a lucky break in the clouds, they saw only a few Jap fighters in the air and sparse flak, though they felt some concussions as they were over the target. After bombs-away, the planes turned east and lost altitude to pick up speed. Later, Cliff told Walt that leaving Balikpapan; they could look back and see smoke from refinery fires climbing 20,000 feet high. Most of the planes made it back to Noemfoor and landed safely in the dark around 1800 or 1900. A few of the B-24's that were running low on fuel diverted to a new airstrip under construction on Morotai Island, to the northwest of Noemfoor.[19]

Everyone on Noemfoor was relieved that the raid had gone as well as it had. The refinery was not knocked out, but substantial damage had been done. The preparations and special procedures had proven the ability of the sturdy B-24's – and the men who flew them – to make a successful bombing run much farther than any had thought possible. The Japanese had been taken by surprise, but they would likely be better prepared the next time.

Everyone was relieved and proud. The 13th Air Force had proven its mettle. To reinforce this pride, the following day copies of the 307th Bomb Group Headquarters Bulletin – a typed and mimeographed single page newsletter was passed around from tent to tent. It included a cable that had been sent to General St. Clair Streett from Lt. General Sutherland, MacArthur's Chief of Staff, which stated:

> "I have just had a letter from Admiral Carney, Chief of Staff, Third Fleet, in which he said that the "Damned 13th Air Force has just about spoiled the war for our carriers, particularly at Yap; Davison's Command, Task Group 38.4 left Yap in disgust after the first day because our old Ex-SoPac 13th Air Force had left no decent targets."[20]

Walt and the rest of the 'Coorssen' crew were earmarked for the next mission to Balikpapan. It started to dawn on the crews in this second wave of attacks that now the Japanese defending

Balikpapan would be on alert and the defense would likely be much stiffer. The weather on Sunday and on Monday morning was stormy, but it was forecast to be good the following day, so Coorssen was alerted to tell his crew to be ready for a mission the following day.

A big briefing was held for all the crews the evening before the mission. Walt and the other men learned that the specific target at Balikpapan was the Pandansari refinery, which supplied 80% of Japan's aviation fuel. The overload of fuel and bombs, with reduced ammunition loads to compensate was described. The detailed instructions worked out during the previous weeks were described again, emphasizing the importance to be able to accomplish the mission.[21]

The procedures used by the first wave were to be repeated. More than 90 planes from the 13th and 5th Air Forces were involved, led by the 13th. Six planes from the 424th Bomb Squadron were assigned, led by Lt. Rodwick in A/C 951. Other crews assigned included Lt. Lewis in A/C 605, Lt. Wheeler in A/C 585, Lt. Rider in A/C 101, Lt. McGinnis in A/C 273, and Coorssen in A/C 547. Lt. Lewis and Lt. Rider's crews each included an eleventh man as photographer. Coorssen's crew included Walt, plus Don Aubrey, Ed Dunne, Jimmie Clark, Boris Hidalgo, Frank Mansir, Jack Riley, and the two other new men, Sgt. Cosgrove and Sgt. Raynor.

What did the six-man core of the crew – Walt, Frank, Jack, Don, Jimmie, and Ed – think about Coorssen? Successful and happy crews depended on trust in each other and confidence in their pilot. They had confidence in Charles McCrea, and had been disappointed five months ago when he developed an ear infection and could no longer fly. They had come to like and trust the Texan, Carl Appling, when they were starting out on Guadalcanal and later on Los Negros. The young West Virginian, Jack Arnett was a competent and courageous pilot for three missions flown from Wakde, even if he was not very friendly to the crew. However, Arnett went missing in action when he had commanded Coorssen's crew just a month earlier, over Palau. Now Walt and the other would learn what kind of man Coorssen was.

After the briefing earlier in the day, the men rested as much as possible on Monday and then hit the sack after dinner to get a few hours of sleep, if they could sleep at all. As they were returning to their tents, they learned that a change in wind direction led to a decision to move the mission up one hour. Well before midnight – at about 2300 - the men assigned for the mission were awakened and dressed quickly. As they assembled in quiet groups down by the road, they could see stars in the sky, meaning good flying weather. Soon trucks arrived which drove them first to the mess tent, where they grabbed a fast bite to eat and some coffee, and then out to the flight line, where they found A/C 547. The men stowed their gear and checked their equipment – mostly by feel, because the planes were darkened to protect against Japanese air raids. Each of the gunners, Walt included, noted the half supply of .50 cal ammunition for their guns. Would they come to regret this short supply over Balikpapan, considering that the element of surprise was likely lost? With the extra load of fuel in the bomb bay and the 10 – 250 pound bombs stowed away, the men took their take-off positions. Coorssen and Aubrey passed through the plane once to ensure that all the gear was stowed forward, and that Walt, Boris, and Frank knew their positions at the ball turret during take-off. The three men settled in and listened in the darkness as one by one they heard the engines of A/C 547, and the other aircraft all around them, sputter and start.

Soon the sounds of individual engines were drowned out by the collective drone of all four engines of A/C 547, with the background of the dozens of aircraft all around them. Right around midnight, the three men felt their plane lurch in to motion, move forward, then slowly turn down the taxiway. Now they could hear the roar of individual planes ahead of them, as their engines sped up as they accelerated down the long runway and out over the sea. Slowly, A/C 547 lumbered along, then finally turned to the left at the end of the taxiway, and then turned left again and stopped. Walt and the others knew that now Coorssen and Aubrey up on the flight deck were staring into the darkness of the runway waiting for the signal from the control tower to go. Just after midnight, they lurched into motion again as the four engines sped up at full power and they slowly accelerated...the nose raised up and within a minute they were in the air climbing slowly. Once they reached their cruising altitude of 7,000' the men could start moving to their normal stations.

Despite the fact that conserving fuel was important for the long flight, their planned route to the target was not a straight line. The flight plan was to follow the Equator, just a little north of Noemfoor, then fly due east. This route would take them north of most Japanese held land where ground observers might see or hear the dozens of bombers and then radio an alert that a bombing raid was in progress. The route along the Equator was designed to fly north of the 'Voegelkopf' peninsula of New Guinea, cross the narrowest southern arm of Halmahera Island, then head across the Molucca

Sea and up the Gulf of Tomini to cross the narrowest neck of the Celebes. This would lead them into the Straight of Makassar, where they would then turn to the southwest towards their IP, just north of Balikpapan. After the bombing run, each plane would turn back towards base by a direct route, as all surprise would then have been lost.[22]

In the darkness, Coorssen, Aubrey, Ed Dunne the navigator, the flight engineer, and the radio operator had to stay awake and alert, but the other men could try to sleep in the cramped and uncomfortable space. But then at 0350, in the darkness, came word through A/C 547 that they were turning back to Noemfoor! Of course each of the gunners had questions. Why? What was wrong? They felt the plane bank and turn. The word was passed back to Walt and the others to keep clear of the bomb bay. The doors opened and Jimmie Clark released the 10 250 pound bombs and they dropped away into the Sea of Halmahera, below. They had reached a point at 00°10'N-128°10'E when they turned back to Noemfoor. The word was passed back to the gunners – the autopilot had failed, and Coorssen had also announced that he was ill. He had made the decision to turn back. Coorssen, as commander had the responsibility for the crew and for decisions like this. Still, what did the rest of the crew think? Could the plane had made the flight without the autopilot? Why not? Could Aubrey had taken over from Coorssen? Why not? Despite their questions, in the early morning light, they touched back down at Noemfoor, at 0650.[23]

This was not Walt's first 'turn-back' – this had happened twice before with Carl Appling on missions to Truk. Walt and the others would get no credit for a combat mission. Back on Noemfoor, the men wondered, just how ill was Coorssen? What really had happened? They headed back to their tents – to wonder and wait to find out how the rest of the mission would turn out.

The mission did not turn out well. Although the bombing of the refinery was rated as 'excellent', as the details came back in to be sorted out later that day and into the next, it was apparent just how bad it was. The 307th Bomb Group, with about 24 planes committed to the attack, lost seven! In addition, 13 planes returned damaged. Two were lost to enemy fighters, two to A/A fire, and three to a combination of both. 63 men were missing, one man died of his wounds, and 27 returned injured.[24] Of the five 424th Bomb Squadron planes that made it to Balikpapan, only one was able to bomb the primary target, but with good results. The other four bombed secondary targets.

The Japanese threw in about 40 fighters, mostly Zekes, to attack the bombers. The fighter attacks started about four minutes before they reached the target and continued about 40 minutes after. The 424th shot down four, with another probable. While over the target, the 424th B-24 (A/C 101) piloted by Lt. Rider was rammed in the right wing by a Tojo. The Japanese plane exploded on impact, which caused the B-24 wing to break off. Lt. Rider's plane then turned onto its back, went into a dive, and crashed. Two bombs were seen falling out of the plane, but no parachutes. Lt. Rodwick's plane was hit in the flap, nose turret, #2 engine, and the tail.

Also from the 424th, Lt. Wheeler's plane (A/C 585) took a hit from a 20mm A/A shell that exploded on the command deck. Wheeler suffered shrapnel wounds in his left arm and leg, as well as lacerations on his forehead, which caused profuse bleeding. Lt. Francis Stepheny, his copilot, suffered a ¾ inch deep by 2-1/2 inch long gash in his face as well as arm wounds. He went into a semi-conscious state. Lt. John Wright, the navigator took a piece of shrapnel through his wrist which fractured two bones. Their plane also lost one engine. With Stepheny not conscious, Lt. White and S/Sgt Byrum, the Flight Engineer, helped the wounded Wheeler get to Morotai on three engines where they were forced to crash land.[25]

One man, Sgt. Thomas W. Pelle later recounted his experience on this mission:

> "I was hit over the target and we had I think about 420 holes in the plane. There was this Japanese plane, we were almost airplane number eight. This Japanese plane—later it occurred to me that he was trying to crash into us—but we were flying along, and he came down like this, trying to crash his right wing into our left wing, but he missed, and he came down, I looked at him right in the eyes, and I could see his goggles, his nose, his moustache, and his teeth—he was smiling, that's how close he was. He was probably 50 feet from me.
>
> "I don't remember much before or after, I got hit about the same time. I heard a loud boom and saw part of the back of the bomb bay open up, and I looked down and I saw part of a G.I. shoe with a piece of a leg stickin' out of it. "Who's is that?" I looked down and it was mine, and I didn't even feel it. Hit me in both legs, took a piece outta the calf of my left leg and took off my right one right below the knee. I didn't even feel it, so

procedure was to call the navigator and let him know you were hit. So I called up there three times, but I noticed on the third time this long gash inside of the plane where my intercom was, and I just had my wire dangling in front of me, holding on the machine gun I turned around to get the starboard gunner to call up front.

"But he was lying on the floor. So I looked back, we had a cameraman on this time, so I motioned the cameraman to come up and take over my machine gun, and I lay down and I could see blood spattered against the side of the plane so I tried to reach down and stop it but my left arm was asleep cause I was lying on it and I passed out. When I came to, I was lying up where the armored gunner had been, but the co-pilot had come back and got the armored gunner and carried him up through the bomb bay and up to the front to make room for them to stretch me out. I don't know what they used for tourniquets, their belts I guess, the tail gunner took care of the tourniquets and the engineer and bombardier and took care of the morphine and the two bags of blood plasma and they gave me everything they had. But they didn't think they could keep me alive until we got back to Noemfoor. And they probably couldn't have because I had ten more transfusions when I got on the ground.

I woke up twice on the way back and saw what they were doing, and I'd throw up and pass out again. I woke up again, the third time as they were passing me out of the waist window and I could see the doors of the ambulance and I passed out again and I woke up once in the operating tent. I was operated on in a M.A.S.H. of the 13th portable army surgical hospital, P.A.S.H. And I came too once and they had this shiny thing above that reflects light and I could hear sawing and I looked up and saw what they were doing and I quietly passed out again. That was my leg they were sawing off."[26]

During the next couple days as the details of this mission came in, morale was not good. It did not help the men of the 307th BG to learn that the 13th Air Force had sent some long-range P-47 fighter escorts with external fuel tanks from the new airstrip on Morotai to cover the bombing run, but because no one informed them that the raid was being moved up one hour because of the wind change, they arrived an hour too late.[27] In all, on October 3, 1944, the 307th Bomb Group lost seven B-24s, with 13 men killed and 41 listed as missing in action. 21% of the men who started out that day never returned. Of all the aircraft put into the air that day, 29% would never fly again.[28]

Because the command on Noemfoor was concerned about morale, MacArthur had his air chief General Kenney fly out to investigate and to give the men a pep talk. He talked with individual crews to get the men to listen to him and let them blow off steam. Kenney had information that the Japanese were flying in reinforcements to the air squadrons defending the oilfields and refineries, which convinced him of the importance of Balikpapan. Kenney told the men that the raids would go on, but that he would also provide better fighter support, time for training in new 'box formation' tactics, and more bombers.[29]

Then a day or two later, as Walt and the other men were hanging around their tent, they were visited by Don Aubrey, Jimmie Clark, Ed Dunne....and Charles McRae. They were now the McRae crew again! They learned that Captain Vanderpoel had been furious with Lt. Coorssen because he had turned back on the mission to Balikpapan. Vanderpoel had him grounded.[30] The autopilot malfunction excuse was deemed dubious. On the October 3 mission, Lt. Rodwick's plane lost its autopilot and he managed to fly to Balikpapan and back without it. Whatever their feelings in the matter, or their opinion about Coorssen, Walt, Frank, and Jack were very happy to see McRae again and they assured the other crewmembers that they were in good hands now!

Noemfoor was now a buzz of activity as preparations were being made for the next raid. So many planes had been shot up or sustained damage on the October 3 raid, that by October 6, there were only 12 flyable B-24's remaining in the entire 13th AF. These were used for training flights for the pilots on October 6 and 7, to practice the new 'box' bombing formation that Kenney had ordered. For these training flights, up to 30 P-47's were used to simulate enemy air attacks on the formation.[31]

The Combat Box tactic had been developed by the USAAF in England, beginning in 1942. The combat box formation was designed to bring in 12 planes in a tight, but staggered formation of four elements of three planes each. Each box consisted of a lead element, two elements behind but to the right and left of the lead, and then a following fourth element behind the first three. Each element flew at a slightly different elevation also. The combat box supported a tight bombing pattern, and also massed the defensive firepower of the gunners in a way that gave them clear lines of fire at any attacking fighters.[32]

While the pilots were practicing, the ground crews were frantically making repairs so that each squadron could put up six planes. New crew replacements were brought forward and there were more transfers. Sgt. Raynor, who had flown with Walt, Jack, Frank, and Boris the week before on the ill-fated flight with Lt. Coorssen, was transferred to another crew, to be replaced by Sgt. Mitchell. They learned that the date for the next Balikpapan raid was set for Tuesday, October 10. They would be flying in A/C 273, which Lt. McGinnis had flown on the raid the previous Tuesday. McGinnis' crew was transferred to A/C 022.

The plan for the October 10 raid would include 106 B-24 bombers supported by 35 P-47 and P-38 fighters.[33] The bombers would come from both the 13th and 5th Air Forces, with all of the fighter escorts from the 5th Air Force. The raid would be led by the 43rd Bomb Group of the 5th AAF. About 30 minutes later, at five-minute intervals, the 22nd and 90th Bomb Groups of the 5th AAF and then the 5th and 307th Bomb Groups of the 13th AAF would make their bomb runs.[34]

The 424th Bomb Squadron would contribute seven planes, led by Capt. Vanderpoel himself in A/C 605. The crews had their final briefing on Monday evening, and then hit the sack for sleep. However, what little sleep the men could get was interrupted late that evening. Anyone awake and listening could hear the sound of the unsynchronized engines of a single Jap bomber, even before the alarm sounded. Walt grabbed his helmet and stumbled into a nearby trench as searchlights lit up the night. Then a couple miles away, in the 13th Bomber Command area, the 'thump, thump, thump, thump' sound of the raider's bombs exploding could be heard, and then he was gone.[35]

With the all clear the men stumbled and grumbled back to their tents and then through the mosquito netting into their cots. Soon enough, the men were awakened again to prepare. The usual routine visit to the latrine, a wash up, breakfast in the dimly lit mess tent, then the drive out to the flight line shortly after 0200. Once again, the planes carried extra fuel and half loads of .50 cal. ammunition for the gunners. Each plane was loaded with five 500-pound bombs. The stars were out as Lt. McRae put A/C 273 into the air at about 0342.[36] The routing to IP was the same as used for the September 30 and October 3 missions: fly due west following the Equator north of the Voegelkopf, cross the narrow southern neck of Halmahera, west across the Molucca Sea and up the Gulf of Tamini, then cross the Makassar Strait to the IP just north of Balikpapan.

A/C 273 made its westward progress through the equatorial night as Walt and the other gunners tried to sleep as best they could. They flew through a weather front for an hour or so, with intermittent rain and mild turbulence. Dawn came as they were crossing Halmahera and heading out over the Moluccas Sea. They flew over the tops of the scattered cumulus clouds for most of the rest of the flight out, with only an occasional towering cumulus to the left or right. They were now deep within Japanese controlled airspace and each plane was on the alert for interceptor fighters.

An hour after dawn the word was passed back that Lt. McRae was worried about making the IP rendezvous on time without increasing power settings on the engines. But if he increased power settings beyond those in the detailed instructions, they would not have enough fuel to make it back to Noemfoor. McRae decided to jettison one bomb to try to keep his speed up without consuming too much fuel. The men set the fuse to safe, secured the other four bombs, and then stood back away from the bomb bay doors. McRae called each man's name over the intercom to ensure that they were all clear of the bomb bay. He then gave Jimmie Clark the order to drop one bomb. The twin doors swung open and one bomb dropped away. The doors closed and they continued on their way westward 500 pounds lighter. Would that get them to the IP on time…and back to Noemfoor OK?[37]

At about 1000, they crossed over the narrow neck of the Celebes and then McRae turned to the southwest and towards the IP. Now Walt descended into his ball turret. Normally he would have been in the ball the entire time during daylight, but to reduce drag, the special instructions specified that the gunners wait until they approached the IP to lower the ball turret. McRae also slowly increased altitude, from 8,000 feet to their bombing altitude of 19,500 feet. The men were all on oxygen now. McRae brought them to the IP on time to find five of the other six planes assembling into formation over Cape Karang, just northeast of Balikpapan Bay and the Pandansari Oil Refinery. One of the 424th B-24's was not to be seen. As they approached the target in the new box formation, Walt could see the other bombers of the 424th Squadron nearby around them, and then more bombers of the 307th BG further away. The 307th was the last group to go in, following the three 5th AAF Bomb Groups and the 5th Bomb group of the 13th AAF. The 43rd Bomb Group had started the attack about 45 minutes earlier, so the Japanese defenders were well alerted.

ENGINE CHANGE AT MUNDA
Ground crews far outnumbered flying personnel in the Thirteenth Air Force. They were critical to ensure the safety and effectiveness of the combat flying crews.

MOBILE REPAIR IN THE SOUTH PACIFIC:
Typical of the Thirteenth Air Force mobile repair shops that kept airplanes, jeeps, trucks, and other equipment functioning.
Painting by Sgt. Robert A. Laessig, courtesy United States Air Force Art Collection.

JUNGLE CAMP
The jungle was home to the Thirteenth Air Force. A cleared area made for a more compact and orderly camp, but trees could provide shade and a picturesque setting.
Painting by Sgt. Robert A. Laessig. Courtesy United States Air Force Art Collection.

TENTS IN THE SOUTH PACIFIC
Sack time was a common relief from the high heat and humidity of Thirteenth Air Force island bases.
Painting by Sgt. Robert A. Laessig, courtesy United States Air Force Art Collection.

KP IN NEW GUINEA
Ground crewmen performed dozens of mundane jobs to keep the Jungle Air Force bombers flying. Many served two years or more before rotation back to the U.S.
Painting by Sgt. Robert A. Laessig, courtesy United States Air Force Art Collection.

A MESS HALL IN THE SOUTH PACIFIC
Far from home-cooking, the men of the Jungle Air Force were fed as well as possible from canned meat, vegetables, and fruit, as well as fresh bread.
Painting by Sgt. Robert A. Laessig, courtesy United States Air Force Art Collection.

AIR BASE IN THE ADMIRALTIES
Mokerang Air Field on Los Negros, in the Admiralty Islands, brought the 307th Bomb Group together for the air assault on Truk and Yap Islands.
Painting by Sgt. Robert A. Laessig. Courtesy United States Air Force Art Collection.

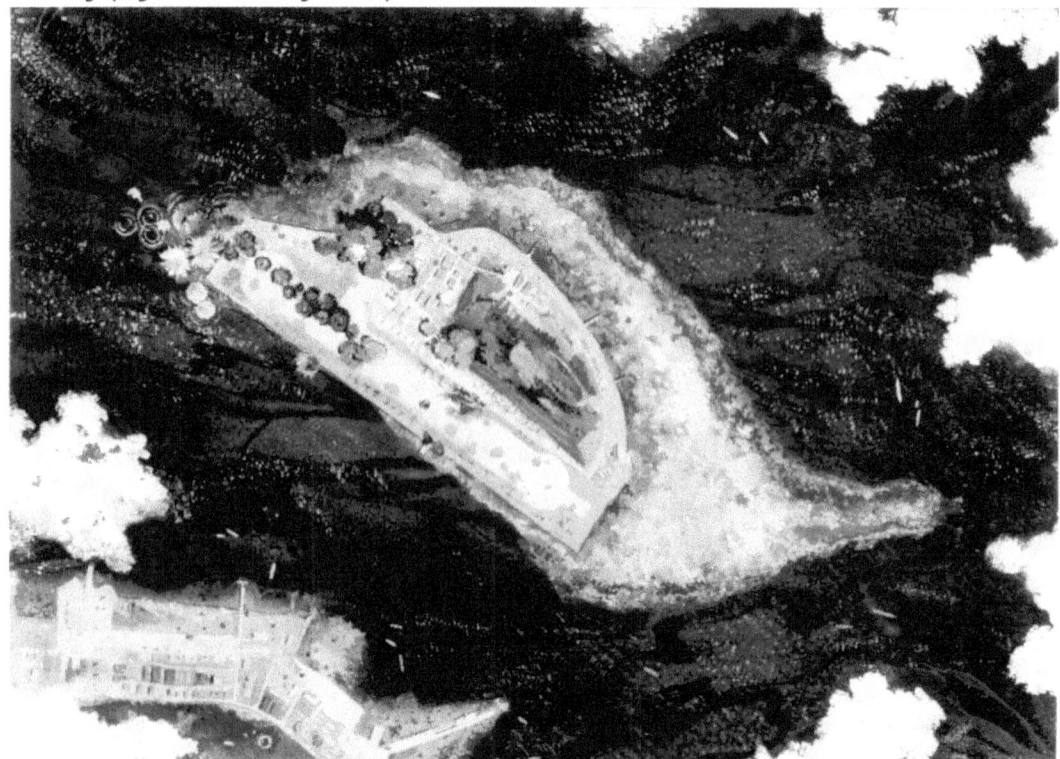

THE BOMBING OF TRUK
Truk was the Japanese 'Gibraltar' - a major heavily fortified naval base that needed to be neutralized to support the two-pronged American advance towards Japan.
Painting by Sgt. Robert A. Laessig. Courtesy United States Air Force Art Collection.

B-24s OVER YAP
A quintet of Liberator bombers withdraw to the south after bombing a Japanese airstrip on Yap Island.
Painting by Sgt. Robert A. Laessig. Courtesy United States Air Force Art Collection.

B-24s OVER BALIKPAPAN
Balikpapan was the longest and toughest, but arguably the most important target that the Thirteenth Air Force attacked, returning again and again.
Painting by Sgt. Robert A. Laessig, courtesy United States Air Force Art Collection.

FLIGHT NURSE
Each Island base had a field hospital, but men with serious injuries or illnesses were flown back to hospitals on Guadalcanal, New Guinea, or in Australia.
Painting by Sgt. Robert A. Laessig. Courtesy United States Air Force Art Collection.

SKYTRAIN OVER BOUGAINVILLE
The C-47 Skytrain was the workhorse transport for the Thirteenth Air Force. Walt flew in C-47s to and from Australia and from Morotai to Biak.
Painting by Sgt. Robert A. Laessig. Courtesy United States Air Force Art Collection.

A RED CROSS LEAVE CENTER IN AUSTRALIA
Facilities like this in Brisbane and Sydney provided relaxation and recreation for crewmen lucky enough to get a short furlough away from the fighting.
Painting by Sgt. Robert A. Laessig, courtesy United States Air Force Art Collection.

BEACH SCENE – SIDNEY, AUSTRALIA
Bondi Beach, about 10 miles outside downtown 'Sidney' provided an altogether different waterside experience for the island-hoping Jungle Air Force crewmen.
Painting by Sgt. Robert A. Laessig, courtesy United States Air Force Art Collection.

A BOMBED OIL REFINERY IN THE DUTCH EAST INDIES
After Balikpapan, the oil refineries at Tarakan were another key target. Tarakan oil was so pure, it could be burned in ship boilers without refining.
Painting by Sgt. Robert A. Laessig, courtesy United States Air Force Art Collection.

B-24s RETURNING FROM A MISSION
For tough missions, men would wait anxiously as dusk fell for their comrades flying on other crews to return safely.
Painting by Sgt. Robert A. Laessig, Courtesy United States Air Force Art Collection.

RESULTS OF A JAPANESE AIR RAID
The Japanese had brave and determined bomber crews of their own who pressed their attacks on heavily defended Thirteenth Air Force bases like Morotai.
Painting by Sgt. Robert A. Laessig, courtesy United States Air Force Art Collection.

OPEN AIR WASHROOM IN THE SOUTH PACIFIC
This is typical of the facilities near the 424th Bomb Squadron's mess tents on Los Negros and Morotai.
Painting by Sgt. Robert A. Laessig. Courtesy United States Air Force Art Collection.

A TENT CHAPEL IN NEW GUINEA
Each Bomber and Fighter Group in the Thirteenth Air Force had its own chaplains and chapels where crewmen could practice their faith and pray for a swift end to the war.
Painting by Sgt. Robert A. Laessig, courtesy United States Air Force Art Collection.

BASKETBALL ON BIAK
Organized sports provided a critical distraction and release of energy for Thirteenth Air Force crewmen.
Painting by Sgt. Robert A. Laessig, courtesy United States Air Force Art Collection.

Looking ahead in his ball turret from 19,700 feet, Walt could see that there was minimal cloud cover – maybe 30% - but there was a lot of black smoke billowing high into the air from fires started by the preceding bombers. Capt. Vanderpoel announced that because the smoke would obscure visibility over the refinery, it was decided that the 307th would bomb the secondary target – the Balikpapan Oil Cracking Plant. This required just a slight adjustment in direction. Looking ahead over the target the men could see the 5th Bomb Group on their bombing run. There did not seem to be any Japanese fighter attacks, but there was a considerable amount of flak. Off in the distance a Japanese observation plane could be seen flying at their elevation. It was obviously reporting their altitude down to the anti-aircraft gun crews below, so that they could set the fuses for the correct elevation. Just after 1100, McRae announced that Jimmie Clark had the control as they lined up for their bomb run behind Vanderpoel's lead. Now the flak started up ahead and the planes started rocking with each blast. Walt, with just plexiglass ahead, could see that it was pretty accurate, moderate to heavy in intensity, but bursting maybe 200 feet below the 424th altitude.

Jimmie opened the bomb bay doors and the four remaining 500-pound bombs were fused. At 1107 they were over the target and the bombs dropped away. While being ever alert for possible fighter attacks, and worried by the heavy A/A fire, Walt could see, and all the men could feel the plane lurch as the ton of bombs fell away. The string of four bombs wobbled when first released, and then the tail fins tipped them down to fall straight away. A half minute or so after bombs away, Walt could see their bomb explosions 'walking' towards the cracking plant, with its complex of pipes and process vessels and storage tanks. One string walked through what looked like a barracks area. Another string went right through the storage tank area. One was dropped close to the shoreline, and another missed land altogether, going into Balikpapan Bay. Two minutes after the bombing run began they were turning to the left, away from the flak, to set their course back to Noemfoor.

Then at about 1115 the intercoms crackled with the news 'here come the Japs'. Now Walt and all the gunners were on high alert. No need for radio silence now. There were no masses of enemy fighters, just small groups of Tojo, Jack, Tony and Zeke types that were milling around. Then small groups of them would make a passing attack on the 424th. The typical attack was from two o'clock or ten o'clock, with some from above or some from below. For the attacks from ahead and below, Walt and the nose gunner would have the best shots and their guns would blaze away in short controlled bursts, as they had been trained and as their limited ammunition load warranted. Then one of the Jap fighters came in from behind, climbed high above and lined up over and slightly ahead of the 424th box formation. It then dropped a phosphorus bomb that exploded and rained sticky white-hot burning gel down towards the planes. Fortunately it missed the 424th formation, but the distraction was the signal for another attack at the bombers in their defensive box. Then one pass by the Japs was made from seven o'clock high and Walt could see the enemy as it dove down below. Then more phosphorus bombs dropped from above – a total of four, with two that dropped their sticky burning streamers right through the formation. More fighter attacks were made, a total of seven in all, but fortunately none of them came in closer than about 600 yards. Then finally, after 28 minutes, at about 1143, they broke off and were seen no more.[38]

Now Capt. Vanderpoel came on the air to give all the crews a 'well done' and good luck getting back to base. Now it was each pilot's job to get back as best he could, generally by a direct route. The crews were also reminded that they should be on the watch for any enemy naval activity or transport shipping. Walt and the other ball turret gunners noted maybe 50 miles to the south, a small warship escorting maybe 10 small cargo ships headed towards Balikpapan, as well as various small cargo ships in Balikpapan Bay.

Within a half hour or so, the men started to relax as the danger of massed interceptor attack declined. They had dropped back down to their cruising altitude of about 8,000 feet, so oxygen was no longer needed. Soon Walt got the order from McRae to retract the ball turret to help minimize wind resistance on the way back. Walt stretched his legs and relieved himself, then grabbed a spam sandwich and can of peaches for his lunch.

Now, because of the takeoff later than the other squadrons, it was likely that they would be making a night landing back on Noemfoor. McRae had been worried about their fuel consumption all along, which had led to his decision to jettison the one bomb before the attack. Now radio reports were being picked up that indicated that a weather front over Noemfoor was quite strong and stretched from there for many miles to the west. McRae was faced with three challenges. Under the best of circumstances, did they have enough fuel to return to Noemfoor? Would the weather front further cause them to burn fuel? In addition, would the thick weather over Noemfoor, the nighttime landing, and the low fuel jeopardize a safe return?

At some point a couple hours after leaving Balikpapan, McRae told Ed Dunne to plot a course for the new emergency landing strip at Morotai. McRae, Aubrey, and Dunne conferred briefly over the navigator's calculations. Some advance echelons of the 307th were already on Morotai to support emergency detours. While the airfields on Morotai were not yet ready for normal bomber operations, 13th and 5th Air Force fighter squadrons were already using the island. A detour to Morotai would eliminate any concerns over low fuel, avoid the heavy weather front further to the east, and allow them to make a daylight landing. McRae decided quickly to change course for Morotai, flying northeast up the Moluccas Sea with Halmahera off to the right. In late afternoon McRae brought A/C 273 on a course over the northernmost tip of Halmahera and over the strait of water leading towards Morotai and its long southern peninsula. Wama airstrip was the original Japanese airbase on the island. Off not too far to the north, Pitoe Airstrip was under construction to be able to handle future bomber operations.[39]

After making radio contact with Morotai, McRae brought their bomber in for an easy landing. The control tower directed him to the appropriate spot where the plane could be serviced and refueled. Soon the crewmen were met by ground personnel who directed them to the base area where they could check in, find a transient tent to spend the night, and get some food. The men learned that they were not to stray too far away, as the Japanese were still on the island in large numbers. The Australian Army had a perimeter around the airbase area, but they had no intention of pushing the Japanese off the island. That would require too many casualties. All the allies wanted to do was secure the base for major bomber operations by the 13th and 5th Air Forces, as well as by the RAAF. The men were advised to carry their .45 cal. pistols with them, in case any Japs infiltrated the defensive perimeter.

They were also warned that many nights the Japanese would send a plane or two to drop a few bombs or strafe the airstrip. On the previous three nights a single Jap plane dropped bombs. At 0300 on October 7, one plane had dropped six bombs that killed one man and wounded several others. There were no injuries or damage from the raids on October 8 or 9, but despite heavy A/A fire and P-61 night fighters, the raiders all escaped unharmed.[40]

Fortunately they spent an uneventful night. After breakfast the following morning, McRae reassembled the crew at A/C 273, which had been checked over, serviced and refueled. Within an hour, McRae had them up in the air and on their way back to Noemfoor, now off to the southeast. At exactly 1300 they landed back on Noemfoor with no damage.[41] Walt had now completed his 16th combat mission. The men were relieved to learn that the mission the previous day had been very successful and that only one B-24 was lost, and that only when it crash-landed on Noemfoor. Only two men in the entire group had been injured, from flak over the target. It would be the most successful mission of the 307th Bomb Group for the entire month.[42]

A few days later on October 14, half of the 307th crews and the rest of the 13th Air Force and 5th Air Force returned to Balikpapan, but the McRae crew sat this mission out. This turned out to be a long but easy mission, with no planes lost due to enemy action. In addition to the P-38 fighters sent to protect the bombers over the target, the U.S. Navy stationed a submarine in Makassar Strait to pick up any downed crews.

The 13th Air Force bomber crews were now being briefed to prepare for single-aircraft shipping prowl and bombing missions. To help disrupt Japanese shipping of fuel and other supplies to the Philippines and to search for Japanese navy ships, single B-24s would fly missions to designated locations where they would fly search patterns to look for shipping.

The 307th was also starting to get an influx of new crews. Walt and the other men of the McRae crew met a couple dozen of these new men on Saturday afternoon, October 14. They had assembled along the road from the flight line when a number of trucks pulled up and new crewmen assigned to the 424th Squadron climbed down and lined up with their duffle bags. Major Vanderpoel stepped forward and addressed the newly arrived men:

"I wish to welcome you to the 307th Bomb Group, 424th Squadron. I am your commanding officer, John Vanderpoel. Gentlemen, for the first time in your Air Force careers there will be no training. If you do not have it, it's too late. There will be no practice formation, no transition. You will soon fly a combat mission.

"But every flight, every day, we improve our techniques, refine our tactics and work towards perfection of our task as a unit. And we do it under enemy fire.

"The good side of our task is that our missions are mostly over water. Actual contact with the enemy will be of shorter duration than other theaters, but it will be much more intense, it will be accurate and you will find the enemy very skilled. We are a tactical unit, which means that we stay close to the front line at all times. Already plans are underway for the 307th to move forward again to a new base.

"Every possible effort will be made by the enemy to keep your mission from being a success. We fight the enemy, we fight the weather, we fight the bugs. Tomorrow you will have a briefing at 08:00. Any questions?"

One of the new pilots asked: "How many crews are there in the 424th Squadron?"

Vanderpoel answered: "I have seven crews, but with you I guess I now have ten crews. Now the men in our Squadron will show you where you can find a cot in a tent, then we will see about getting you something to eat."[43]

On October 16, a number of the new crews were sent to bomb Namiea Airfield on Boeroe Island. This was considered an 'easy target' similar to the combat training bombing runs to Rabaul that Walt made when he first arrived on Guadalcanal, back in May.[44]

The following day McRae informed the crew that they were scheduled to fly another mission the following day. At the briefing that afternoon Walt and the other men learned that they were headed back to Balikpapan. This operation would be a 13th Air Force show, with the 5th Air Force starting to focus its missions on the Philippines themselves. For this mission the McRae crew would be the same as their last mission on October 10, except that Sgt. Mitchell was replaced by another new man, Sgt. Sellers.

For this mission on Wednesday, October 18, the 307th Bomb Group would be led by Major Vanderpoel and the 424th Squadron. The 424th put seven aircraft in the air, led by Major Vanderpoel in A/C 381, then Lt. McMillan in A/C 424, Lt. Youngmark in A/C 605, Lt. Hunter in A/C 802, Lt. Woodward with A/C 236, Lt. McGinnis in A/C 540, and Lt. McRae in A/C 958. Three of the B-24's included a photographer, so a total of 73 men were on the mission. Each plane was loaded with six 500-pound G.P bombs and half loads of .50 cal ammunition for the defensive guns. Operational procedures were unchanged from the previous missions. The 424th Squadron planes took off from Noemfoor at about 0230 on the 18th. The route west was the same as previous missions, although the designated IP was a little further to the north of Balikpapan than before. The route out was flown at elevations between 5,000 and 9,000 feet. The weather at first was very good, but the farther west they flew, the clouds increased in both coverage and in thickness.

At about 1015, the squadrons started to assemble at Cape Karang on the Borneo coast north of the target. Climbing to elevation in box formation the 307th Bomb Group headed towards the target. Walt and the other gunners were on oxygen and at their stations. At about 20,000 feet elevation, it was cold in the drafty planes. Then up ahead they could see that they were flying straight into a thick heavy cold weather front right over Balikpapan. Soon the group was in the middle of the front with zero visibility, moderate turbulence, and…snow! After a minute or so, Vanderpoel broke radio silence to give the order for the entire group to spread out and to bomb the target on 'ETA.' With zero visibility, the planes had to spread out to reduce the risk of mid-air collisions. It was now the job of each navigator to do his best job to calculate their plane's estimated time of arrival over the target, then drop their bombs blind at that point. Now the planes started reporting icing on the wings and flaps, so many started to slowly drop to lower elevation.

At about 1035, Sgt. Augsberger, tail gunner in Lt. Hunter's plane looked out and all of a sudden saw A/C 540, piloted by Lt. McGinnis, in a steep bank and peeling away from the formation. Augsberger thought the plane seemed to be under control after it disappeared into the thick clouds of the heavy weather front. Now it was Ed Dunne's job as navigator to put A/C 958 over the target. Some of the planes circled around to ensure that the group was well spread out to minimize the risk of dropping bombs on their own planes in the zero visibility and heavy turbulence. At least the thick weather meant there was no flak and no Japanese interceptors to deal with. Finally, at around 1100 Dunne gave the word. McRae flew a steady course; Jimmie Clark opened the bomb doors and on Dunne's word, toggled the six 500 pound bombs down to the target. Walt saw them disappear in an instant, as they raced down, hopefully near the refinery.

Now each plane was free to head back on its own. However, the crews had been instructed to use an intermediate base to refuel, unless absolutely certain that they could make it back to Noemfoor safely. Rather than Morotai, tiny Middleburg Island near Sansapor on the Voegelkopf had been designated. The 13th Air Force Fighter Command used Middleburg as well as Mar Airfield on the mainland near Sansapor. The 42nd Bomb Group (M) flying B-25 bombers used Mar on a regular basis. McRae brought A/C 958 down safely on the short, 5400-foot long runway on the tiny 240-acre island. The 13th Air Force ground crews there were efficient though. Walt and the other crewmen had just an hour or so to stretch their legs, get a bite to eat and talk to crews of other nearby B-24s headed back from Balikpapan. Then they were back into the plane for the short hop back to Noemfoor. It was after dark when McRae made a safe landing on Neomfoor. Walt had finished his 17th combat mission, credited with flying 2,485 miles for 14 hours and 55 minutes.[45]

Even though they were tired, the crew headed to the intelligence tent for debriefing. There was not much to report, considering the zero visibility over the target. But there was a question that the crew was asked...had they seen A/C 540 at any point over Balikpapan or afterwards? Lt. McGinnis' plane did not return to Middleburg Island, Mar, nor Morotai. Eventually the men would learn that Sgt. Augsberger's sighting of A/C 540 banking away steeply from the formation would be the last time anyone saw the plane. No radio report, no nothing. It had just disappeared from the face of the earth with ten of their friends. The 424th Squadron had lost another crew missing in action.[46]

Noemfoor – October & November 1944

Balikpapan had been severely damaged by the 13th and 5th Air Forces during the previous three weeks and more missions were planned. Although the oil fields and refineries had not been completely knocked out, it was decided to shift the focus to support the invasion of the Philippines.

Even though Japan had been pushed back steadily across the Southwest Pacific, from Guadalcanal in mid-1942, to the edge of the Philippines now in late 1944, she still possessed powerful forces. Hundreds of thousands of fanatical troops were supported by courageous Japanese pilots in excellent fighters and light bombers. Most worryingly, lurking somewhere in an arc from Japan, Formosa, the Philippines, and south to the East Indies and Singapore, the enemy still possessed naval forces that included at least 10 battleships, 20 cruisers, dozens of destroyers and submarines, and at least four powerful aircraft carriers. But where were they?[47]

The U.S. Navy was much more powerful, but it was spread out over millions of square miles of ocean. If Japan could concentrate its naval forces at a key point, it might still overwhelm the Americans and cause the invasion to falter, or even to fail. Now the 307th Bomb Group would be tasked with looking for elements of the Japanese Imperial Navy.

As the 13th Air Force bomber crews rested, the ground crews repaired their damaged and worn out planes. On October 19, a first 'shipping prowl' was conducted by a single B-24. It searched the Ceram-Ambon area, but saw no shipping. On October 20, the 424th Squadron was tasked to search the Makassar Strait, finding and attacking an armed transport vessel.[48] Later that day, news of the invasion of the Philippines was received on Noemfoor. MacArthur had returned, as he had promised in 1942, personally coming ashore on Leyte Island. Later that night at the group outdoor theater there was a band that played some 'swell music' followed by a movie.[49]

Over the next two days, more 307th B-24's were sent on shipping prowls. On the 21st Lt. Woodward from the 424th took his crew back to the Ceram-Ambon area and saw shipping, but they could not get their bomb bay door open for an attack. The following day two bombers were assigned, one searching Makassar Strait and another back to the Ambon- Ceram-Southern Celebes area. After a day off, another 307th B-24 was sent to search the Makassar Strait- Eastern Borneo area.[50]

Early on the 23rd McRae alerted his crew that they were assigned to a shipping prowl the following day. A few days earlier, Sgt. Sellers moved over to another crew, to be replaced by Gene Schreiner. Walt, Boris, Frank, and Jack had flown with Schreiner before. Back in late August, they had flown three missions with Schreiner on the Arnett crew...including two missions to Yap and the August 29 mission to awful Palau.

At their mission briefing that Monday they learned why Schreiner was joining the crew. They were assigned to A/C 951, a radar equipped B-24, and Schreiner was a trained radar operator. He had trained on operating the radar sets at a secret base in Florida, where all the men had been sworn to secrecy. The radar was in a small dome that protruded below the right side cockpit window. Within the dome was the radar itself that rotated once per minute and sent a signal to a scope that the radar operator monitored. The scope could pick out the shape of islands and coastlines, and most importantly, it could pick out ships in the distance, even in the dark or through clouds. When a radar equipped B-24 came under attack, the radar operator would man one of the waist guns.

During the briefing the men learned that earlier that day, a U.S. submarine operating in the South China Sea, northwest of Borneo, had radioed a sighting of a large force of Japanese ships headed towards the Philippines. This was the First Strike Force/Center under Vice Admiral T. Kurita, with five battleships, 12 cruisers, and 15 destroyers. It was known that two other Japanese Strike Forces were approaching from the north. However, there was one other Japanese Navy Strike Force that was unaccounted for. This was Force C/Southern, under Vice Admiral S. Nishimura, with two battleships, one cruiser, and four destroyers.[51]

Nishimura and his powerful force was thought to be somewhere between the Celebes and Borneo. This force had to be found to keep it from moving north to threaten the landings on Leyte. The 307th was tasked to put four B-24s in the air on the 24th to search for Nishimura and Force C/Southern. Two planes from the 371st Squadron and one each from the 372nd and 424th Squadrons would conduct the search. Each plane would fly a different search pattern, scouring the Ceram Sea, Banda Sea, Flores Sea, Straits of Makassar, Celebes Sea, and the Moluccas Sea. If Nishimura was sighted, the rest of the 307th would spring in to action to try to attack his fleet. The 424th contribution to the reconnaissance mission would be the McRae crew and A/C 951.[52]

There would be no movies for the men of the McRae crew on Monday night. They needed to try to get as much sleep as possible. Sometime shortly after midnight, a crewman from operations came to their tent to waken Walt, Frank, Jack, Boris, Gene, and Cosgrove. In the officers' tent, McRae, Aubrey, Dunne, and Clark were also getting their wake-up. Walt and the others hit the latrines, and then they headed with their gear to the mess tent. They were served by a bleary-eyed cook and his helper: reconstituted scrambled eggs, toast, and coffee. The tent was lit by a dim red bulb, so there would be no light visible from the air that might attract the occasional Jap bomber. As the men ate their breakfast they grumbled about the bread...made from very course flour that was often infested with bugs. They would hold the toast up to the dim red light, to search for any of the little critters, which they would tap out onto the table if they found them.

Breakfast finished, the men grabbed their gear, as well as a greasy cardboard box with more loaves of bread, cans of spam, and cans of peaches. This was the typical meal for a bomber crew with the 307th. They hopped onto a waiting truck, which drove them the couple miles through the quiet camp out to the airstrip, where A/C 951 was sitting, with its little put-put humming inside and a few ground crew standing around, greeting the men. These were the mechanics, armorers, and fueling crew. Walt and the others quickly moved to their positions to check out what they were responsible for before the ground crew left. In addition to extra bomb bay fuel tanks, they were loaded with five – 500 pound GP bombs. This was not a bombing mission, but if any target of opportunity were sighted, they could be ready.

The four 307th BG B-24's assigned to the reconnaissance were flying independent search patterns, so as each plane was ready it was cleared for takeoff. As the men finally climbed aboard, they could see stars in the sky and very few widely scattered clouds. Soon McRae had A/C 951 rolling down the runway and in to the air.

In the dark, McRae turned the plane to the south, headed for a point on the narrow neck of the Voegelkopf at 2°30'S, 134°14'E. Everything about this mission had been planned to maximize their time in the air and to search possible places where the IJN might be found. Their designated slow speed, power settings, and moderate elevation were intended to maximize their range and time in the air and the effectiveness of their search by radar and visual sighting. McRae and Aubrey would spell each other at the controls, Dunne of course was busy keeping them on course, Cosgrove and Hidalgo had responsibility as radio operator and flight engineer, and Schreiner had to monitor the radarscope. In the dark though, Walt, Frank, and Jack had little to do except doze, if they could with the drone of the four engines.

Once Dunne reported they were at the neck of the Voegelkopf, McRae turned the ship to the west. The tropical sun came up as they were over the Ceram Sea, headed almost due west. Frank and Jack lower Walt into his ball turret and they moved forward and aft to man the nose and tail turrets. Each of them would search the sea below and report any sighting to McRae. Slowly they passed Misool Island on the right, Ceram on the left, Obi Island on the right, then Boeroe Island on the left.

As they left sight of Boeroe behind, the weather changed and they flew into a mild front with light, intermittent rain and moderate turbulence. The rain and a haze layer on the water surface reduced visibility to about three miles. After Dunne determined that they had reached the next designated point at 3°00'S, 123°00'E, about 75 miles east of the Celebes, McRae turned towards the southwest. They were now headed towards a point located at 3°45'S, 122°30'E, off Kendari, a port on the southeast Celebes coast where the IJN might be found. However, they saw nothing. Their next waypoint was at 5°30'S, 122°°00'E, just south of Kabena Island.

McRae decided to do a sweep of four medium sized islands south of the southeast neck of the Celebes: Wowoni, Boetoeng, Moena, and Kabena Islands. Then approaching Dwaal Bay on the east coast of Boetoeng Island, Walt saw small specks ahead on the water, which he reported to McRae. Soon others on the crew picked up the sightings too. As they flew closer, it became apparent that they were small two-masted inter-island schooners. They were estimated to be in the range of 50-60 feet long, with about a dozen scattered around the bay and another three headed north out of the bay under sail. Although they were certainly carrying supplies to aid the Japanese, coastal schooners were not what they were looking for, and so McRae flew on. Farther south, in and around Pasarwadu Bay near the southern end of Boetoeng, another 20 or so of these small schooners were seen. McRae turned west around the southern end of Boetoeng. As they neared the southern end of Moena Island, another half-dozen schooners were sighted heading east under sail. At the southern end of Moena, at 5°05'S, 123°00'E, a possible emergency fighter airstrip was noted in the jungle, but no planes were sighted.

Their orders were to radio any sightings back to Noemfoor. But try as he might, Cosgrove could not get a response on the primary radio frequency. He tried the secondary frequency, but still no response. He kept trying but was never able to make radio contact throughout the entire mission.

Next their orders were to fly up the eastern side of the Gulf of Bone, between the two southern peninsulas of the Celebes. Finding nothing, McRae then turned back south, along the western side of the gulf, past the southern tip of the Celebes, to search round the southern end of Salajar Island. They had been flying more than 10 hours now and they were in the middle of the Flores Sea, when Schreiner came on the intercom. He had sighted a small blip on his radarscope, 20 miles off in the distance that was clearly not a wooden schooner. McRae changed course to close with the location that Gene indicated: 6°10'S, 119°50'E. Was it a Japanese naval vessel?

However, as they approached, they discovered that it was an American seaplane! It was a Catalina PBY-5 amphibious patrol and rescue aircraft on the water. It was seen taxiing on the water, on one engine on a 180° heading. They fired a recognition flare, and then McRae dropped down to 200 feet. They tried to raise the PBY via radio, but there was no response. For 20 minutes, McRae circled. The aircraft was seen to be towing two large life rafts, with what seemed to be more than just crewmen aboard. The Cat kept sending a Morse code signal of 'a – a – a' but there was no other attempt to communicate with the circling B-24. No one knew what it meant. They had to move on, so McRae ordered that they drop a recue kit, with a note stating that rescue facilities would be notified at the first opportunity. The kit was seen to be recovered from the water by the Catalina.

McRae then headed north and slowly gained altitude, to continue their patrol, which now led up the Strait of Makassar and along the western coast of the Celebes. Up the entire coastline of the Celebes, clouds were built up along the landmass. Visibility ranged from one to five miles, making any sighting difficult. On they flew, 12, 13, 14 hours. No shipping at all was seen in the Strait.

Finally at Cape Simatang, on the northwest coast of the Celebes, McRae turned back east. Their orders now were to bomb the Japanese base wharf area at Manado, at the far northeast corner of the Celebes. However, as they approached, they entered full overcast, with towering cumulus with tops over 12,000 feet and visibility reduced to two miles. Fuel was a concern now. McRae flew around and ordered Dunne to calculate an ETA over the wharf area at Manado. Jimmie Clark opened the bomb bay doors. Clark found a landmark down below, calculated a course and in they went. At 1617 at 7,300 feet elevation, at Dunne's mark, Clark toggled the five 500 pound bombs away down through the clouds and towards the target. About two miles beyond, Walt in his ball turret could finally see below that they had missed the wharf but had hit a barracks area about a half-mile north. Two small fires could be seen rising two hundred feet into the air.

That was it, now time to head back as daylight was fading away. But then through the clouds, at the very northeastern tip of the Celebes, Walt and the others noticed a small Japanese airstrip where they counted five Zero fighter and two Betty bombers. If the Zero's took off, they would easily catch the lone B-24. McRae decided that they would drop down through the clouds and surprise them on the ground. Back at Muroc, earlier in the year, McRae had them practice low-level strafing against old cars parked out on the dessert floor, as well as against the 'Muroc Maru.' Walt in the ball turret would have the best shots, but the two nose guns and the waist gunner would have good shots too.

They came in vary low and slow over the tops of the trees and Walt and the others opened up with their .50 cal. Browning machine guns at the planes parked on the ground. They took the Japanese completely by surprise. All the planes were either destroyed or damaged on the ground, plus the barracks area was shot up.

However, as they started to pull away at the far end of the airstrip, a single 20mm Jap A/A gun started firing at them and hit the #3 engine. As they gained altitude, McRae was able to feather the prop and shut down #3 with no further damage. They were now very short on fuel, with just three engines, and unsure of their fuel situation. There had been little likelihood of making it back to Noemfoor before, but now the only option was to try for Morotai. They had now been in the air more than 17 hours...farther and longer than any of them had flown before. Boris, the flight engineer, calculated the fuel and it did not look good. McRae ordered them to lighten the plane any way that they could. With a sigh, the gunners heaved their beloved .50 cal Brownings out the windows, one by one, and then the remaining ammunition went, as well as flak jackets and tools. Finally, Cosgrove was able to raise Morotai and alert the base of their situation. They were cleared to land.

It was dark when McRae finally approached Wama airstrip on Morotai. As he descended on A/C 951's three good engines, they touched down safely. Then just as they turned to taxi off the runway, all three remaining engines died. They were completely out of fuel! A/C 951 had been in the air for 18 hours and 40 minutes. A duration in the air that was never matched by any other B-24. They had not seen any evidence of the IJN, but they provided valuable information about where the sea was clear of the enemy.[53]

As tired as they were, the crew had time to visit the latrines, clean up, get a little chow at the mess tent, then they were off for a vital debriefing. They would repeat the debriefing again the following day on Noemfoor, but the Intelligence officers had the men pour over charts of their route, as they made notes of what had been sighted at each point. Finally, late at night they stumbled away in the darkness to a transient tent area where they hit the sack for some well-deserved sleep. Walt had now completed his 18th combat mission, an amazing circumnavigation of the Celebes, deep inside enemy territory.

As Walt and the rest of the crew were dozing off for some well-deserved sleep, at about 2100, all hell broke loose. A/A fire started as searchlights lit up the sky. Within a minute there were two big explosions near the docks. Then the sound of a diving plane and machine gun fire could be heard as a single Jap raider made a strafing run, then departed unharmed. The attacker strafed and dropped two phosphorus bombs, which injured about 20 men on the ground, but no one was killed.[54]

The following day, with A/C 951 being judged in no condition to fly, the crew hopped a ride back to Morotai on one of the C-47's that was busy ferrying advance ground echelons and equipment to Morotai. Returning to Noemfoor, they reported to the 424th Intelligence Officer to go over the details of their mission. However, it soon became apparent that something big was up for the following day. Having flown such an epic flight the day before, Walt and the rest of the McRae crew would sit this one out, but there were reports that the Japanese Navy had been sighted headed towards the Philippines. The 307th would try to catch them the following day, on their way north.

Thursday, October 26, as Walt and the rest of the dog-tired McRae crew were resting up, eight B-24 crews of the 424th were assigned to combat. One, commanded by Lt. Heillie, was sent west on a lone shipping prowl to Moetang near Borneo. The seven others were assigned to join other 307th BG planes to search for and attack a major Japanese Navy formation that was threatening MacArthur in the Philippines. In total 28 B-24s were to seek out and attack this taskforce, which had been sighted just south of the northeast point of Panay Island in the Sulu Sea. Each plane carried three bomb bay fuel tanks and just a single 1,000-pound bomb. They would attack at 9,000 feet elevation, which was above the range of light and medium shipborne ack-ack, but low enough to have a reasonable chance of success.

The Japanese taskforce, which had participated in the Battle of Leyte Gulf, consisted of three battleships (two Kongo Class, and the Yamamoto, the largest battleship ever constructed, with 17.1-inch main guns), five heavy and light cruisers, and four destroyers. The 424th was the lead attack squadron. A/A fire began ten to twelve minutes before the attack by the heavy antiaircraft guns and by the main battleship guns. Men later reported seeing shells the size of bathtubs (and weighing over one ton) fired by the Yamamoto. When asked to pick a ship to target by the group commander, Lt. McMillan, the 424th Squadron leader replied, 'I'll take the big bastard' – meaning the Yamamoto.

But the attack was very difficult. The flak was intense. Many men later reported that they thought their plane was on fire, because of the thick black smoke from the A/A. As the 424th pressed their attack, Lt. Houston G. Hick's plane was hit between the bomb bay and the tail. It completed its bomb run, but then the tail fell off and the plane went into a spin, disappearing into a cloud and never to be seen again. Another plane flown by Lt. Joseph W. Jones was hit by flak just after it dropped its bomb. It burst into flames and spun into the ocean with all crew. Lt. Russel Sutphin's B-24 was hit by flak in the #3 engine after its bomb run. Sutphin cut the engine and feathered the prop and continued to fly. However, an hour or so later he lost #4 engine; the plane went into a spin and hit the water. Three men bailed out, but one man's chute was tangled in the plane and he went down with it. Another man got tangled in his cords and sank. The third man hit the water and was seen floating on top of his chute, but without movement. Lt. Balovich dropped a raft and radio to him and circled several times at 50 feet, but the man made no movement before Balovich left.

The 424th made one hit on the Yamamoto and one of the Kongo class battleships was hit twice by the 371st Squadron. But the 424th Squadron lost three of the seven plane sent on the attack![55]

October 26 marked the second anniversary of the 307th BG being overseas. The attack on the Japanese Navy represented another first for the Group – an attack on a battleship group by Heavy Bombers. The 307th BG crews on Noemfoor received a special dinner with a small ration of beer to mark the anniversary, but the mood was somber considering the losses the group had suffered in October.

By the end of October, 424th Squadron manpower had declined to 92 officers and 138 enlisted flying staff and 15 officers and 368 enlisted men in the ground support.[56] Conditions with the 424th Squadron were very bad. At the beginning of the month the Squadron had 13 B-24's. During the month, five were lost in combat, six old planes were transferred out, but only seven new planes were received. This was a net decrease of four, for an end of the month total of just nine planes. The 424th personnel were spread out – a few were still back on Wakde, most on Noemfoor, but some were also on Morotai. Many of the crewmen, especially the ground crews, were beginning to show strains. Many of the ground crew had been in the South Pacific for the entire 25 months that the 307th BG had been in combat. There was no rest leave for ground crews, and only 19% of the men had been granted R&R leave to Australia or New Zealand. At Noemfoor, the facilities were all very temporary in nature, with no comprehensive sports or recreation facilities, poor laundry services, and the quality of food was not consistent. Petty arguments, flare-ups, and griping was reported. Technical Inspectors noted a marked increase in maintenance and repair discrepancies. Men were supposed to be rotated back to the U.S., but none had done so yet. A large number of written requests started coming in from the long-serving ground crews for 21-day furloughs back to the States, but none could be granted. All of this was reported by Major Vanderpoel to 307th BG HQ, but the conditions were common to all of the squadrons.[57]

To help alleviate the shortage of aircraft, on Saturday, October 28 Walt was assigned to a small crew to retrieve A/C 951, which had been left on Morotai after the October 24 shipping prowl reconnaissance mission. This was an easy 500-mile flight each way, taking about five and a quarter hours of flying time total.[58]

At the very end of October or beginning of November, there was a change in Walt's crew. It was announced that Carl Appling would again be replacing Charles McRae in the pilot's seat, but other than that, the core crew would remain the same, with Don Aubrey, Jimmie Clark, Ed Dunne, Jupe Bretherton, Frank Mansir, Jack Riley, and Boris Hidalgo. They were all sad to see McRae go, but they all liked the cheerful Texan Appling.

On Thursday, November 2, Walt and the Appling crew were assigned another combat mission. A new man, Sgt. Roth, joined the crew as one of the waist gunners. Major Vanderpoel led seven B-24s from the 424th on a search for a Japanese Naval Task Force in the Philippine Islands. The seven bombers, each loaded with five 1,000-pound GP bombs, took off from Noemfoor at about 1030. Once in the air, Vanderpoel received an order by radio to search for the enemy naval force in the Mindanao Sea, to the west of Camigain Island. He led his seven-plane formation to the northwest, to a point at 9°N, 124°E, to the west of Mindanao. The weather enroute was altostratus clouds above 20,000 feet, with scattered cumulus at 6,000 feet. One plane, piloted by Lt. Youngmark (A/C 236), was seen flying in a tail-heavy attitude. It jettisoned one of its bombs to try to correct this. An hour later Youngmark reported excessive fuel consumption. He radioed that he would return to Morotai separately, searching for enemy shipping along the way.

Vanderpoel then turned the rest of the squadron north, crossing Mindanao Island and into the search area to the west of Camigain Island. The weather was good for the search, with scattered broken cumulus clouds from 6,000 to 10,000 feet. They spread out and ranged across the sea with

Walt and the other ball turret gunners scanning the water below, but nothing was to be seen. At one point, off in the distance towards southern Leyte Island, where MacArthur was fighting inland, intense A/A fire was seen. Finally, when they had reached a point at 9°50'N, 125°50'E over the Siargao Islands, dusk began to fall and Vanderpoel turned the group back. At 1804, Vanderpoel ordered the planes to salvo their bombs from 13,000 feet. Two minutes later, off in the distance, a warship was identified in the ocean. However, because of the dark, it could not be identified. Well after midnight, the squadron returned to Morotai (not Noemfoor) landing safely in the dark after 13 hours and 50 minutes in the air. Walt had completed his 19th combat mission.[59] Fortunately, there were no Jap air raids on Morotai that night, so the men got a good night's sleep.

Four days later, on Monday, November 6, the Appling crew was again assigned to a combat mission by Vanderpoel. The Appling crew had one change. Sgt. Roth, who had flown on the November 2 mission, was replaced by Sgt. Shaughnessy, another newcomer at the waist gunner position. Walt had a little chat with Shaughnessy to ensure that he knew the correct procedure for lowering Walt into the ball turret and getting him back up and out. For this mission they would be going back to the Philippines to support MacArthur's invasion force. Their target was Alicante Airfield, a Japanese air base, on Los Negros Island.

This mission would be commanded by Lt. McMillan in A/C 381. Lt. Balovich would fly old A/C 951 that the McRae crew had flown around the Celebes for almost 19 hours on October 24. Cliff Llewelyn was back now, flying on A/C 179 with Lt. Raue, and Boris Hidalgo was on this mission flying on A/C 236 with Lt. Heille. The Appling crew was assigned to A/C 276. Each plane was loaded with 30 – 100 pound GP bombs, designed to destroy or damage aircraft and buildings.

The crews were roused around 0100 on Monday. After chow and a ride out to their armed and ready airplanes, they took off at about 0245. The planes flew independently to a rendezvous at Flecha Point on Mindanao in the Moro Gulf. As dawn broke, they found full overcast skies with altostratus clouds to 15,000 feet. From there the planes formed up and flew north over Mindanao, crossing Los Negros and then the Panay Gulf, before reaching their IP over Guimaras Island. Alicante is at the northern end of Los Negros. As they flew up Guimaras Strait, Walt and the other ball turret gunners could see a large Japanese cargo ship at a wharf in Bacolod harbor on Los Negros.

As they approached Alicante, the cloud cover thickened to full overcast over the target area. Their target was the intersection of the taxiways south of the shop area, with aircraft dispersal areas and the supply and personnel areas. McMillan led them north of the target, then turned south on a bomb run heading of 180°. Shortly after 1040, as they were approaching the target, between 12 and 15 Zekes attacked. For the next 20 minutes or so, before, during, and after their bomb run, these Jap fighters made from 15 to 20 passes against the 424th B-24's in the defensive box formation. With the low cloud cover, there was no A/A fire to speak of, but this made finding the target difficult.

At 1043, a Zeke attacked A/C 951 from three o'clock high, knocking out the #3 engine. The plane stayed in formation until 1049, when Lt. Balovich peeled off to the right. Walt watched as it descended to about 4,000 feet, where it leveled off and went into a cloud escorted by two P-47's. Shortly after Balovich radioed that he was trying to reach the Negros coastline with injured men on board, but that they had to bail out.[60]

Lt. Raue's A/C 179 took fire from an attacking Zeke, which damaged the rollers for the bomb bay door. The crew was not able to open the doors so Raue broke off their attack. Later the crew was able to open one door so they could jettison 20 of their bombs before returning to Morotai. Later, 28 holes were found in A/C 179, including the antenna shot out and holes in #2 and #3 props.

McMillan's plane, in the lead position, used its automated flight control equipment to sight for range and the other planes dropped their loads at 1057 in a 2,900-foot long bomb train. It was believed that some of the bombs hit the shop area, but it was difficult for Walt and the other ball turret gunners to confirm results. The attacking Jap planes made multiple attacks, always using the same approach from behind at between five and seven o'clock level. One Zeke coming in at six o'clock got concentrated bursts of fire at close range from Jack Riley in the tail gun, as well as from gunners in two other planes. This Jap plane started to smoke badly. It passed under A/C 276 with Walt firing away at it. The enemy then swerved away to the right and went into a steep dive into the clouds at 4,000 feet with black smoke pouring out of its fuselage. Later it crashed into the sea. The 424th Bomb Squadron gunners got a 'probable' for this plane.[61]

Finally, the four remaining Liberators left the attacking Jap planes behind. They made their way back towards Morotai, and then returned to Noemfoor at about 1840. Walt completed his 20th combat mission after 15 hours and 55 minutes in the air.[62] However, A/C 951 with 10 men of the Balovich crew was never seen again.

What was the mental state of these young men who saw a B-24 with their friends disappear into a cloud, never to be seen again? Walt's thoughts were likely similar to those of Charlie Dowdy, a gunner who Walt knew. Walt overheard him talking with some of the other men who saw Balovich disappear, including Sgt. Gootee, Lt. Heille's flight engineer in A/C 236 and Sgt. Dague, McMillan's flight engineer in A/C 381. It was difficult for them to understand why some people caught hell, while others seemed to go through day after day unscathed. The men of Balovich's crew were their friends. None of the men who flew that day would forget this devastating mission.[63]

Two days later, during a return mission to Alicante, the Japanese struck the 424th again. A/C 381, with Lt. William R. Hunter Jr. as pilot, was hit by a Japanese fighter in the No. 4 engine. He feathered the prop but the engine was in flames and the plane slowly drifted down. Four men were seen to parachute from the camera hatch. At the same time, four men were seen to parachute from the bomb bay – one with his chute in flames. At that moment, the right wing buckled upwards and the plane went in to a spin. One more chute was seen before the plane went down.[64] In just 48 hours, the 424th had lost two planes with 21 men missing in action.

The next couple of days the 424th Bomb Squadron was a flurry of activity as everyone was busy packing up to leave Noemfoor. Two days later, on Friday, November 10, it was the turn of the Appling crew. Walt made the 500 mile, three hour and 15 minute flight to Morotai, his home for the next three and a half months.[65]

- 13 -
BACK HOME 1943-1945:
BEYOND A WIDE AND SPACIOUS SEA

The Home Front

The Babinski and Kurzyniec families had followed the progress of the war through 1943 on the radio, in newsreels, and in the *Detroit Times*, *Free Press*, and *News*. As the year began, they read about the Casablanca Conference between Roosevelt and Churchill under a headline in the *Detroit News* announcing 'Total Destruction of Axis Pledged at Roosevelt-Churchill Africa Parley'[1] in January. On May 13 they read the details of the Russian Victory at Stalingrad: '16 Generals Among 175,000 Axis Captives: How Von Arnim Gave Up.'[2] Later in the year, after the invasion of Sicily and then of Italy itself, people were cheered to read, 'ITALY GIVES UP.'[3] At the end of the year, the *News* published a story about the Teheran Conference between Roosevelt, Churchill, and Stalin: 'Triple Front Drafted by Big 3 to Smash Reich.'[4]

Walt had left Detroit for the last time after a two-week furlough at the very end of 1943. This was before travelling to Muroc for his Bomber Crew Phase Training. After that came his assignment with the 307th Bomb Group in the South Pacific. California and Hawaii had seemed exotic and exciting to a young man who had spent his whole life in bustling, industrial Detroit. However, since mid-April his life was confined to a string of remote and tiny tropical islands: Guadalcanal, Los Negros, Wakde, and Noemfoor, which were absolutely nothing like home. Walt had a short interlude in Australia on a two-week furlough to Sydney, where he was able to see his brother Johnnie. But the B-24's that he flew from those tiny islands every few days with his comrades was the single most concrete reminder of industrial Detroit. As Walt travelled still farther from Detroit – from Noemfoor to Morotai – he must have longed to be home again.

Walt had seen in 1943 that the United States was being transformed to win the war in Europe against Germany and Italy, and in the Pacific and Asia against Japan. Agriculture, transportation, mining, and manufacturing – all the components of the American economy and society – were realigned to win the war. More than any single city, Detroit was being transformed into what would be known as the Arsenal of Democracy. It would equip and supply all the nations fighting the Axis, from the Red Army still struggling across Eastern Europe, to the British and French and Poles who were now fighting in France and up the Italian peninsula. It would equip and supply British India, and Australia, and New Zealand, and the Chinese Nationalists fighting Japan on the Asian mainland and across the Pacific.

What was life like for his family and friends back home in Detroit, Walt wondered. Just as they wondered, what his life was like so far away.

Walter senior was working long hours at the Ford Rouge plant. Tom Kurzyniec was likewise working long hours as a foreman at the Ternstead Steel Division of General Motors, just the other side of the tracks down on Livernois. Ford had ceased making cars a couple years before. Now they were turning out thousands of M-4 tanks and M-10 tank destroyers for the Army and Allies. They were also manufacturing gliders that would be used in the airborne assaults in Normandy and Holland. General Motors had turned out 854,000 military trucks by the end of 1943, while Chrysler was building trucks and tanks, as well as radar systems and anti-aircraft guns. By November 1943, the Ford Willow Run plant turned out its first 1,000 B-24 bomber and the pace of production would soon quicken.

One of Ford's most famous employees, Charles Lindbergh, left Detroit in March 1944. He had been a technical advisor to Ford on the development of the Willow Run plant. As part of his work, he had piloted a Ford B-24 to 43,020 feet elevation – an amazing feat in an unpressurized airplane. But now he was being sent as a civilian advisor to the South Pacific, to help the Army Air Corps get maximum performance out of its B-24's.[5]

Thomas and Walter, every day on the front line of the Arsenal of Democracy, were contributing to the success of the Allies just as Walt and Johnnie were across the Pacific. The Babinski and Kurzyniec families continued their special friendships. In August, Dorothy Kurzyniec was able to get away for a few days to Detroit Beach, near Monroe, Michigan. She wrote a post card to her friend Genevieve Babinski:

Hi Jenny,
Thanks for the rackets. It's nice here. Hope to remain here till Thursday. I like you. Do you still love me?

Dorothy[6]

As was common across America, the Babinski house on Pulaski displayed a Service Flag with two stars in the front window, to indicate that the family had two sons on active duty. When John went into the Navy, the family displayed a flag with one and after Wally was drafted, they changed it for a flag with two stars. How did the family feel about John being in the Navy and Wally in the Air Corps? At that time in the neighborhood, almost every young man was gone or leaving. Ma and Pa worried of course, but it was a shared experience.[7]

Walt's sister Genevieve helped their Pa and Ma write out letters in English that they would dictate to her. They could speak English, but only wrote Polish. They would send letters to Walt and Johnnie and to some of Walt's crewmates.

In June 1944, Walt's sister Genevieve, her friend Dorothy Kurzyniec, and another friend, John Besek, graduated from Southwestern High School. In the summer of 1944, Walt's, older sister Anne spent a few weeks at St. Mary's Catholic Girls Academy, in Monroe, Michigan. Anne wrote from the Academy to her sister Genie back in Detroit:

Dear Jenny,
Won't write much. Just to say hello. Have you been riding? That's about all. Just that I'm having a swell time.

-Anne[8]

John and Wally tried to keep in touch with their family by writing often. Johnnie wrote that in Australia he was first stationed on a minesweeper, but that made him very nervous. He was eventually transferred because of his education out of the minesweepers. That made him happy, but he still hoped to be assigned to a major warship. In Brisbane, where Johnnie was stationed in Australia, he 'stood up' at the wedding of one of his shipmates, who married an Australian girl. Because Ma & Pa Babinski did not read English well, Genie would always have to read them their letters. One time, Walt sent Gene a package with a 'hula skirt' from the South Pacific made from old parachute silk. Another time he sent Leona a bracelet made from Australian coins. Wally had written to them from the places in the US where he had been in training, and also from overseas.[9] It was more difficult for him to describe the specifics of targets and bases and actual combat, because letters were censored before they went home.

One way for the Babinski family to get an idea of what the war was like for Walt was from periodic magazine articles. Every family subscribed to weekly photo magazines, like Look, Life, and The Saturday Evening Post. In the late summer of 1944, The Saturday Evening Post published a photo essay covering a single 13th AAF bombing mission to Yap. The essay showed the work of the ground crews preparing a B-24 for the mission. It showed the gunners inside the airplane over the target, and then the return home, with Red Cross girls serving refreshments before the crew debriefing in the intelligence tent.[10]

In late October 1944, the magazines were reporting on the October attacks on Balikpapan. The article described the importance of the Balikpapan Oil Refineries for supplying Japan. It included dramatic photos showing B-24 bombers flying in formation over burning oil facilities below.[11] Another article was written in the first person, describing one mission to Balikpapan in the words of Lt. Richard Reynolds, a 13th AAF bombardier: 'Here is what it's like to fly 16 hours unescorted against one of the toughest targets in the Southwest Pacific. Here's what it's like to fly into a Japanese hornets' nest and out again.'[12]

Before he shipped overseas, Wally provided the Babinski family with the addresses of the other crewmen's families. Gene would write letters dictated by Ma Babinski to the families of Cliff Llewellyn in Minnesota and Jimmie Clark in Louisiana.[13] Frank Mansir's wife, Virgie, wrote to the Babinskis that she thought all the men looked too thin in their photos and she worried that they were not getting enough to eat while overseas.[14]

In addition to frequent postcards and mail from Walt and Johnnie, other friends of the family would write. For example, Walt's friend Joseph 'Tex' Jekielek was also a friend of the Kurzyniec family. At about the same time Walt had arrived in St. Petersburg for the first time in 1943, 'Tex' had just arrived at his training base in Miami Beach. He was assigned to the 598th T.S.S., Flight B of the Army Air Force Technical Training Command. He wrote to Mrs. Kurzyniec:

> *Hello Ma!*
>
> *How's everything? Store closed yet? Sure nice out here. Quite a change from Detroit climate. Everything went swell. I'll be a soldier yet.*
>
> *-Tex*[15]

Walt, Johnnie, and other friends in the service would also send V-Mails home. Joe 'Pudgy' Prus wrote a V-mail from the 67th General Hospital where he was stationed in New York in April 1943 addressed to the 'Kelly' family. The 'Kelley' family was the Delray nickname for the Kurzyniec family. This V-mail depicted three bunnies labelled Martha, Dottie, and Stasia, with the note:

> *Hello Everyone: Happy Easter to you all.*
>
> *-Pudgy*[16]

Later Pudgy was shipped to England with the 67th General Hospital, which took over a British Naval hospital near Taunton, in Southwestern England. This hospital was being readied for the D-Day invasion planned for 1944. While the hospital buildings were snug brick designs, Pudgy and the other men were quartered in 108 six-man pyramid tents, even through the winter. Although each tent had a coal stove for heat, it could not have been comfortable. At the end of 1943, Pvt. Prus sent another V-mail to the 'Kelly' family:

> *Season's Greetings for Great Britain. Merry Christmas and Happy New Year.*
>
> *-Pudgy*[17]

Walt's friend Joseph Stoklosa, now a Corporal with the Third Tank Battalion, U.S. Marines, sent the Kurzyniec Family a V-mail dated December 10, 1943 from 'in the field.' It depicted a marine laying against a palm tree with a tropical sun setting, while dreaming of a pretty blond girl and a decorated Christmas tree.[18]

Walt sent V-mails to the Kurzyniec family also. One addressed to Martha showed a soldier sitting and looking at an island beauty with the caption:

> *The Jap's here have surrendered, but the natives are still causing us considerable trouble.*
>
> *-Wally*[19]

Another V-mail addressed to Miss Dorothy Kurzyniec showed a soldier kneeling in prayer on a tropical island, while thinking of his sweetheart safe in bed at home. The caption read:

> *Dear God,*
> *Watch over her for me,*
> *That she may safely guarded be.*
> *Help her each lonely hour to bear,*
> *As I would lord, if I were there.*
>
> *When she is sleeping, watch her then,*
> *That fear may not her dreams offend.*
> *Be ever near her through the day,*
> *Let none but goodness come her way.*
>
> *Sweet, faithful girl that waits for me,*
> *Beyond a wide and spacious sea.*
> *Be merciful, oh God, I pray,*
> *Take care of her while I am away.*
>
> *That's no kidding.*
> *Love,*
>
> *-Wally*[20]

The occasional letters from family and friends put a human face on the war news as the year wore on. June 6 brought an exciting headline: 'Beachhead Won on French Coast'[21] as the D-Day invasion commenced. But nine days later headlines seemed alarming: 'German Attack Checks Allies' and 'U-Boats Prowling Off Canada.'[22] Reports about the war in the Pacific always seemed to take second place to the war in Europe. Roosevelt had made clear the 'Beat Germany-First' strategy agreed with

Stalin and Churchill, but this added to the anxiety about Johnnie and Walt off in the Pacific Theater. The headline for July 21 read 'Troop Mutiny Reported, Berlin Cut Off Again' while down below there was an article titled 'Marines Win Beachhead on Guam.'[23] In September the *News* had a big headline blaring '2 Yank Columns Pound Opening in West Wall' while down below a tiny headline reported '2nd Blow Dealt in Philippines.'[24] Finally, on October 20, there was some major news from the South Pacific as the *News* headline read 'M'Arthur Lands 250,000 on 3 Philippine Beaches.'[25]

In early November, as Walt was moving from Noemfoor to Morotai to support the Philippines invasion, his brother Johnnie was on his was home to the U.S. When they saw each other in Sydney back in September, Walt had learned that his brother was going back to the States on leave and then transfer. After some delay in securing a berth on a ship, Johnnie left Australia sometime in October. Finally, on Thursday, November 9, 1944, he arrived in San Francisco, where he received orders granting him 21 days annual leave plus six days of travel time to and from Detroit.[26]

Assuming that Johnnie made good time crossing the country by train from San Francisco to Detroit, he likely returned home to Detroit late on Sunday, November 12. Johnnie likely had little gifts for his Ma and Pa and all his sisters and brothers. Despite rationing, there was a big Polish dinner for Johnnie that night, with talking and drinking into the night. But not too late. His Pa and Anne had to go to work the following morning, and it was a school day for Genevieve, Leona, and Henry.

The next three weeks Johnnie savored sleeping late in a warm bed, with the smells of his Ma's good Polish cooking filling the house. With his family and neighborhood friends, he had stories to tell about the Navy and Australia and his hopes for the future. Of course, everyone wanted to hear about Walt and the time Johnnie and Walt shared together in Sydney just a couple months earlier. Johnnie was able tell the family about Walt's life on remote island bases and about the long and dangerous bombing missions he flew – things that the censors would not pass in letters or cards home. Johnnie would have also talked about his frustration with his assignment to a maintenance depot in Australia. He had put in for transfer to a major warship and that was what he was hoping for after he reported back for duty.

His three weeks home would have passed quickly. On Thanksgiving Day that year the Babinski family was very grateful that Johnnie was safe, but worried about what the coming year held for him and Walt. Johnnie enjoyed time with some of his friends from childhood, school, and work, but many of them him were away in the service also. Then about Tuesday, December 12, Johnnie's allotted 21-day leave was at an end. Saying his goodbyes to Ma and Pa, Anne, Halen, Gennie, Leona, and Hank, he was off again to Michigan Central Station and his train back to San Francisco.

Johnnie arrived on Friday the 15th and spent the next couple of weeks in San Francisco, not a bad place to spend the holidays for a sailor, though still not as good as home. Finally, on Tuesday, January 2, 1945 he received orders to proceed to the Small Craft Training Command (SCTC) on Terminal Island, San Pedro, California. He arrived there the following day and settled into his new home for the next several months.[27] It was here that he would be processed for reassignment. His records show that he was then 5'-9" tall at a solid 175 pounds. A few days later he failed a night-vision test.[28] Would that affect his chances for the assignment he was hoping for, he wondered.

Later in January, the Babinski family would receive a post card from Johnnie in Los Angeles, showing the intersection of Broadway and Sixth Street, postmarked January 22, 1945 from Terminal Island, San Pedro, Cal. Johnnie's message was short and maybe indicated a little frustration: "...I'm still waiting for assignment."[29] His wait would go on for many more weeks. At around the same time, Geenie received a postcard from Johnnie showing a night view of the NBC Studios at Sunset and Vine in Los Angeles:

> *Hello, It is quite nice over here, but it sure is cool at night, cold in fact.*
> *-John (John J. Babinski, MM1/C, APO Pool Terminal Island, San Pedro, Calif.*[30]

Throughout the war, the Babinski family would try to send packages of food, medicine, soap, or other scarce item to their Babinski and Gebolys relatives in German occupied Poland. Cora Babinski would put small items of medicine and food in packages, and then make a small parcel out of old cut-up bed sheets, which she would then sew closed. Then young Henry would address the packages in indelible ink. The packages would then make a slow journey to Europe via neutral countries, assisted by the Red Cross. Cora would include letters written in Polish. In one, she mentioned that her son Walter was in the American Airforce. Months later a return letter arrived from Poland saying

"...if Walter is in the American Airforce, could he please fly his plane to Poland, land nearby, and then take us all back with him to America?"[31]

There was one family member who was making significant contribution to the war effort in Europe: Lt. Julian Gebolys. He was a cousin of Walt's mom Cora. Lt. Julian Gebolys had served as an instructor at the Polish parachute training school at Bydgoszcz, Poland before the war broke out in 1939. Lt. Gebolys had escaped from Poland to England in 1939, after the German invasion. As an expert in parachuting techniques and he was sent by the British to the RAF Parachute Center at Ringway, near Manchester. Later Gebolys served as a flying officer and in 1941 he requested to be transferred to the Indian Army Parachute Training School in Chakala, India. But the Polish General Staff considered him too valuable to approve this request. He was promoted to Flight Lieutenant in the RAF. Through 1941 and 1942 Gebolys trained a large numbers of Polish parachute troops.[32]

German and British parachuting technique was to just float down with the wind, fall forward and roll. Gebolys had developed a technique in Poland of using the shoulder harness webbing to control the direction of fall. Gebolys and his comrade, Lt. Jerzy Gorecki, also trained RAF flyers and British and Polish special operations men who would be dropped into France, Czechoslovakia, and other occupied countries.[33]

For the Kurzyniec family, in addition to Thomas working at Ternstead Steel, Martha and the three girls kept the store on Pulaski going. In addition, there was another member of the family for part of this time. Martha's sister Seal Skrabut had suffered post-partum depression after her first son, Norman, was born. A variety of treatments were tried on Seal, including electric shock therapy. Conditions were not conducive to raising a little boy, so it was agreed that Norm would live with the Kurzyniec family. For Thomas, Norman became like the son that he never had.[34]

Dorothy Kurzyniec and her sisters continued at school. Dorothy continued her close friendship with Genie Babinski, who was a year older at Southwestern High School. Dorothy worked part-time during the war as a file clerk in a warehouse at nearby Fort Wayne. She remembered it as a nice pleasant job. The old fort was operated as a POW camp for Italian prisoners during the war. By 1943, Italy was out of the war and actually changed sides to join the Allies. Some of the POWs worked in the warehouse. On Sundays, the POWs would be marched down Fort Street to attend Catholic mass at a Hungarian church in Delray.

At some point in late 1943, the Kurzyniec family moved into a house on Livernois, one block off West Jefferson and very close to Thomas' job.[35]

As always for Detroit, professional sports provided a focus that could unite everyone. The Tigers has finished the 1943 season with a disappointing 78-76 record, in fifth place 20 games behind the Yankees. In 1944, the Tigers started to turn the team around. Detroit would finish the season with an 88-66 record, just one game behind the St. Louis Browns.

In hockey, the Red Wings won the Stanly cup in 1943. The following year they finished the season in second place with a 26-18-6 record, but they were eliminated by the Black Hawks in the first round of the playoffs. At the end of 1944, the Red Wings were in second place in the NHL, pursuing the Montreal Canadiens.

The Detroit Lions were less successful. They had gone 0-11 in 1942, scoring only five touchdowns and never more than seven points a game all season! 1944 saw the team improve to 6-3-1, but it did not inspire the kinds of fans that the Tigers and Red Wings did.

Detroit's economy was booming and pay packets were fat for those working in war industries. However, rationing changed consumption patterns. After Pearl Harbor, daylight savings time was instituted in 1942. By April, sugar, red meat, butter, and canned vegetables were being rationed. Meatless Tuesdays, powdered milk, and Spam became common. Liquor was in short supply, which boosted wine and beer sales. Cigarette shortages resulted in a resurgence of pipe smoking. Because Japan controlled most sources of natural rubber, car tires along with gasoline were rationed, limiting the use of personal autos. Most families received only 5 gallons of gasoline per week.[36]

Every member of a household, including children and infants, was issued a ration book. Typical of a Detroit family were the ration cards and books issued to Thomas and Martha Kurzyniec. Ration books were issued in series – Ration Book One, ration Book Two, and so on. Martha Kurzyniec's War Ration Book No. 3[37] was issued in October of 1943, and Thomas Kurzyniec's War ration Book Four[38] was issued at the end of 1943. Each book showed the name and address of the ration book holder and included various restrictions and the proper use of the stamps bound into the ration book. Each ration book contained sheets of tiny perforated stamps of different color and with different images. The stamps were not designated for any particular type of commodity or product. Local communities would publish in the newspapers and in stores how many stamps and of what kinds were required to be able to purchase food, clothing, fuel, and other rationed items. Items purchased with ration stamps could not be resold, although there was an active market for the stamps in most communities.[39]

Thomas and Martha were also issued ration cards by the Michigan Liquor Control Commission.[40] These cards were to be presented when purchasing liquor in any State liquor store or other licensed outlet. The cards have a note referencing No. 2 Gov't Ration Book, with Thomas' and Martha's specific ration book number. They are interesting because of what they tell us about Thomas and Martha. Thomas was listed as being 55 years old, with blue eyes, dark brown hair, 5'-8" tall, and weighing 165 pounds. Martha is listed as 35 years old, also blue eyes and dark brown hair, 5'-2" tall and 145 pounds. There may have been rationing in place for two years, but no one in the Kurzyniec household seemed to be going hungry.

The Babinski family also faced the limitations that rationing imposed. But Kunegunda Babinski was resourceful. She had a few distant Polish relatives and other family friends who lived in Windsor, Canada, across the Detroit River. It was possible to buy some items in Canada that were hard to come by in Detroit. However, bringing them back into the U.S. was 'restricted.' When Kunda could get a ride into Windsor with a Delray friend, she would stock up on fresh meat, kielbasa, beer, and wine. Then she would put her purchases under her seat in the car. When they came up to the border crossing and were asked '...are you brining anything back from Canada?' Kunda would smile, mumble something in Polish, and inevitably the crossing guard would wave them through.[41]

Despite the key role that Detroit played in helping to win the war as the Arsenal of Democracy, there was a dark blot on the city's history during the war. In addition to attracting Poles, Italians, and other Europeans to the opportunities that Detroit created in the Twentieth Century, the city was also the focus of a huge demographic shift within the United States. In 1900 there were fewer than 5,000 blacks living in Detroit, many in Delray. As the century progressed, two key factors increased the proportion of blacks in Detroit's population. First, the growth of the auto industry created thousands of jobs that attracted people from all across the country. Blacks though often were relegated to the most difficult and dangerous jobs. At the same time, across the South, a subtle reign of terror comprised of black lynching and the growth of the Ku Klux Klan provided another motivation for Black Americans to move north. By 1940, there were almost 150,000 Blacks living in Detroit, over 9 percent of the population.

Then with the coming of the war, the pace of people moving from the South to Detroit accelerated. The black population of Detroit would double during the decade, but joining them on the move north were also many tens of thousands of southern whites. This group stood out from the existing Detroiters, with their folksy culture and the distinctive twang to their speech. Known as 'Hillbillies' they sometimes met discrimination in house and job opportunities. However, many of the Hillbillies also brought their prejudice against Black Americans. There was certainly prejudice against blacks by other whites in Detroit, including some Poles and Italians. But the Hillbillies, competing for the same entry-level jobs as the black, stoked the fires of racial hatred. On a hot Sunday in June 1943, a number of fights broke out on Belle Isle between groups of Blacks and Whites. The fighting spread to other part of the city, and fueled by rumors; gangs attacked innocent people, looted shops, and burned cars. After a day and half, 6,000 troops were needed to restore calm. 34 people were killed, 800 injured, and about 1,000 were arrested. These riots involved only a minority of the population, but represented a dark stain on Detroit's history.[42]

In November 1944, war news seemed to indicate that things were going better for the Allies. The Red Army was in Poland. The Western Allies were slowly moving north up Italy. In France, Paris had fallen and the American, British, French, and Polish armies were making good progress towards Belgium and the Rhine River. Of most interest to the Babinski family, knowing where Walt was based and what he was doing, MacArthur had returned, as promised, to the Philippines. But how long would the war drag on? Another year? Another two years, or longer? No one knew. But everyone hoped and prayed for the best.

- 14 -
MOROTAI: WITH COURAGE AND DEVOTION TO DUTY

On to Morotai
Walt and the rest of the Appling crew flew from Noemfoor to their new home base on Friday, November 10.[1] Morotai Island is located 15 miles northeast of Halmahera Island. It is 45 miles from north to south and 32 miles from east to west. It is mountainous, with peaks up to 4,100 feet, and mostly jungle covered, except for some flat coastal strips of land in the southeast and southwest. Before the war, about 12,000 people inhabited the island, one-quarter Christian, one-quarter Muslims, and half pagans. The islanders were all either fishers or farmers. They ate fish, pig, wild deer, and they grew vegetables, fruit, sago, and rice. Their villages were all located along the coast with buildings of planks and corrugated iron and native homes of reed grass and palm leaves.

Outside the flat coastal area, there were no roads. Only jungle trails led inland, mostly along streams, into the densely forested interior. Being just two degrees north of the Equator, the yearly rainfall averaged 80-90 inches, with year-round average temperature of 80 degrees, and humidity ranges from 60-90%. The flat and narrow, four mile long Gila Peninsula is located in the southwest of the island, adjacent to Pitoe Bay. In the native Malay language, 'gila' means 'man.' A glance at the map shows why the peninsula was so named.[2]

To the northeast of the peninsula were the two parallel airfields: Wama and Pitoe. Wama is located close to the beach and was used only for fighters and transport aircraft. Pitoe had two parallel coral asphalt runways, each 8000 feet long with 50-foot shoulders and 500-foot overruns at each end. They are aligned E/W with no obstructions at either end.[3]

MacArthur had ordered the capture and development of Morotai into a major base in late summer 1944 to support the invasion of the Philippines. For much of the war it had been garrisoned by the Japanese with only 100-250 men. But as the Allies threatened the Philippines, the garrison was reinforced to about 1,000 men, dispatched from nearby Halmahera. There had been a small airstrip developed by the Japanese at Dobera, but it was later abandoned due to poor drainage. Nearby Halmahera and other of the North Molucca Islands had extensive but scattered Japanese garrisons and many small airfields. Beyond the North Moluccas – on Ceram, the Celebes, and other islands, the Japanese had about 500 bombers and fighters that could be deployed forward to attack Morotai.

Most of the Japanese defenders on Morotai were located in the relatively flat coastal plain in the southwest, north of the long narrow Gila Peninsula. However, they were no match for the invasion force sent by MacArthur: the entire 31st Division supported by the 126th Regimental Combat Team from the 32nd Division. More than 40,000 troops were supported by 15,000 men of the 13th and 5th Air Forces, as well as No. 10 Operational Group of the Royal Australian Air Force. The 5th Air Force and the RAAF provided tactical support for the invasion, while the 13th Airforce provided more distant covering support and continued to harass the Japanese in the Philippines.

The first battle of Morotai began with a two hour naval bombardment on Friday, September 15 (while Walt was in Sydney), the allied invasion force landed on the west side of the peninsula. There was little opposition and few casualties on either side as the Japanese defenders retreated inland in the face of overwhelming American strength. The Americans made landings on several other points around the perimeter of the island, to establish radar and observation stations. The only significant Japanese counter-attack occurred on September 22, but it was easily repulsed. North of the Gila Peninsula, the American infantry continued its advance to predetermined objectives, which were reached by October 4. No attempt was made to proceed into the rugged interior of the island, where Japanese forces remained. The Americans lost 30 killed, 85 wounded, and one missing. The Japanese lost over 300 men killed and 13 captured.[4]

MacArthur visited Morotai and confirmed that it was to be developed into a major base. A PT boat base had been established the day after the invasion and work quickly started on the two new airfields. Plans to use the small Japanese airstrip were dropped when it was realized that it would interfere with flying operations from Wama and Pitoe. The plan was to develop Morotai to house 60,000 allied army, navy, and air personnel. The base would also include a 1,000-bed hospital. To begin construction, the invasion force included 7,000 American and Australian army engineers, who were later assisted by several hundred native recruits. The construction included the two new airfields, as well as docks and jetties to unload fuel and other supplies, an extensive tank farm area, warehouses, a water supply, and a road network. The engineers also cleared bivouac areas, which were assigned to the various units that would be based on Morotai, but it was the responsibility of the units themselves to do much of the base development work.

Originally, the 307th Bomb Group had been assigned an area to the north of the airfields as its designated camp area. However, General Matheny did not like the location and had chosen a site in a coconut grove that stretched for about 2,000 feet on the west side of the Gila Peninsula. The men all agreed it was the best site the Long Rangers ever had. There was space for sports fields and the location along the beach provided cool breezes, swimming, and sailing opportunities.[5]

The Australian Army Engineers were responsible for overall base design and operation. To assist crews to complete well laid-out and uniform camps, the Thirteenth Air Force published a simple large format booklet to help the men with their work. The prototypical camp area would have been used by each of the four 307th BG squadrons, as well as by the 307th BG HQ group. The typical camp included the squadron's own electrical generators and water supply tower. Also in a central area would be a mess hall, a dispensary, a supply room, orderly room, and separate enlisted men and officer's showers. These would all be connected to the water supply tower. Also centrally located would be a mess kit washing area and garbage area. The camp would have separate areas for officers' tents and enlisted men's tents. Urinals and latrines would be scattered around the periphery of the tent areas. There would also be a parade ground, and maybe a baseball field and another area of hard packed coral that could be used for tennis courts or basketball. At one end of the camp was located the squadron motor pool with a small vehicle maintenance shop.

To help ensure an orderly layout of the camp area, simple instructions were provided for field surveying. There were also simple diagrams to help the men fabricate various camp facilities, like shower rooms, mostly from local material. Showers could be improvised by hanging large tin cans with holes in the bottom from overhead bamboo water pipes with holes above each tin can. Water was stored above the showers in big black 'Lister Bags' which would be fed by a pipe from the water supply and then heated by the sun. The most personal structure for the men would be their shared tent – a standard design to accommodate either four officers or six enlisted men. A 22 by 22 feet square area was prepared with hard packed coral fill. The tent itself was 16 by 16 feet square with three-foot overhangs. The men's cots would be located along the perimeters with a small table and some stools in the center. Each man would have his own cot with a storage locker and shoe rack beneath, and a shelf and net hanger for clothes and other small items above the cot. An important item for each tent was sufficient mosquito netting to protect the men from flying pests.[6]

However, the Appling crew did not really need the Air Force telling them how to erect their tent. They had lots of experience already. Walt's crew tent on Morotai was very good because Frank Mansir and Jack Riley were good scroungers. They always managed to find lumber. The Australian Army Engineers had set up a small sawmill north of the Gila Peninsula. The crewmen pooled part of their whiskey ration to barter for lumber. So they were able to build a tent foundation a foot off the ground to minimize the snakes, lizards, centipedes, and various 'black bugs'.[7]

The Australian Army engineers were responsible for laying out the basic road pattern. They ran pipelines from the water source at the north end of the Gila Peninsula to a small number of water spigots where individual unit would have to drive trucks to obtain water and take it back to their own camp area. For the hundreds of men in the 424th this was a major daily operation. A few weeks after arriving on Morotai, it became apparent to the men in the 424th that there was a spigot about a half mile away where they obtained their water, as well as another spigot on the other side of their camp area. Some of the men discovered the location of the pipeline under the 424th camp area, but the Australians would not allow units to tap into their pipeline.

The 424th CO, John Vanderpoel was not happy with this Australian bureaucracy. Vanderpoel arranged for some of his men to dig a concealed trench down around and below the pipeline in the 424th camp area. He then had a tent set up over the trench and late one night (when water was not normally flowing in the pipe), a 424th mechanic entered the trench with welding tools to cut into the line and then weld in a tap with a shutoff valve. The tent was to hide the sparks and light from the welding from any Australian QM or other officers who might try to stop the tapping of the line.

The following day, again under cover of dark, a small diameter pipe was welded onto the tap to a spigot run into a water tank in the trees in the 424th camp area. After the welding was completed, the trench was covered over. The spigot in the trees provided 50,000 gallons of water per day to the 424th; eliminating the long, hot round trip to retrieve water. Later other squadrons found out what the 424th had done and made plans to install their own taps. Vanderpoel warned them to weld under a tent at night.[8]

Despite the best plans and intentions, all the camp areas took on a somewhat chaotic appearance. Many years later, one 307th Bomb Group vet would depict a tented camp area in a famous film based on his experiences on Morotai.

Thirteenth Air Force Together on Morotai

Eventually the entire Thirteenth Air Force would be based on Morotai. By December 1944 it would include the Fifth Bomb Group with four B-24 squadrons and the 307th Bomb Group with its four squadrons equipped with B-24s: the 370th, 371st, 372nd, and 424th. These eight heavy bombardment squadrons represented the chief offensive punch of the 13th AAF. There was also a group equipped with shorter-range B-25 medium bombers – the 42nd Bomb Group with five squadrons. There was also an independent squadron of secret radar-equipped B-24s: the 868th bomb Squadron known as the 'Snoopers.' The 868th specialized in nighttime anti-shipping patrols, to try to interdict Japanese tankers and supply ships from the Dutch East Indies to the Philippines and farther north. To defend their base the 13th AAF also had two fighter groups: the 347th FG with three P-38 equipped squadrons and the 18th FG with a further three P-38 squadrons. There were also two independent night-fighter squadrons, the 419th and 550th, with P-61 night fighters. These were intended to defend Morotai from Japanese nighttime bombing raids. There was also the 4th Photo Group with two F-5 photo-recon squadrons, the 403rd Tactical Cargo group with C-47 Skytrain cargo planes, and an amphibious rescue squadron, the Second ERS, equipped with OA-10 (Catalina) search and rescue airplanes.[9]

One of the 424th Bomb Squadron's non-flying officers at this time was Charles McRae, Walt's pilot from back at Muroc and bomber phase training. In the South Pacific he had been plagued by ear infections, which made it difficult for him to fly. McRae was a great pilot. Walt had flown with him in October on his two completed missions to Balikpapan and on the epic 18 hour and 40 minute shipping prowl around the Celebes on October 18. Now McRae was grounded again and serving as the 424th BS Assistant Operations Officer, helping to plan the missions that he could no longer participate in.[10]

The mission of the 13th Air Force was to protect MacArthur's invasion of the Philippines beginning with the landings on Leyte. This was to be accomplished by attacking Japanese airfields across the Philippines, patrol the sea-lanes from the south to prevent supplies reaching the Philippines, and also to attack and destroy oil installations at Tarakan on Borneo.[11]

The Japanese command on Halmahera realized that Morotai was being developed into a major Allied base, so large reinforcements had been dispatched to the island in September and then more again later in November. These reinforcements were usually sent by barge at night. The US Navy PT boats based on Morotai tried to intercept the reinforcements, but many snuck through. These included the Japanese Army's 211th Regiment and the Third Battalion of the 210th Regiment. However, supplying the Japanese forces was difficult and many would die from disease and hunger. However, attacks on the Allied perimeter were a constant threat.[12]

Give Us a Goal

Walt had now been in the South Pacific for seven long months. He had progressed to the west with the 307th BG from Guadalcanal, to Los Negros, to Wakde, to Noemfoor, and now to Morotai (with a two-week interlude in Australia). However, in seven months he had completed just 20 combat missions. There was no apparent definitive policy on how many combat missions a man had to complete to qualify for rotation back home. At one point, 25 missions had been assumed to be the 'magic number'. Each man accrued points for completed missions flown. This is why Walt and the other crewmen got no 'combat credit' for the few times their plane 'turned-back' for various reasons, before completing their assigned mission. But by November, the few men who had been sent home had each finished about 35 combat missions.[13]

However, the need to maintain operational efficiency came first – meaning a unit needed enough experienced crews to man its bombers when and as assigned to combat missions. When replacement crews arrived and were checked out and available to keep the unit's planes flying, the men with the highest number of points would be given the opportunity to rotate back home. It was not a magic number – the operational needs of the unit came first.[14] Therefore, Walt had a way to go before he could start thinking about rotation home. But soon the operational tempo would pick up for the 307th BG and for Walt and his buddies.

As early as May 1942, the Air Force had noted that fighter and bomber crews in Australia were 'getting burnt-out.' The implication is that 'burnt-out' crewmen were less efficient and also more likely to become combat casualties because of fatigue. In addition to the stress and danger to the individual crewmen there was the fact that experienced combat crewmen would make excellent instructors for new crews being trained in the United States. These experienced combat crews (like Walt and the Appling crew) could provide superior instruction to trainees on the combat use of the equipment and on the unique characteristics of the various theaters of the war and the enemy opposition encountered. [15]

Around the same time, the Air Force was recommending that combat crews be given periodic rest leave. In mid-1942, it was recommended that after 100 to 125 hours of combat duty, a crewman should be given one week of leave in a designated rest area. However, there was a disclaimer to this and later refinements: The rest leave was subject to local conditions at the discretion of the unit commander. And in the South Pacific, even if the commander was amenable, there was a lack of adequate transportation to get men to suitable rest areas in Australia and New Zealand. The Air Force also directed that the combat tour for aircrews be limited to one year, after which they were to be returned to the U.S.[16]

In the 13th Air Force, concern for the well-being of their combat crews started at the top. When Gen. George Kenney took over command of the allied air forces in the Southwest Pacific, he found some of the crews 'punch drunk.' In December 1942, he sent home the entire 19th Bombardment Group (H), a B-17 unit that began the war in the Philippines. Air Corp Commanding General Hap Arnold called some of these men to Washington, DC, to assess combat conditions and the impact on aircrews. One commander said 'Our crews have been going a long time. They are getting tired. We can give them no reasonable assurance as to how long they will have to carry the ball. To them there appears no end – just on and on until the Japs get them.[17]

Conditions in the various U.S. air forces were very different from theater to theater. Aircrews in England or Italy could get away for a weekend to a town or city for R&R. Even men in India or Africa had relatively easy access to rest facilities. However, there was no such possibility in the Central and South Pacific. Brig. General Nathan F. Twining, commander of the 13th Air Force since its inception in January 1943, established a formula to determine eligibility for rotation home. Each crewman's position on the eligibility list was determined by his total points based on the following formula:

Score = T/100 + M/10 + A/3, where T= total hours of flying time, M = total number of combat missions, and A = total months in the South Pacific Theater of Operations.[18]

Within the 307th Bomb Group, the operations office of each squadron maintained a record for each flying crewmember that listed their name, rank, duty, army serial number, date left for foreign duty, date arrived in the South Pacific, and a record of each flight that they were credited with. Each individual flight record indicated the date, flying time for that date classified by type of mission (strike, search, combat, or other), with total flying time for the day and remarks (usually the target).[19]

However, this was the policy for the 13th Air Force only, and the policies and eligibility for rotation home varied from Air Force to Air Force and from theater to theater. By February 1944, before Walt left California for overseas duty, General Arnold in Washington decided that, for maximum operational efficiency, each unit was two have two complete crews for each heavy bomber and one and a half crews for each medium bomber and fighter, in all theaters. This would run into conflict with rotation plans like the 13th AF's based on a formula. Arnold also intended that rotation home would not mean that a combat crew airman would not serve another tour of combat duty. Arnold sent a letter dated February 16, 1944 to all Air Force commanders laying out his requirements.

General Kenney immediately pushed back on March 24, 1944, stating in a letter to Arnold, "Our policy out here in regard to returning combat personnel is based entirely on combat fatigue. The number of months or years that a man has spent in the Southwest Pacific Area has nothing to do with his returning home." Kenney did offer to take back veteran crewmen rotated back home who would volunteer to return for another tour of duty in the Southwest Pacific. Other Air Force commanders became emboldened by Kenney's push back and started to chime in. General Ira Eaker, commander of all Allied Air Forces in the Mediterranean wrote to Arnold '....the thing that makes it most difficult to maintain morale is to have no policy, leaving clearly in the mind of the combat crewman the belief that he must go on until he cracks up and becomes a jabbering idiot or an admitted coward, or until he is killed. We are all of one mind that the effectiveness of tactical units cannot be maintained under this condition.'[20]

In mid-1944, a plan was put into place for combat crewmen from the European theaters to be rotated home for 30 to 60 day rest periods before being returned to duty. In July 1944, the Seventh Air Force (operating in the Central Pacific) began to send combat crewmen home for 30 days of rest leave after completing 30 combat missions. However, medical evaluation of men who were returned to combat or to instructor duties in Hawaii, found that one-third were in worse emotional condition than before being sent home. General Kenney had been planning a similar furlough rotation program for the Southwest Pacific Area, but AAF headquarters vetoed this plan because of the poor results.

By September 1944 however, conditions began to change. The flow of newly trained crewmen arriving in all theaters began to increase. At the same time, the war was turning definitely in the Allies favor and combat aircrew losses were decreasing. The pressure from Arnold on Air Force commanders to lengthen combat duty tours began to ease. General Kenney delegated responsibility to his units on the guiding basis that the local commander should determine a tour of duty adequate for rotation and that a replacement should be available. The Fifth Air Force set 400 hours in a heavy bomber as its criteria, while the 13th Air Force kept its formula-based point system.[21]

Walt and his buddies knew little of this high-level discussion and debate about rotation policy, and rumors abounded.[22] It is very likely that Walt, Cliff, Jupe, Frank, Jack, and their friends had conversations about tour of duty and rotation something like this.

> *'Give us a goal'*
> *'Well, we will fly as many missions as you think we should fly – but, by God, after that number of missions, or months, we want a chance to go home.'*
> *'I don't want a tour fixed by number of months...I want a definite number of missions or a positive number of combat hours.'*
> *'As it stands now, we have nothing to look forward to – no incentive.'*[23]

The Pace of Battle Quickens

The first night on Morotai, as Walt was trying to settle into his new camp, the Japanese paid a call. At about 0300 early on Saturday morning, the air raid signal went off. In the distance, a lone raider dropped two phosphorus bombs. Things quieted back down, and then another raider dropped two GP bombs. Finally about 0500, a third Jap raider made a strafing run. Fortunately, there was no damage or injuries from any of these attacks, but there were many bleary-eyed men that morning.

The following night the Japs were back again, starting about 0330. On the first attack a heavy GP bomb was dropped and then a phosphorus bomb. Intense A/A fire boomed away and searchlights scanned the sky. On a second pass, more bombs were dropped and the Jap raider escaped.[24]

Just two days after arriving on Morotai, the Appling crew was called out on their first mission. When men were scheduled to fly the following day, they would usually go to sleep early. Some would shave during the evening so they would not have to in the morning. Very early in the morning – sometimes as early as two or three, men would be awaken by the CQ who would say 'time to fly sergeant – wake up.' After dressing, the men would get breakfast at the mess tent, which would be lit at night by a dim red light, so that the occasional Japanese bombers would not have a beacon to aim at if they raided the base in the dark. Breakfast was usually scrambled eggs, toast, and coffee.

Reporting to the briefing room, the details of the mission would be spelled out for the crews. The weather officer would report the weather conditions to be expected over the route. The armament officer would tell the men the size and number of bombs that had been loaded on the planes. The planned location of PBY rescue planes or submarines would be indicated, in case a plane went down to or from the target. The engineering officer would tell the men how much fuel had been loaded on each plane. The communications officer gave the radio operators the 'codes of the day' with separate codes for strike, weather, and general communications information, as well as the radio frequencies. Lastly, the intelligence officer would announce the primary and secondary targets, likely anti-aircraft guns, enemy fighters, and anything else of importance to the mission.[25]

Then the crew would ride in a truck several miles out to Pitoe airstrip where maintenance crews and armorers would have the planes checked out and loaded with fuel, bombs, and ammunition for the guns. Each crew would also have a cardboard box with cans of Spam, loaves of bread, and canned peaches, which was their typical meal during long missions.[26]

On Monday, November 13, as Walt was going through this pre-mission routine, another Japanese air raid began. The Jap plane could be heard, but there was ground fog that morning which made it difficult for the searchlights to spot the plane. Fortunately, the fog also made it difficult for the raider to see his target. There was sporadic but ineffective A/A fire. Finally, two phosphorus bombs were dropped, but they were far wide of doing any damage.[27]

The Appling crew in A/C 276 joined five other B-24s of the 424th Bomb Squadron to attack the dispersal, supply, and personnel areas of Fabrica Airdrome on the north end of Negros Island in the Philippines. Lt. 'Mac' McMillan in A/C 619 was in command. The Appling crew on this mission included Don Aubrey as copilot, Ed Dunne and Jimmie Clark, plus Walt's friends Jupe Bretherton, Frank Mansir, Jack Riley, Boris Hidalgo, and Gene Schriener. On this mission, there was an eleventh man on the crew – a Sgt. Walker as combat photographer. Each plane was loaded with 30 100-pound frag cluster bombs.

Take off from Pitoe airfield was at about 0600. The weather was overcast at 10,000 feet with light intermittent showers. The route was to the northern tip of Raoe Island, just off the west coast of Morotai, to Tuna Point at the southern tip of Mindanao, to Capton Point, to Paulino Point to the IP. On the way out, Walt and the other ball turret gunners spotted a large Japanese transport ship (code name Fox Tare Baker) grounded on the southern tip of Talaud Island at 3°55' N – 126°50' E. From this point north, the weather cleared up a little with visibility two to eight miles.

Shortly after 1030, they reached the IP north of the target and the six planes formed up in the group box formation with McMillan and A/C 619 in the No. 3 position. Approaching the Japanese airfield at 10,200 feet, the weather was 70% cumulus cover tops at 9,000 feet and visibility 10 miles. As they approached, everyone nervously looked for enemy fighters but none were seen. Then the clouds thickened up and the target was obscured. McMillan ordered the formation to circle around and try again. At about 1052, they approached on a heading of 130° magnetic. Still no enemy fighters and no A/A fire. They followed the course set by McMillan's plane and each bombardier dropped his load when the target was in range. Walt watched as the squadron's 180 bombs drifted down then began exploding in a line that extended about 2,900 feet. The bomb pattern started east of the personnel camp then progressed southeast through the center of the runway, and then through the southeastern half of the dispersal area. Three fires were seen in the dispersal area. The attack was over by 1054 and amazingly there was no A/A fire! Sgt. Walker managed to get 11 photos of the attack at 1053.

The planes then each set a direct course back to Morotai. The weather was essentially the same as earlier, except that over southern Mindanao they encountered heavy turbulence and rain as they penetrated a weather front. At 1520, after 9 hours 20 minutes and 1,367 miles, A/C 276 returned safely to Pitoe. Walt completed his 21st combat mission. They secured their plane in the dispersal area and made their way back to the 424th camp for their coffee and donuts. Later they reported to HQ for mission debriefing with Lt. John Shields. This mission would be rated as excellent.[28]

Early in the morning on Tuesday, just before dawn, Walt's sleep was disturbed by an air raid alert. Off in the distance he could hear light to heavy A/A fire as the raider made three passes, dropping bombs south of Pitoe strip with no damage.[29]

Two days later, Wednesday, November 15, the Appling crew was assigned another mission. Six bombers from the 424th Squadron would join another six from the 372nd Squadron to bomb the dispersal area at La Carlota Air Field on Negros Island in the Philippines. La Carlota was inland, about two-thirds of the way up the length of the island. They would fly A/C 276 again and the crew would be the same as the last mission on the thirteenth, except that the photographer Sgt. Walker would be replaced by Sgt. Mario Campora. Walt had flown with Campora before, back on August 10 on a mission to Yap with Lt. Jack Arnett. Lt. 'Mac' McMillan would again lead the attack in A/C 619. Their friend Cliff Llewellyn would be on this mission too, flying with Lt. Raue in A/C 236.[30]

Walt and the rest of the crewmen were awakened at about 0400. This gave them time for breakfast, collect their gear, attend the briefing, and then catch a truck for the drive to Pitoe. At about 0700, the 12 Long Ranger B-24s took off one by one into the tropical morning. The weather was good with scattered cumulus with tops at just 3,000 feet. Walt and the other men test fired their guns and then settled down at their stations for the flight north. Their route took them from the navigation point at the north tip of Raoe Island off the Morotai coast, then across the Molucca Sea to Tuna Bay on the southwest coast of Mindanao at 6°24'N, 124°05'E. A life raft was sighted in the water, but there was no one in it and no apparent life in the area. Then they turned north to Tagulo Point on the north tip of the Zamboanga Peninsula, before heading to the westernmost tip of Negros

Island. Turning north into Panay Gulf, at about 1130, Walt and the other BT gunners reported seeing a 'Sugar Charlie' – a small Japanese freighter - at the IP southwest of La Carlota.

Just before noon, the dozen Long Ranger Liberators arrived at the IP southwest of La Carlota and formed up at between 9,200 and 9,500 feet in the group box formation. The weather was good, with widely scattered clouds and unlimited visibility. No enemy fighters were sighted as they headed in for the bombing run on a course of 30°. Fortunately, no A/A fire either. Walt had a great view of the target area and then at 1204 Jimmie Clark called out 'bombs away.' Each of the dozen planes dropped 30 100-pound cluster bombs. Walt watched as the bombs from the 424th dropped and then began to explode, walking north-northeast across the airstrip and then into the northern revetment area for about 3,300 feet. Walt saw four planes hit and burning on the ground and another large fire started by the 372nd Squadron bombs. Campora managed to get 14 photos during the attack, and Sgt. Walker in A/C 236 got five photos. Walt and the other observers noted that there were six twin-engine bombers on the ground in a northern dispersal area that had not been part of the target.

After the attack, Appling had Ed Dunne plot a direct course back to Morotai. There had been no A/A fire and still no enemy fighters to challenge them. They crossed back over Negros and then Cebu Island. In the Cebu Sea, on small Naburos Island, Walt and others sighted a small building with a white flag flying from a pole in front. The flight back was otherwise uneventful. Appling put A/C 276 back down at Pitoe at about 1702. After 10 hours in the air and 1,368 miles, Walt completed his 22nd combat mission. After their debriefing, Lt. Shields would rate the mission as excellent![31]

Early on Thursday morning there was another air raid. A single Japanese plane dropped a cluster of small bombs near Wama airstrip.[32] In addition, the men heard machine gun fire and saw flashes of light offshore in the dark. The rumor was that the Navy PT boat squadron based on Morotai was intercepting barges and small boats trying to bring reinforcements and supplies onto the island.

The next day, on Friday, November 17, Walt and a few of the other crewmen were ordered to report to 307th BG Headquarters in good clean uniforms. Arriving at HQ, Walt saw Lt. Aubrey and Lt. Clark, as well as a few other enlisted men he recognized, including S/Sgt. Richard Cosgrove and Pvt. Mario Campora. The men were told to form up in ranks, and then called to attention as Gen. William Matheny, 307th BG CO appeared. Matheny addressed the men telling them:

> "By direction of the President under the provision of Executive Order No. 9158 as amended, you are each awarded a Bronze Oak-Leaf Cluster in lieu of an Air Medal, by General St. Clair Streett, Commanding General, Thirteenth Air Force."

General Matheny continued his address:

> "These awards are for meritorious achievement while participating in sustained operational flight missions in the Southwest Pacific area, during which hostile contact was probable and expected. These operations consisted of bombing missions against enemy airdromes and installations and attacks on enemy naval vessels and shipping. The courage and devotion to duty displayed during these flights are worthy of commendation."

Walt's own citation stated that it was for "operational flight missions from 17 May to 9 June 1944."[33] With that, the men were dismissed and headed back to camp, where they learned that there would be little rest for them.

The following day, Saturday, November 18, the Appling crew was called for another 424th Bomb Squadron mission. Lt. McRae, grounded again for several weeks because of his recurring ear infection, gave the mission briefing.[34] McRae told them that their target was the Pamoesian Oil Refinery complex at Tarakan, in northeast Borneo. Tarakan is several hundred miles north of Balikpapan, which had been the target for some costly attacks by the 13th Air Force in October. Tarakan is a marshy island that produced about seven percent of the oil from the Dutch East Indies and was a key fuel source for the Japanese Navy. The quality of the oil was so good, that in a pinch it could be burned without refining by Japanese Navy ships. Tarakan had to be knocked out to help neutralize the enemy fleet by starving them of fuel. While much of the oil headed from the Indies to the Philippines and Japan was destroyed by sinking tankers and small transport ships, the Long Rangers were now tasked with striking at the source.

McRae told them that each Long Ranger squadron would put up six or seven Liberator bombers for the attack. Major Vanderpoel would lead seven of the 424th Squadron B-24s. The 424th would be the lead squadron in the attack, followed by the 371st, 372nd, and 370th Squadrons, and then four squadrons of the Fifth Bomb Group. In total the 13th Air Force would commit 49 heavy B-24 bombers,

54 P-38's, and even five medium B-25 Mitchel bombers.[35] Over the target, they were to rendezvous with three P-47 squadrons for fighter protection. The 424th would be the lead squadron, and Lt. Appling and his crew would be the number two plane behind Vanderpoel. Each of the planes was loaded with nine 500-pound bombs: seven general purpose and two AM76 incendiary bombs. The bombs were set to be released so that they would fall at 85-foot intervals on the ground.

Considering the importance placed on this mission, Maj. Vanderpoel as C/O of the lead squadron would fly the lead plane (A/C 619) itself, with Lt. Col. Robert D. Hinton, 307th BG Deputy Commander in the Co-pilot seat, and a Capt. Terry as navigator, and Major Mitchell as bombardier!

As usual, the briefing instructions included many details. The pilot and his assigned A/C was indicated, as well as their position in the formation and radio call letters. The primary target was named with planned elevation and time of attack, in this case 10,750 feet at between 1235 and 1240. The route was spelled out: from base on Morotai, they would fly independently to the Group assembly point at Dum Dum Island (04°27'N, 118°2'E), where they were to circle around to the left from 1124 to 1200 until all squadrons had formed up. Vanderpoel would then lead them to their turning point at 03°56'N, 117°21'E, then to the IP at 03°22'N, 117°31'E. Then they would make a turn of 42° to the left, putting them on a magnetic heading of 118° to the target. The other squadrons would follow the 424th in trail, with each squadron 250 feet lower than the squadron ahead. Designated attack speed was 160 MPH. After bombs away, the planes were ordered to break away to the left at 165 MPH, and then reform in the group box formation for defense against any possible fighter attacks. If needed, two secondary targets were named: nearby Sandakan and Mapanget A/D's.

Each plane was to be loaded with 3100 gallons of fuel. Various radio frequencies were specified for each A/C, as well as for base, other intermediate radio stations, the P-47 fighter squadrons, as well as their code names (Outcast, Leper, and Beaver). The location of two rescue points was designated, where Air-Sea rescue planes would be circling: Southern tip of Sibutu Island at 04°00'N, 19°30'E between 1245 and 1330, and at Siroe Island at 2°40'N, 125°20'E, from 1515 until dark.

After their briefing and breakfast, Appling's crew assembled out at Pitoe field at A/C 276. Knowing that the lead plane with Major Vanderpoel included brass from Group HQ, the men were on their toes. In addition to their bomb load, considering the fierce defense encountered at Balikpapan, there was no scrimping on ammunition for the .50 cal guns. It was about 0630 that Appling piloted A/C 276 into the air, headed out over the Molucca Sea, then turned to the west. The weather was good all the way the turning point at Dum Dum Island, to the northeast of the target. From there they flew southwest to the IP at 03°22'N, 117°31'E, which is actually northwest of the target, up a wide meandering river in the jungle with a distinctive crook in its course, chosen for easy identification. Many of the planes were late or early to the IP, so the attack was carried out by multiple ad-hoc groups of planes as they assembled. The bomber crews were relieved to learn that the P-47s and some P-38s from the 13th Air Force showed up to provide fighter protection.

At about 1206 Carl Appling announced that the target was in view. They were at about 11,500 feet elevation. Walt and the other gunners remained alert for enemy fighters, but none was seen. As they approached the target, medium anti-aircraft fire opened up from below. Fortunately, most of it was a thousand feet or more below them. Walt watched as the bombs from six of the seven planes dropped, but for some reason the lead plane, with Vanderpoel and the Lt. Col. Hinton, failed to drop their bombs. As they turned to the left to leave the target, Walt and the other gunners could hear over the intercom as Appling cursed and announced that they were going back, in formation, so that Vanderpoel in the lead plane could drop his load. They were lower now after the turn, at about 10,800 feet. The earlier bombing run a few minutes before had already started some fires below and in places thick black smoke billowed up to their elevation and higher. After they lined up for the second pass over the target Walt kept an eye open for enemy fighters as he watched the bombs fall away the second time. He was able to get a second look at the detonations from the 424th planes the first time around, that had walked through the refinery installations and tank farm for about 680 feet. The bombs from the second run extended only about 340 feet, because they came from a single plane. The detonations started in the oil separation plant, where three small fires were started, and then into the tank farm, where three oil storage tanks were set ablaze.

The A/A fire shifted a little then, but now it burst high above by about 200 feet or more. Walt could feel A/C 276 shake as the pressure waves from the explosions above rattled the retreating planes. As they left the target at about 1303 there was radio chatter that indicated a P-38 had been hit and the pilot was seen to parachute down. Then Walt and the other ball turret gunners saw the P-38 crash into the sea, while other P-38s circled around the fallen flier.

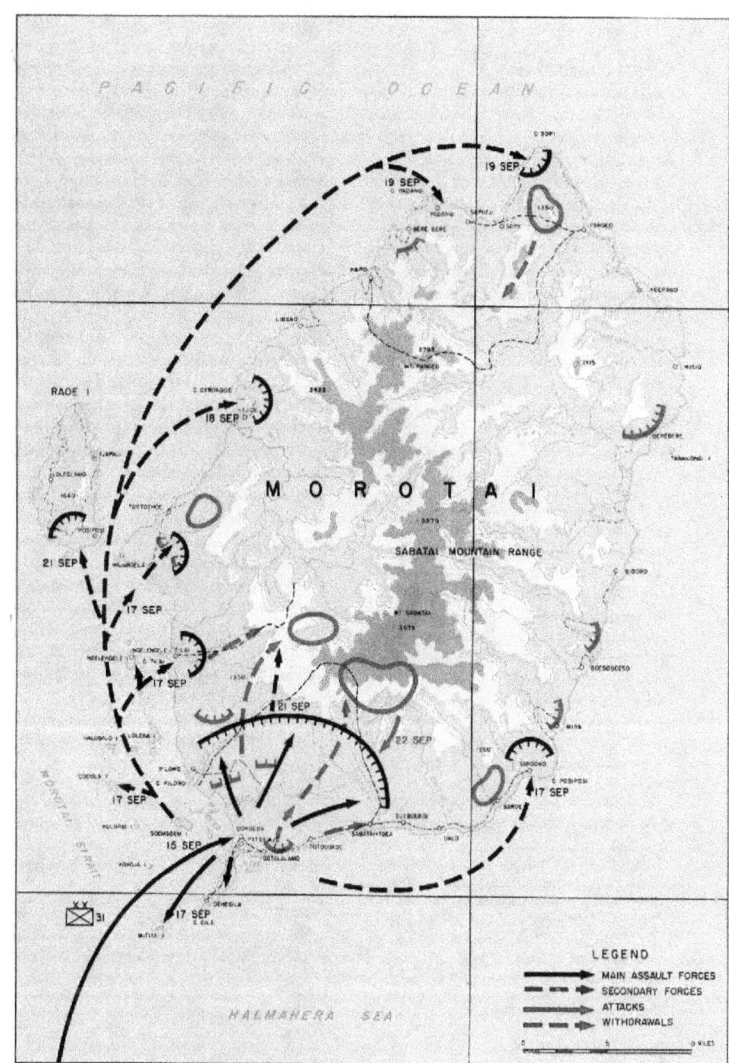

Invasion of Morotai: Morotai Operation, September 1944. US Invasion in black, Japanese defensive moves in red.
Wikicommons

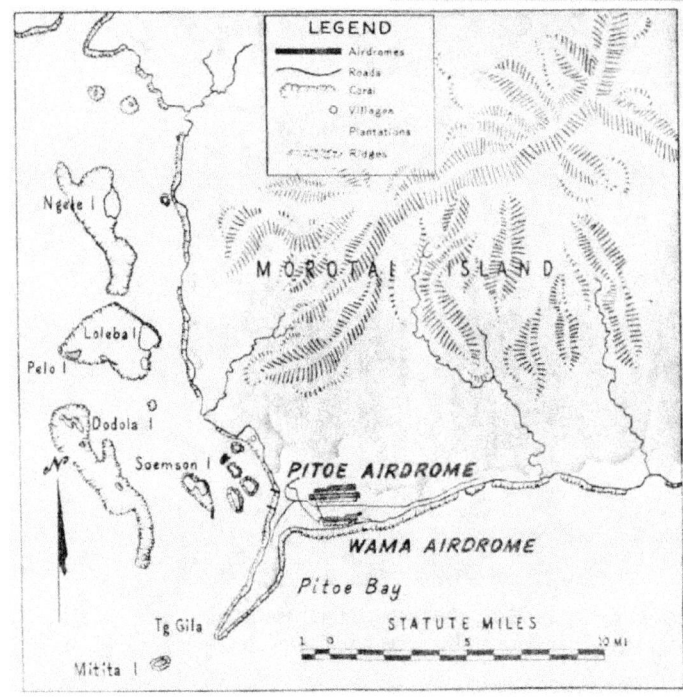

Morotai: Gila Peninsula where most crew camps were located with Pitoe and Wama Airdromes.
Airdromes Guide Southwest Pacific Area, P. 89

Morotai Base Layout Map: The 2/1 Australian Army Survey Company mapped the base layout at Morotai in late 1944. Note the perimeter line north of the engineering and dumps area. Route 3 (Skyline Drive) runs the length of Gila Peninsula.
Australian National Archives

Setting Up Camp on Morotai: After the first few Japanese night-time air raids, each crew worked on a bomb dugout near to their tent.

Setting Up Camp on Morotai: There was a mill on Morotai run by the Australian Army Engineers where whiskey and beer could be exchanged for lumber for tent floors and frames. *Missing Aircrew Project*

Setting Up Camp on Morotai: The 424th Bomb Squadron tent area known as 'Beer Bottle Alley' (right) and Walt's camp area (below).

Setting Up Camp on Morotai: Two views of Skyline Drive, running the length of the Gila Peninsula, here passing through the 424th Bomb Squadron camp area. *307th Bomb Group Association*

Flight of the Malfunction

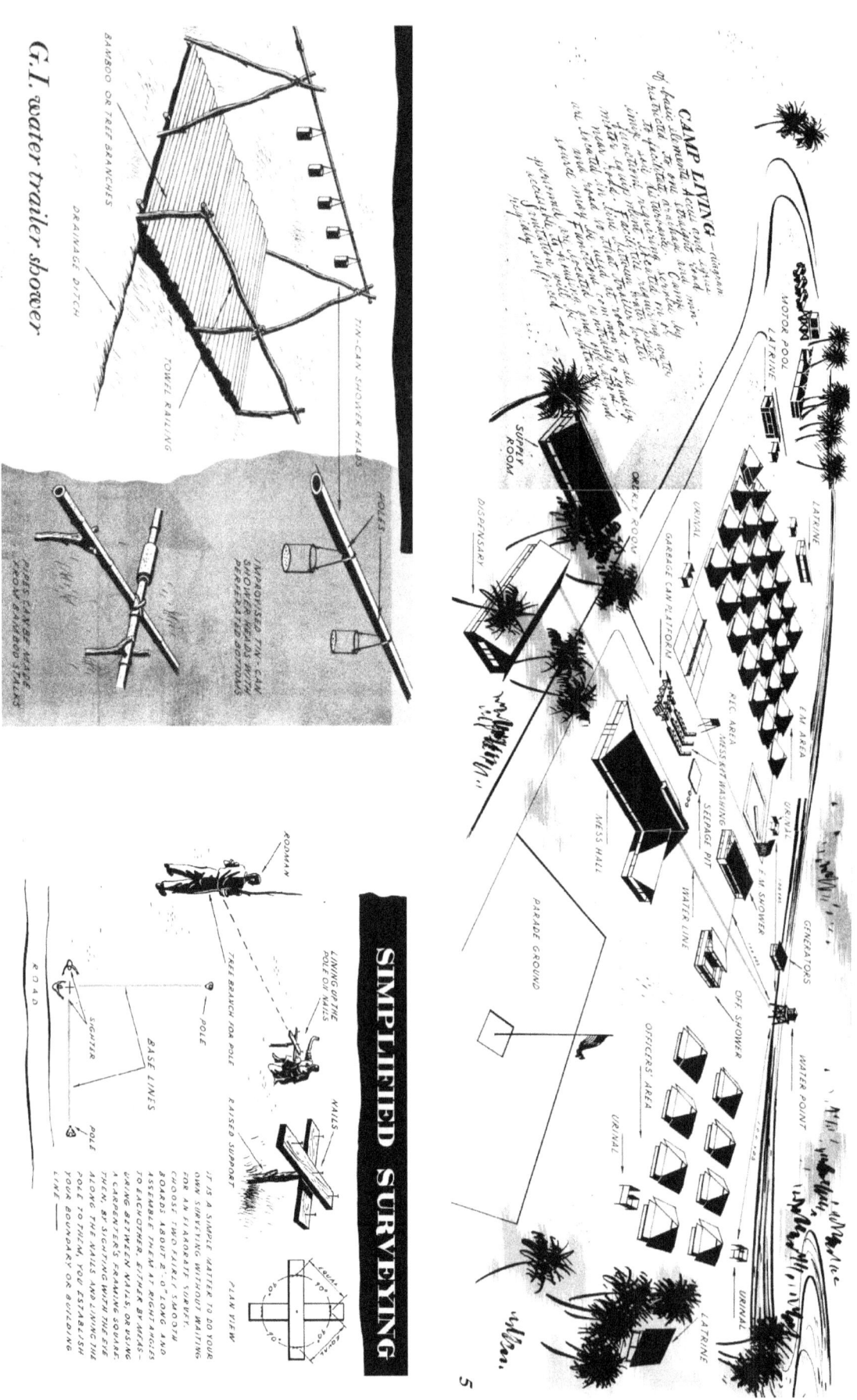

Camp Living: The Thirteenth Air Force issued a booklet to advise crewmen how to lay out effective and orderly camp facilities. This included simple field surveying and how to construct a field shower for local materials. *Camp Living*

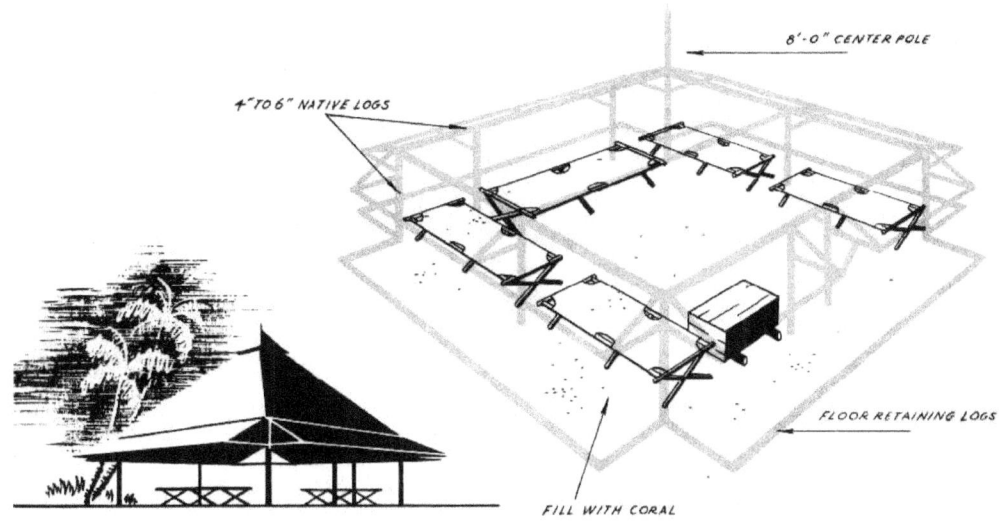

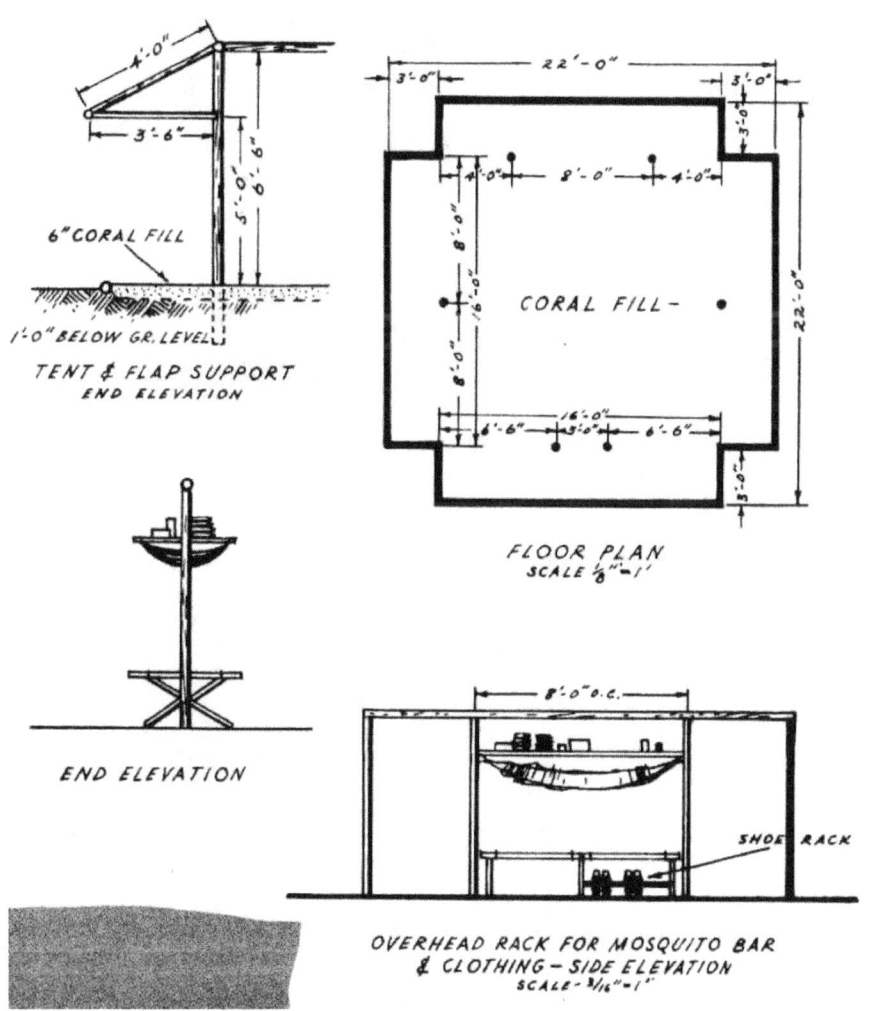

Camp Living: A Thirteenth Air Force booklet advised crewmen how to construct a comfortable tent frame.
Camp Living

Walt on Morotai: Walt relaxes outside his tent.

Crewmen on Morotai: Walt and Jupe Bretherton (above) and Walt and Jack Riley (below).
Bretherton Family Collection

Crewmen on Morotai: Walt (front row, left) with very informal crewmates posed in front of A/C 276. Walt flew at least six missions from Morotai in this machine in November, 1944.

Crewmen on Morotai: Cliff Llewellyn near the crew's bomb-dugout.

Morotai Base Facilities: The Long Ranger's HQ area was well marked.

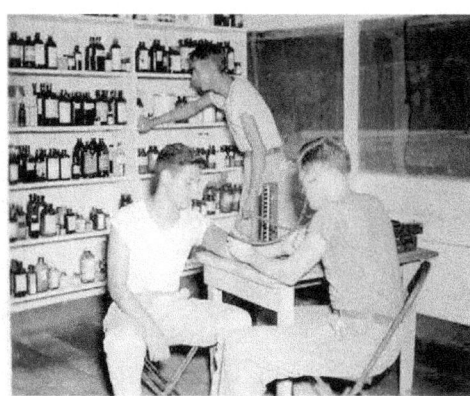

Morotai Base Facilities: Long Ranger's Dispensary. *Missing Aircrew Project*

Morotai Base Facilities: 424th Squadron Intelligence Tent and staff after mission debriefing.

Morotai Base Facilities: 424th Squadron Mailroom (left) and Orderly Room (right). *Missing Aircrew Project*

Morotai Base Facilities: Awarding medals. *307th Bomb Group Association.*

Morotai Base Facilities: 424th Squadron EM Latrines (above) and movie theater (below). *307th Bomb Group Association.*

Keeping the Planes Flying: 50 hour maintenance work on a Liberator in the hot tropical sun.

Keeping the Planes Flying: Walt, Jupe, and crewmates clean their guns after a mission in the shade of a wing (above, right, and below).

Keeping the Planes Flying: Armorers (right and below right) worked through the night loading bombs and ammunition, while other men unloaded an endless supply of 55 gallon drums with aviation fuel to feed the Long Ranger B-24s (below left).
307th Bomb Group Association

Combat Tempo: Each Long Ranger mission began with planning by Operations (left) based on targets selected by XIII Bomber Command. The following evening or morning crews were briefed with detailed orders (below).
307th Bomb Group Association

Combat Tempo: In early morning the bombers would be signaled one by one to begin moving from their revetments (above) and fall in line (right) waiting for the signal from the control tower to begin their take-off run down Pitoe airstrip (below).
307th Bomb Group Association

Combat Tempo: Once in the air each gunner test-fired his guns as the aircraft proceeded to the IP via pre-determined route, which might pass over high volcanic islands. *Left, 307th Bomb Group Association*

Combat Tempo: The ball turret gunners had the best view as they flew west or north towards the target. But each crewman always scanned below for ships or other signs of life. They were always happy to sight other Long Ranger bombers off in the distance.

SUGAR BAKER
1500/2500 G.T.

1. Hatches, no catwalk
2. Bridge amidships
    Speed, N.C., 10
    Speed, A.M., 13
Two exceptions, three goalposts-4500 G.T.

SUGAR CHARLIE
300/700 G.T.

1. No bridge amidships, only a mast
2. Two hatches
    Speed, N.C., 10

SUGAR CHARLIE ABLE
70/150 G.T.

1. No bridge amidships, only a mast
2. One hatch
    Speed, N.C., 10

Combat Tempo: Any shipping sighted would be reported using a standardized identification nomenclature.
Photographic Interpretation Handbook

Combat Tempo: More friendly aircraft would appear and then at the IP, form up in the designated bombing formation as the target approached. *Below, 307th Bomb Group Association*

Combat Tempo: The sky looks clear and unthreatening (left) but experienced crews knew there would be long minutes of extreme danger as the target approached. The flak begins as the bombardiers salvo their bomb loads (below). *Below, 307th Bomb Group Association.*

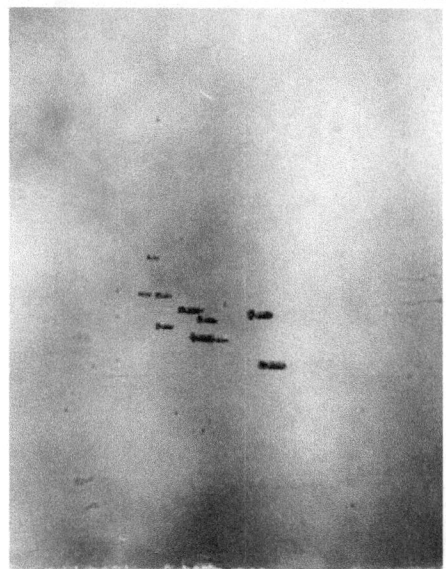

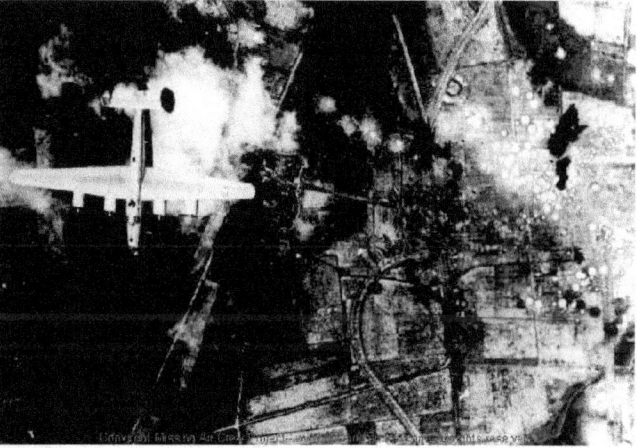

Combat Tempo: Walt would see 'bombs away' many times (above), then watch the bomb pattern 'walk' across the target (right). *Right, 307th Bomb Group Association.*

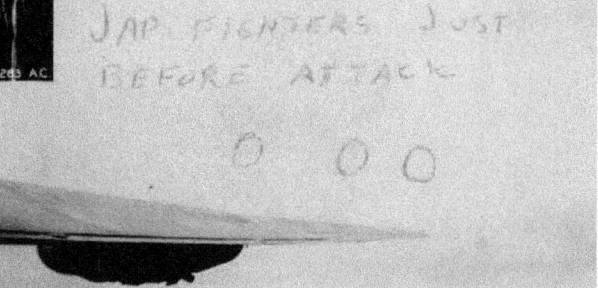

Combat Tempo: Jap fighter attacks often began with phosphorus bombs dropped from above the bombers (above), then the fighters would turn in to make their attacks (right). *Above, 307th Bomb Group Association Right, Bretherton Family*

Danger on Morotai: Japanese pilots were brave and persistent. They took a steady toll in lives and equipment for almost the entire time Walt was based there.
Above and right, 307th Bomb Group Association

Danger on Morotai: A scant five miles from Pitoe, the second Battle of Morotai began on December 26, 1944 after a two-day but ineffective artillery bombardment. Here a squadron of the U.S. 33rd Division advances cautiously through the dense jungle.

Danger on Morotai: In addition to the danger of bombing raids, a system malfunction or a moment of pilot error could lead to disaster on take-off or landing.
Missing Aircrew Project

Danger on Morotai: Aircraft damaged beyond repair were quickly stripped to provide spares to keep other machines flying.

Neighbors on Morotai: Some 12,000 natives inhabited Morotai in 1944-1945, one quarter Christian, one quarter Moslem, and half pagan. Almost all lived along the coast.

Neighbors on Morotai: Elderly natives weaving mats learned to smoke pipes from the Dutch and were happy to barter for American tabaco.

Neighbors on Morotai: Two young Morotai men you would not want to meet walking down a dark jungle path at night (left), and two children who surely melted the hearts of GI's who offered them candy and chewing gum (below).

Neighbors on Morotai: Idyllic scenes of Morotai's native communities provided contrast to the battles being waged nearby by young American, Aussie, and Japanese men, far from their homes.

Relaxing on Morotai: Pickup baseball game.

Relaxing on Morotai: Walt and Frank Mansir navigate the Moluccas Sea with the Gila Peninsula in the background.

Relaxing on Morotai: 424th BS Crewmen Walt flew with Elliot Lott (left) and Dick Roth (center holding lizard)

Relaxing on Morotai: 424th BS Crewmen enjoy Christmas Day. *307th Bomb Squadron Association*

Relaxing on Morotai: Richard Cosgrove with his dog.

Relaxing on Morotai: Watching a movie in the rain as searchlights seek a Jap raider. *Australian Women's Weekly*

Relaxing on Morotai: Above: Walt (left) and two friends relaxing on top of their bomb dugout. Right: Walt tending a fire at a camp along the beach.

Walt got a good look at Tarakan. The little A/A fire that there was seemed to be coming from a beach at the south end of the island. He also noticed the airfield on Tarakan, which had two runways, seemingly in good condition. He saw three enemy aircraft, including two two-engine types, parked along the runway. At about 1310, the observers sighted three little Japanese 'Fox Tare Dog' freighters on a heading of 350°. For a half an hour after leaving the target, Walt reported that he could still see the smoke billowing above the refinery complex.

The route back to Morotai was direct, with continued good weather, although over the islands there was usually a buildup of clouds and slight turbulence over the mountains. The men kept alert reporting any minor shipping or unusual sighting observed below. At about 1645, Appling put A/C 276 down on Pitoe airstrip after 10:15 in the air and 1,485 miles flown. This mission was rated as 'excellent' with 31,500 pounds of bombs dropped on target by the 424th BG alone.[36] Walt now had 23 completed combat missions under his belt.

The following day, November 19, while the Appling crew rested, the 424th returned to attack Alicante. 26 planes from the 307th BG were on this mission. There was no A/A fire over the target, but they were attacked by about 15 to 20 Japanese fighters: a variety of Zekes, Hemps, Jakes, and Tonys. The men of the 424th encountered a new tactic. The Japs would drop strips of metal, about two inches wide, half an inch thick, and two feet long, above and in front of the bomber formation. These would then float down and many hit the approaching planes. Many planes returned to Morotai later with some of these strips embedded. The Jap fighters were very aggressive. One Zeke flew directly into a B-24, embedding itself between the number one and two engines of the unlucky Liberator. 424th crewmen looked on in horror as both planes descended to the left locked in a death grip as four chutes were seen to emerge from the bomber. Lt. Ihne's A/C 276 was hit by fighter fire. The nose gunner, Sgt. Bryce, and the bombardier, Lt. Finney were both hit. Charlie Dowdy replaced Bryce in the nose turret as other men tended to Bryce and Finney. Unfortunately, Finney died before they returned to Morotai but Bryce survived. It was tough flying the damaged A/C 276 back to Morotai through a rough storm, but Ihne managed it. On the ground, Finney's body was removed by the Graves Registration Unit and Bryce was taken away in an ambulance. Sgt. Robert Nugent, the 424th ground crew chief came up and looked over the damaged A/C 276. One of the metal strips dropped by the Jap fighters was found embedded in the right wing. They also counted 117 bullet holes in the plane![37]

That same day there was a small reshuffling of crewmen. Cliff Llewellyn, Walt's old friend from Muroc and quite a few earlier missions with the Appling crew, had returned from his two-week R&R to Australia on November 3 and then made the move from Noemfoor to Morotai on the 10th. As a radio operator, he had been very busy helping to ferry 307th BG crews and equipment to the new base, as well as his share of combat missions with other crews. He was a little behind Walt, Frank, Jupe, and Jack with only 19 combat missions. Gene Schreiner and Boris Hidalgo had both fit in well with the other crewmen, but Cliff was like family. After a little negotiating and the OK of both pilots, Gene agreed to switch to the nearby Raue crew, with a couple men that he knew from before. Walt, Jack, Jupe, and Frank asked Cliff if he had heard anything about his brother Lorian, missing in action many months before, but he had no news. They helped him settle into the nice tent that they all shared and pass the day reminiscing, talking rumors, and sharing news from home.[38]

The following day, Tuesday, November 21, the Appling crew was assigned another early morning mission. At the briefing, the men learned they would fly in A/C 180 with 'Mac' McMillan in command of the six 424th Baker-24's assigned. The target was another Japanese airstrip – Bacolod - on Negros Island in the Philippines. The goal was to keep pressure on the Jap Airforce, to aid MacArthur as his invasion force battled its way onto Leyte. Each plane carried 24 – 260 pound fragmentation bombs, instantaneously fused. The 424th would be joined by the other three Long Ranger squadrons.

Take off was later than usual – Carl Appling had A/C 180 in the air at about 0740. Then he headed north into scattered clouds. The gunners test fired their guns, and then Walt settled in for the long flight. The route took them over Mindanao Island. As they reached the north end at Macajalar Bay, a small Japanese 'Fox Tare Dog' freighter was sighted in harbor at the mouth of a river.[39]

However, as Appling proceeded north past Togolo Point to the IP at 10°34'N, 122°57'E, the clouds thickened into turbulent frontal conditions. Assembling any number of planes in thick weather like this would be difficult and dangerous. Appling tried flying southwest into the Sulu Sea and then back around to the north, to look for a break in the clouds that could be used to assemble the group, but no luck. Then over the radio came a message from McMillan in A/C 619 with orders to proceed instead to the fourth designated target. This alternate target was Lumbia Aerodrome on the north side of Mindanao Island. Several planes had noted better weather conditions there and this seemed like a better plan than dropping their bombs blind.

At about 1257, as they returned south and passed Togolo Point again, Walt and others noticed three parachutes in the air just north of the point, with two planes circling in the vicinity. McMillan kept the six planes going and soon they arrived at the fourth target IP at 10°34'N, 122°57'E. The cloud cover was about 40% cumulus with tops at 5,000 feet and unlimited visibility. The six 424th BS liberators formed into the group box formation at between 7,800 and 8,400 feet on a bearing of 210°-220°. No enemy fighters were seen, but Walt and the other gunners remained alert. They began their bomb run at 1312. No A/A fire was encountered as the bombardiers sighted on a clump of trees in the storage area, east of the center of the runway. The bombs fell away striking the ground in a train of over one mile. Walt and the other observers saw the explosions cover the runway and then destroy two large buildings in the personnel area.

After the attack, as the planes headed south, two of them dropped propaganda leaflets over Cagayan Town. As A/C 180 headed over the island on a beeline back to Morotai, they passed over Del Monte A/D, a dozen or so miles south of Lumbia. Walt and the other observers could see through the clouds that there were two or three twin-engine planes parked on the ground. Other observations that would be passed on during the debriefing were a radio station on the beach of Mindanao, at 06°30'N, 124°02'E, and a possible fuel tank farm at 05°50'N, 125°00'E, near Sarangany Bay. At about 1810, after ten and a half hours and 1,014 miles in the air, Appling set A/C 180 back down safely at Pitoe. Walt had completed his 24th mission. Later, at the mission debriefing, Lt. Shields would rate the results as 'excellent.' The experienced pilots and crews had responded well to the heavy weather over the primary target and achieved good results.[40]

There had been at least one Jap air raid every night since Walt arrived on Morotai. However, on Wednesday, November 22, as the Appling crew was trying to rest before their mission the following day, the Japanese launched their most successful air raid on Morotai. They attacked with nine planes at night. Across the Allied base area, 15 airplanes were destroyed on the ground and eight damaged.[41] The Australians were hard hit, losing nine of their Boston bombers at Wama airfield. Three men were killed and 11 wounded on the ground and just one Japanese bomber was shot down. There had also been frequent sounds at night, far offshore, as the Navy PT boats patrolled to intercept reinforcements from Halmahera for the Japanese on Morotai.

The entire Thirteenth Air Force got some bad news on November 22. Earlier that day the 347th Fighter Squadron had sent 23 P-38's to attack Mandai airfield and to attack an escorted convoy. The attack was successful, with planes destroyed on the ground and four small ships sunk. However, three P-38's were shot down by A/A fire. Lt. Col. Robert E. Westbrook, the 13th Air Force's top ace, credited with 20 enemy planes destroyed, piloted one of the three P-38's lost.[42]

Later that same evening, all hell broke loose across the island as the men were watching their evening movie. At about 2000, a single Japanese plane – with its lights on - swept over Pitoe airstrip dropping bombs. Finally, the American and Aussie A/A fire got a kill as the plane went down in flames. But then Japanese planes kept on coming in ones and twos – at least a dozen in all, bombing and strafing the airstrips, dispersal areas, and camps. One of the Jap planes was claimed as a probable kill. However, it was a devastating attack. Three more men were killed and eight wounded. At Pitoe, four B-24s were destroyed and seven damaged. Three B-25s were destroyed and a further nine damaged. One Lockheed Ventura was destroyed and another damaged. At Wama airstrip, seven A-20s were destroyed and six damaged. A total of 15 aircraft destroyed and 23 damaged.[43] It seemed to Walt that maybe being in his ball turret in a B-24 flying over a distant Japanese held island was safer than spending the night on Morotai!

The next day was Thanksgiving. However, before any Thanksgiving dinner that the 424th cooks could arrange, the Appling crew was tasked with another mission. They had little rest because of the air raids the night before. Nevertheless, early on Thursday morning, November 23, Walt, Cliff, and the other crewmen rolled out of their cots, ate, and attended the pre-mission briefing. Their target was Bacolod, a major Japanese Air Field on Negros Island. Mac McMillan would lead six 424th Bomb Squadron Liberators in A/C 619, with the Appling crew in A/C 276. Each plane had been loaded with six 1000-pound bombs. Take off from Pitoe was at about 0645. The men settled into their normal routine after test firing their guns. Walt in his ball turret watched the clouds and ocean pass by, scanning for anything unusual to report. Their route took them to Flecha Point to Tagolo Point.

At Tagolo, an American airstrip was under construction and Walt saw two P-38s on the ground beside the strip.

The weather north gradually thickened until they encountered full overcast but no real turbulence as they approached their turning point off the west coast of Negros at 10°16'N, 122°50'E. From there it was a short flight to the IP, inland and to the south of Bacolod. Then the clouds opened up with 30% cumulus tops at 6,000 feet and unlimited visibility.

McMillan formed up the six planes at between 10,400 and 10,600 feet in a squadron javelin formation, on a heading of 0° to 10°. About five miles from the target, the bombardiers opened the bomb bay doors. Then Walt was startled when he saw the bombs from A/C 958 piloted by Lt. Rodgers drop away. There had been a malfunction that caused their bombs to drop as soon as the door was opened. Walt saw them drop harmlessly into an open field. Rodgers kept his plane in formation because he also carried one of the combat photographers. At about 1212, the attack began. Again, Walt and the other nervous gunners scanned the sky for Jap defenders, but none was seen. Closer they came and no A/A fire was encountered. The attack formation had the six planes in parallel, with an aiming target on the east side of the runway, about 300 feet from the southern extremity. At bombs away, Walt felt A/C 276 lurch as he watched the fall of the bombs. The five parallel strings of bombs fell across the southern end of the airstrip and into the dispersal area. The airstrip sustained several direct hits. As they passed the airstrip and made their turn back to the south, Walt noted several barges in Bacolod harbor.

The weather on the return flight was essentially the same and there was nothing of note observed. At about 1715, with a little remaining late afternoon light, Appling eased A/C 276 back down at Pitoe after ten and a half hours and 1,393 miles in the air. Walt had completed combat mission 25. At the post-mission briefing, the results of the day were again rated as 'excellent' by 424th Squadron Intelligence Officer Capt. Edward G. Ginnane.[44]

That evening, the men lined up outside the 424th BS mess tent for their special Thanksgiving dinner. The cooks did what they could with some turkeys flown in from Australia and the best dressing they could cook up from leftover bread. It was not an elaborate meal, but at least a little change from the normal routine. Walt and Cliff certainly wondered what Thanksgiving dinner was like that day back home. The men all missed their families and friends that night, but they recognized that they had a job to do. A year earlier Walt had celebrated Thanksgiving at Laredo Air Force Base in Texas. He hoped that he would be able to celebrate next year at home.

During Thanksgiving dinner, the men talked about the increased frequency of combat missions. It was strange that they were encountering little or no fighter or A/A defense over their targets in the Philippines, because the Japanese were making an increased effort to attack Morotai itself. For the week after Thanksgiving nine B-24's were lost to nighttime bombing raids on Morotai. The Japs appeared to be bringing in planes by night from Ceram, Celebes, and Mindanao to nearby Halmahera Island in various dispersal airstrips, from which to launch these raids.[45] Most of the raids concentrated on Pitoe and Wama, about eight miles away from the 307th BG camp area on the Gila Peninsula.[46]

However, one night a few days after Thanksgiving, Walt, Cliff, and the others were watching a movie in the early evening when the air raid siren sounded. They all ran to their air raid dugout and waited, but no Japs came. The men drifted back to the movie theater and started watching again, when the air raid siren sounded a second time. Again, they ran to their dugout, but still no Japs. Once again, the men drifted back and the film began again. A third time the air raid warning sounded, but this time most of the men stayed where they were. About five minutes later two Jap fighters came down low and started strafing the camp area. All hell broke loose as men started scrambling in every direction. Walt, Cliff, and some others found a nearby truck to dive under. For an hour and a half, nine Jap raiders bombed and strafed the base. None of the 424th men was hit, but there were some injuries and the mess hall area was shot up. At Wama, a B-25 and an OA-10 were destroyed. Despite heavy A/A fire and six P-61 night fighters, only one for the enemy planes was destroyed. Something had to be done about these frequent nighttime raids![47]

The Appling crew was called for yet another mission, their fourth in eight days, on Saturday, November 25. Just after midnight though, a single Jap plane broke the tropical quiet as it strafed and dropped a string of anti-personnel bombs. An hour and a half later, another Jap plane attacked, dropping three bombs and machine-gunning with no effect...other than more lost sleep.[48] After the all-clear, the men tried to get a little sleep before their early wake-up. Walt and the others learned that they would be returning to Fabrica Airfield on Negros, which they had bombed back on the 13th.

The Appling crew remained unchanged. They were assigned to A/C 318, loaded with eight 1,000-pound bombs.

Appling lifted their plane into the air at about 0640 and set a course north. The weather began very well, with widely scattered cumulus. Their route took them to Tabolo Point and then to Tuburan Bay, off the west coast of Cebu Island. From there the cloud cover became increasingly dense. Ed Dunne navigated Appling west to an assembly and turning point at Malocaboc Island, off the north end of Negros. With the six planes formed up trailing McMillan, the Squadron leader in A/C 619, they made their way to the nearby IP just off Suyac Island at 10°57'N, 123°27'E. But from here south to Fabrica, the ground was obscured by thick clouds. The six planes from the 424th were in a loose group formation with planes from the other Long Ranger squadrons. A squadron of P-38s also appeared to provide the bombers with fighter protection.[49] However, the target could not be seen through the clouds and so for 45 minutes they circled, looking for a break in the weather. They passed over nearby Alicante airfield, where the Japanese put up some slight, medium, but inaccurate A/A fire. They were all glad to get back into the clouds then. Finally, making another pass at about 10,100 feet and on a heading of 240°-250°, there was a break in the clouds over Fabrica at about 1138. McMillan radioed the other planes telling them to drop their bombs if they had a clear view of the target. With the clouds clearing just a little, Walt tensed up looking for enemy fighters. At 1140, Jimmie Clark yelled that he saw the target and toggled their bomb load away. About this time, there were ten bursts of A/A fire around them, but no damage.[50] With bombs away, Boris Hidalgo also tossed a package of propaganda leaflets out the window. Walt saw that each of the four planes following Appling dropped their loads, but that A/C 619 up ahead did not. Walt saw five seemingly operable aircraft on the ground at Fabrica, but then the clouds closed in again, obscuring the results of their bombing.

McMillan in A/C 619 announced that he was going to take the Squadron to the secondary target at Seravia A/D, a few miles southwest of Fabrica. There must have been some grumbling about having to do another bomb run as the other five planes followed A/C 619. Then McMillan found an opening in the clouds at about 1203 at 10,000 feet elevation and dropped his eight one-thousand pound bombs. Fortunately, there was no A/A fire at Seravia and still no enemy fighters to be seen, largely thanks to the thick cloud cover. The six bombers then spread out and set their course back to Morotai. As they passed over Bizal A/D on Negros, Walt and the other BT gunners saw 15 apparently operative aircraft near the airstrip. About a half an hour later, after skirting the Bohol Sea and flying up Ilagan Bay to Mindanao, Walt sighted two possible signal fires burning below near the northern end of Lake Linao, at 08°N,124°17'E. The rest of the flight back to Morotai was uneventful, as Appling set their Liberator down at Pitoe at about 1625 after 9 hours, 45 minutes, and 1,367 miles of flying. This was Walt's 26th combat mission. Later, after their debriefing, Capt. Edmund Ginnane, 424th Intelligence Officer, would record the results of the mission as 'unobserved, due to almost full undercast.'[51]

That evening, just at dusk, a high-altitude Jap reconnaissance plane flew over Morotai but dropped no bombs. The next night three Jap raiders were back, first bombing Pitoe and strafing the nearby tent areas. Two B-24s were destroyed and others slightly damaged. The A/A fire seemed meager and the P-61 night fighters had no success bringing down any of the planes.

The Appling crew was in a fast-paced routine now. Walt flew his first combat mission, from Guadalcanal to Rabaul, back on May 20. From Guadalcanal, over the next five months he had progressed to Los Negros, Wakde, and (after his two weeks of R&R in Australia) Noemfoor, flying an average of just four combat missions per month. However, since arriving on Morotai, Walt and the rest of the crew had flown a mission every two or three days. That pattern continued. After a day of rest, the Appling crew was called for another mission on Monday, November 27. Just as they awoke, at 0430, two Jap planes scattered clusters of small bombs around the runway, but caused no damage. Again there was moderate A/A fire and the P-61 night fighters scrambled, but with no luck.[52]

At their briefing there was no surprise that they were headed back to Negros Island to help knock out Jap airfields that threatened MacArthur's advance on Leyte. Walt learned that he and the Appling crew would be assigned to A/C 180 and they would be joined by Sgt. Walker as combat photographer. The target was Malogo A/D in the flat plain on the northwest coast of Negros. Each of the six 424th Squadron B-24s assigned was loaded with five 1,000-pound GP bombs. One challenge in scheduling missions from Morotai was the somewhat predictable arrival time over targets in the Philippines. Daylight near the Equator was around 12 hours every day of the year. For maximum safety, it was preferred to take off near dawn and return before dusk. A nighttime landing would require the lights at Pitoe to be turned on, which would aid Japanese night attacks. Consequently, the Japanese could almost always anticipate when the heavy bombers would arrive over their airfields on Negros Island. This would allow them to prepare their defenses and ensure that their aircraft were in the air then, either attacking MacArthur's ground forces or ready to pounce on the heavy bombers.

The mission hoped to achieve an earlier than normal arrival to surprise the Japs on the ground unaware. This meant an earlier than normal departure from Pitoe. At about 0545, Appling lifted A/C 180 off the ground as the six 424th Squadron bombers began their flight north. Walt and the other gunners got into position and test fired their guns. The weather was clear to the Talaud Islands, mid-way between Morotai and Mindanao. Then they passed through a mild weather front before emerging again into clear weather as they approached Mindanao. They made their way to the turning point at 10°08'N, 122°20'E, out in the Panay Gulf. Then they formed up with the other bombers and headed northeast to the IP at 10°45'N, 122°54'E, off Talisay in the Guimaras Strait. At the IP, a squadron of P-47's arrived to provide them with welcome fighter cover.

The weather was good, with widely scattered clouds and unlimited visibility, as they approached the target in group box formation at 11,000 feet on a heading of 55°. Approached the target at about 1059, A/A fire began bursting around them at their elevation and course over the target. At least their escorting squadron of P-47s would keep any fighter attacks at bay, but the flak rattled all of the planes. McMillan in A/C 276 was in the number one position, followed by Appling in A/C 180 and then the other machines behind. Within three minutes their loads of bombs were dropped and they turned away from the target. Walt noted that the A/A fire was coming from one position at the southeast corner of the runway and from two more positions at the extreme north end. Then Lt. White in A/C 318 reported on the radio that they were hit. A piece of flak had gone through the nose turret and another piece caused a three-inch hole in the right wing behind the #3 engine.

Their early arrival had caught 11 Japanese aircraft on the ground: two in a dispersal area near the southern end, three along a taxiway running east-west from the southern end, and six along the eastern edge of the strip, south of its center. Walt watched as the thousand-pound bomb detonations began about 200 feet west of the central edge of the airstrip, then walked for about 700 feet northeast across the runway and into the northeast dispersal area. At least 13 to 15 hits were on the runway itself. As they made their withdrawal, Lt. White reported that A/C 318 was leaking hydraulic fluid, but otherwise seemed to be OK.[53]

As they headed south, they passed Silay airstrip, inland from Bacolod, where Walt saw a large oil fire, with smoke billowing up to about 3000 feet. Ed Dunne gave Carl a heading for a direct flight back to Morotai which took them south across Negros and then Cebu Island. They crossed the Bohol Sea towards Mindanao with generally good weather and widely scattered cumulus. As they approached the southern end on Mindanao, Walt reported seeing two distinctive signal fires together. Ed noted their position at 06°15'N, 125°05'E. After the long crossing of the Molucca Sea, the now-familiar sight of Morotai and the Gila Peninsula came into view. Boris helped Walt back up into the fuselage and out of his turret and Carl piloted A/C 180 down for a landing on Pitoe Airstrip at about 1545. After ten hours and 1,393 miles in the air, Walt completed his 27th combat mission. After their debriefing, Lt. Shields, 424th BS Intelligence Officer would rate this mission as 'Excellent.'[54]

Tuesday was a day of rest for the Appling crew, but they learned that once again they were assigned to combat the following day. The air raid gun went off again that night at 0100. Out at Wama airstrip a small Japanese plane came in behind a B-25 that was making a late night landing. The Jap plane dropped a cluster of small bombs, but there was no damage. The A/A guns opened up on the plane as it departed but without success.[55]

At their briefing early on Wednesday, November 29, they learned that the Long Rangers had been assigned a new target: Puerto Princesa airfield on Palawan Island. Japanese aircraft based at Puerto Princesa did not directly threaten MacArthur's troops on Leyte, but they provided fighter cover for Jap shipping from Borneo making its way up the South China Sea to Manila and north to Formosa and Japan itself. The 424th would contribute six aircraft for the mission, led again by Mac McMillan in A/C 619, with the Appling crew in the number two position flying in the squadron namesake B-24, A/C 424. Each plane was loaded with 12 500-pound GP bombs. They learned that a total of 56 B-24's from the 307th and 5th Bomb Groups would join together on the attack and three squadrons of P-38's would rendezvous with them to provide escort over the target.[56]

Pitoe was noisy in the early morning. Certainly, there were Japanese observers just beyond the perimeter who could hear and see the 13th Air Force activity and radio a general warning to their forces on nearby Halmahera and the more distant Philippines. At about 0720, Appling piloted A/C 424 into the air up to cruising altitude and set a course to the northwest. The crew remained unchanged, with Don Aubrey at his side in the copilot seat, Ed Dunne and Jimmie Clark as navigator and bombardier, with Cliff as radio operator, Jupe Bretherton as flight engineer, and Boris, Jack, Frank as the other gunners with Walt.

Their route took them to Olutanga Point, deep in the Moro Gulf at 07°17'N, 122°45'E and then across the Zamboanga Peninsula. The weather was fair, with 50% stratus cover at 14,000 feet and 40% cumulus at 3000 feet. Over Mindanao, the cloud cover thickened up. This routing over Mindanao was not unusual and would likely lead any Japanese observers to report another bombing raid to Negros Island. But once across the island they did not turn north, but kept on their generally northwest heading to North Islet[57], a remote reef in the middle of the Sulu Sea. The weather conditions now changed, as they penetrated a moderate weather front with slight turbulence, which would persist to the target. Walt was having a rough ride in his glass ball slung beneath A/C 424. From North Islet there was a slight course change taking them to the assembly area and turning point at 10°01'N, 119°21'E. This was about eight miles southeast of two small but distinctive islands, South Verde and North Verde, off the east coast of Palawan. Once the 424th and the other bomber squadrons had formed up with the three P-38 squadrons, they headed off to the IP at 09°56'N, 118°51'E, just off Barlas Island. The cloud cover increased to 80% cumulus at 8,000 feet. They were flying in at about 10,000 feet and based on the weather over Puerto Princesa, the group commander decided to approach the target from a reciprocal heading from the original plan. After maneuvering the group first to the south, they then approached on a heading of 26° with the 424th bombers following 'squadrons in trail' at between 9,600 and 10,100 feet elevation. The gunners, ever alert saw no enemy fighters, thanks to either the element of surprise or the protection from the P-38 escorts. Then the clouds opened up over the target at about 1244. They approached the target over the little harbor of Puerto Princesa, where Walt saw two little 'Sugar Charlie' class freighters of 300-700 tons, as well as a barge. Appling turned A/C 424 over to Jimmie Clark who engaged the AFCE to control the plane for the bomb run. Already the squadron ahead had begun dropping its loads, but there was still no A/A fire from the ground. It was just a couple minutes for bombs away and Carl Appling had the controls again to turn away from the target.

Walt as usual had a great view of the results. Walt saw five twin-engine and six single-engine planes along the western half of the airstrip, on both the north and south edges. Their bombs started detonating just north of a road running east-west south of the runway. The bombs walked north for 1100 feet, across and up the airstrip. Two-thirds of the western half of the airstrip was covered by the bombs, with at least 13-15 direct hits on the concrete runway. Walt let out a cheer as he saw a bomb hit and set afire one of the twin-engine planes on the north edge of the runway, then a single engine plane nearby had its tail blown off, and finally another twin-engine plane on the south side was hit by a bomb and completely destroyed. Some of the bombs also targeted another ungraded strip nearby. Six single-engine planes seen parked there were not hit, but both airstrips were left in an unusable condition.

After a half hour, the P-38 escorts departed as the Liberators made their way back in a loose group on a reverse course. The weather was about the same as earlier in the morning and the five and a half hour return flight was uneventful. Nothing but clouds and ocean, scattered little coral islands and the jungle covered Zamboanga Peninsula down below for Walt to gaze at and ponder his future. At about 1825, Lt. Appling eased A/C 424 back down onto Pitoe airstrip. After 11 hours and five minutes and 1,691 miles in the air, Walt completed his 28th combat mission.[58]

The following night, at 0310 on November 30, the Japanese raided Morotai again. Four enemy planes made the attack, bombing Pitoe, Wama, and the camp areas. They destroyed two PV-1 Ventura patrol planes at Pitoe and two P-47 fighters at Wama. There were no injuries though, and the P-61's managed to shoot down one of the Jap planes.[59]

MacArthur had made good progress against fierce resistance since his landing on Leyte Island on October 20. The Fifth and the Thirteenth Air Forces were doing their part to provide support to the invasion. The Japanese were trying to get reinforcements and supplies to the defenders on Leyte, but the incessant air raids from the big bombers from the south made this difficult for them. Their easiest source for fresh troops and material was via Samar Island to the north. In early December, the Long Rangers would be doing their part to support MacArthur's invasion of Samar.

The 424th Bomb Squadron and the entire 307th Bomb Group (HV) was getting a reputation as a very effective and efficient fighting force. At the end of November, 424th Bomb Squadron personnel had increased to 96 officers and 142 enlisted men in the flying group and 15 officers with 403 enlisted men in the ground support group for a total strength of 656.[60] The four squadrons of the 307th totaled about 3000 men plus about 100-200 in the HQ units.[61] The entire 13th Air Force probably totaled over 15,000 men. The 424th had completed the move of all air echelons from Noemfoor on November 8, and the last rear echelons arrived from Wakde on November 10 via the S.S. Owen Summers, one of thousands of Liberty ships that were streaming out of American shipyards. The 424th started the month with 12 B-24's.

Two 424th planes and their combat crews were lost during the month. On November 6, over Alicante, Walt saw A/C 961 piloted by Lt. Donald Balovich go down. Two days later, again at Alicante, A/C 381 piloted by Lt. Hunter with a crew of 11 went down. These two crews were not heard from again. Alicante would be bad luck for the 424th. On November 19, Lt. Curtis Finney, a bombardier and Cpl. H. E. Bryce were both wounded by enemy fire. Bryce and the navigator, Lt. Ihne tended to the wounded Bryce for hours, but he died less than an hour from Morotai.[62] The 424th more than made up for the loss of two planes, receiving four replacements during the month, bring their total to 14 of the Liberator heavy bombers. With these resources, the Squadron managed 22 missions, dropping 546,540 pounds of bombs on target.

During the month, the 424th developed very suitable camp facilities – judged the best they ever had. The camp was situated in a cultivated grove of coconut palm trees on the western shore of the Gila Peninsula, about three miles from Wama and Pitoe. The location on the shore provided frequent cooling ocean breezes. Semi-permanent buildings were erected for offices and the mess hall. Coral roads and walkways were laid out to help the appearance of the camp and to provide dry footing when there was rain. Flights crews were given lectures during the month on the geography of the Philippines and the location of emergency landing areas.[63] There were also two wooden rafts tethered offshore by the 307th that were used to instruct groups of men in the proper method of getting out of a parachute harness in the water, inflating their Mae West flotation jacket, and staying afloat. They were also taught how to get into a life raft. This training combined pleasant time in the water with useful knowledge. Most flight crew had seen planes go down, so everyone knew that these were practical skills that they might need to put into use someday.[64]

Life on Morotai

Life for the men of the 424th Bomb Squadron settled down to a routine on Morotai. Despite the frequent enemy air raids, this seemed like much more of a permanent base, more like Guadalcanal or Los Negros, and not at all like Wakde or Noemfoor. For a typical mission Walt and the other crewmen would be awakened early in the cool of the morning by the CQ rattling the crews' mess kits hanging from a hook at the entry to each tent. Crews at Morotai would assemble after breakfast in a nearby open area were they would board trucks for the five-mile ride to the airfield itself. At other bases (Wakde, Los Negros, Noemfoor) the airfields were closer and crews had been able to just walk to their waiting plane if they chose.[65]

Once on Morotai, Don Aubrey was taxiing in a B-24 lined up with others in preparation to take off for a mission, when the air raid sirens went off. All crews shut down their planes, got out and ran into ditches at the side of the runway...then there was a big explosion – the B-24 he had been in was hit by a bomb, exploded and destroyed. Later that night Tokyo Rose reported that the "Japanese Airforce hit Morotai today and destroyed two 307th Bomb Group B-24's!"[66]

After a typical mission, 'Red Cross Girls,' would meet the crews and have coffee and doughnuts waiting for the men after they had driven back to their camp. Then the men would have to meet with an intelligence officer who would 'debrief' them, to find out all the results of the mission, what they saw, and any other important combat details. Later, crewmen could report to the group medic who would issue each man two ounces of whiskey, supposedly to calm their nerves. Some men would drink the whiskey immediately, while others would save it to trade later for food from packages sent to their friends from home, for furniture to make their tents more comfortable, or other things of perceived value.[67]

Often the men would return to their plane for a little maintenance work and cleaning. Once Walt and Cliff's crew was out near the airfield, cleaning their guns the day after a mission. The gun parts were cleaned in gasoline. One of the Appling crewmen was smoking while doing this! Just by chance, a general came by and saw this man smoking near the gasoline. He ordered the man to attention, 'dressed him down', then the general ordered the entire crew to report early the next day before dawn for a 'refresher' safety film and instructions on proper maintenance procedures training.[68]

Back on Wakde, men of the 307th could hear the firing on the New Guinea mainland as the Army battled the surviving Japs in the jungle. Some men even went over to the mainland to try to get near the 'front lines' for some souvenir hunting, although this was discouraged. But on Morotai, the Japanese surrounded them. They still occupied 90% of the island and they would try to sneak men and supplies onto Morotai by barge from Halmahera many nights. Many nights Walt and the other men would hear U.S. Navy PT boats firing at and attacking small Jap craft trying to sneak over to Morotai. There were also frequent rumors of possible Jap mass attacks to try to drive the Americans and Aussies off Morotai. It was a very nervous situation and most men kept their guns handy.[69]

Then there were the air raids. Morotai was raided by the Japanese more frequently than any other base in the South Pacific. Each crew dug a deep bomb shelter near their tent and covered it with coconut tree logs and sand bags. Three shots from a 90mm anti-aircraft gun was the signal that a Jap raider had been sighted and then the men would make a mad dash for to the nearest hole. Even after the Jap planes left, the men would wait in the hole until after the shrapnel from the anti-aircraft bursts had all rained down on the base.[70]

Most of the air raids occurred at night, but that was when the men enjoyed Hollywood films shown in open-air theaters scattered around the island. Walt and Cliff's tent was located about six 'blocks' from the 307th BG open-air film theater. The theater was excellent so it sometimes attracted men from other units. Sometimes 307th BG men could not find a place to sit. So a fence was erected around the theater and a rule was made that only men with a 13th Air Force patch could enter. But most men did not have a shirt with a patch, so men would pin a patch to their shirt, then once inside pass it to their buddies on the outside who could then use it to enter. The men usually brought a poncho, so if it rained they could continue watching the film. When it rained and the movie was not very good, some bored men would collect rainwater in their poncho, then stand up and dump it down the neck of the man in front of them.[71] Mosquitoes and other pests were a continual problem. By now, Walt and Cliff were used to shaking out their shoes to make sure there were no scorpions before they put them on.[72]

The Appling enlisted crewmen built a covered foxhole - their 'bunker' in case of an air raid. It was covered with wood and logs, but was not very deep and probably would not have provided much protection in the case of a nearby bomb hit. However, it did provide a feeling of safety during a raid.[73]

On Morotai, the enlisted men on the Guild and Appling crews had their tents side by side and shared the nearby bomb shelter. One day Jack Riley got word that his wife had had a baby – so he was a father. All the men congratulated him and he spoke of his desire to get home safely and see his son. That same night the 'Rrrrrrr-Rrrrrrr-Rrrrrrr!' sound went off which all the men knew was an air raid. Men yelled 'turn out the lights!' Before going into their dugout, the crew would always try to send one guy into the shelter first to check for snakes or scorpions. That night it was Walt's turn and after a few seconds he said that it was all clear, so they all went in. But in the dark, for a joke, Walt 'walked his fingers' gently across Riley's back like a bug. Jack screamed and jumped out of the shelter, leaving the rest of the men laughing hysterically.[74]

There was an alternative to the free movies. On Morotai, there was a tent in the 370th BS enlisted men's tent area where big time gambling went on.[75] Sometimes these games lasted three days and wound up with just a couple men with most of the crews' pay.[76] Walt and a few of his crewmates liked to play cards. But Wally was always prudent. He would decide, "OK, I'm going to risk $5 tonight". If he lost the $5, he would stop. If he started winning, once he won $10, he would put $5 back in his pocket and not touch it. Then he would risk only his winnings as long as they lasted.

Most men had some money to burn on gambling, because there was not much else to spend your money on. There were several PX's on Morotai but they were never well stocked and open only limited hours. They sold items like toothpaste, soap, cigarettes, soda, and warm beer. The 424th Squadron Mess was one of the few semi-permanent structures: The food was usually based on spam, dehydrated eggs, potatoes, vegetables, and beans - and strong coffee. The squadron baker was very good – he made great sweet rolls that some men would remember many years later. Sometimes there would be fruit, vegetables, and fresh meat from Australia. If they were lucky the meet would be beef, but usually it was what they called 'strong mutton,'[77]

There were many funny incidents that would make the rounds for the amusement of the crews. Not everyone tolerated the food the same. Dick Reis, a 424th BS gunner on the 'Smokey' Guild crew, told Walt the story of man on his crew who went 'jungle crazy' on Morotai. This man was drinking far too much liquor and went to Guild with a tommy gun and said 'sir, I'd like permission to shoot the cook.' Guild had a great sense of humor and he did not want to argue with a deranged man with a tommy gun, so he said 'OK, but use short bursts to save ammunition.' The drunk man saluted Guild, turned, and left. After he left, Guild's co-pilot, Vern Michel said 'You know – I think he's going to do it –we should stop him!'[78]

One day, a number of completely bald enlisted men were seen walking around the 424th camp area. The story spread quickly. A ground crewman, Herb Keffler, appeared in his tent the afternoon before with nothing but a towel wrapped around his waist and a bar of soap. He announced that he was going to take a shower and then come back with a bottle of whiskey. His tent mates were certain that this was impossible and egged him on. One of his friends, D. D. McQueen, said that if Keffler came back with a bottle of whiskey, he would have his head shaved. All of their friends agreed that they would shave their heads too. McQueen also said that if Keffler came back with a bottle of whiskey he would draw a circle in the dirt floor and take a crap in it, and another man said that he would carry it out with his bare hands. Well, somehow, Keffler did come back from his shower with a bottle of whiskey…the crap was taken…the whiskey consumed….and eight men visited the barber to have their heads shaved.[79]

The Japanese disrupted the natives on Morotai when they constructed their airstrip back in 1942. However, relations between the natives, Americans, and Aussies were good. Men could buy things from the natives, like fish, native crafts, carvings, and bow and arrow sets. They would accept American dollars, Dutch guilders, or Aussie pounds, but they preferred to barter for things like cigarettes, liquor, canned food, and candy. One of the most prized trading items the Yanks had, in the eyes of natives, was a 'DuPont Spinner.' These were half sticks of DuPont brand dynamite the natives used for fishing. The lighted stick would be thrown into the water and the explosion would stun fish for many yards around. The natives could then easily scoop up the fish as they floated to the surface.[80]

Most of the 13th Air Force men on Morotai were not flying crewmen like Walt. Ground crews made up the bulk of the personnel. They included a few non-flying officers in each squadron. Very few of them got R&R to Australia and it was not clear when they would be rotated back to the States. Many of them had been in the South Pacific now for 18 months or more. Morale sometimes was bad for these men. They did mundane but necessary jobs like cooks and kitchen helpers and clerks and drivers. Others were mechanics or plane armorers or medical orderlies. Some were photo-interpreters or ground radio operators. They too each had a story.

For example, Cpl. Glenn Norwood served as an armorer. He served on a crew that would load the bombs on each plane after the mission plan had been completed. The armorers also loaded each plane with the correct amount of .50 cal. ammunition. The two waist guns were fed ammo from belts in cans at the gunners' feet. But the four turret gunners needed to have their ammo fed into intricate containers that were an integral part of the turret.

One time, one of Norwood's friends got a lister bag – a large rubber bag that held about 30 gallons of purified water to drink from. This guy got all the raisins that he could from the mess tent and put them in a lister bag. He added a few coconuts and canned fruit and then added some yeast that he got from the camp baker. He made an alcoholic beverage that Norwood swore would 'make your hair stand straight up.' Another time Norwood and his buddies built an oven out of rocks for a new baker. This new guy would then bake fresh bread every day.[81]

Another ground crewman was Sgt. Robert Robinson. He had graduated from Boston Latin School for high school then completed two years at Boston University Law School when the war broke out. He had been recruited to become an officer, and then for two years he had gone from army school to army school. Finally, because of his education he was enrolled in a radar school. Because radar was such a critical and secret weapon, Robinson was induced to remain an NCO. Eventually shipping out with the 307th BG to maintain and repair airborne radar sets.

Years later, Robinson would remember that month after month….

> "…that's what I did. We made them, we calibrated, we repaired them. We were over-trained for what we had to do. I spent more time with Q-tips than with any real tools. Q-tips right out of the hospital, well, because we used vacuum tubes and there were magnetrons, small ones as long as your little finger. And, in the tubes there were pins, well you get about four pins in a small little thin tube, the pins are very close together. Now the Japanese had mica. Their radar was not as good as ours but they did have mica, and their tubes were set in the mica, The platform for us was made out of pressed fiberboard which had been impregnated with a fungicide. Yet, because, you see you get fungus in the

humid conditions that we had there. And, they know they had to kill the fungus. The fungus would eat the fiberboard, that's edible stuff to them, I don't care for it myself. But the little fungi died in the tubes, right there and with the little bodies crowding up they shorted out between the pin sockets, they made a connection, I mean, it was like as if you wired the damned pins together and blew the tube. I tell you I spent more time cleaning those damned tubes, to make sure!"[82]

Another typical enlisted ground crew man was James V. Walsh from Texas. He had worked as a truck driver for a lumber company and then a steel barrel manufacturing company along the Gulf Coast of Louisiana and Texas. He had been inducted early in 1942, and because of his truck driving experience he was sent to the Pacific with little training. From 1943 into early 1945, he worked as a truck driver with the 307th. By 1944, he worked on a crew fueling planes to prepare them for the next day's mission. He remembered....

"We would fuel our airplanes up during the day which was probably for 4-6 hours. In the morning the mechanic who takes cares of the planes; he would come out in the early morning when they were getting ready to take off for a long bombing mission and crank the engines on the airplanes, when he turned them off we would come along and refuel the tanks with all the gas it would need to make that mission. And at night when it came back in we would do the same thing. I guess you could say we never were overworked. Nobody bothered us, it was just like doing a job, we would do the job and nobody would bother us."[83]

Recreation for the men included volleyball and basketball courts, and of course swimming and boating from the beach. However, there was only one baseball field on Morotai. There were many organized teams made up of the best players from each Squadron. Walt wanted to play baseball in one of the many leagues, but this was difficult for flying crewmen with many days on combat missions. But he played in frequent informal pickup games.[84]

Hijinks were not limited to the enlisted men. One time a 424th Bomb Squadron officer borrowed Major Vanderpoel's personal (and brand new) jeep to 'romance' a Red Cross nurse. The officer parked the jeep with her at dusk on the beach. However, it turned out that the 'beach' was really just a sand bar. The tide came up silently but steadily as they were 'otherwise occupied' with romance. The jeep was soon surrounded by water on the sand bar...and the tide was continuing to rise! The officer had to carry the nurse through waist-deep water back to shore. Then he had to report the problem to Vanderpoel, who chewed him out for a good hour. Vanderpoel sent the unfortunate officer and the 424th Tech Sgt to the beach with a crew and a winch – to winch the jeep out of the water. Then the tech sergeant had to have his crew strip the engine, flush the salt water, and then clean, dry, and oil all the parts before reassembly.[85] Stories like this spread everywhere and kept everyone amused.

Around the beginning of December, Charles McRae was back flying but with a crew of mostly new men. By this time the Appling crew, with Walt, Frank, Jack, Jupe, Don Aubrey, Ed Dunne, Jimmie Clark, and Cliff was one of the best in the 424th. There were friendly rivalries with other crews. There were many friendships between the men of the Appling crew and of the nearby crew of Allen 'Smokey' Guild. Walt was particular friends with Dick Reis on the Guild crew.[86] The crews competed to see which was best at most anything, including bombing, flying the best formation, poker, dice, drinking, and having fun.

Carl Appling was the subject of one particular story that all the officers and enlisted men spread throughout the entire Long Rangers. Once Appling was on a 'fat-cat' mission flying the squadron's C-47 down to Australia for R&R and then to return with booze, beef, and produce. At the big airbase complex at Nadzab, on New Guinea, he met a man who introduced himself as Lt. Kelso. Kelso said that he was an A-20 pilot who had been shot down, survived the crash, and was then sent to hospital in nearby Lae to recover. Kelso said that he was looking to hitch a ride to Australia. Carl was a friendly fellow and he did not ask many questions so he agreed. He even let Lt. Kelso fly part of the way to Australia in the left-hand co-pilot's seat.

However, when they landed in Brisbane, Kelso was spotted by some MP's and arrested. It turned out that Lt. Kelso was an imposter. His real name was Beck, Pvt. Beck! Apparently, Beck was variously in an ack-ack unit and also had some time as a flight crewmember. Then at some point he went AWOL. Beck had flying experience as a civilian in California and knew his way around planes.

That is when he started impersonating 'Lt. Kelso'. His story was that he had no identification when his plane was shot down, so he managed to pull off his scam.

After being apprehended in Brisbane, Kelso – or Beck - was returned to a major Army Air Corps brig at Nadzab. But a couple months later, he escaped again and made his way by various ruses to Morotai and the 424th camp area. He made his way to Major Vanderpoel's office and convinced him that he was a B-24 pilot who had crashed a B-25 offshore and was rescued at sea near Morotai – again with no identification and just the clothes he was wearing. Some of the other officers gave him some clothes to wear and he was assigned as co-pilot on one of the 424th Squadron crews. [87]

However, at his first mission briefing, Carl Appling was one of the pilots. He recognized Kelso (private Beck) in the dim light and slipped out to inform Vanderpoel. Vanderpoel called some MP's and they returned to the briefing in the operations tent. They all snuck up behind Beck, and in his gruffest military voice Vanderpoel yelled out 'Private Beck!' Beck sprung to his feet and snapped to attention, giving himself away. The MP's hauled him off and he may still be in a military brig somewhere. It turned out that he had been impersonating officers all over the South Pacific and he had a list of charges as long as his arm and then some!

The story of Lt. Kelso was one of the great scams of the war, as far as the men of the 13th Air Force knew. Everyone gave Appling a hard way to go and teased him at every chance they got. They also teased Vanderpoel, who had also been taken in by Beck and had him scheduled to go out on the next day's mission as a co-pilot. The officers and men made up an imaginary crew, named "Kelso's Raiders." On it as ball turret gunner was "Legs" Looker, the 424th BS operation officer at the time, who was 6 feet 5 inches tall. Every position was filled by the most ridiculous names. They all had a lot of fun with it and gave Appling and Vanderpoel a hard time in jest."[88]

December Missions

As December began, there were reports that the Japanese were planning an invasion to drive the Americans and Aussies off Morotai. With thousands of Japanese soldiers still in the jungle-covered hills of Morotai and thousands more on nearby Halmahera Island, everyone was nervous. The Japanese were short of food and there were many unconfirmed reports of shadowy men lurking around the camp at night, when lights were out to guard against air raids. The squadron mess tents were lit by dim red lights in the morning and many flight crewmen imagined seeing what they thought was a Japanese soldier in fatigues with a hat pulled down in the early morning chow line. It seemed plausible that a starving enemy soldier would risk capture for the chance of some food.[89]

Once, a group of bombers that had just taken off from Morotai and began heading towards their target in the Philippines received a radio message ordering them to turn around and return to base. However, the message turned out to be fake. The Japs had figured out the frequency and code for the day and they were trying to stop the mission from proceeding. After that, the procedure for assigning codes and frequencies was changed to prevent similar problems.[90]

And then a Japanese corporal was captured. Under interrogation, he said that the Japanese were planning a combined attack with a parachute landing, seaborne assault, and ground attack on the Allied perimeter. Around the same time, detailed plans were found on the body of a dead Japanese officer that seemed to confirm the attack plans. Everyone was put on a state of alert and instructed to keep their .45's and any other weapons handy at all times. Suspicions were heightened when Tokyo Rose began reporting the name of the nightly movie to be shown in the 307th Bomb Group Theater! She also said that the Japanese were going to bomb the mess tents of each of the four Long Ranger Bomb Squadrons.

The Long Ranger Bomb Squadron camps were along the beach, facing Halmahera. The beach was suitable for an assault from the sea. Each squadron set up a couple machine gun nests along their stretch of the beach, which were manned from dusk to dawn, as armed men patrolled the beach. At the sign of an invasion, everyone was instructed to don their helmet and go to predetermined defensive positions. One night the Japanese did try to bomb the mess tents, but all the bombs fell just off shore. Non-flying personnel began to volunteer in increased numbers to go along on bombing missions as observers, assuming that being in the air was safer than remaining as a sitting duck on Morotai.[91]

One pilot recounted how a few of the black men from the service corps wanted to go fly as observers on a mission. Each man was given a parachute and told where to sit in the waist of the plane as they took off. After they were in the air, the flight engineer came screaming up to the flight deck saying that the service corpsmen were trying to bail out of the plane, because one of the engines was off. However, they were just going through a test and had feathered the prop. After that, 'observers' were always included in the briefings so they would know what to expect in the air.[92]

On Friday, December 1, the Appling crew was called for another mission. Major Vanderpoel in A/C 619 would lead six 424th BS Liberators. The Appling crew would remained unchanged, with Don Aubrey, Jimmie Clark, Ed Dunne, Jupe Bretherton, Cliff Llewellyn, Frank Mansir, Jack Riley, and Walt in A/C 180. Six machines from the 424th would be joined by more B-24's from the other three squadrons to bomb Bacolod airstrip on Negros Island. Each plane was loaded with 40 – 120 pound fragmentation bombs.

Charles McRae was back with a new crew on this mission. His recurring ear infection, which had plagued him off and on ever since he arrived in the tropics back in April, had cleared up again well enough for him to fly. Some new crews were starting to arrive in increased numbers, and as an experienced and respected pilot, McRae could help give them some guidance and experience. McRae would pilot A/C 958 with nine of these new men.

Takeoff from Pitoe was at about 0612 into clear skies with widely scattered cumulus clouds with tops at 8,000 feet. Their route took them northwest to Flecha Point on Mindanao at 07°20'N, 122°20'E, then a little east of north to Tagolo Point, then northwest again to the IP at Sojoton Point on Negros at 10°00'N, 122°53'E. The weather remained good, but after they formed up with the other planes and headed to the target with the squadrons following in train, a rainsquall hit. They were approaching on a heading of about 18° and the word came down from Major Vanderpoel for each plane to drop its bombs independently only if they could see the target – the airstrip at Bacolod. They approached at 11,200 feet at 1132, with no A/A fire and no enemy fighters seen. Only McRae in A/C 958 and his new bombardier Lt. Plummer, felt they had a good enough view of the target to drop their bomb load. The other five planes held off for a second run at the target.

Vanderpoel in A/C 619 led them around in a wide circle for another run at Bacolod on a reciprocal heading. At about 1142 at 10,500 feet they approached the target, the rain having cleared but still with haze over the ground. Still no resistance from the Japanese in the air or from the ground. Walt watched as the 200 little bombs from the five planes floated down towards the target. The bomb pattern began about 1000 feet south and west of the runway and then walked NNE across a river, through the dispersal area, and then about 200 feet the other side of the strip – a total of about 2,335 feet. Walt could see the bomb bursts, but no other explosions, fires, or destruction on the ground. By 1144, the attack was over and they headed for home.[93]

One of the planes in the 372nd Squadron had a freak incident over Bacolod on this mission. When one plane dropped its bombs, two of them contacted each other in the air and exploded, damaging the following plane piloted by Lt. John Jonas. Fortunately, it was not serious enough to prevent them from returning to Morotai.[94] The route back to Morotai was uneventful with good weather. Carl Appling had A/C 180 back on the ground at about 1612, after ten hours and 1,393 miles in the air. At the debriefing, the results were rated as just fair, but Walt had successfully completed his 29th combat mission.[95]

The 'just fair' rating of the December 1 mission was primarily due to the rainsquall that obscured the target and the mist that they flew through on the second attempt at Bacolod. The following day the danger of flying in formation through thick weather hit home for the men of the Long Rangers. The target for the group that day was Carolina A/D on Negros Island, but it was socked in with thick weather. The group flew on to the secondary target at Fabrica A/D, but it too was obscured by complete cloud undercast making accurate bombing impossible.

As the group proceeded towards the tertiary target at Dumaguette A/C in loose formation, it approached very heavy weather with no visibility. The group started a slight turn to the right as it entered the rain clouds, when a loud explosion was heard and flaming debris was seen through the clouds falling to the sea in the Tanon Strait. A 424th BS plane with a new crew of 11 men piloted by Lt. Jack K. Mitchell had collided in the air with a plane from the 372nd Squadron piloted by Lt. James York. A wing section with a flaming engine, a wing section with two engines, a tail section, and a section of fuselage were seen going down by various crews. Another plane from the 372nd was hit in the tail and fuselage by falling debris. No parachutes were seen.[96] More than 20 men were lost in the twinkling of an eye, missing in action. The Long Rangers had encountered little opposition from the Japanese in the air or from the ground the last few weeks, but danger was still there. To add insult, that night the air raid warning went off as the Japanese launched a bombing attack on Morotai.

The following day, Walt and some of the other crewmen were summoned to 424th HQ. There Major Vanderpoel met them. Lt. Appling, Lt. Aubrey, Lt. Dunne, Lt. Clark, and S/Sgts Bretherton, Hidalgo,

Schreiner, Mansir, Riley, and Babinski had been formally credited with the destruction of one Japanese Zeke on November 6, during the attack on Alicante A/D on Negros Island. Each man had a letter placed in his file that stated:

> *"The following named officers and enlisted men, Air Corps, United States Army, are officially credited with the destruction of one enemy fighter type aircraft in aerial combat over Negros Island, Philippines, at 1053/I on 6 November 1944....This crew was part of a formation of B-24's that was intercepted by a formation of Japanese fighters. Sergeant Riley, tail gunner, fired on a Zeke attacking from 6 o'clock and the enemy fighter began smoking and went into a dive. It was observed to crash into the sea."*
> *By Command of Major General STREETT.*[97]

The Thirteenth Air Force had a very strict policy on awarding credit for the destruction of enemy aircraft, so the men of the Appling crew that day were very gratified to have this recognition.

A new rumor spread across Morotai – it was said that the following day a suicide battalion of Japanese paratroopers would be dropped onto the base, while other Jap troops infiltrated across the perimeter. All of this was supposed to happen during a Japanese bombing attack.[98] It was decided that all of the bombers would be evacuated back to Noemfoor until the threat of an attack had passed, but they would take a shot at the Japanese along the way. That night there was another nuisance Jap air raid to rattle everyone's sleep.

On Sunday, December 3 the Appling crew was scheduled for this mission. During the usual briefing that morning, Walt learned that a new man, Sgt. Walter Blawicki, would be joining the crew to take the place of Boris Hidalgo. Boris had gotten the news that he was headed home! The influx of new crews was starting to increase. Blawicki was from Newark, New Jersey, where he had been raised by a foster father.[99] Boris would still be around on Morotai for a few weeks, as he waited for transportation back to the states. This would be an unusual mission for a number of reasons. First, the 424th BS would put eight planes in the air for this mission, each loaded with 30 – 100 pound GP bombs. They would be led by MacMillan in A/C 619. The Appling crew in A/C 180 would be in the number 2 position. In addition, Lt. Charles McRae was flying with a mostly new crew in A/C 179.

The target was a new one – Malimpoeng A/D, on the Celebes. Walt, Jack, Frank, Don Aubrey, Jimmie Clark, Ed Dunne, as well as Boris and Gene Schriener remembered the October 24 shipping prowl around the Celebes they flew with McRae at the controls: 18 hours and 40 minutes of flying! No one had met anyone who had flown longer in a B-24. This flight would not be as far, because they had just a single target. But then the biggest surprise was that after the attack, they would not be returning to Morotai. Because of the threat of a Japanese attack on Morotai, the plan was to return to Noemfoor. This way, if there really was a Japanese assault on the island, at least the bombers and their crews would be safe. Malimpoeng was a staging and repair base for many of the Japanese planes that were bombing Morotai. For the attacks on Morotai, the Japs would fly the planes to one of the many small airstrips scattered around nearby Halmahera Island a day or two before. Then one or two nights later they would make their attack. Bombing Malimpoeng was intended to disrupt the nightly Japanese attacks. In all 84 B-24's from the 13th Airforce would participate in this mission.[100]

Take off for the group was very early, to accommodate so many planes being launched into the air in case the Japs attacked. Carl Appling had A/C 180 in the air by about 0540. Their route took them southwest across Halmahera and then into the Molucca Sea. Off to the left Walt could see little Ternate Island, one of the original 'Spice Islands. Later they crossed Peleng Island and the Gulf Van Tolo, before the Celebes came in to sight. Their first way station was across the southeast peninsula to 02°40'S, 121°07'E, near Malili Bay. From this point they continued across the Gulf Van Bone to a turning point on the east coast of the southwest peninsula at 03°40'S, 120°25'E. Their course then turned west to the IP at 03°46'S, 119°45'E. The squadrons formed up here then turned north to the target, Malimpoeng A/D at 03°43'S, 119°55'E. The airfield was located in a flat agricultural area.

Enroute to the target they had flown through variable scattered clouds, with a moderate front encountered midway with cumulus clouds at 9,000 feet and 12,000 feet and some towering cumulus to 12,000 feet. A/C 236, with Lt. Sasner as pilot, emerged from the weather front finding the rest of the squadron too far ahead to catch up. Sasner radioed that he would proceed independently to bomb the number 4 alternate target. As the rest of the squadron approached the target on a heading of 280°, they encountered 70% stratocumulus from 5,000 to 9,000 feet and a 30% stratus layer at 11,000 feet. However, the target was clear as they approached.

At 1139, the seven remaining 424th bombers approached the target in Javelin down formation, with A/C 180 in the number 2 position with McMillan in the lead. The planes ranged from 9,985 to 10,200 feet in elevation. Ever alert for enemy fighters, but none appeared and there was no A/A fire encountered. At 1142, they were over the target following the lead plane. As Jimmie Clark saw the target come into range, he salvoed their load of 30 – 100 pound bombs. Walt watched as they dropped away towards the target. A beautiful pattern of explosions started about 500 feet west of the target, then the bomb pattern walked over the southern half of the main runway, ending about 500 feet on the other side. Walt saw many hits scored on the runway, as well as a small fire on the south end and a large explosion and fire – either an airplane or a fuel truck to the west, near the center of the number 2 runway. One of the planes also dropped a load of psychological warfare leaflets over the target – something for the Japanese to read in their spare time. By 1143, Appling was turning A/C 180 to the right. A few minutes later, Ed Dunne confirmed a direct heading to Noemfoor.

The route to Noemfoor was mostly obscured by thick clouds, so there were few chances for observation reports. At about 1640, Appling had found the old familiar airstrip at Noemfoor and set A/C 180 down on the ground safely. Walt had now completed his 30th mission after eleven hours and 1,431 miles in the air. There was no 424th BS intelligence officer to report to on Noemfoor – the ground echelons were still back nervously waiting on Morotai.[101]

Conditions that night on Noemfoor were somewhat rustic. Most of the old camp facilities had been moved to Morotai, but there were enough tents to accommodate this reverse surge in aircrews. The food was adequate too, but there was no nighttime movie. The following day the men had to do some of the maintenance of their planes themselves. The guns did not need to be rearmed, because there had been no enemy aircraft encountered. However, late in the afternoon of the fourth the enlisted men had to help the few armorers there were load their planes for the next mission. Others had to help fuel the planes. Walt's rating as an airplane mechanic, earned many months ago at Keesler, came into use again here on Noemfoor. That evening the men were told that they would find out the following morning if they would return to Morotai.

Sure enough, Tuesday morning December 5, they learned that conditions on Morotai were deemed safe enough for them to return. But they planned to give the Japanese threatening Morotai from Halmahera a little greeting from the Long Rangers. They would attack every known airfield on Halmahera. The 424th, which now had nine planes on Noemfoor, would send six to bomb Hatitabaco A/D and one each to bomb Lolobata, Hiti, and Djailolo A/D's. The Appling crew in A/C 180 was assigned to bomb Lolobata, on the northeast central peninsula. Lolobata was only a single runway, but it was used for some of the nightly raids on Morotai. The bombing was to be conducted at very high elevations, so the men had to load their thick jackets and gloves, despite the tropical location.

The Appling crew was the same as on the previous mission, with Sgt. Blawicki in place of Boris Hidalgo. It took a long time for so many planes to get into the air, due to the limited ground support facilities remaining on Noemfoor. They were loaded with 15 – 250 pound GP bombs. It was not until about 0900 that Appling piloted A/C 180 into the air. The weather just north of the Voegelkopf was full overcast of altostratus at 15,000 feet, with 30% cumulus with tops at 6,000 feet. The route to the IP at 01°30'N, 129°50'E took them on a heading just north of west. The IP was just a spot in the Halmahera Sea where the 424th bombers would rendezvous, then head off to their targets.

The weather over Halmahera was generally good, with 30% cumulus with tops from 6,000 to 9,000 feet. A/C 180 was instructed to make its attack at a very high elevation – 17,500 feet. Even near the Equator it was very cold at this elevation so the men had all donned their cold-climate flight jackets and were on oxygen. Making a solo attack at this very high elevation would be a challenge for the bomb aiming skills of Jimmie Clark. Lolobata was just a small airstrip. Appling maneuvered A/C 180 to the south, over open water, and then turned north for their bombing run on a heading of 359°. There were no enemy fighters to be seen, no A/A fire coming up to harass them and fair weather. Appling gave Jimmie control of the plane at 1111. He sighted on the south end of the airstrip where a taxiway intersected, and when he thought he was on target toggled the load of 15 – 250 pound bombs away. Walt watched for a long long time as they fell away, and then he saw them hit. The detonations began about 4,000 feet south of the airstrip and then walked north towards the dispersal area, but never reached it. Just detonations, no explosions and no apparent damage on the ground. From there it was a short flight, descending the entire way back to Pitoe, and Appling had them back on the ground on Morotai at about 1310. The results were disappointing, but Walt had completed his 31st combat mission after four hours and ten minutes in the air.[102]

Later that afternoon the crews reported to Lt. Shields to be debriefed on both the mission they just completed, as well as the mission two days before to Malimpoeng that terminated on Noemfoor. The December 3 mission to bomb Malimpoeng on the Celebes was rated as 'excellent'.[103] The solo mission they had just completed, to Lolobata, was rated as 'unsatisfactory.' The result was due to the difficulty of accurate bombing from the extremely high elevation they flew that day.[104]

The threat of a Japanese assault from the air, sea, and land still seemed real, so the next few days were nervous for everyone. Every night there was another Japanese air raid. The A/A crews seemed to be getting better. In the wee hours of Wednesday, November 6, at 0330, heavy A/A fire opened up. A single Jap raider decided to jettison his bombs offshore in the sea and not chance coming closer. The next night, at 0230, very intense (and loud) A/A fire opened up on a Jap plane flying very high overhead. It eventually withdrew without dropping its bombs.[105]

Despite the frequent air raids and the threat of a Japanese attack by land or sea, the 13th Air Force was also expected to continue hitting their assigned targets, so they had to get back to preparations for the next mission. On Thursday, December 7, the men of the 424th were called to assembly on the parade grounds and Major Vanderpoel addressed them. He reminded them that it had been three years to the day since the Japanese sneak attack on Pearl Harbor, but the 424th Bomb Squadron and the entire Long Rangers bomb group were taking the war closer and closer to Japan itself every day. A commendation letter had been received from Major General St. Clair Streett addressed to each man of the 13th Air Force. Vanderpoel read:

"HEADQAURTERS THIRTEENTH AIR FORCE
"Office of the Commanding General
APO 719, 7 December 1944

"I wish to extend my person congratulations for your superior performance as a combat crew member of a magnificent air team during the recent South and Southwest Pacific campaigns.

"Your devotion to duty and your achievements have been noted by our commanders. Lieutenant General Kenney has radioed the following message:

'THE EXCELLENT JOB ACCOMPLISHED AND MAINTAINED BY YOUR COMMAND IN NEUTRALIZING SATAWAN, TRUK, YAP, PALUWAT, WOLEAI AND PALAU WITH REPEATED AND AGGRESSIVE ATTACKS RECEIVES MY HEARTIEST COONGRATU;ATIONS....A DIFFICULT MISSION WELL ACCOMPLISHED.'

"In the operations against Balikpapan you made Thirteenth Air Force history by your courage and skill in one of the longest over-water flights ever attempted. These missions prompted General MacArthur and General Arnold to state:

'KINDLY PASS TO GENERAL STREETT AND PERSONNEL CONCERNED MY ADMIRATION OF THEIR UNSHAKEN DETERMINATION IN PRESSING HOME THE BALIKPAPAN ASSAULTS IN THE FACE OF DIFFICULTY AND HAZARD. THE FINEST TRADITIONS OF THE AIR FORCE ARE AGAIN LIVED UP TO. SIGNED MacARTHUR.

'WHIPPING THE JAP ANY TIME AND ANY PLACE, AS INDICATED IN THE BALIKPAPAN ATTACKS, SHOWS AGAIN THE INDOMINITABLE COURAGE OF THE THIRTEENTH AIR FORCE TO WADE IN AGAINST THE JAPS. AGAINST BALIKPAPAN THE FIGHT WAS PRESSED HOME IN THE FACE OF KNOWN STRONG FIGHTER OPPOSITION. A 'WELL DONE' TO YOU AND YOUR MEN FOR CONTINUING TO SLUG THE JAP UNDER MOST DIFFICULT CONDITIONS. SGD ARNOLD.'

"Through the skill and determination of men such as yourself the "Thirteenth" has earned the respect of all fighting men. Your service has brought honor to yourself, your organization and your country. I am proud that I can claim service with you during these critical operations, the success of which has so appreciably contributed to the shortening of the war.

Signed...
ST. CLAIR STREETT
Major General, U. S. Army, Commanding"[106]

Each man felt pride in knowing that the difficult and dangerous job they were doing was recognized and appreciated. Vanderpoel added a few words of his own expressing his pride in his men and the job that they were doing. Then he saluted and dismissed them back to their duties. Walt, Jack, Cliff, and the other tent-mates wandered back to their camp. On the way, Appling made sure that they all knew that they were assigned yet another mission the following day, so they should turn in early. Again, their sleep was disturbed for an hour or so by another Jap air raid that night. It began at 0345 with five planes. Again there was very heavy A/A fire and the P-61's got into the air and at them. Four of the Japanese planes dropped their bomb loads without causing damage and the fifth one dropped its load in the sea.[107]

Early on Friday, December 8, Walt and his buddies went through the familiar pre-mission routine. At briefing they learned that they were headed back to the Philippines. Their target was La Carlota Airfield on Negros Island, which they had bombed back on November 15. Six planes from the 424th would be led by Capt. Looker, Major Vanderpoel's deputy, in A/C 619. Looker would be leading the entire group, with planes from the other three squadrons as well, for a total of 56 bombers. The Appling crew would fly in the number two position in A/C 318, with one change in the crew. Sgt. Blawicki, who had flown with the Appling crew the previous two missions would be replaced by Sgt. Roger Williamson. Each of the trusty Liberators was loaded with eight 1,000-pound GP bombs.

As the day was just dawning, they made their way out to Pitoe Airfield to prepare. At about 0653, Appling had them in the air and on their way to the northwest. As they started out, they flew above 70% cumulus with tops at 8,000 feet and below full overcast of altostratus clouds. Visibility started out at seven miles, but steadily decreased the farther north they flew. Their planned route took them to a series of check points, at Flecha Point, to Tagola Point, Sojoton Point, and then to the IP at 10°17'N, 122°53'E, south of La Carlotta. The four squadrons formed up with Capt. Looker in A/C 619 in the lead. The approach was made at about 10,000 feet on a northerly heading with 80% cumulus and seven miles visibility.

The plan was for the four squadrons to follow in Group combat box formation and drop their bombs after the lead plane. At about 1153 they headed towards the target. No enemy fighters were seen in the air and no A/A from the ground. However, there was cursing by Capt. Looker in A/C 619, because they could not get their bomb bay doors open. The order went out to the group to follow Looker around for another run at the target after the crew in A/C 619 managed to get the bomb bay doors opened manually. Everyone was cursing at this point. No one liked to have to make multiple runs over a target, but fortunately, there was still no A/A fire from below. At 1157, spread out between 9,500 and 10,500 feet they were over the target again on a heading of 170° but cloud formations interfered with the bomb sighting. A/C 619 dropped its bomb load and the other planes in the group followed suit. Walt watched as the bomb patterns appeared. One group of bomb strings hit within 500 feet of the runway, but off to the north. Another four strings of bombs landed between the runway and a major highway to the south. There was no apparent damage that Walt could see.

There was grumbling amongst the men by the poor results, but there was nothing to do but head back to base. Off the southwest coast of Negros Walt reported that he saw a Sugar Baker class small freighter in harbor. Then as they approached Mindanao Island, several of the observers saw a PBY Catalina on Dipolog Runway, but it did not look like the plane was serviceable. The rest of the return to Morotai was uneventful and by 1618, Lt. Appling had A/C 318 back on the ground at Pitoe. After nine hours and 25 minutes and 1,368 miles in the air, Walt had completed his 32nd combat mission. Later, at the debriefing, the 424th BS effort that day would be rated as 'unsatisfactory' due to the malfunction with the bomb bay doors on Looker's plane during the first pass, and poor bombardier aiming on the second pass at the target.[108]

That night, at about 0300, there was yet another Japanese air raid on Morotai. The P-61 night fighters managed to shoot down one of the two planes – a 'Rufe' floatplane – about five miles offshore. The following night, beginning about 0315, three JAF planes attacked for about an hour. There was heavy A/A fire. One raider dropped bombs on Pitoe; one dropped three bombs into a nearby tent area, and the third scattered small bombs on Wama, all with no damage or casualties.

Despite the continued air raids, the 'black alert' that they had been under was called off. It was plausible that the Jap's would still try an attack on the Allies on Morotai by ground, air, or sea, but the imminent threat seemed to have receded. Walt had that Saturday and Sunday off, although the Japs kept up their nightly bombing raids. Late Sunday Walt was notified that he was called for another mission the following morning.

At the early morning briefing on Monday, December 11, the Appling crew learned that their target would be a new one for them: Ilo Ilo Airfield, on Panay Island in the Philippines. Panay Island was beyond Negros, where they had been many times before. Lt. William M. (Bill) Hughes in A/C 318 led the six 424th BS bombers assigned. Walt and the Appling crew would be in the number 2 position behind Hughes. Each of the six bombers was loaded with twelve 500-pound GP bombs.

Take-off was at 0652. Appling headed off to the northwest, gradually gaining altitude. The weather was good, with 30% stratocumulus above at 15,000 feet, and 30% to 50% cumulus with tops at 5,000 feet. Walt and the other gunners test fired their guns in preparation for the long flight. The route led first to the familiar waypoint at Flecha Point. Then they encountered a mild weather front with towering cumulonimbus clouds from 3,000 to 15,000 feet. From there they flew to Baliango Point, then to Sojoton Point and then to the southern tip of Guimaras Island. Here they formed up with the other 424th BS bombers and flew to the IP at 10°35'N, 122°35'E, in the center of Guimaras Island. The weather was good with 50% cumulus tops at 5,000 feet and unlimited visibility.

Hughes led them to the target, located across a strait of water just beyond Guimaras Island. They were at about 11,200 to 11,900 feet elevation with squadrons in trail formation on a heading of 25°–41°. Walt was on the lookout for enemy fighter opposition, but none was in sight. At 1152, they were over the target and one by one, the bombers dropped their loads. Walt saw no A/A fire from below as he watched as the strings of bombs fell away. The bomb patterns began about 1,500 feet west of the southern tip of the runway and then walked three-quarters of its length and through the area west of the strip. None of them fell on the runway, but a few of the bombs hit and destroyed a service building. Walt saw one twin-engine bomber parked off the center of the runway.

With that, at 1154, Appling turned A/C 180 away and headed for home. They flew across Negros Island on their return, and then crossed the strait between Negros and Mindanao. Nearing the coast, Walt observed two P-38's on the runway at Dipalog, near Tagolo Point. As they crossed Mindanao, the weather front had become stronger with wind and rain buffeting the plane and reducing visibility. After a couple more hours, they were nearing Morotai and Appling gave the OK signal for Walt to come out of the ball turret. At just about 1637, Carl Appling set A/C 180 safely back down on Pitoe airstrip. After nine hours and 45 minutes and 1,430 miles in the air, Walt had safely completed his 33rd combat mission. After some coffee and donuts from the Red Cross nurses, they made the drive back to the camp area. A little later, they reported to the Squadron Intelligence tent. After their debriefing, Lt. John H. Shields completed his report, rating the mission as 'Fair' because they missed the runway, their primary target, but managed to destroy one building.[109]

As they were finishing though, Shields told them some news. Carl Appling and Ed Bretherton had gotten orders to cease combat flying and await return home. Because of Carl's earlier missions flying B-25's with the 42nd Bomb Group, before he replace Lt. McRae on Guadalcanal, he had more than enough points accumulated for rotation back home. Walt, Cliff, Jack, Fred, Don, and Jimmie congratulated Carl and Jupe and wished them all the best. The men had come to like and respect the Texan Appling and wild, funny Jupe. Carl had proven himself a great pilot and brought them all back safely after tens of thousands of miles flying over lonely seas and enemy infested islands.[110]

What would that mean for the crew? They soon learned that Don Aubrey would move to the pilot seat for the crew. He would be replaced by Lt. James J. Fielding as copilot. Jim was from Bridgeport, Connecticut. Lt. Fielding had served as copilot for the McMillan crew. He spent time over the next two days getting to know the enlisted men and he made a favorable impression on them.

The nightly Japanese air raids continued. The men grumbled a little. The 13th Air Force now had two P-61 night fighter squadrons, but they seemed unable to stop the Japs coming in their ones and twos to bomb Pitoe and Wama.[111] There were new crewmen arriving each week. At this time, the 307th BG policy was that any man who had accumulated 35 combat missions was eligible to return to the States on rotation and become an instructor.[112]

On Tuesday the 13th, some great news spread through the 424th Bomb Squadron Camp. The day before, the 424th had sent six machines to bomb Bacolod A/D on Negros. While returning from this mission, Dick Reis in the lead plane, piloted by Lt. Alan 'Smokey' Guild, sighted mirror flashes from the ground.[113] Guild radioed the 13th Air Force Catalina PB-Y flying boat assigned for rescue duty during the mission, informing them of the precise location of the mirror signals. Nine hours later, the Catalina returned to Morotai with nine men from Lt. Balovich's and Lt. Hunter's crews, who had been shot down on November 6 and November 8.

It turned out that when Lt. Balovich's plane went down on November 6 four of the men had survived, parachuting into the water. However, only one, Lt. Primsberger, the bombardier, had a one-man raft. The other three men were Sgt. Minier, Sgt. Gannon, and Lt. Yuschak, the navigator. They could see Panay Island, but the current was taking them away from the island. The four men took turns resting for one hour in the raft, while the other three stayed in the water, supported by their Mae West jackets. After resting for one hour, Sgt. Minier decided to try to swim to Panay, but he was never seen again. At one point, a small Japanese patrol boat came within 50 yards of the raft, but they were not seen. After 40 hours in the water, the life raft with the last three men washed ashore on Negros Island.

They dragged themselves and the raft into some bushes and slept. The next morning they crept to the edge of a small fishing village. After explaining that they were American fliers, the Filipino villagers fed and cared for them for a few days. They then turned the men over to a group of partisan guerilla fighters. The partisans led them to their headquarters in the jungle, where they found other rescued American fliers.[114]

When Lt. Hunter's plane had gone down on November 8 over Negros, a number of men were seen to bail out. Lt. Willard McGee saw Hunter's body on the plane before he bailed. The seven crewmen who survived were all picked up by Filipino partisan fighters. In addition to Lt. Hunter, three other men died: Lt. Dawkin the copilot, Sgt. Witham and Sgt. Wilson. The Filipino Partisans took the airmen to the same HQ where the survivors of Balovich's crew were found. One man from Hunter's crew, Sgt. Maximillian Augsberger, had a bad infection in both of his feet. He was taken to a partisan hospital to recover. Then the partisans led the men from the two crews to a spot at Sojoton Point, where they used signal mirrors to flash every American plane that passed by.[115]

The men had survived many days crossing steep and rain-swept mountains, dangerous journeys by small boat, and evading Japanese patrols. However, they returned in good mental and physical condition, despite their harrowing experiences. Their return provided a morale boost for all the men of the 424th Bomb Squadron.[116]

On Wednesday the crew, now the Aubrey crew, was assigned another mission the following day. Early on Thursday, December 14, Walt and his buddies reported for the pre-flight briefing. They learned that their mission was to support the invasion of another of the Philippines Islands. MacArthur had decided to invade Mindoro Island, bypassing Negros and Panay Islands. Mindoro was just 100 miles south of Manila, and would provide fighter airbases for the eventual invasion of Luzon Island. The job of the 13th Air Force was again to suppress Japanese fighter planes by attacking their airfields on MacArthur's flank.

Capt. Looker would lead the six planes assigned for this mission in the Squadron's namesake, A/C 424. Don Aubrey would fly A/C 180 in the number 2 position, a credit to the well-honed crew he now commanded. Seven of the crewmen: Aubrey, Dunne, Clark, Mansir, Riley, Llewellyn, and Walt, plus Don Aubrey, had now been together for almost a year. Walter Blawicki would come back to the crew, with Sgt. Williamson going over to the Raue crew in A/C 546. Replacing Jupe would be a new man: Sgt. Rafko – one of the replacements starting to arrive in greater numbers. As co-pilot was James J. Fielding, who usually flew with the MacMillan crew.[117]

The target would be Malogo A/D on Negros. Half of the plane were be loaded with twelve 500-pound bombs with .1 nose and .01 tail fuses. The other half were loaded with eleven of the same 500-pound bombs, but their twelfth bombs were fused with a 24-hour delay. Once out at Pitoe, Don Aubrey was very active, checking everything out and ensuring that everyone, especially Blawicki and Rafko, were properly prepared. He did not want any foul-ups on his first mission as pilot. At about 0645, Don had A/C 180 up in the air and on its way to the north.

The weather north was good with 70% stratocumulus clouds above at 12,000 feet and unlimited visibility. As Walt settled into his ball turret, he had a great view of the empty ocean. Their first waypoint was Flecha Point. As they crossed Mindanao, they started to encounter widely scattered cumulus clouds with tops at 5,000 feet. As they crossed south of Negros Island, at about 1109, Walt let out a whistle. There below was a large convoy. He counted three cruisers, two destroyers, and 23 landing craft: 28 ships in all. They were headed on a course of 340°.[118]

From here, the route of the 424th BS planes took them around Negros Island, first up Panay Gulf, then to their IP in Guimaras Strait at 10°46'N, 122°53'E. As the 13th Air Force bombers formed up, they headed off to the northeast. The weather was excellent for bombing – 20% cumulus with tops at 4,000 feet and unlimited visibility. For the attack, they flew between 9,800 and 10,300 feet on a heading between 60-68°. They were in squadrons in trail formation. At about 1145, they approached the target. No fighter opposition was seen and no A/A fire from below. The planes followed Captain Looker in A/C 424 and as each plane reached the designated bombing point, the center of the western edge of the airstrip, it was bombs-away. Walt watched the bombs drop in their mesmerizing pattern. Then they began to explode in a beautiful rectangle, about 1000 feet wide, which walked diagonally towards and then over the southern portion of the north half of Malogo runway. At least ten of the bombs hit the runway itself. Walt also saw between eight and ten single engine planes in the revetment area that seemed to be operational.

By 1149, the attack was completed and Aubrey turned left and then south as Ed Dunne set a course back to Morotai. Over Alicante A/D Walt saw about 15 aircraft – about two-thirds single engine and one-third twin-engine types. A few minutes after they turned away from Malogo they crossed the coast again and Walt observed a large two-stacked 'Tare Able' freighter tied to a jetty at Talisay.[119] Later, on the southwest coast of Negros, Walt and the other observers noted a fire with white smoke, about 500 yards in from the beach, appearing to be a signal fire. They continued south and crossed over Mindanao. Later, at about 1417, on the southern tip of the island Walt observed a large brush fire with white smoke.

The rest of the flight south there were no observations. As they passed the Talaud Islands, they flew into a heavy weather front, with turbulence and heavy rain. Despite visibility reduced to three miles, with Ed Dunne's navigating Don Aubrey found the Pitoe Airstrip without trouble. Don eased A/C 180 back down safely at about 1635. After nine hours and 50 minutes and 1,420 miles in the air, Aubrey had completed his first mission in the pilot seat. Everyone congratulated Don as they rode back to the 424th Camp area. For Walt, it marked his 34th combat mission. He wondered how much longer it would take him to follow Boris Hidalgo, Jupe Bretherton, and Carl Appling back to the states. Later, at their post-mission briefing, Lt. William N. Vincent Jr., Intelligence Officer, would rate their results that day as 'excellent'.[120]

Later that evening, back at their tent, Cliff came back with more news. Ed Dunne had gotten his orders with permission to stop combat missions and to await shipment back to the states! Cliff, Frank, Jack, and Walt stopped by the officer's tent to shake Ed's hand, slap his back, and congratulate him. And of course, they were all hoping and waiting for their own orders to come through soon. Walt and his crewmates had the next two days off, although each day the squadron put six machines in the air to participate in bombing raids on Puerto Princesa A/D on Palawan Island. Each night sleep was interrupted by the continuing nuisance Jap air raids that seemingly could not be stopped. For example, at 0300 on Friday morning, in drizzling rain, a Japanese plane dropped three bombs beyond the infantry perimeter north of Pitoe with weak A/A fire.[121] Late on Saturday, Walt learned that he was called for another mission the following day.

Very early on Sunday, December 17, at 0315 there was another air raid alert. Three Japanese planes attacked Wama through moderate A/A fire. Two of the planes bombed from high altitude, while the third dived down before dropping two bombs. Luckily there was no damage or casualties.[122] Shortly after the air raid ended, Walt and his buddies were awakened by the orderly rattling their mess kits hanging from the tent pole. Grumbling, the men went through their familiar routine: get dressed, hit the latrines, grab their gear, get some chow, all in the early morning dark. Then thy assembled in the briefing tent to learn what lay ahead for the day.

First the mission: Their target was the runway at Jesselton, on the northwest coast of Borneo, on the South China Sea. Jesselton was the capital of British North Borneo and had become a major Japanese administrative center. The squadron would put six machines in the air to support the attack on the runway there. Don Aubrey in A/C 180 would be flying with a new co-pilot, Lt. Farman. Farman was a new replacement who had just recently arrived on Morotai. Lt. Charles Kaiser, Jr took Ed Dunne's place as navigator. He was relatively new too, although he had flown as a replacement on a few other crews earlier in the month. They would depend on Kaiser a lot, because they were flying to a completely new target and an unfamiliar stretch of ocean and Borneo coastline. Jimmie Clark remained as bombardier, and Walt, Cliff, Frank, and Jack were in their usual spots on the crew. The other gunners would be Walter Blawicki and Robert Williamson. They would also carry a combat photographer – Sgt. Bauer. Aubrey would fly A/C 180 in the number two position, behind Lt. Bill Hughes in A/C 318. Each plane was loaded with twelve 500-pound GP bombs, each with .1 nose fuse and .01 tail fuse – except one bomb on each plane that would have a 24 hour delayed fuse.

The briefing completed, the crews loaded into waiting trucks that ferried them out to the revetments surrounding Pitoe. Don Aubrey spent some extra time going over the expected routines with the two new men – Farman and Kaiser. After everything checked out and everyone had settled into their places, Don got the signal to start his taxi onto the side runway. Dozens of big Pratt and Whitney engines hummed as the planes made their way in line for take-off. At about 0535, A/C 180 was at the head of the line. Aubrey and Farman throttled the engines, released the brakes and they started rolling down Pitoe strip. Soon they were up and climbing towards the west into 70% cumulus clouds with tops at 11,000 feet.

The route took them just north of due west, first to the Sangi Islands, midway between the northern tip of the Celebes and the southern tip of Mindanao. Their next waypoint was Hog Point, at 5°19'N, 119°16'E, on the east coast of North Borneo. From here, the cloud cover thickened to 100% alto-cumulus with tops at 12,000 feet, with 30% to 50% cumulus tops. Don Aubrey called back to Lt. Kaiser to ask if he was doing OK at his chart table. Their route took them across North Borneo to a turning point on the northwest coast at 06°25'N, 116°25'E. This route led them within a few miles of Mount Kinabalu, a 13,455-foot high volcano. Walt and the other men gave out a whistle as they caught a glimpse of this monster mountain off the left side of the plane as they made their way by it safely. Their turning point was on the coast near the mouth of a meandering jungle river.

From here they turned north out over the South China Sea to the IP, at 06°03'N, 116°03'E, situated between two islands, Sepangar and Gaya, that formed the bay that Jesselton lay on. As the planes circled at the IP the weather was good – 70% alto-cumulus at 12,000 feet above and 20% cumulus with tops at 4,000 feet below and unlimited visibility. Then Lt. White, piloting A/C 276, reported that he had blown a cylinder in his No. 4 engine. He reported that he had feathered the engine and activated his fire extinguisher to douse the fire. White reported that he was OK to go ahead with the attack on three engines. The 424th planes were staggered at 8,200 to 8,800 feet as the 307th began its bombing run with the squadrons in trail on a heading of 180° to 185°. The runway was aligned with the shoreline, south of the town.

The bombardiers sighted on a point 100 feet short of the center of the runway. Walt scanned the skies below for enemy fighters and A/A fire, but none was seen. As they approached, Walt counted eight small vessels in Jesselton harbor. Then at 1135, the bombs away began as the planes crossed the target one by one. Walt watched as the bombs covered the northern half of the runway and 250 feet on either side, for a distance of about 990 feet. Sgt. Bauer took seven photos from A/C 180 at bombs away. At least 20 of the bombs hit the runway. By 1137, Don Aubrey was turning away to the right in a big circle, following Hughes in A/C 318.

From here, their orders were to fly along the coast of North Borneo to the northeast, observing along the way. Walt reported seeing a single 70-foot long vessel in an inlet at 06°22'N, 116°20'E, southeast of Usakan Island. Before reaching Banggi Island, just off the northern tip of Borneo, they turned to the southeast to continue their coastal reconnaissance. Just off Hog Point, Walt spotted a submerged freighter in shallow water – a small Sugar Charlie class. Now Aubrey headed southwest around Hog Point and into Darvel Bay to check out a large fire that Walt and other observers sighted at about 1309. Nothing was seen except a column of gray smoke that leveled off at about 2,000 feet. From here, Aubrey turned A/C 180 to the east and had Lt. Kaiser plot a direct route back to Morotai. They flew on an east by southeast heading out over the Celebes Sea. Walt scanned the ocean below as they flew slowly back to Morotai. As they approached home, Walt was helped out of the ball turret and back in to the belly of plane for their descent. At about 1705, Don Aubrey eased A/C 180 down onto the landing strip at Pitoe. After eleven and a half hours and 1,773 miles in the air, Walt had completed his 35th combat mission. Later, during the briefing, Lt. Vincent, the intelligence officer, rated the mission results as 'excellent.'[123]

For the next two days the 307th Bomb Group was 'on alert.' The invasion convoy that Walt had seen south of Negros Island on Thursday, December 14 had landed on Minoro Island. They were making good progress, but there were reports that a Japanese naval task force was headed to Mindoro to attack the U.S. transport ships on the beach. This task force did in fact shell the American beachhead, but the 307th never got the call to deliver a bombing strike.[124] After the alert was cancelled, the 424th Bomb Squadron began routine bombing missions again.

Every night, the Japanese continued to send a small number of planes to bomb the base area on Morotai. Early on Wednesday, December 20 at 0430, one plane dropped several bombs along the shore, but with no damage. Moderate A/A fire drove the plane off. That evening at 1945, as the men were watching their nightly movie, a twin engine Japanese plane came zooming over the island strafing and dropping a cluster of bombs at Pitoe. The A/A fire was light and late, but fortunately the raid caused no damage or casualties.[125]

After four days with no flying, the Aubrey crew was called for another mission. This time they were called much later than usual. It was Friday, December 22, and it had rained through the night and was raining still in the morning. At their briefing, Walt learned that the target would be a couple Japanese A/D's over on Halmahera. They would try once more to disrupt the staging airstrips that were being used for the nightly nuisance air raids. Except they were not just nuisances. Every few days a plane would be destroyed at one of the airfield dispersal areas, or a crewman would be wounded or even killed by the bombs.

The 424th Squadron would put eight machines in the air. The Aubrey crew would target the large personnel and supply area that was located between Hate Tabako and Lolobata airstrips. These two airstrips were located parallel to each other and shared this large crew and supply area. Their location was maddeningly close to Morotai – just about 50 miles by air, located at the tip of a small peninsula on the Kaoe Baai that separates the two northern Halmahera peninsulas. However, they had a nasty reputation for having a fair number of widely dispersed heavy A/A guns manned by skilled crews. Lt. Hughes in A/C 318 would lead the Squadron. The Aubrey crew in A/C 180 would fly in second position with Jim Fielding in the copilot seat again.

The navigator on this mission would be Lt. John A. White IV, an experienced man who was well respected throughout the squadron, because of his exploits on the October 3 mission to Balikpapan.[126] Walt, Cliff, Frank, and Jack were the other old-timers on the crew. Sgt. Bauer would be combat photographer again, and Sgt. Walter Blawicki was on the crew. One last new man was Sgt. Roy Buchleiter from Denver, an experienced gunner who had been credited with a kill back in June. Each plane was loaded with 20 – 250 pound GP bombs, instantaneously fused.

It was still raining when the crews drove out in the early morning daylight to the Pitoe dispersal area to find A/C 180, armed and waiting for them. It was about 0630 when Don Aubrey lifted off from Pitoe airstrip and began a slow steady climb through 100% cumulus and stratocumulus clouds up to 20,000 feet with continuing heavy rain. However, as they turned south, low and behold, they emerged from the weather front into sunny clear skies with 20 miles visibility. It was a fast flight to their IP at 01°20'N, 127°59'E, just the other side of Kaoe Baai from their target. The IP was close enough to the target that their attack could not be a surprise. The squadron flew at between 12,400 and 12,700 feet on a heading of 115° across the bay to their target.

As they approached, with all the bombardiers and ball turret gunners on the alert, Walt saw the A/A fire begin. Walt had never seen A/A fire from so many guns. The Japanese gunners seemed to have gauged their elevation, because the fire was heavy and accurate. Walt curled up inside his ball turret, trying to get smaller, as he heard shrapnel ping the aluminum skin of A/C 180. At 0801, they were over the target and one by one, the bombardiers dropped their bomb loads and the planes then peeled away. By 0805, the last of the eight planes had dropped and they soon were clear of the ack-ack. Walt watched as at least 100 of the bombs landed in the target area, causing two large secondary explosions and many fires. Walt counted at least five groups of three heavy A/A gun emplacements – one group southeast of Lolobata R/W, one in a wooded area between Lolobata and hate Tabako, one group just to the east of Hate Tabako, one close to the coast and midway between the two airstrips, and one further down the coast near the mouth of the Titilegan River in Wasile Bay.

As a ball turret gunner, Walt was expected to be always alert and report anything significant seen below. Near the last gun emplacement along the shore, Walt saw a Japanese destroyer at anchor, with a Sugar Charlie class freighter laying close-by. A minute later, at 0812, he saw a Fox Tare Charlie class freighter at anchor along the southern shore of Wasile Bay at the mouth of the Sohrain River. From there it was a fast short hop north to Morotai. By 0845 Don Aubrey had A/C 180 safely back down on the ground. After maybe 115 miles as the crow flies and just two hours and 15 minutes in the air, Walt had completed his 36th combat mission. As Don brought A/C 180 to a stop in their assigned revetment, the crew climbed out with their gear and started counting the many little holes from the A/A fire they had encountered over Halmahera. Fortunately, there was no serious damage and none of the crew had been hit. An hour or so later back at the intelligence tent, Lt. William Vincent, the Assistant Intelligence Officer, rated the mission as 'excellent.'[127] Lt. Vincent let them know that their mission was just the start of a big daylong attack on the Jap airbases on Halmahera. In addition to the heavy B-24 attacks, both US and Australian medium B-25 bombers, fighters, and fighter-bombers would attack the Wasile River and Goeroea areas, Lolobata, and Hate Tabako Airfield on Halmahera Island.

But even after the pounding the Jungle Air Force gave to Halmahera that day, the Japanese managed another attack on Morotai that very night. Two Jap planes came in at 5700 feet at 0300. Despite sporadic A/A fire they dropped their strings of bombs, but with no damage or casualties.[128]

About this time, it became apparent that the Army was planning a big attack against the Japanese on Morotai. American troop and supply transports had been arriving for several days and unloading men, trucks, artillery, and stores of all kinds.

More and more of the experienced old hands were getting their orders with permission to halt combat flying and await shipping back home. Walt, Cliff, Jack, and Frank, as well as Don and Jimmie Clark were wondering when their time would come. Late the following day Walt got the orders for another mission on Christmas Day. But Major Vanderpoel was also looking for volunteers to help fill out crews for a mission on Christmas Eve. Walt, Cliff, and Walt Blawicki each volunteered.

Early on Christmas Eve 1944, Walt, Cliff, and Blawicki woke early. At the briefing, Walt learned that he would be flying in A/C 179 piloted by Lt. Vern Michel. Michel, from Oakdale, California, had been Smokey Guild's co-pilot and he was now getting his chance in the pilot's seat. Michel's copilot would be a new man, Lt. Clark (not Jimmie Clark), with Lt. Irving as navigator and Lt. Neeker as bombardier. Along with Walt in the ball turret, the other EM included Sgt. Dick 'Junior' Reis, Sgt. Vernon Gentry from Davenport, Iowa, Sgt. Edward M. Rees, Sgt. Elliot Lott, from St. Augustine, Florida, Sgt. Thomas Cottone, and as combat photographer, Cpl. Worley. [129] The target for the day was Puerto Princesa A/D, on Palawan Island. Walt had been on a mission to Puerto Princesa before, back on November 29. The 424th would put six machines in the air for this mission, with the Michel crew in the number two position, behind Lt. Kasser in A/C 546. Cliff was assigned to the Kasser crew for this mission. Walt Blawicki was flying with the Lt. Christian crew in A/C 275, and Lt. McRae was flying again as pilot of A/C 958. Each B-24 was loaded with 20 – 250 pound GP bombs with .1 nose and .01 tail fuses. They learned that there would be no fighter cover available for them, so the gunners needed to be ever alert.

Out at Pitoe field, Walt and Dick Reis helped coach the new men. Lt. Michel got the crew going and by 0640, he had A/C 179 into the air and headed northwest. The weather started out with 70% stratocumulus at 5,000 feet, with 20% alto cumulus at 10,000 feet, and a 20% cirrus layer above at 20,000 feet. The route took them over the Talaud Islands and across Moro Gulf. Walt spotted many small luggers and barges in the Gulf. Then they flew over the western arm of Mindanao, where Dick Reis spotted very distinct mirror flashes that looked like a rescue signal. Michel had the navigator make a careful note of the position, so they could search more on their way back to Morotai. From here Michel flew to a waypoint at North Islet[130], where Walt observed a radio tower, and then on to the IP at 09°40'N, 118°52'E, over open water about 20 miles south of the target. Here the six planes from the 424th rendezvoused with others from the 370th, 371st, and 372nd Squadrons. At about noon, they formed up, with the 424th in the number three position with squadrons in trail as the Long Rangers approached the target.

The weather over Puerto Princesa was good – 40% stratocumulus at 4,000 feet with a 20% stratocumulus formation towering to 12,000 feet. The 424th planes flew between 10,400 and 10,600 feet on a heading of 296° to 306°. Walt and the other gunners scanned the skies for enemy fighters but none was seen. At about 1223, the 424th planes were on the target with no A/A fire observed. Walt saw the strings of 20 bombs drop away from each plane ahead and then felt A/C 179 lurch as its load was dropped. The seconds ticked away as he watched them first disappear in the distance and then the detonations began…first on the apron on the south side of the runway, then over the western half of the runway proper, and then some hit the apron on the north side of the runway. Walt saw at least 50 hits on the runway itself and Cpl. Worley took a dozen photos of the target. By 1225, Lt. Michel was turning away from the target. As they turned away, Walt noted two Sugar Charlie freighters anchored off shore at Puerto Princesa with three floatplanes nearby.

Enroute back to Morotai, Michel flew to the spot where Junior Reis had noted the rescue mirror flashes earlier in the morning. He dropped down to about 400 feet and they all spotted a crew that had been shot down. Michel rocked his wings to signal that they had been seen, and then they made another pass and dropped emergency rations. Michel radioed the position back to Pitoe.[131]

The return to Morotai was uneventful. Lt. Michel touched down back on Pitoe strip at about 1730. After 10 hours and 50 minutes and 1,691 miles in the air, Walt had now completed combat mission number 37. Later at debriefing, Lt. Vincent would rate their results as excellent.[132]

Walt, Cliff, and Blawicki would be flying again tomorrow. Back at their tent, they met Frank and Jack and they all agreed to get dinner together and then get some sleep. On Christmas Eve, every man's thoughts were on home and on how they had celebrated Christmas in the past. Walt talked about being back home on furlough the year before, enjoying Christmas with his family and friends. They all hoped to be back home again soon. Christmas Eve – silent night, holy night.

However, it was not to be a silent night on Morotai. The Army had other plans and the Japs had other plans. That night an intensive Army artillery bombardment opened up on Japanese positions. These were targeted on known Japanese camps and supply routes. Then by ones or by twos, the Jap air raids started at about 2200. The searchlights would light up the sky and once they caught an enemy plane in their lights the A/A guns would open up. There were to be at least four separate raids that night, but for all of the banging away at them the A/A gunners could not hit anything.

Fortunately, the target for the air raids was the two airstrips – six or more miles away, but they could never know if one of the raiders would bomb or strafe the camp areas.[133] At one point, two of the raiders came in at 10,000 feet from Halmahera and the searchlights were on them for six or seven minutes, but still the A/A couldn't hit them. Fortunately, one of the P-61 night fighters got one as it was withdrawing after its bomb run.[134]

Finally, Walt and his friends got a little sleep before their early morning briefing. They learned that the raid the night before had destroyed one 424th machine – the Squadron's mascot Liberator A/C 424! Three other 424th planes were damaged and required repair before they would be airworthy again. In addition, an RAAF Spitfire being used as a night fighter was shot down.[135]

The squadron men were angry about the loss of A/C 424. Still they had a mission to fly. The 424th would put up every serviceable plane they had – five in all. The target for Monday, December 25 was Sandakan R/W on the north coast of British North Borneo. Sandakan served as a major Japanese administrative center. There was also a prisoner-of-war camp where over 2,000 Australian and British soldiers were imprisoned.[136]

Walt was assigned to A/C 619 with Don Aubrey in command. Cliff, Frank, Jack, and Walt Blawicki were on the crew, with a few new men as well. Sgt. Rafko was flight engineer, and Sgt. John W. Gilchrist from Springfield, Massachusetts, was combat photographer. Jimmie Clark was the bombardier, with Lt. Beam as copilot and Lt. Charles Kaiser Jr. from Los Angeles as navigator.

At about 0645, on a cloudy Christmas Day, Don Aubrey got A/C 619 in the air with its crew of 11 men and 20 – 250 pound GP bombs. As he circled around to head west, the men were able to look down and assess the damage done by the Jap air raid the last night. All the men grumbled about A/C 424, still a smoldering wreck from a direct hit by a Japanese bomb. But soon they penetrated the low 80% cumulus, from just 2,000 feet with buildups to 13,000 feet. Walt eased himself into his ball turret and looked for the scattered openings in the clouds below. After several hours flying, their route took them northwest to Hog Point, the easternmost tip of British North Borneo. This was the assembly point, and Don circled A/C 619 until the other four 424th BS bombers, and more planes from the other squadrons arrived.

From here the route turned due west, over the jungle covered wilderness, for about 80 miles, to their designated turning point. The Long Rangers were flying in trail, with the 424th bringing up the rear behind the 370th, 371st, and 372nd Squadrons. From here they turned north. Sandakan is located on the northern shore of a large estuary. The crews had been briefed to take care because the POW camp was only about one mile from the runway. At about 1150, the Long Rangers began their bomb runs on a heading of about 15°. The weather was good for bombing – 30% cumulus cover with tops at 8,000 feet. Walt was scanning the sky for any Japanese Zeros, but none was seen and no A/A fire from below. Finally, at 1155 the five 424th bombers were over the target at between 10,100 and 10,300 feet when Jimmie Clark toggled away their score of 250-pound bombs. Walt watched them descend, disappear with the distance, and then the explosions began. The bombs made a tight pattern extending about 900 feet, covering the northern half of runway #2. Walt saw two fires at the west edge of the southern half of the runway. Gilchrist snapped ten photos before Don took control back from Jimmie Clark and headed in a loop back to the east. Walt counted barges in the Sandakan Bay as quickly as he could – at least 10, maybe 12. Looking back, he saw no Jap planes along the runway. A little up the coast, Walt saw two barges near the mouth of the Kabon China River.

The flight back to Morotai was uneventful. The weather was much the same going back, although by the time they reached Morotai, the cumulus cover at 7,000 feet had become completely overcast. At about 1535, Don Aubrey put A/C 619 back down onto Pitoe airstrip. Walt had now completed his 38th combat mission after eight hours and 50 minutes and 1,507 miles in the air. Later, after the crew gave their debriefing report at the 424th Intelligence Tent, Captain Ginnane rated the results of the mission as excellent.[137]

What was the rest of Christmas day like for Walt and his friends there on Morotai, thousands of miles away from family and friends? Here is what the 424th Bombardment Squadron History said about that day:

> *"The enemy, Japan, faced an open declaration of war by the 424th Bombardment*
> *Squadron when on Christmas Eve the unofficial flagship of the squadron, A/C # 424, was*
> *destroyed on the ground by an enemy raider. Despite the season, it was quite obvious that*
> *the enemy's heart was not filled with the peace and glad tidings of Christmastide – but*

then neither was George Washington on a social call when he visited the Hessians at Trenton on Christmas Eve of 1776. The loss of A/C # 424 was our only loss during a month of almost nightly air attacks that diminished in frequency and intensity as the month progressed."

Walt and his buddies must have laughed long and hard at the thought of the 'declaration of war on Japan' by the 424th Bombardment Squadron. As the day ended, the officers and men of the 424th Squadron made their way to the mess tent for what served as Christmas dinner and celebration in their little corner of the Gila Peninsula on Morotai.

"Miniature Christmas trees, crumpled by their long journey in gift packages from the States, were smoothed and set in the center of tables in a small, but powerful, effort to hold back the nostalgia of the third Christmas Season overseas. Bits of gift wrapping and ribbons were used as decorations and greeting cards were set upon shelves. The squadron sat down to its Christmas dinner of turkey (white or dark meat, please?) and cranberry sauce, the high point of Christmas Day on Morotai Island."

"As an aid to the attempts at close harmony that were as inevitable as Christmas Eve itself, a song sheet was hastily compiled and distributed among the squadron members at evening mess. The primary purpose of including this song sheet as a part of this historical report is to direct attention to the first song, fittingly, or otherwise, entitled "Saga of the Long Rangers."[138]

The Japs tried to disrupt Christmas on Morotai again though. At about 2230, a single plane approached the base. There was meager A/A fire and the raider dropped a few small bombs out west of Pitoe without damage.[139] That night the Army artillery bombardment of the Jap positions began again, to add to the nighttime noise. The artillery bombardment moved very close to the Allied front lines – within 25 to 35 yards, to hit the Jap soldiers in their dugouts and bunkers. The next morning, 35 Japanese bodies were found by Army patrols.[140]

Walt and his crewmates were talking more and more about home. At this point, Walt had credit for 38 combat missions, and Cliff had 39. The supposed rule was that crewmen were supposed to be rotated home to become flight instructors after 35 missions – but with the caveat that operational considerations of the unit came first. Cliff had it even worse, because as a radio operator, he was flying many more days on supply missions in C-47's and B-24s used as transports. The fatigue was beginning to show on him. Finally Jupe and Walt convinced Cliff that he should go and ask for fewer missions. He went and talked to Major Vanderpoel. Vanderpoel said that it wasn't his call to give him his shipping orders back home, but he did say 'I'll get you assigned as radio operator on a C-47 taking men to Australia for R&R. They can't fly you if you aren't here.'[141]

Second Battle of Morotai

That next day a big infantry battle began on Morotai, just a few miles away to the north. Despite the bombing raids earlier in the month on Halmahera by the 13th Air Force heavy and medium bombers and fighters, and similar attacks by the Fifth Air Force and the Aussies, little by little the Japanese had managed to bring in reinforcements and supplies to Morotai Island. The bulk of the Japanese 211th Infantry Regiment and the Third Battalion of the 210th Regiment were in place around a hill to the northwest of Pitoe. Colonel Kisou Ouchi had assumed command of the Japanese forces in October with the express purpose of retaking the Island. The Japanese plan was to launch a suicide attack on the American-Australian perimeter, break into the base, and then destroy planes and supplies and eliminate all allied forces from the Island. To support this plan, three Japanese Suicide Units, each with about 120 men, had been sent to reinforce the Japanese on Morotai.[142]

The U.S. 33rd Division had arrived from New Guinea on December 21 to attack the Japanese before they launched their attack. The artillery barrages the two previous nights were preparation for the attack. The Army planned to attack the Japanese with overwhelming force, with landings up the east and west coasts, a strong defense in the center of the front, and two deep encircling moves to trap the Japanese before they could retreat into the rugged central mountains. The goal was to encircle the Japanese and kill or capture as many as possible and eliminate the threat to the airbase.

The attack was launched the morning of December 26. Everything went according to plan, except movement in the thick jungle was much slower than anticipated. Within one mile from the beach, the forces landed on the east and west coasts lost radio contact because of the dense jungle. Later light scout planes were used to maintain radio contact with the American columns making their way through the jungle. The jungle through which the Army attacked was oozing with thick mud and dripping with rain. There were many signs of small Japanese bivouacs as the Army advanced little by little. One combat commander reported: 'Japs appear to infest the area. They are well fed, but seem disorganized, travel in small groups. Many small bivouacs were encountered holding 2 to 60 Japs.'[143] There were many small actions as an American and a Japanese patrol would encounter each other in the jungle followed by a short intense firefight.

Despite this battle beyond the perimeter, life for Walt slowed down. He had not gotten his orders to cease combat yet, but there were many new crewmen arriving on Morotai assigned to the 424th. Tuesday, Wednesday, and Thursday passed with no combat for Walt. There were still almost daily Jap air raids. On Tuesday, at 2035, a single twin-engine bomber flew over the base at treetop height with a few .50 cal. A/A guns firing at it. In the early hours of Wednesday morning at about 0045, another Jap plane made a treetop pass over the base. Neither of these planes seemed to drop bombs.[144] On the 27th, Cliff flew a mission to Talisay R/W in the Philippines. He returned to some good news though – his orders had come through for the R&R in Australia that Major Vanderpoel had promised. He would fly to Australia as radio operator on a C-47 ferrying men on R&R.[145] Walt and Jupe wished him well, and asked him to bring back some fresh fruit or Australian beer. Early that morning, at 0400, a single Jap plane raided the base, diving from high altitude to drop a cluster of small bombs then escape, despite heavy A/A fire.[146]

Unknown to the 13th Air Force rank and file, MacArthur was planning a landing on Luzon Island, north of Manila, in early January. The focus of the 307 BG shifted to a stand-by status in the event Japanese Naval forces from the Dutch East Indies attempted to sortie north to threaten MacArthur's landing. Each day from December 27 to 30, the 307th Bomb Group was on alert – planes armed and ready to take off at a moment's notice, but awaiting word of a Japanese naval movement.[147]

There was no mission scheduled on the 28th, but Walt learned he was assigned to a mission the following morning. Walt, Jack, Frank, and Blawicki were wakened early in the morning for their normal pre-mission routine. They were flying with Don Aubrey and felt reassured to have someone they knew well as their pilot. At the briefing, they learned that they were the only plane assigned from the 424th Squadron that day. They were flying a shipping prowl – out to Cape Mangkalibat, on the east coast of Borneo, about mid-way between Tarakan to the north and Balikpapan to the south. Their mission was to search for any Japanese naval movements through the Straits of Makassar, which would then be radioed back to Morotai and the waiting 307th BG heavy bombers.

They were loaded with 2500 pounds of GP bombs. After their early morning take-off, Walt settled into his ball turret and was lowered below the belly of the plane. Their route took them across the Moluccas Sea to Noord Kaap on the northeast tip of the Celebes. From here, Don flew just offshore, along the curving north coast of the island. Walt was of course the primary observer, but the other crewmen were tasked with scanning for shipping as well. Manado was the only sizeable town, but beyond that, there were many inlets and bays for ships to hide along the mountainous coast. Not much was seen though except for a few beached hulks of Jap freighters. The Japanese at this point were reduced to using very small wooden freighters and they went to extreme lengths to camouflage them with palm fronds and hide them in little inlets during the day.

At Strooman Kaap, Aubrey turned to the southwest to make for tiny Simatang Island. Dondo Baai off to the southeast would be a good hiding place for a Japanese flotilla, so Walt scanned for any sign of ships. At Simatang, Aubrey turned west; across the Straat Makassar for Cape Mangkalibat on the extreme eastern tip of Borneo. He flew south and then further east and up Sangkoelirang Baai – an excellent hiding place, but no sign of the Japanese Navy. From here Aubrey turned back to cross Straat Makassar, but further south so that Walt and the other crewmen could search far out to the south for any sign of ships. A few more times Aubrey crossed the strait again but always further south to extend their search area, but still no luck. Finally, their fuel just adequate for their return, Aubrey had the navigator plot a straight-line return back to Morotai. Somewhere along the way on their return, they dropped their 2500-pound bomb load – maybe on an airfield or on some coastal freighters. Finally, after 11 hours and 45 minutes in the air, Don landed back on Morotai. Walt had now completed his 39th combat mission. Later at the debriefing, the results of the mission were rated as excellent. They did not find where the Japanese Navy was, but they were able to verify where it was not, and their bombing had good results.[148]

Walt's mission on the 29th was the very last mission for the 424th Squadron in 1944. The Squadron was seeing a steady inflow of new men. The flying personnel saw an increase in officers from 96 to

120 while the enlisted men increased from 171 to 190. This resulted in rotation home for more and more of the old hands. The ground personnel only increased by one to 15 officers and 375 enlisted men. The 424th BS had exactly 700 men at the end of 1944. The Squadron started the month with 12 B-24's. It lost two to enemy action, but received three new planes for a net 13.

The December 29 mission was the last for more of the old Muroc crew. Jack Riley, Frank Mansir, and Jimmie Clark all received orders relieving them of combat duty by the end of the year. Only Walt, Jupe, and Cliff from the old McRae crew were still waiting the same orders every day.

The following night the 13th Air Force had a little success over Morotai. A Japanese 'Jake' dive-bomber was intercepted by a P-61 night-fighter. Captain Richard D. Stewart of the 419th Night Fighter Squadron got credit for bringing down the raider. This was the last successful kill for the 13th Air Force in 1944.[149]

The morale of the Squadron was considered high. There were only two major gripes the men had. One was the long-standing uncertainty about the rotation system – 'when the hell am I going home?' Replacements were arriving, but for some enlisted men the new arrivals were at higher ranks, which blocked promotion for some of the older men. But by and large, things were OK considering the situation. At the end of 1944, the condition of camp life for the 424th was described as:

> *The squadron is now well settled in its routine of carrying the war to the enemy, both operationally and administratively. As is so typical of the American soldier, it is being done in the most pleasant and comfortable manner that can be contrived. Porches have been built as an addition to the living quarters, small gardens have been industriously hoed and raked, tropical ferns have been transplanted to line walkways and door-ways. There are many radios throughout the area and, at times even the hum of an electric razor can be heard.*[150]

However, there had been a grim toll. At the end of the year, the 307th Bomb Group totaled its losses for 1944. 23.3% of flying personnel were killed, MIA, or wounded during the year![151] 13th Air Force crewmen began to use a new catch phrase. They had all seen or been near to a lot of death during the previous year. The cold statistics proved that. The phrase they now used to greet or say goodbye to each other was 'Home Alive in '45.' It was part question and part hope. The phrase recognized that they had survived, and maybe, just maybe they were going to pull through the war OK.

Fate though sometimes interferes with hopes and dreams. On January 1, A/C 461 was on a routine crew transition flight with Lt. Earnest C. Bean at the controls. As it was landing on Morotai, the left tire blew out and the plane swerved violently, left the runway, and crashed into a large mound of earth and coral. All crewmembers escaped with minor scrapes and bruises, except for Lt. Bean, who was seriously injured. He was taken to the hospital where on January 8 it was found necessary to amputate his left arm. At 1815, he died from complications and he was buried the next day with full honors in the Morotai American Military Cemetery.[152]

There were a couple special events to celebrate New Year Day. The men of the 371st BS put on a talent show called 'Mile High' and each man was issued a case of beer – a first![153] While the men of the 13th Air Force on Morotai rang out the old year and rang in the new, there was a grim deadly fight going on just a few miles away in the thick Morotai jungles, as the attack of the U.S. 33rd Division against the Japanese continued. The going was much more difficult than originally imagined. The Army intended to build roads inland from the landing beaches to support the attacking troops. However, a D7 bulldozer landed to build this road became mired within 10 yards of the beach. It became apparent that roads could not be built fast enough to resupply the combat troops. There were not enough native porters available to supply them either. So it was decided to supply the forces by air. By the time the campaign ended, 31,675 pounds of supplies were airdropped.

By the first of the year as the American forces penetrated inland, most of the Japanese were cut off from retreat into the mountains. They began falling back to an area of high ground called 'Hill 40', which had a view of the Pitoe and Wama airfields. From December 30 to January 2, the Americans advanced towards the hill via deep gullies and ravines. The vegetation was thick with visibility up the trails usually limited to only 20 feet. The trees, many 100 feet tall made it difficult to use mortars. The Japanese defenders were armed with rifles, grenades, light machine guns, and knee mortars.[154]

The Japanese were tenacious defenders, but they were pushed back little by little. Conditions for the Americans were difficult. A major problem was evacuation of wounded men. It took two days for two men to carry a single casualty to the beach. The poor radio contact made it difficult to call in artillery support. Eventually a stream was found that could be used to float casualties to the beach on improvised rafts, which cut evacuation time to just one day. Eventually also, a wire line was run to the beach which would allow better communications with the artillery.

The attack on Hill 40 commenced on January 4 after a nightlong artillery barrage. It was later learned that most of the artillery fragments were deflected by the thick trees on the hill. As the dawn attack was launched, there was little possibility to use mortars and automatic weapons and most of the fighting was done by men with M1 carbines and hand grenades. By nightfall, slow but steady progress had been made. It was assumed that the Japanese would counterattack at night, so the American troops were ordered to fall back about 100 yards to clear the Japanese front lines for another artillery barrage. By January 8, Hill 40 was cleared of Japanese defenders. 264 Japs were killed and seven taken prisoner. The Americans lost 27 killed and 73 wounded.[155]

On January 2, as the Hill 40 attack began, Walt and a number of other men were summoned in their best uniforms to 424th Bomb Squadron HQ. There Walt and the others were awarded a Second Oak Leaf Cluster to their Air Medal for 'Missions flown between August 29 and November 1, 1944'.[156]

The following day Walt was ordered to the infirmary where he stood in a long line with other men to receive triple typhoid, typhus, and cholera vaccinations.[157] This was a good sign for Walt, because these inoculations were required for any man returning to the U.S. from duty in the South Pacific.

The Japanese air raids on Morotai tapered off but did not cease altogether. Late on Saturday the 4th, a single Japanese plane approached the airstrips as harassing A/A fire opened up. It was enough to cause the plane to jettison its four or five bombs into the ocean as the pilot turned away back home. Two days later, on January 6 at 1600, a single Jap recon plane sped over the Island, catching everyone by surprise.[158]

As January proceeded and the men of the 13th Air Force went about their business, the Army continued to harass and kill the Japanese and drive them further north. After Hill 40 had been cleared, the Army continued working its way up the east and west coasts and sending strong patrols inland to seek out Japanese retiring into the interior.

There was no single big battle, but many small quick and deadly firefights deep in the dense jungle. On the morning of January 5, the Japs made a desperate Banzai attack on Company B. Eight Japanese were killed; the last man was the Jap officer leading the charge, who threw his samurai sword into the American positions as he was cut down by BAR fire. Elsewhere another squad of Company B was advancing when they came over a small rise. Eight men were ambushed and cut down by Jap machine gun fire in just 10 seconds. Another squad on the right flank attacked the Japanese and finished them off with hand grenades.[159]

On January 7, a short distance up the Toetoehoe River a force of 75 Japanese was encountered. Five enemy were killed with the loss of a single American officer. However, his body could not be recovered until the following day, when it was discovered that the Japs had been mutilated his body beyond recognition.[160]

That same day a Japanese soldier was captured who reported a plan to land reinforcements from Halmahera at the mouth of the Toetoehoe River on January 10. Patrols continued to crisscross the jungle search for enemy units. By January 10, 'G' Company had been moved to the mouth of the river and laid in wait. North of the river five barges came ashore with about 250 Jap reinforcements who slipped into the jungle unharmed. On the morning of January 11, 40 Japs landed at the mouth of the Toetoehoe River but Company G killed 20. The next morning, January 12, a prisoner was captured who reported that 5,000 Japanese reinforcements were going to land on Morotai on January 15 to destroy the garrison! Based on this intelligence, strong elements were redeployed to oppose the threatened attack on January 15, but it never materialized. The following day the 31st Division was ordered to relieve and hold the advanced positions achieved by the 33rd Division. This relief was completed by January 19 and the operation was considered complete.

At the beginning of the operation on December 20, 1944, the Japanese strength on Morotai was 1,439 men. By January 18, 1945, 804 Japs had been killed or captured. Although the enemy was able to land some additional reinforcements, they never again posed a significant threat to the Allied forces on Morotai. Total U.S. Army losses were 46 killed and 111 wounded.[161]

January 1945 Missions

While this second Battle of Morotai was concluding, time passed by slowly for Walt. A week had passed as new crews were getting flying time, but without Walt. Cliff was back from Australia and had flown a couple supply flights between Morotai and Biak. Biak – another Island off the north coast of Dutch New Guinea had first been invaded by MacArthur back in May, but it had not been until the middle of August that the island was secured at the cost of 3,000 casualties. The Japanese lost 6,000 men. Cliff brought back the news that Biak had become a shipment point for American servicemen in the Southwest Pacific being returned to the U.S.

Later the following week there was a slight increase in Japanese air raids on Morotai. At 0145 on Monday the 8th, two Japanese planes launched a dive-bombing attack. One released four bombs that hit A/A batteries and a hospital area near the beach. 10 men were killed and 31 wounded. The second plane dropped its bombs near a large gas storage tank, but fortunately with no effect. The following night four planes attacked at 0300. Their bombs dropped harmlessly on the beach and in the water. A 419th FS P-61 night-fighter, piloted by Lt. Ralph R. Levitt, shot down one Betty twin-engine bomber.[162] Two days later, at 0300, a single dive-bomber attacked from high altitude before releasing three heavy bombs. There was meager A/A fire and the Japanese bombs destroyed one B-24 and wounded six men.[163]

A week and more had gone by in the New Year and Walt had yet to be called for a combat mission. The Squadron had been busy though, with bombing missions to Halmahera, to pin down the Japanese while the ground combat was concluding on Morotai, and to Borneo and Palawan, to harass any Japanese that tried to interfere with U.S. operations.

On January 6, the 424th began its first bombing raids on the airfields around Manila: Nielsen, Batangas, and Grace Park. These raids on Manila were timed to coincide with the arrival of strong U.S. Naval forces and more than 200,000 soldiers in transports in the Linguyan Gulf, about 120 miles north by northwest of Manila. A naval bombardment of the landing beach area began on January 6, as the Japanese began to counter with bombing and kamikaze attacks on the ships. Three days later, the U.S. Sixth Army, with 68,000 men under Gen. Walter Krueger landed in Linguyan Gulf and begin driving inland, towards Manila. By January 9, all the men had landed ashore. The role of the Thirteenth Airforce was critical during these landings. By January 12, the Japanese managed to sink 24 American ships and damage 67 more American and Australian ships.

Finally, on Friday, January 12, Walt was assigned to another combat missions. He would be flying with a mixed crew piloted by Lt. William M. Hughes III[164]. The target was Grace Park Airfield, north of Manila. Prior to the war, Grace Park had been laid out as a subdivision of Manila. The airport was the first commercial air facility in the Philippines. It was turned into a major Japanese airfield and its destruction was key to the American invasion. Walt was somewhat excited after the briefing to be going to see Manila. He had not seen a 'big city' since his two week rest leave in Australia. Morotai to Manila is a long flight, so take-off was before dawn. As Walt was lowered into his ball turret and Hughes headed north by northwest, Walt again saw the Pacific flow by, the clouds come in to view and then disappear, and islands large and small. The route took them over Mindanao, Cebu, Negros, and Panay, islands that Walt and the 13th Air Force had visited on many past missions. Off to the left was Mindoro Island, and then Luzon, the main island of the Philippines came into view.

There was a special routine that the 13th Air Force used for all bombing missions to Manila. The routine was to always approach the city from the southwest so as to fly over the Santo Tomas Internment Camp. Since 1942, over 4,000 American civilians and military were held here under near-starvation conditions by the Japanese. This was intended to demoralize the Japanese captors and give their prisoners hope that they soon would be liberated.[165]

Finally sprawling Manila came in to view below, with a dense old Spanish central city and meandering rivers winding towards Manila Bay. Grace Park came in to view and the mass of bombers made their attack with no opposition other than some sporadic A/A fire. The 13th Air Force was good at what it did and Walt observed the excellent bomb patterns spread over the airfield and nearby barracks and shop facilities. With that, Hughes turned their B-24 for home. The return was uneventful and by about 1700 they were back on the ground at Morotai. Walt had clocked another 11 hours 40 minutes in the air and a 2,000-mile round trip to complete his 40th combat mission.[166]

Walt had the following day off but learned that he was slated for a mission the following day, Sunday, January 14. Cliff had great news – after 40 combat missions, he had received permission from 424th BS Operations to stop combat. Walt and Jupe were happy for him, with hearty congratulations. Cliff was told to turn in his flak-suit, parachute, and other government-issued combat gear. This was still not shipping orders home, but that would be coming next for Cliff. Now he had nothing to do but laze the day away, swimming, making necklaces from shells, and of course watching a movie every night.[167]

But Walt was going back to Manila to support the Army on Luzon. The target was Grace Park Airfield again.[168] Walt was assigned to fly with Lt. Alan 'Smokey' Guild, a well-respected pilot that he knew of by reputation and from having flown with one of Guild's regular crewmen before, Dick Reis. They would be flying in old A/C 180 in the number one lead position with six other 424th BS B-24's. Other men flying with Guild, Reis, and Walt were Lt. Willie, the copilot, Lt. Charles Kaiser from Los Angeles – Guild's navigator, and Lt. William Fawcett from San Antonio[169], Guild's bombardier. The other enlisted men with Walt and Dick Reis were Sgt. Vernon Gentry from Davenport, Iowa, Sgt. Edward M. Rees, Radio operator Sgt. Glenn F. Scott[170] from Richmond, California, Sgt. Klein, and Sgt. Elliot Lott, the combat photographer. Ed Rees was the ball turret gunner on Guild's crew, so Walt agreed to man one of the waist guns on this mission and relieve Rees, if needed.[171]

Take off was very early, to allow them to make their return to Morotai while there was still light. By 0400 the planes were taxiing at Pitoe and by 0415, Guild had A/C 180 in the air and climbing to cruising altitude. Guild set a northwesterly course in the pre-dawn darkness. It first became light as they passed over the Talaud Eilanden, about half way to Mindanao. The early morning cloud cover was dense, with 90% stratus at 13,000 feet, and visibility two miles. Their route took them to Glutango Point in Sibuguey Bay at 7°28'N, 122°45'E. From this point, an observer on the ground might think they were headed to Palawan or even further west to French Indo China. But they crossed over Mindanao and then turned north. After a couple more hours of flying through 60% stratus at 12,000 feet with 40% cumulus tops at 8,000 feet and 20 miles visibility, they reached Naso Point, the southern tip of Panay Island. From here they proceeded first to Dihanic Point, at 14°36'N, 121°42'E, on the east coast of Luzon, opposite Manila on the west coast, and then yet further north to Uniray Town, a small fishing village up a broad meandering river at 15°13'N, 121°25'E. The weather over the water east of Luzon Island was clear, but as they turned southwest to the IP at 14°47'N, 121°00'E, they encountered thick clouds built up over the land.

Despite the thick weather, Guild had managed to gather the other six 424th BS machines together. However, as they headed south towards their target at Grace Park, they flew in to 90% strato-cumulus with tops at 6,000 feet, and one of the planes became separated. It later joined up with another Long Ranger squadron and returned safely to Morotai. Guild lowered the altitude to 8,500 feet to try to penetrate the clouds, even though this would make any Japanese A/A fire more dangerous. As they approached Grace Park, Guild found it completed closed in with thick cloud cover and he ordered the six planes to proceed to the secondary target at Nielsen Filed.

Nielson Airfield, located on the southern outskirts of Manila, had been built before the war with two good intersecting runways and taxiways connecting them with extensive dispersal areas. It was used before the Japanese conquest in 1942 by the American Far Eastern Air Force and by Philippine Airlines (PAL). After the war began, the American FEAF enlarged the base with hangers, workshops and facilities. This was FEAF's Manila Air Depot where new aircraft were assembled and equipment stored. In early 1945, the Japanese Unit based at Nielson Field was 51 Hikoshidan Shirebu Hikohan (51st Air Brigade) Ki-49.[172]

However, Guild found Nielsen Field socked in too so he headed the Squadron south to the tertiary target at Batangas A/D. Enroute, the clouds parted over Lake Lawa ng Taal, and as they passed by West Lipa Airfield, six unidentified aircraft were sighted at 1030. Walt saw one take off and the others were dispersed along the western edge of the runway. Approaching Batangas, the cloud cover thickened again. Guild told the planes to bomb the target if they could see it through the shifting clouds. At 1110, at 8,500 feet one of the following planes saw a little seam in the clouds and dropped its load on target. 3,600 pounds of bombs from this one plane spread over the length of the strip, from north to south. There were other nearby targets, but now Guild made a fast assessment of his remaining fuel and asked for fuel reports from the other planes. He concluded that there was not enough remaining fuel to try another target and ordered the squadron to return to Morotai via direct route. Staying in formation, at a point about midway between Luzon, Mindoro, and Marinduque Islands, the six planes jettisoned 21,600 pounds of bombs into the sea. Then they headed southwest over Mindoro. Just as they cleared the coast at 1151, near the little seaport of San Jose, an unidentified ship was observed through the clouds. It was under attack by six aircraft and the ship was on fire, but no clear identification of ship or aircraft could be made.

On the return to Morotai, A/C 847 piloted by Lt. MacCluskey went low on fuel. MacCluskey decided to land at Tacloban A/D on Leyte to refuel. The route back for A/C 180 was otherwise uneventful. The weather was 60% to 80% towering cumulus with tops at 14,000 feet over landmasses, but clear over the water. At about 1735, Guild brought A/C 180 safely back down at Pitoe after about 2,000 miles and 13 hours and 35 minutes in the air. This completed Walt's 41st combat mission. During the debriefing the results of the mission was rated as excellent despite the fact that only one bomber dropped its load on the target.[173]

The next week passed quietly for Walt, Jupe, Blawicki, and Cliff. The Squadron was busy with combat missions every day, to Jesselton A/D in Borneo on the 15th, Miri A/D in Sarawak on the 16th, and Talisay A/D in the Philippines on the 17th. On the 18th, the target was a Jap airfield on Miti Island, just 40 miles south of Morotai, off the coast of Halmahera near Tobelo. The next day the target was La Carlota on Palawan, then Fabrica on Negros Island on the 20th. On the 21st two targets were hit: Zablan A/D near Manila and Silay A/D on Negros near Bacolod. Late that day, Walt, Jupe, and Blawicki learned that they were scheduled for a mission the following day.

At about 0530 on Monday morning, Walt and his buddies were awakened for the normal pre-mission routine. After breakfast and a latrine stop, they grabbed their gear and assembled for the briefing. Their target for the day was Nielsen A/D near Manila. The 424th would put seven planes in the air along with like numbers from the 370th, 371st, and 372nd Squadrons. Each plane was loaded with nine 500-pound GP bombs, except for one, which carried nine 300-pound GP bombs. This plane (A/C 236, Lt. Vaneiel, Pilot) would serve as a spare, if a plane from any of the Long Ranger squadrons had to drop out for any reason.

Walt, Blawicki, and Jupe were assigned to A/C 958, flying in the #2 position. This mission was partly a test for some new pilots or co-pilots who had not flown in the pilot seat before. The lead plane, A/C 275, was piloted by Lt. Evans, but the co-pilot was Alan 'Smokey' Guild to check out Evan's skills and abilities. Piloting A/C 958 was Lt. James L. Raue, from Elgin, Illinois.[174] On this mission, his co-pilot was Capt. George Luketz. Luketz was commander of the 394th Bomb Squadron of the Fifth Air Force and was observing Long Ranger combat practices.[175] Other crewmembers were Lt. Lee W. Birkenfield, the navigator, and Lt. Loren Patten the bombardier, from Waupun, Wisconsin. Patten was the old bombardier for the MacMillan crew.[176] The other enlisted men were Sgt. Robert S. Riddle from Centerville, Mississippi, Sgt. Ted V. Samples from Murtagh, Idaho, Sgt. Hansbery, and the combat photographer, Sgt. Wilson.

It was not until about 0800 that Raue and the crew on A/C 958 took off from Pitoe and climbed into the tropical morning. The flight north began with intermittent lite rain and moderate turbulence with one-mile visibility. Their route took them first to Olutanga Point on Mindanao, at 07°20'N, 122°46'E. Once southern Mindanao was reached, conditions cleared with 50% stratus at 15,000 feet and 30% cumulus tops at 7,000 feet, and unlimited visibility. Their next waypoint was Naso Point on the southern tip of Panay Island at 10°30'N, 121°55'E. From Naso Point to the southern coast of Luzon Island the clouds were higher still, with 60% cirrus at 25,000 feet, 60% stratus at 20,000 feet, and 30% cumulus tops at 7,000 feet and continued unlimited visibility. Over Negros, at 09°31'N, 122°36'E, Walt spotted mirror flashes and a column of smoke. The location was noted so that a search could be conducted later, to see if it was a rescue signal. Then they flew on to Nuestre de Campe Island at 12°56'N, 121°45'E, where they then formed up into squadrons and then into a group box formation. More than two dozen Long Ranger heavy bombers then headed north to the turning point at 14°20'N, 121°26E, located at the mouth of the Pagsanjan River where it empties into Laguna de Bay, a large fresh water lake to the southeast of Manila. Here the bombers formed into squadrons in trail formation and flew on to the IP at 14°35'N, 121°07'E, just over the eastern suburbs of Manila.

Once over the Manila area though, the cloud cover changed. Three decks of clouds made sighting the target difficult, with 30% stratus at 17,000 feet, 60% stratus at 13,000 feet, and 70% cumulus at 9,000 feet. The visibility was unlimited, but the cumulus cover at 9,000 feet made it tricky to get a clear view of the target from the attack elevation at about 10,000 feet. From the IP they turned left to Nielsen, on the southern outskirts of the city. Walt got occasional views of Manila and the Bay to the north, but the lead planes had no good clear view of the target at Nielsen Field. They tried the secondary target at Zablan A/D on the eastern outskirts of the city, just a few miles away. Still no good view of the target suitable for accurate bombing. For an hour or so, Raue followed Lt, Evans in the Squadron lead plane, who followed the squadron ahead, as they flew back and forth in formation looking for a clear sight of either the primary or secondary target. Fortunately, there was no A/A fire and no Japanese fighter planes to interfere with this extended tour of the Manila area.

The decision was made to head to the tertiary target, at Fabrica on Negros Island. This was about 300 miles distant on the route back to Morotai, so a good choice considering the fuel that had been expended over Manila. The direct route took the Long Rangers southeast down the length of the Sibuyan Sea, providing Walt a great view of little islands and blue water and scattered clouds. By about 1400, they were nearing Fabrica at the northern tip of Negros. Walt had flown on missions to Fabrica twice before, but this time they were approaching from over the sea, flying south. The 424th was in the number four position, behind the three other squadrons in the group. The weather was looking good for the attack, with 40% cirrus at 20,000 feet, 30% cumulus at 4,000 feet and unlimited visibility. As long as there was no cumulus cloud cover directly over Fabrica A/D when they arrived, they would have a good chance to launch the attack.

At about 1415 at an altitude ranging from 9,900 to 10,600 feet and a heading of 165° to 180°, the attack began. Walt scanned the skies front, sides, and to the rear for any possible Japanese fighters in the air. He braced for any possible A/A fire. Then he saw the explosions below as the bombs from the first squadrons began to hit the target. Finally, it was the turn of the 424th Squadron and by 1428, Walt felt the lurch as A/C 958 dropped its bombs. The bombs were targeted at a point one-third of the way from the south end of the runway and they walked about 400 feet along its length. At least three bombs made direct hits. Walt saw black smoke from three fires rise from the taxiway and revetment area – the smoke from two of the fires rose 1,000 feet into the air. Then the bomb bay doors closed and Raue turned the plane away to head back to Morotai.

The route back took them southeast, over the length of Mindanao. As the four engines roared and Walt gazed down at the strange mountainous green land below, what did he think about his recent past in the Southwest Pacific? And what were his hopes and dreams for the future? Just as the plane cleared the southern tip of the island at 1740, flying at about 6,500 feet, Walt reported a barge below headed on a course of 195°. Soon the tropical night descended over A/C 958 and Walt was able to leave his ball turret and relax as best he could in the belly of the plane. It was late – at about 2020 – when Raue made a successful night landing at Morotai. After 12 hours and 20 minutes and another 2000 miles in the air, Walt had completed his 42nd combat mission. It was very late when the post-mission debriefing was completed. Lt. William Vincent, the new 424th Intelligence Officer rated the results as excellent, with 29,700 pounds of bombs dropped on target.[177]

It was almost midnight when Walt, Jupe, and Blawicki made it back to their tent for a well-deserved sleep. Walt awoke the following day, Tuesday, January 23. After breakfast, he checked the Squadron duty board and seeing that his name was not listed, he was free for the rest of the day, and then the following day, and the day after that. Jupe, Blawicki, Walt and other old hands started having a lot of free time. There was a little excitement when the Squadron enlisted men were assembled on the parade ground to be read the 'Articles of War.'[178] This was just a formality that reminded the men of common-sense rules of conduct. The rumor was that the rules were being read to them now to remind men of expectations once they were rotated back home and in a military environment much less informal than a remote combat base thousands of miles away.

On Sunday, January 28, Walt and the rest of the camp was awakened at about 0615 with the sound of loud explosions from the direction of the airfields. A bomber had cracked up on takeoff for a mission. The plane started burning at the end of the runway as rescue crews approached. Then all of a sudden the flames flared up and one of the 1,000-pound bombs on board exploded. Men went running in all directions and diving for cover as shrapnel started raining down. Then another bomb exploded, and then another. Half of the crew perished in the unlucky plane.[179]

The old hands like Walt in the 307th became more and more interested in the point system used to decide when a man was eligible to go home. The common understanding was that a man received a point for every 3 months overseas, a point for every 100 hours of 'South Pacific' flying time[180], and a point for every 10 missions. The lists were kept by specialty; so there was a list for pilots, navigators, engineers, bombardiers, and gunners. The point system was carried out to three decimal places and it was posted on a bulletin board. Everyone kept track of his name on the list. When there were sufficient trained replacement crewmen, a man at the top of his specialty list would be eligible to return home.[181]

Walt did the math: He had arrived on Guadalcanal on April 11, 1944. As of January 23, he had been overseas for nine months and 12 day, or 9-2/5 months. He had now been credited with 42 combat missions. He had been assigned to combat 45 times. However, three missions were 'turn-back' to base and not credited as combat. Then he carefully totaled his hand-written tally[182] to compare against a running total kept by the Squadron Operations Office for each man. This official list included the formula.[183] Walt discovered an error on his FDFR. It was missing his December 24 mission to Puerto Princesa flying with Lt. Michel. He ensured that the operations clerk checked the mission report and added that mission to the bottom of his record. Walt's calculations on his personal hand-written record contained on error. Through October 24, he had correctly tallied 211 hours and 40 minutes of combat flying time. However, when he added the 13 hours and 50 minutes for his next mission on November 2, he came up with 224 hours and 30 minutes: one hour short. He carried this error forward for the rest of his calculations. As of January 23, he calculated 458 hours and 45 minutes of combat flying, plus 32 hours and 5 minutes of other South Pacific flying, for a total of 489 hours and 50 minutes. Walt added up the time, checked and double-checked it. Then he ran the calculations:

9-2/5 months / 3 months =	3.133 points
42 combat missions / 10 mission =	4.200 points
489 hours and 50 minutes, or 489.833 hours / 100 hours =	4.898 points
Total:	**12.231 points**

Was that good? Walt and Jupe compared their calculations with what was posted for gunners in the 424th Operations Tent. Yes – their names were getting closer to the top. They hoped that their day would come soon, but they knew from past experience that some men waited many weeks before the Squadron finally and conclusively determined that any given man could be rotated back home.

For Walt and Jupe, time seemed to drift by slowly. On occasion, they would help instruct newly arrived aircrew members on combat flight routine. The Japanese air raids were infrequent now and the second battle of Morotai had moved the remaining Japanese troops on the island many miles distant into the rugged and thick jungle-covered interior. Baseball, softball, swimming, and volleyball during the day and movies at night were their chief diversions.[184]

By the end of the month, the 424th Bomb Squadron had lost one plane because of wear and age, but received two new planes, bringing the total to 14 B-24's. Some replacement ground personnel were arriving and some of the old hands were being sent home. Some of these men had been in the South Pacific for two years and no one begrudged them the transfer back to the states. By the end of the month, the ground echelon was reduced by one officer to 14, and by ten enlisted men, for a total of 388. Of the aircrews, there was a net reduction of two officers to 188 total, and increase of five enlisted men for a total of 184. Total strength of the squadron was 704 officers and men.[185]

The month of February began with heavy rainfall, and almost every day of the month there would be rain. On Monday, February 5, Walt and many other old hands were ordered to report to the Squadron parade ground when the rain cleared and there was a short spell of clear skies. There Major Vanderpoel presented awards for many of the men assembled there. By order of HQ-FEAF APO 925, Walt was award the Air Medal, 'For operations June 11, 1944 through August 10, 1944' plus a First Oak Leaf Cluster 'For operations August 29, 1944 through November 15, 1944' and a Second Oak Leaf Cluster for 'Operations November 11, 1944 through December 8, 1944.'[186]

Another week drifted by and it had now ben three weeks since Walt had been scheduled for combat. February 13 marked two years since the Long Rangers first began combat operations. That morning Walt was assigned to check out a couple new crewmen on ball turret operations. New gunners and even some of the older men were become lax in their efficiency and alertness. Mid-morning they reported out to Pitoe where a B-24 was waiting for them. In addition to Walt, a few other old hands were there as well, to check other newly arrived men out on operation of the nose, tail, and top turrets. After an hour or so of briefings, Walt and the other men settled into position for take-off. Soon the training flight was in the air and headed out over the ocean. Walt assisted the two new men assigned to the ball turret one by one. Each man test fired his guns, rotated the turret through all of its various ranges, then helped raise the turret again and help the man out. The same routine was followed with the other turrets. Soon everyone was checked out and the plane returned to Pitoe after just an hour in the air.[187]

Later that day there was a memorial program at the Long Ranger Chapel to commemorate all the men who had lost their lives in two years of war.[188] Chaplain Frank Dennis presided. Sgt. Mellville Miller sang the hymn *Sleep, Comrades, Sleep*, with words by Longfellow:

> Sleep, comrades, sleep and rest
> On this Field of the Grounded Arms,
> Where foes no more molest,
> Nor sentry's shot alarms!

All is repose and peace,
  Untrampled lies the sod;
The shouts of battle cease,
  It is the Truce of God!

Rest, comrades, rest and sleep!
  The thoughts of men shall be
As sentinels to keep
  Your rest from danger free.

Your silent tents of green
  We deck with fragrant flowers;
Yours has the suffering been,
  The memory shall be ours.

Taps closed out the ceremony.

But the truth was that the job of the aerial gunners was becoming easy. Of course, every man on every crew had to be always alert, but as February progressed, there were no attacks by enemy fighters or by ground A/A fire, at all. The 424th Bomb Squadron continued to bomb the Manila area, including Corregidor Island and the Canacao Peninsula base facility until the middle of the month when American airborne troops landed on the island, finding almost no effective Japanese resistance.

On Wednesday, February 14, Cliff Llewellyn and Jimmie Lynch said goodbye and left Morotai at 1000 via C-47.[189] The rain continued – every day heavy rain. Fortunately, the camp area had been well constructed with crushed coral side roads and walkways. The crushed coral drained quickly, so that the camp area remained in good shape. The main road up the length of the Gila Peninsula was a problem though. Known as Skyline Drive, it saw heavy truck and jeep traffic, which churned up the road into a quagmire of thick, slippery mud and cratered in many areas. Some men said that the road looked like the Long Rangers had done a good job bombing it. It was a rough, gut-wrenching five mile drive from the 424th BS Camp area to the flight lines, and one man said 'it ain't the missions I'm worried about, it's that damned trip back and forth from the line.'

The Squadron was equipped with 14 B-24's, but the ground crews were having difficulty with maintenance. Not because of qualified mechanics though – in fact, now in its 29th month overseas, 59% of the Squadron personnel were from the original contingent! The problem was supplies and parts – the engineering department was short of spare aircraft engines, fire suppression equipment, and parts of all kinds. The motor pool was short of vehicles, with half its authorized number and many were three years old and broke-down frequently. Major Vanderpoel made sure that his superiors at the Group level were aware of this situation. Still, the Squadron was operating at a high level of efficiency and bringing destruction to the Japanese in the Philippines and on Borneo.[190]

On Thursday, February 15, Walt and many other men were again paraded and then presented with the 'Good Conduct' medal by Major Vanderpoel.[191] Three days later, on Sunday, February 18, Walt and a few dozen other men were told to report to the Squadron Operations office. Jupe was there, as well as Dick Reis and some other men that Walt had flown with before: Roy Buchleiter, Angelo Caputo, Elliott Lott, Robert Riddle, and Ted Sample. There Walt and the others got the news they had been waiting and hoping for. They each received orders from Headquarters, Far East Air Forces, stating that they were 'Released from present assignment and from further duty in the Southwest Pacific Area and assigned for processing and transportation to the United States.' Their next stop would be the 11th Replacement Battalion at APO 920 (on Biak Island). There they would get orders to return to the U.S. via either air or sea. They might be assigned to military or civilian aircraft or transport ship. Once they reached the U.S., they were instructed to report to the nearest port AAF Liaison Officer, who would then give them each orders for the next AAF redistribution station where they were to report. Any man traveling by air would be restricted to carrying 65 pounds of personal baggage. Any excess personal baggage would be transported by ship to arrive later.

The men were told that over the next few days they would each have a physical exam, have their immunization record checked and if necessary receive any required vaccinations. They would have their service record, pay record, and qualifications card checked and brought up to date. They were also told to write to their families and friends back home and tell them to stop sending any letters to them, until they were notified of their new permanent station.[192]

Each man was given a sheet of paper titled 'Suggestions for Returning Combat Crew Members' some of which were:

> *"You, a returning veteran, will be beset with requests for interviews, statements and public appearances. Often, through other veterans, military information of value to the enemy has been disclosed, inadvertently.*
>
> *"Here are a few pointers which it is suggested you remember:*
>
> *"1. What you state to the press may reach the enemy. Any pertinent news published in the U.S. newspapers reaches enemy hands within 24 hours.*
>
> *"2. Stick to personal experiences….Avoid opinions and predictions…remember, your <u>comrades</u> are still here; <u>protect</u> them.*
>
> *"6. The Army Air Forces, and the Thirteenth Air Force, desire to have the public as fully informed as possible on what their personnel are doing in the combat theaters. Your part in this is to tell your story accurately, without embellishment and with regard for the fact that there are several thousand other men doing every day what you have done – and who, perhaps, will be affected by the report of your story.*
>
> *"7. You are a member of the Thirteenth Air Force. You owe a debt of loyalty to your comrades. The outstanding achievements of your Air Force, and an interest in its continuing success to which you have contributed so much should be a matter of personal pride."*[193]

Walt also received a copy of his combat flying record signed and certified as correct by Captain Carl S. Looker, Operations Officer of the 424th Bombardment Squadron.[194]

Walt spent another week on Morotai. He wrote home telling his family that they should stop writing to him at APO #719 (Morotai) and await further news from him before they wrote again. However, he was able to tell them the good news that he was on his way home. Walt started sorting through his things to get down to 65 pounds of luggage in a single large standard duffle bag. He found his good uniforms that he had brought from the U.S., but had only worn in Hawaii and Australia. At some point, he had bought a leather jacket made in Australia, which he festooned, with Thirteenth Air Force and 424th Bombardment Squadron patches. He packed a few other souvenirs – a tropical skirt made from an old parachute, shell bead necklaces, and odd native carvings.

Walt carefully packed away his military service records. He would need to present his records and orders once he returned to the U.S. for future assignment and orders. There was also a frenzy amongst the returning men to exchange home addresses and promises to 'look you up' after the war was over. Walt, like most every man, had compiled a collection of photographs of many things – friends, camp scenes, group photos, photos of natives, photos taken in the air, and copies of combat photos showing bomb damage to Japanese airfields, or in some cases of Japanese fighter attacks. He also had a collection of currency acquired in the islands: Dutch, Japanese, Philippino, even Chinese. These were all carefully packed away.

On Sunday, February 25, Walt received orders to proceed by air to Biak Island. In the morning, he went to the operations office where he was given a Battle Participation Credit – Awards & Decorations letter, signed by Capt. David J. Davis. This listed the awards and decorations he was authorized to wear.

After an early lunch, Walt retrieved his duffle, said his last good-byes, and headed to the operations tent where a dozen other men were waiting to leave. A truck arrived and each man heaved his duffle up and climbed aboard. One last time Walt made the bumpy five mile drive up the Gila Peninsula to Wama airfield. They were driven out to a C-47A Skytrain transport plane of the 13th AAF – 403 Transport Group, 13th Troop Carrier Squadron. Walt showed a copy of his orders to Cpl. John C. Waker, who welcomed him aboard. Waker, the pilot, Lt. Elbert Lesh and co-pilot Lt. William Goodman had arrived a couple hours earlier from Biak Island. The other passengers were Cpl. Bernedict Gold and S/Sgt Frank J. Barry, from both the 13th TC Squadron, Sgt. Elliott Lott, S/Sgt. Ted V. Samples, S/Sgt. Roy A. Hinkle, S/Sgt. Buchleiter, S/Sgt. Frank J. Doblekar, S/Sgt. John W. Gilchrist, and S/Sgt. Robert S. Riddle. Walt had flown before with Lott, Buchleiter, Gilchrist, and Riddle. He looked back, down the runway and at the trees swaying in the breeze, and then climbed aboard. He tied his duffle down in the middle of the airplane and then took a seat on one side of the plane. At 1330, the door was closed and Lesh was cleared to take off down Wama airstrip.[195]

As Walt felt the plane accelerate down the runway and lift off for the last time from Morotai he reflected on luck. Although none of the men knew the statistics, during 1944, 23% of Long Ranger flying personnel had been killed, MIA, or wounded. Except for a bad cut on his hand from a can of peaches, Walt survived the year unscratched. All of the old original Muroc crew – Charles McRae, Don Aubrey, Jimmie Clark, Ed Dunne, Cliff Llewellyn, Len Sherman, Frank Mansir, Jack Riley, Ed Bretherton, and Walt – had survived unscratched. Cliff had kept track of the crews that had graduated from Muroc. Of the 13 crews in the IV Bomber Command - 382nd Group - 539th Squadron, only three had suffered no casualties at all.[196] Most of the other men Walt served with survived too – Gene Schreiner, Dick Reis, Boris Hidalgo, Carl Appling, Walter Blawicki, Allen Guild, Jim Fielding, and many others. One crew Walt knew, but that was not lucky, was the one that flew with Lt. Jack Arnett five and a half months earlier and was later lost over Palau.

Walt and his friends survived danger on the ground too. By the time he left Morotai, the Japanese had raided the island 82 times with 179 individual sorties. 54 of the raids caused no damage, but 42 allied airplanes were destroyed on the ground and 33 damaged. 19 men were killed by these raids and 99 wounded. Life on Morotai had not been a picnic and for many it was the end of the road.[197]

However, for Walt – the future lay ahead – first to the southeast and then home.

- 15 -
1945: THE OLD MALFUNCTION BROUGHT US IN

Biak

C-47A 43-15236 flew on to the southeast. It is 585 miles from Morotai to Biak and by 1650, Cpl. Walker told the passengers that Biak was in sight and they would be landing soon at Sorido Airdrome. At just about 1700, Lt. Lesh landed and taxied to a few small tents that served as the terminal area. Walker helped get their baggage unloaded and then directed the eight men towards the tent were they were to report and present their orders.[1] The ATC operations clerk told them where to find a tent and bunk, where the mess tents were, and the daily routine. When they asked how soon they would have to wait before a plane back to the States, the orderly laughed. 'There are hundreds – thousands of men like you waiting to return.' They learned that there were C-54's that left from Biak for the U.S., but the priority was for senior officers, and for pregnant nurses who needed to return so that their child would be born a naturalized American citizen. At that point, there were about 750 men ahead of them on the list for transportation back home. There were hundreds more men who had just arrived in theater and were waiting for transportation to Morotai, the Philippines, or other SWPA destinations. They would just have to be patient and settle in to their new home on Biak – and wait.

Biak is similar in size to Morotai, but elongated – 45 miles long and about 23 miles at its widest point. It is midway between Wakde and Morotai, located in Geelvink Baai. Noemfoor is about 75 miles to the west. The native population was mostly Christian. MacArthur had wanted Biak taken to help protect his flank, as he proceeded to invade the Philippines. The island had been held by 11,000 Japanese troops, who also had nine small tanks. On May 7, the Americans had taken Wakde, which was close enough to Biak to provide air cover for the assault. The U.S. 41st Division landed on Biak on May 27, 1944 to little opposition. The Japanese commander, Colonel Kuzume Naoyuki, decided to allow the Americans to land unopposed and advance inland into prepared battle zones, where he hoped to destroy them. The Japanese tanks tried to attack the Americans coming ashore, but a small number of M4 Sherman tanks destroyed them.

Then, as the American infantry advanced the dug-in Japanese opened fire. The topography of Biak includes many limestone caves and low ridges, which dominated the landing area. It took the Americans until August to clear out the bulk of the Japanese, who fought from many small isolated bunkers and strong points. By August 7, the Japanese suffered 6,100 killed and 450 captured, The Americans suffered 474 killed, 2,428 wounded, and an additional 3,500 with bush typhus.[2] Walt was warned that just like on Morotai, there were still Japanese stragglers who would sneak into the base areas to scrounge through the garbage for a little to eat.

There were three airfields on Biak, all adjacent to each other: Sorido, Borokoe, and Mokmer. Connected together by roads, encampments, repair facilities, supply depots, and fuel dumps; they were like a little temporary city in the jungle. There was another airstrip on Owi Island, about three miles away.[3]

Walt and the other new arrivals were directed to rows of canvas-covered platforms that were lined with bare cots, 50 yards from a rocky beach. For the first days the men mostly woke up, ate, and then sat on their cots under the shade of the tarps and talked to other transients waiting to go home. The area was mixed with men from both the Thirteenth and Fifth Air Forces, and sometimes there were heated arguments about the merits of each. The men shared their experiences and compared stories of their missions to Balikpapan, Palau, Truk, and Yap.

Soon Walt found that Cliff and Jimmie Clark were still on Biak. In fact, many of the crewmen that Walt had flown with were still there, including Dick Reis, Gene Schreiner, Richard Cosgrove, Jupe, Frank, Jack, Carl Appling, and others – all waiting for transport. Sometimes the men would go swimming. Across from the rocky beach, there was a coral reef and there was a sunken ship on its side that could be used for diving. Men hunted for shells. There was little information from the outside world, because as transients, they had no recent magazines, newspapers, or letters from home.[4]

Walt and his friends made the most of their leisure time on Biak. They toured a little native village along the coast called Borokoe, where the men wore old Dutch uniforms, handed down from

314

government officials. The food was not bad and there were several outdoor theaters, so the men watched a movie or two after chow every evening. Even though there were warnings about Japanese soldiers who might still be wandering in the jungle, the men started following roads and trails into the jungle back of the airfields. One time, back in the jungle up a bulldozed road, they came across a fenced area with warning signs. They saw a mountain of chemical gas bombs stacked high – to use in retaliation if the Japanese ever used chemical bombs.

One day, Walt and Cliff discovered that they were assigned to KP duty. A T/Sgt and a S/Sgt on KP Duty! They were angry, but what could they do? With all transients and the men returning to the U.S. mostly promoted, it was inevitable. Well, at least because they out-ranked most of the other men on KP, they were treated well and got some of the best food for themselves.[5] On March 1, Walt and most of the other men were summoned to the administration area, where they each had their pay brought up to date.[6] Now they just had to avoid gambling it away at cards and get shipped to some place where they could spend American money.

With many mosquitoes and the risk of malaria in the area, all the crew men were required to take Atabrine tablets to suppress the disease. However, the Atabrine eventually began to turn the men's skin yellow. Before returning home men would work on a suntan, to hide the yellow tint.[7] They had found a fresh water-swimming hole back in the jungle with crystal clear water where they could swim and sit on rocks in the sun. Once Cliff and Gene were swimming there alone. There was a rock with an underwater tunnel that you could swim through to the other side. Cliff dived down and headed for the hole; about 8 feet down, but missed it and bumped his head. He tried again and bumped his head again. Now he was disoriented and starting to panic. However, Gene saw his shape down below and realized something was wrong. He dived down and pulled Cliff up, saving his life.[8]

Across the Pacific

A week later, on Thursday, March 8 a troop transport ship arrived at the pier adjacent to Sorido A/D. This was the U.S.S. Puebla. The men learned that this ship had arrived from the Philippines, where she had picked up several hundred recently liberated POWs and civilian prisoners who had been held in captivity by the Japanese since early 1942. The men learned that the Puebla's next stop was San Francisco, and they had two options – wait for who knows how long to reach the top of the air transport list for a quick flight back to the States, or leave on the Puebla for a long, slow, but sure crossing of the Pacific to San Francisco. The Puebla would be leaving in a few days, so they had to decide quickly. Walt, Cliff, Dick, Gene, and the other men debated the pros and cons of each option.

First, they sized up the ship. The Puebla, they learned, had been built in 1928, by Vulcan Bremen, Vegesak, Germany. Originally named the Orinoco, it served between Hamburg and Bremen and ports in South America.[9] The Orinoco was in Mexico when, on May 22, 1942, Mexico declared war on Germany, Italy, and Japan. The Orinoco's German crew was interned and the Mexican government turned over the Orinoco to the United States, as a contribution to the war effort. It was renamed for the Mexican city Puebla, originally founded in 1531 by Spain and the site of important battles during the Mexican-American War and during the French occupation of Mexico in 1862. The War Shipping Department put the ship into troop transport service beginning in 1943. She was 9,660 gross tons, 484 feet long, with a beam of 60 feet and she drew 24' 6" of water. She was powered by two German diesel engines and could do 13-1/2 knots with a range of over 16,000 miles.[10] She could accommodate over 2,000 passengers.

Soon, Walt and his buddies decided. There were still hundreds of names ahead of them on the air transport list. Walt and 38 other men from the 307th would take passage home on the Puebla. It took a couple more days to get their paperwork completed and orders cut. On Sunday, March 11, they boarded the ship and found places to bunk. The main deck of the Puebla still had the look of a classic passenger ship, with an ornate colonnaded ballroom on the main deck. But down below the ship had been remodeled to accommodate more passengers and the bunks were seaman hammocks. The ship was taking on more passengers, so they were lucky that they had boarded early. They were able to claim bunks on the first deck below the main deck, where, they were told, there would be less ship motion. Plus further below, the stink of thousands of men crammed together would be worse.[11]

The men had been warned when they left Morotai, that no one would be allowed to bring home pets that they had adopted. Many men had acquired little monkeys, parrots, and stray dogs, to help with the loneliness of jungle life. Nevertheless, many of these animals had been carried to Biak. Once again, the men had been warned as they boarded the Puebla, no animals would be allowed into the United States, so they had better release them on Biak. But many were smuggled aboard anyway. Walt's friend Richard Cosgrove was one of the men who had befriended a stray dog on Morotai and he had brought him to Biak and then aboard the Puebla. Despite the warnings, Cosgrove asked his friends to help him keep his dog hidden. This would be difficult, because space was at a premium. The hammocks were slung in sets of three, a top one that looked up at the ship piping and steel girders above, and then a middle and a bottom one that each allowed two feet clearance below the man above. Between every two rows of hammocks was a narrow aisle. The men had to store their dufflebags under the bottom hammock and in the aisle ways, so they would always be stumbling in and out of their bunks. The mess hall where the food would be served was not big and they learned that the food, though adequate, would be monotonous.

For two days Walt, Cliff, Dick and Gene watched as the Puebla loaded more returning troops and released POWs. Walt and Dick noticed a sign posted on the ship asking for volunteers to serve as waiters for the officers' mess. They would be fed anyway, but they learned that the waiters would get to eat the same food prepared for the officers and they would be able to eat in the mess after the officers' meals were finished. All of the galley staff – the cooks and cooks helpers – were Philippinos, so it was not a question of KP duty. It would just be serving the food. Dick and Walt agreed that they would volunteer together.[12] Finally, at 1230, on Tuesday, March 13, the Puebla cast off its mooring lines, turned around in the narrow channel, and headed slowly east.[13] Walt and his buddies watched as the south shore of Biak slowly receded in the distance and they headed out to sea. They learned that they would be sailing alone and non-stop now to San Francisco. Even though they were headed east, away from Japan, they could not be certain that there was no Japanese submarine lurking beneath the water. They would sail a zigzag course, changing direction frequently to make a torpedo attack more difficult. Their route would not be direct to San Francisco, but first they would skirt the north coast of New Guinea, then head southeast across the Solomon Sea until they turned northeast, passed through the Solomon Islands, and then headed direct to San Francisco.[14]

By about 1800 on the first day out from Biak the tropical night descended and the men learned a new routine as an announcement came over the loudspeaker: 'the smoking lamp is out on deck. Smoking below deck only in designated areas.' Smoking on deck at night was forbidden so it would be more difficult for any Japanese ship on or below the surface, or airplane above, to spot the ship. All of the portholes and windows were blacked out and the ship ran without running lights. However, on deck during a clear night, in a warm breeze, the intensity of the stars overhead was dazzling for any of the men who cared to gaze up into the tropical night sky. They could also watch the phosphorescent glow of the sea in the wake left behind, as the ship churned small luminous microorganisms as they passed.

The first nights sleeping on board, Cliff, Walt and many of the men felt a little woozy from the motion of the ship, but they all quickly got their 'sea legs.' It was easier to get used to the motion of the ship for returning aircrew men like Walt and Dick, because of the 'iron stomach' developed from hundreds of hours flying in a B-24. But for ground crewmen or ground troops, the first days on board the ship found many of them with chronic seasickness. The airmen learned to avoid these men because at any moment they could be hurling the contents of their stomach over the side rails…if they were fast enough.

Walt slept well, although the crowded bunk area was sometimes noisy, with steel decks, bulkheads, and scores of snoring and farting men. As men first bedded down at night, and then in the morning as men started coming and going, it could be even more raucous. On warm nights, Walt often chose to sleep on deck and watch the stars slowly pass by, or see the moon reflected in the ocean.[15] Walt and Dick quickly fell into their routine of reporting to the officers mess for their server duties, then when the meal was done, enjoying their own leisurely meals in comfortable seats.

The Puebla plodded along at about 13 knots (maybe 15 miles per hour) powered by its two efficient but old German diesel engines. As the days passed by, they would sometimes see little islands or the north coast of New Guinea in the distance. Walt tried to follow their route on the map of the Pacific Ocean that he had purchased the year before at a bookstore in Sydney. Sometimes they would see schools of dolphins riding the bow wave of the ship. Sometimes schools of flying fish would appear. For entertainment, for two hours every morning the ship would pipe popular music over the loudspeakers. There was a small library where men could check out books. Crossword puzzles and checkers were popular, and of course cards. Men played poker, bridge, and Walt always liked to find a pinochle game. Every few days, men would be assigned to KP duty, to help in the enlisted men's galley, but Dick and Walt would be excused because of their volunteer duty in the officers' mess.[16]

After a day at sea they reached a point north of Hollandia, the Dutch capital of New Guinea. Often they saw planes flying far overhead – some they could identify as B-24s headed west to reinforce the Fifth or Thirteenth Air Force. They remarked on how tiny a plane looked a couple miles up in the sky, with the realization that to spot anything or anyone on the water or on the land from that height was largely pure luck. On Thursday the 15th they were opposite Wakde, now all but abandoned as an American air base after the war had moved farther west half a year earlier. The following day they passed through Vitiaz Strait, between New Guinea and New Britain. Walt saw on his map that at the other end of New Britain was Rabaul, where he had flown his first combat missions 10 months previous. Now for two and a half days they sailed further to the southeast through the Solomon Sea. On Sunday, March 18, church services were held in the Puebla's ballroom. There were four chaplains aboard; one was an ex-POW who had lost one leg in captivity on the Philippines Islands.

One day, as Dick and Walt were walking in to the galley to prepare to serve the officers mess, they heard a big commotion. The Philippino cooks and helpers were cursing in Tagalog and slashing at each other with butcher knives! They did not know what the fight was about, and the cooks were not coming at Walt or Dick, but they didn't like the looks of it. Walt and Dick agreed that they would always stick together in the galley and watch out for each other.[17] Except for Walt and Dick, the enlisted men got much poorer food than the POWs and officers. Jim Fielding thought this was not right, since the officers and enlisted men had served together as a team and shared the same dangers and hardships. Fielding complained to the ship's crew about this, but nothing changed.[18]

Boredom set in for many men and there were many practical jokers. To accommodate so many men, additional heads had been constructed in various locations throughout the ship. These were in a large communal room. To one side was a long metal trough with salt water continually flowing in one end and slowly draining out the other end. On top, the seats were made from a long thick wooden plank, with holes spaced at appropriate intervals. The room was poorly ventilated and dimly lit. Men would sit and smoke, waiting for nature to call. Sometimes a man at the end of the trough near the water source would take a wad of crumpled toilet paper, set one tip of it on fire and then set it into the water to float with the flowing water. As the smoking wad floated under each man, they would leap into the air cursing, while the jokester and other men would laugh at their surprise.[19]

For days now, they had been sailing east, but farther and farther south. Now on Monday, the Puebla changed course – northeast – as it made its way through the Solomon Islands just north of Guadalcanal, with Santa Isabel Island on their port side. Now their course was straight home (with anti-submarine zigzags) all the way to San Francisco. Slowly, as the days rolled on, Walt, Cliff, Gene, and Dick, would mingle more with other men on the ship. The released POWs from the Philippines were a sorry lot. It had been a couple months since they had been liberated, but most of them were still thin and frail looking from two and a half years of harsh captivity. They looked emaciated, even after several weeks of freedom, medical care, and ample food. Jim Fielding remembered one man – his '...eye pupils had fused with the whites of his eyes due to malnutrition. Another man had bones so rubbery, that he could visually bend even his shinbones with his fingers.'[20] All of the POWs were extremely well treated and fed by all the ships' crew.

Cliff recounted a story told by one of these POWs who was tortured by the Japanese. This man's friend was a Philippino POW whose father and brothers had been killed and his mother and sisters taken by the Japanese as 'comfort women'. When he was released, he vowed to torture and kill any Japanese soldiers that he found. The men felt pity for these released prisoners and it reminded them of the purpose that they had risked their lives for the past 10 months.[21]

Time passed by slowly – so slowly – as the Puebla kept at its steady pace. The Pacific Ocean is vast, and even though they passed Narau to the north and Tarawa to the south, they never caught sight of these islands. On Sunday, March 25, they crossed the International Date Line and then it was March 24 all over again. A few days later, they briefly sighted Johnson Atoll off on the port side. This tiny island was still maintained as a seaplane base. A few days later, they passed to the northwest of Kauai, westernmost of the inhabited Hawaiian Islands. April 1st was Easter Sunday – 5800 miles out from Biak, and all the men had a little better food for their Easter dinner, but still eaten in cramped and uncomfortable conditions for most of the men.

As the Puebla passed Hawaii to the south and headed east towards San Francisco, it started to become cooler. After many many months in the hot humid tropics, the cool springtime air felt refreshing. A thousand miles out from port, the Puebla developed some mechanical problems. The ship could not keep both of its twin shafts going, which slowed them down. They did not have to cruise the zigzag route, now that they were so far from any Japanese naval activity. Then the ship's supply of fresh water ran very low, because of the slow speed. To save fresh water, the men had to shower with salt water. They lathered up with big bars of lava soap, but after rinsing with the salt water, their skin felt sticky.

Then a storm hit the ship. Rain and high winds pelted the ship as it sailed into high swells. At one point, the men sighted a freighter steaming behind them. The Puebla would seem to be in a low valley between two big wave crests, then it would slowly reach the top of a swell, the stern of the boat would lift clear of the water and the propellers would race, creating a disquieting rumble and strong vibrations throughout the ship. Later, one of the screw shafts froze up completely.[22] Some men worried that they might be lost at sea. A couple hours later, the men noticed that the freighter that had been following them was now well ahead. So much for their 'luxury liner!'

A few days later the ship's crew found out about a parrot that one of the returning airmen had snuck aboard the ship and managed to keep hidden. It was forbidden to bring animals back to the U.S. Shortly afterwards a ship's loudspeaker announced, "If the parrot that is hidden on this ship is not produced, everyone on the ship stays in quarantine after we reach dock in San Francisco!" All of a sudden, through an open hatch, a parrot in its cage came flying out and overboard![23]

The storm continued and the Puebla's speed was slowed, at some times to just four knots. They commented on the name 'Pacific Ocean' and that some parts, at least, were not always peaceful. The men got used to the storm and felt somewhat snug inside the ship. Finally, after three days the storm died down. At around noon on Tuesday, April 3, men started seeing gulls flying overhead. Gulls – that meant that they were nearing land. No one slept that night and in the wee hours of the morning, they started picking out flickering lights in the distance to the east. In the morning, hills could be seen and then the twin towers of the Golden Gate Bridge came in to view. Men cheered.

They went down to pack up their things. Many men looked at the worn and tattered clothes they had brought back and realized that they did not need them anymore. Some men found sacks to stuff unwanted clothes into, and then they took them onto deck and tossed them overboard. Soon a line of floating bags stretched off behind the Puebla.[24]

At 0630, on Wednesday, April 4, 1945, the Puebla passed under the Golden Gate Bridge with its horn blaring. The ship was met by a pilot boat that guided them to Fort Mason.[25] After 7,760 miles and 23 days at sea since leaving Biak, the Puebla docked at Pier 9. It was two days short of one full year since Walt had departed from Hamilton AFB. He was back home in America.[26]

They were surprised that there was a small band playing for the returned POWs and service men, although the sound of the band was drowned out by the cheering from the ship. There were Red Cross girls waiting for them too.[27] As they came down the gangplank, each man was careful to step onto United States soil with his right foot first. They wanted to start their new lives 'off on the right foot.'[28] After a lot of lining up and waiting on Pier 9, they were soon directed to board a ferryboat, the General Frank M. Coxe[29], for a short trip to Fort McDowel on nearby Angel Island.

Fort McDowell[30] was the processing and quarantine station for troops and civilians back from the Pacific Theater. Walt learned they would spend the night on Angel Island and go through standard routine for returning servicemen. At Angel Island, everyone was given physical exams.[31] Later they had wonderful warm fresh water showers. Each man could send a telegram home, letting his family know that they had returned safely and might be home soon. And the food was great and plentiful. They were served dinner and breakfast by German POWs. At one point there was a scuffle. One of the men off the Puebla had just learned that his brother had been killed by the Nazis in Germany. This man jumped the first German waiter that he saw, and had to be restrained by his friends.

The next morning each man was issued new clothes. Then they were directed to a place for final inspection of luggage, before leaving the island. Jim Fielding, Carl Appling, Cliff Llewellyn, Jupe Bretherton, Richard Cosgrove, Gene Schreiner, Jimmie Clark, Dick Reis, Frank Mansir, Jack Riley, Ed Dunnne, and Walt were all smiles and happy to be headed home. Jim Fielding had decided to spend an extra day in San Francisco, but the others all wanted to get home as quickly as possible.[32] Then as they moved along in the inspection processing line there was a big commotion. There was a Lieutenant and an MP and Sgt. Cosgrove – and his dog! They were insisting that Cosgrove's dog had to go into quarantine, but he was insisting that it was coming with him. Walt, Cliff, and the others began giving the Lieutenant a hard time. More officers were called. Cosgrove had a look of crazed determination that after so many months and so many miles, he was not going to be separated from his dog! Finally, after conferring, the officers and the MP turned their backs. Cosgrove walked out the door and onto the waiting ferry, with his dog, to the cheers of everyone there.[33]

Late on Thursday morning, the delay with Cosgrove's dog cleared up, the little ferry left Angel Island for the Southern Pacific Railroad Pier across the Bay in Oakland. The men had each been issued train tickets and meal vouchers for transportation to a processing center. Walt was headed to Fort Sheridan in Chicago and Cliff to Fort Snelling, near St. Paul, Minnesota. At these depots, they would get furlough orders and then orders on when and where to report back to the Air Corps for further assignment. The rumor was that experienced crewmen like Cliff, Walt, and Gene would be assigned as aircrew instructors at some base in the U.S.

At the Southern Pacific Railroad Pier in Oakland, the men were deposited at Depot. The Southern Pacific's main station was at 16th Street in downtown Oakland. However, trains originated at the Oakland Pier, to accommodate passengers from San Francisco and elsewhere in Marin County. It was a short easy walk for the GI's from the ferry and into the big train shed at the depot. They found the correct platform and made their way onto the waiting train. Walt said goodbye to his friend, crewmate, and ship galley partner Dick Reis, who was headed home to Pacific Grove, near Monterey. Walt, Cliff, Gene Schreiner, Jimmie Clark, Carl Appling, and Jupe Bretherton all found seats together.

Soon the train departed and headed east, then north with the Bay out the windows on the left side of the train. There were quick stops in Berkeley and Richmond. Walt looked across San Pablo Bay, thinking about Hamilton Airfield, where he and Jack had left for Hawaii the previous year. The train sped on to Martinez, where a bridge carried them slowly across Carquinez Strait. Then the train turned to the east, across the Sacramento delta flats. Walt and his buddies looked for Fairfield-Suisun Air Field, half way to Sacramento, where they had spent a few days after leaving Muroc the previous March. The train reach Sacramento in the early evening, where they had about 15 minutes to dash out and grab some beers, sandwiches, donuts and newspapers. Then the train began to ascend the western slope of the Sierra Nevada towards Donner Pass and then Reno. At each stop, one or two men would rush off to buy beer and food to eat. Bread, baloney, and beer was standard fare for much of the journey.[34]

The stations passed by in a blur as night came, and then sleep as best they could in their seats. On Friday, April 6, daylight found their train on the D&RGW Railway in central Colorado. Walt gazed at the rugged landscape that he had loved from the western movies as a kid. In the late morning, the train climbed the western slopes of the Rocky Mountains, then into six-mile long Moffat Tunnel. At about 12:15 pm they arrived at Denver's Union Station. Then it became a little emotional as Gene Schreiner, Carl Appling, and Jupe Bretherton said their goodbyes. Gene was headed north to Fort Collins, Colorado to his wife and the baby daughter he had never seen before. Appling and Frank Mansir were headed to their homes in Texas. Jimmie Clark was headed home to Louisiana. There were emotional goodbyes and hopes they would all meet again. At 1:00pm, their train to the east was set to depart and so Walt, Cliff, Jupe, and Jack boarded and continued their journeys home.

Their train arrived late at night in Kansas City, where most of the remaining men boarded different trains as they headed home. Walt said goodbye to Cliff and Jupe.[35] Early next morning Jack and Walt would continue on together to St. Louis. Late the next day, Jack and Walt parted ways – Jack to head home to North Carolina, and Walt to Chicago, and then home. Walt arrived in Chicago late on Sunday afternoon. His orders were to proceed to Fort Sheridan, the Army Administrative Center for men drafted from Illinois, Indiana, Michigan, and other nearby states. He learned that the Fort ran trucks to carry men to and from the Chicago stations. He had time to grab a sandwich and coffee before his truck arrived to drive him north to the pleasant lakeside base. He arrived too late to be processed that Sunday. Walt was anxious to get home, but it took time for the base to bring his records up-to-date the following day. Late in the day he had his new orders. He was granted 21 days furlough. He was ordered to report to Santa Ana Army Air Force Base in California on May 4. His furlough assumed 5 days of travel time between Detroit and Santa Ana, so he had to leave by April 30. Walt was told that at Santa Ana he would be reassigned as an aircrew gunnery instructor.[36]

Early the next morning Walt caught one of the early Fort Sheridan shuttle busses to Chicago, to catch a NYC train back to Detroit. He had time to send his folks a telegram to let them know when he would arrive. Soon, Walt was on his way east again, around the south end of Lake Michigan, then back into Michigan. Niles, Paw Paw, Kalamazoo passed by. At Battle Creek, Walt remembered his first days away from home at nearby Camp Custer in January 1943. Albion, Jackson, Chelsea, Ann Arbor passed by. As he passed through Ypsilanti, Walt thought about the Willow Run B-24's that had carried him many thousands of miles. Passing through Dearborn, he could see the belching smoke from the Ford Rouge plant and from hundreds of other factories spread across the Arsenal of Democracy. Then in mid-afternoon, his train pulled in to Michigan Central Station. Tuesday, April 10, 1945, Walt was home.[37]

Home

Pa was waiting for Walt as he entered the huge Michigan Central Station arrival hall. Henry and Leona had come to the station too. Walt, in his best uniform with his S/Sgt stripes and all of medals on his chest looked handsome and happy. His Pa shook his hand and then hugged him. Leona and Henry looked in awe at their big brother. Pa grabbed Walt's duffel and led them to his car, as Henry and Leona followed, holding Walt's hands. Soon they were driving down familiar Delray streets. The towers of St. John Cantius Church came in to view and they pulled up in front of the house on Pulaski Street. Walt's Ma was waiting there with tears in her eyes, but a smile on her face. 'You're too thin!' she said. 'What's for dinner Ma?' Walt asked. Soon Walt was seated at the dinner table with his Pa and Ma, his sisters Anne, Helen, Genevieve, and Leona, and little Henry. Only Johnnie was missing. They talked long into the evening, but it was the middle of the week, so his Pa, Anne, Helen and Gene had work the following day, and there was school for Henry and Leona too.

Walt learned some sad news. His good friend 'Tex' Jekielek had joined the Air Force also. He became a P-47 fighter pilot and was assigned to the 460th Squadron, 348th Fighter Group in the Philippines, supporting MacArthur's invasion. On December 4, 1944, on his first combat mission, he was killed in action.[38]

The next day was a lazy one for Walt. He woke up late and then ate the big breakfast his Ma had waiting for him. Later Walt walked around the old neighborhood, stopping at the Kurzyniec Store to say hello and enquire how Dorothy was. Walt picked up a couple newspapers. Baseball season would be starting soon and he wanted to see what the prospects were for the Tigers this year.

On Thursday, Walt had to report to his draft board out at 250 West Lafayette downtown, to register for a ration card. Even though he would only be home for three weeks, his ration card would help his family buy the extra food they would need to feed him. Walt's Ma had to fatten him up before she let him head back to the Air Force. Walt went downtown on the streetcar with his Ma. At the draft board, he presented his furlough orders. A clerk stamped them with a little ink stamp and initialed for receipt of a food ration card and a 23-gallon ration of gas for the family car.[39]

On the way home, Walt helped his Ma with her shopping. That afternoon, Ma was cooking dinner and Walt was relaxing at home waiting for the rest of the family to return from work and from school. He was listening to the radio when an announcement came over the airwaves: President Franklin Roosevelt had died. Walt felt shock and sadness. As the rest of the family returned home, everyone talked about the news. The radio reported that the president had died of a cerebral hemorrhage at Warm Springs, Georgia. Vice President Truman had taken the oath of office and requested the FDR cabinet to remain in their posts. Everyone expressed grief and regret that FDR would not see the conclusion of the war in Europe and the eventual victory against Japan. For more than 12 years, FDR had guided America through turbulent and difficult times. Every face was etched with grief.[40]

A day later, a letter arrived from Johnnie. He had heard that Walt was back home on leave. Johnnie had news too. He had been assigned to a ship! His new ship was the USS Pine Island, a new 540' seaplane tender. The Pine Island had been launched in February 1944 and she was just going through commissioning when Johnnie was assigned. Johnnie would be shipping out within a month to two months. He wished Walt well and hoped that the entire family would be together again soon.[41] A few days later, Leona got a postcard from Johnnie showing dozens of boats at Fisherman's Wharf in San Francisco:

> Hi Leona, How are you? I didn't get one of these things for Lize. Did he send you one and do you like it? There's a lot of nice palm trees growing here and oranges growing on trees. I am fine.
>
> -John[42]

Leona and Walt laughed at Johnnie's joke about sending fishing boats to Lize and Leona.

Walt went to mass at St. John Cantius on Sunday with his Ma and Pa, sisters, and little Henry. The sermon focused on the death of Roosevelt and the hope that the war would end soon. There were also prayers for family and friends still in the 'old country' – the captive nations being liberated from the Nazis. The priest and many of the nuns from the school remembered Walt and shook his hand after mass. Everyone wanted to hear about Walt's experiences and what he thought about the war ending soon. Sadly, almost everyone knew of a young man who had been killed in the fighting.

The grief everyone felt at FDR's death was tempered as it became clear that Truman was firmly taking over the conclusion of the war. Unknown to Walt, Truman would soon make an important decision that would affect him, Johnnie, and millions of others in every corner of the world.

Days went by. Tuesday, April 17 came and with it the opening of the 1945 baseball season for Detroit's beloved Tigers. Walt had studied the scouting reports in the sports page of the *Detroit Times* and *Detroit Free Press* every day. The Tigers had a solid team, with Bob Swift at catcher, Rudy York at first base, Eddie Mayo at second, and Bob Maier at third. Skeeter Webb was the shortstop, and the outfielders were Roy Cullenbine, Jimmy Outlaw, and Doc Cramer. Hank Greenberg was a pinch hitter now, but still very potent at the plate. The Tigers had solid pitchers too, with Hal Newhouser, Dizzy Trout, Stubby Overmire, and Al Benton the starters. Walt went to a bar to listen to the opening game against the St. Louis Browns, with Harry Heilmann broadcasting the play-by-play on WXYZ Radio. Ah, sadly, the Tigers lost the opening game 7-1. But they won the next two games, 11-0 and 1-0. On Friday, the Tigers returned home and Walt took the streetcar out to Michigan and Trumble to Briggs Stadium to watch them play the visiting Cleveland Indians. Stubby Overmire was pitching for the Tigers, but he gave up four runs in the first three innings. The Tigers managed one run in the sixth inning, but there the scoring ended and the Indians won 4-1.

As the days passed, Walt stopped by the Kurzyniec store each day, especially when Dorothy Kurzyniec was home from her job. In a year, they had both changed a lot. Dorothy thought that Walt looked so handsome in his uniform with his sergeant stripes, 13th Air Force patch, and medals and ribbons. Walt thought that Dorothy looked even more beautiful than before he left 16 months ago. Walt and Dorothy enjoyed a lot of time together. Walt was sometime the dinner guest of Thomas and Martha Kurzyniec, with their three daughters and cousin Norman.

Time passed by so quickly and Walt neared the end of his furlough. His orders were to report to Santa Ana Air Force Base in California on May 4. He thought that if he left early on May 1, caught a train to Chicago and then a fast Santa Fe Railroad train to California, he could arrive on time. On Sunday, April 29, the Tigers spilt a double header with the Indians in Cleveland and improved their record to 6-3. On Monday, April 30, there was no baseball, but more news on the radio. Hitler was dead in Berlin. The end of the war in Europe seemed in sight. American and Russian soldiers had met in Germany, cutting the once-might Third Reich in two.

Walt had planned to leave on Monday, to begin his cross-country trip to report to Santa Ana AFB in California. However, his friends talked him in to staying another night, for a big party that was planned to celebrate Hitler's death and the imminent end of the war in Europe. The party wound up and down Pulaski Street. Late in the evening, Walt and a group of his friends took a record player up to the second floor porch of a house on Pulaski. They put on a record of a Spike Jones song called *In The Feuhrer's Face* – then Walt and his buddies sang along with the song, with Walt singing lead:

> *When der fuehrer says we is de master race*
> *We heil heil right in der fueher's face*
> *Not to love der fuehrer is a great disgrace*
> *So we heil heil right in der fuehrer's face*[43]

Everyone thought it was very, very funny – everyone had a good time on Pulaski Street that evening, as the tension of the war, at least of part of the war began to ease.[44]

Walt was supposed to report back by May 4. So now, on Tuesday, May 1, he could wait no longer. Early in the morning, after a big breakfast and his bag stuffed with Ma's kielbasa and rye-bread sandwiches and a loaf of her poppyseed cake, Walt said his goodbyes to his family. His Pa loaded Walt's duffle into his Ford and then drove him to the Michigan Central Station. After a fast goodbye, Walt strode into the station to present his travel orders and catch the first train to Chicago. Walt's plan was to catch a Rock Island and Southern Pacific Railroad train – the Golden State Limited from Chicago, direct to Southern California.

Santa Ana AAB

Santa Ana Army Air Base did not exist at the end of 1941, but in early 1942, in just 120 days, 177 buildings were completed and the base became operational as the major Air Corps classification and officer training center on the west coast. SAAAB did not have runways or hangars, but by the end of 1944, it had 800 buildings and more than 150,000 men and women had been processed and trained there.[45] Unknown to Walt and his crewmates, the role of SAAAB was changing.

When Walt arrived in Chicago, he could not get a seat on the train to Los Angeles, so he had to wait one more day. Very late on Wednesday, May 2 he finally managed a seat on the Golden State Limited to Los Angeles. This train left Chicago at 8:30pm and was scheduled to arrive in Los Angeles three days later. All he had to do now was sit back as best he could and enjoy the scenery, talk with other travelers, read the news, and play cards. The only thing he had to worry about was getting to Santa Ana in time, and it looked like he might be late. The train traveled via the Rock Island Railroad through Kansas City, across Kansas, the Oklahoma and Texas Panhandles, and New Mexico, before arriving in El Paso, Texas. Here the Southern Pacific pulled the Golden State Limited across southern New Mexico, Arizona, crossing the lower Colorado River into California before the route turned northwest to Los Angeles.

Meanwhile, Cliff Llewellyn, Gene Schreiner, and a few others from the old crew were already at Santa Ana. But they were worried about Walt. As the men arrived at Santa Ana, they were processed, given physical exams, and did a lot of waiting. SAAAB was a huge base, so a couple times they asked if Walt had arrived. They grew worried that he would be in trouble, reporting in late. Finally, early on Sunday, May 6, Walt reported in at Santa Ana Army Air Base.[46]

He managed to find Cliff and Gene in time to share dinner together. Walt explained that – yes, he left Detroit a little late, but he thought he could make it in time. He got off the train not in Los Angeles, but at a flag stop in Pomona, just about 35 miles north of the base. But then he discovered there were no Army buses at Pomona and he had a difficult time hitching a ride through the farm and orchard county between Pomona and Santa Ana. Anyway, he was there now and Walt did not seem too worried. What could happen? Walt believed that he had enough points for discharge. Cliff and Gene had been thinking about staying in the service to become instructors, but Walt was ready to return to civilian life if he could.

The day after Walt arrived, SAAAB changed from a training and aircrew classification center to become an Air Force separation center.[47] Walt did not get in trouble, but he missed being offered the chance to become a gunnery instructor. That was where Cliff was headed – to Laredo AAF, where Walt attended gunnery school a year and a half earlier.[48] As Cliff thought about it, he decided to apply for discharge too, but he learned that even though he had enough points, he was too late to apply at SAAAB. He would have to wait until he arrived at his next assigned base – Laredo AAF.

That same day early in the morning, cheers went up across the base. The news arrived that Germany had surrendered. Soon a truck drove up with stacks of *Los Angeles Times* special editions with the headlines: 'V-E DAY: Nazis Surrender Unconditionally to Allied Powers.'[49] Work at the base slowed to a crawl as everyone shared the news and celebrated.

The next day Walt put in his application for discharge and started going through various processing. He had a physical exam and then had his military awards and decorations confirmed and signed by Major James A. Smith, Jr.[50] Walt learned he had been approved for discharge, but there was even more processing to go through.

On May 15, Cliff & Gene said goodbye to Walt. They were both headed out on the same train - to Texas at Laredo AFB for Cliff, and to Williams Field in Chandler, Arizona for Gene. Now all Walt had to do was wait for the Army Air Force bureaucracy to work. But the waiting wasn't too bad. The food was good, and for men who were not scheduled for processing, there were always free passes off the base available.[51] More than once, Walt took the opportunity to take a bus into Los Angeles for sight-seeing or to visit a nightclub.

And there was a lot of time. It was not until Friday, May 18, that his pay was brought up-to-date[52] and he was issued an Air Force Veterans Identification Card[53]. It was not for another week before Walt finally received the orders he had been waiting for. Walt (and 31 other Michigan and Illinois men) received orders to proceed back to Fort Sheridan for discharge as had been authorized on May 11. They were issued vouchers for travel and meal allowances.[54] Walt sent a letter home announcing that he would soon be discharged. Finally, early on Monday, May 28, the 32 men departed SAAAB by rail enroute to Chicago.[55]

Home for Good

Walt retraced his route to Chicago via the Southern Pacific's Golden State Limited across Arizona and New Mexico to El Paso. There the Rock Island took the train on to Chicago. He was back at Fort Sheridan on Saturday, June 2. He was asked to report for a brief ceremony where he was presented the Air Medal.[56] The next day, young S/Sgt Walter A. Babinski had his final payroll certified. He received one last payment of $143.60. During his time in the military, he had been paid a total of $14,099.53. He received his honorable discharge and indicated his civilian occupation as Power Saw Operator.[57] Walt grabbed his duffle, hopped a bus to downtown Chicago, and got a train for Detroit. Late that evening he arrived at Michigan Central Station, got a taxi, and was home. Home for good.

Once again, after Walt came back home there were big parties with family and friends, as there would be again later when Johnnie finally returned.[58]

At some point in June, a secretary from WXYZ Radio in Detroit contacted Walt. The station had a feature show, which interviewed a few returned service men each day. They wanted to have Walt on the air for an interview. Kenneth Koppitz would announce the show and Johnny Slagle[59] would do the interview. A WXYZ staff reporter interviewed Walt the day before and they worked out a script, so that the interview was both interesting and moved along at a good pace. Of course, everything would be broadcast live.

Walt made sure everyone in his family and all of his friends would be listening. Walt went to the WXYZ studio on Cass Avenue, near Warren, wearing his best uniform, which his Ma had pressed in the morning. A few friends made the trip downtown with Walt and sat in the studio audience.

Koppitz and Slagle said hello and asked him if he was nervous. Walt said, 'well if there is no flak or enemy fighters, I think I will be OK.' Walt was the third interview and soon he was on the air!

Interview Number Three

"SLAGLE: Hello Staff Sergeant Walter; ; ; I understand you are a native Detroiter and went to school at Southwestern High is that right?

WALTER: Yes Sir:

SLAGLE: Tell us when you entered the service and when you were released.

WALTER: I entered the service in January 1943 and was released June 3rd 1945.

SLAGLE: What branch of the service did you see action with?

WALTER: I was a Ball Turret gunner on a B24 and went overseas in April, 1944 with the 13th Air Force in the Southwest Pacific.

SLAGLE: Being a ball turret gunner for your size is kind of unusual isn't it?

WALTER: Yes! But I've gained a lot of weight since June...but even at that I didn't have much room in the ball turret.

SLAGLE: Tell us about your space in the ship...

WALTER: We were in the ball turret and we were let down after the ship took off and stayed there until we came in for a landing. We manned two 50 calibre machine guns and were there to fight off any enemy attacks. By the way we did have some protective covering... But the armor plate was at our backs, and we had nothing in front of us except the blue sky.

SLAGLE: Did you have a lot of room to move around?

WALTER: No we didn't, in fact we rode out the missions usually with our heads between our knees.

SLAGLE: Cramped quarters I'd say. By the way Sergeant, did you have any actual fighting experience with Jap zero's?

WALTER: Quite a lot of fighting experience.

SLAGLE: Tell us about a run you consider a standout in your memory.

WALTER: My ship made one of the longest runs by any B24 in this war, if not the longest and that trip is a standout for my money. We were sent on a bombing run that took 18 hours...and that's a long time to stay cooped up in the ball turret. We were going over to bomb a big oil center, I can't tell you just where I guess, but it was considered the Ploeste oil center of the Japs. It was a big field and it was really protected. This run had a lot of action...And we lost a lot of planes and crews....The flak was terrific and the fighter opposition was the greatest we had ever ran into. Our tail gunner knocked down a sure zero and I had four probables...

SLAGLE: What do you consider a probable?

WALTER: Well: you see a probable is a zero that goes down and you think you hit him. The fighter pilots could always tell when they got a Zero as they had cameras working with their guns, but we could just guess and with more than one gunner firing at them its kinda hard to say just who gets him. But we're happy if he just goes down.

SLAGLE: Did your ship ever get hit by flak or fighters?

WALTER: Mister we sure did, one run we were on we were hit 85 times, but the old Malfunction brought us in.

SLAGLE: Did you say Malfunction?

WALTER: Yes sir! That's the name of the B24 we flew.

SLAGLE: That's an odd name how come you christened the ship the Malfunction?

WALTER: We didn't pick out a name right away and we wanted one a little different from the rest, most of them have girl names and girl pictures but one day we remembered a dog we knew back at the training base called Malfunction...This dog was a mongrel and was strictly on the dumb side and hard to handle at times....But usually reliable....So we called our ship the Malfunction.

SLAGLE: What insignia did you use with that name, a dog picture?

WALTER: No we didn't we painted a wolf on the side of the ship....

SLAGLE: Walter, now that you're a civilian again with a splendid war record as evidenced by your collection of medals....The Air Medal, with four oak leaf clusters, 5 major battle stars, the Philippine Liberation Medal, and the Asiatic-Pacific theatre ribbon, what do you want to do in the post-war world?

WALTER: Well, although I'm only 23 years old, I figure its time to learn a trade, I always wanted to work with my hands, I had a little experience before the war working a saw in a wood working shop and I want to do something with my hands and learn a good trade with a future.

SLAGLE: Sergeant, I'm sure that someone in Detroit can find a spot for you to learn a trade....Thank you from Koppitz and Myself – Staff Sergeant Walter of the Army Air Corp...."[60]

Walt did not have to look far for his first job as a civilian. Before he left the station, WXYZ radio offered him a job doing odd jobs around the studios.[61]

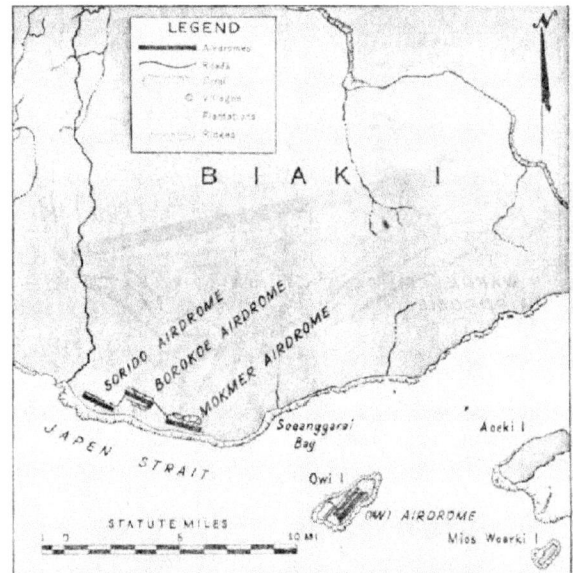

Biak: Walt's home for 17 days, awaiting shipment back home to the States. *Airdromes Guide Southwest Pacific Area, p. 75.*

Across the Pacific: The German SS Orinocco (above), transformed into the USS Puebla (below), Walt's home for 22 days, crossing the Pacific from Biak to San Francisco.

Across the Pacific: Life aboard the USS Puebla was cramped and monotonous (left). *Missing Aircrew Project*

Back in the States: After returning to the States and a three-week furlough in Detroit, Walt travelled to Santa Ana Army Air Base, where he received discharge orders. Here, in May 1945, he enjoys leave in a Los Angeles nightclub with two friends (below).

USS Pine Island: As Walt returned home, his brother Johnnie was finally in combat aboard the seaplane tender, USS Pine Island, here shown near Okinawa.
Wikicommons

Home: Walt poses with his Ma in front of their house on Pulaski. Note the Babinski family Blue Star Banner, one star for Johnnie and one for Walt.

Home: Walt in his leather bomber jacket poses with sister Leona (above left) and in civilian clothes with sister Genevieve (above right).

Home: Walt with Dorothy Kurzyniec, left in his uniform, and right at the Kurzyniec store on Pulaski.

Home: Johnnie, Cora, Walter Senior, and Walt pose in the front yard

Home: Martha and Thomas Kurzyniec relax in the back yard of their new home on Livernois in Detroit, late 1940's.

Home: Walt and Dorothy stand-up at a friends wedding in 1946 (left) and pose on a cruise on the Detroit River to Bob-Lo Island in 1947.

Lines of Life: Walt and Dorothy married in 1948 at St. John Cantius Church.

Lines of Life: Stella Kurzyniec married John Besek.

Lines of Life: Martha Kurzyniec married Anthony Tranchida with Martha's mom to her right and her father Thomas to Tony's left.

Lines of Life: Pa and Ma Babinski with son Henry at their feet and grandsons Edwin and Greg, early 1950's

Lines of Life: Walt and Dorothy Babinski, with Stella and John Besek. Later they would buy a duplex in Dearborn together.

Lines of Life: Walt and Dorothy's first house—6888 Penrod in Detroit. Walt and Gregory pose on the porch.

Lines of Life: Walt and son Gregory in the backyard of their house on Penrod in Detroit, early 1950's.

Lines of Life: 19 Babinski, Tranchida, and Besek children on the basement stairs of the Tranchida house in Detroit for their 1964 Christmas picture.

Lines of Life: The author with his guide at the Morotai WWII Museum (above left) and nearby beach scene (above right), both at the site of the 307th Bomb Group camp on the Gila Peninsula in 1944-1945. *Author photos, 2013.*

Lines of Life: The author on the Indonesian Air Force Base on Morotai. His hand rests on a static display of a WWII 50 cal machine gun with Pitoe Air Strip in the background. *Author photo, 2013.*

Lines of Life: Teruo Nakamura statue (above left) on Morotai, one of the heavy A/A guns (above right) at Hate Tabako and Lolobata air strips on Halmahera that fired on Walt on December 22, 1944, and a Morotai souvenir hunter's collection (left) of WWII artifacts found on the Gila Peninsula. *Author photos, 2013.*

Lines of Life: Alan "Smokey" Guild at the 2004 Reunion of the 307th Bomb Group Association—Nashville, TN. *Author photo.*

Lines of Life: John Vanderpoel at the 2006 Reunion of the 307th Bomb Group Association—Seattle, WA. *Author photo.*

Lines of Life: Mark Babinski, Dick and Mina Reis, and Dorothy Babinski at the 2004 Reunion of the 307th Bomb Group Association—Nashville, TN. *Author photo.*

Lines of Life: Three Long Ranger vets tour Ephrata Airport, before the 2006 Reunion of the 307th Bomb Group Association. *Author photo.*

Lines of Life: Cliff Llewellyn and the author at Cliff's house in Thiensville, WI, 2002. *Author photo.*

Lines of Life: The author and Gene Schreiner at the 2002 Reunion of the 307th Bomb Group Association—Salt Lake City, UT. *Author photo.*

Relics: Souvenir pennant that Walt brought back from Australia with his service ribbons (above), Walt's wings—Combat Crewman, Aerial Gunner, and Aircraft Mechanic, (below left), and Air Corps, 13th Air Force, and Sergeant's patches (below right). *Author photos.*

- 16 -
NEW BEGINNINGS - DREAMS OF LIFE

The Ending of the End

By the end of 1945, slowly, much of the world awoke from the nightmare of the previous six years to get back to the dreams of life. For a while, Walt wrote to some of his buddies from the service. Cliff Llewellyn got a typical letter from Walt at the end of June announcing that he was 'a civilian again!'[1]

By the end of 1945, Johnnie would be back. He had shipped out across the Pacific as a MM 1/C on June 16, 1945, one of 684 men assigned to the USS Pine Island.[2] By July, The Pine Island was off Okinawa deploying seaplanes for air-sea rescue operations. The Pine Island was part of Operation Gunto, the US invasion of Okinawa, from July 10 – July 25, 1945.

At about 3:00 pm on Saturday, July 21, Walt, Henry, and his Pa started listening to the Tigers game versus the Philadelphia A's on the radio. It was a real pitcher's duel and by the end of the ninth inning, it was tied 1-1. The radio was on as Ma served Saturday dinner. Still the game went on. Walt and his Pa drank a few beers into the evening. Still 1-1. Tiger's pitcher Les Mueller went 19-2/3 innings, allowing only one run. In the 10th inning, Jimmy Outlaw threw an A's runner out at home plate to preserve the tie. After 24 complete innings and four hours and 48 minutes of play, the umpire Bill Summers called the game at 7:48 p.m., saying: "I'm sorry, I just can't see the ball anymore." A's third baseman George Kell had the worst game of his career, going 0-for-10.

The headlines the following day reported this game in detail. The *Detroit News* also reported that the navy and air force continued to pound the Japanese home islands. Walt wondered how Johnnie was doing so near to the fighting. At the same time the *News* reported continued fighting on Borneo, as Americans and Australians under MacArthur took the key ports and oil facilities one by one. Walt remembered his contribution to that fighting, but was glad that he was home with his family and friends listening to the baseball game.[3] But how many thousands more Americans and other allied military would die before the war was over?

On August 6, 1945, news came over the radio and in the papers that something called an 'atomic bomb' had been dropped on Hiroshima, Japan. The *Detroit News* explained that this single bomb was equivalent to 20,000 tons of TNT. Four days later the papers reported that Japan was asking for peace terms. Things moved quickly. By September 1, MacArthur was in Tokyo Bay, where the Japanese signed the surrender that ended World War Two. A few days later, the USS Pine Island, with Johnnie aboard, entered Tokyo Bay to assist in the occupation of Japan.

A few months earlier world peace seemed a distant dream. Now it had arrived. Life slowly returned to normal. Baseball was normal and for the Tigers, 1945 was a great year. They finished the season in first with an 88-65-2 record. The Tigers played the Chicago Cubs in the World Series. The first three games were at Briggs Stadium in Detroit, but the Cubs went ahead 2-1. The next three games in Chicago, the Tigers took the first two, but the Cubs won the last one to tie up the series 3-3. The last game, in Detroit on October 10, the Tigers won 9-3 giving them the 1945 World Championship.

Baseball reminded Walt about his baseball card collection. Before he left for induction into the Army back in 1943, he had boxed up his collection and placed it in the attic of the home on Pulaski. However, went he went upstairs to find it, the box was nowhere to be seen. He asked his Pa and Ma, he asked his sisters and brothers. But no one knew what became of it.[4]

Johnnie, the family soon learned, was now in Sasebo, which had been the main Japanese naval base. Sasebo became the center of U.S. Naval operations in Japan. Now that the allies had access to the many air bases available in Japan, one thing that they did not need was a seaplane tender. The short mission of the Pine Island was over. At Sasebo, Johnnie had his records brought up to date. He had now been in the service for three years. His mechanics mate proficiency rating was 3.72 and he was verified to have 42-1/2 points, sufficient in the Navy system for discharge. This is what Johnnie wanted and by early October, he had left Sasebo and was on his way back the U.S.[5]

By November 1 he was back at the Great Lakes Naval Station north of Chicago, and not too far from Fort Sheridan. On arrival at GLNS, Johnnie was 'informed of the benefits of reenlisting in the US Navy.' Johnnie's records state 'He does not wish to reenlist.'[6] So close to Detroit, but so far. On November 10, Johnnie had a mandatory performance rating: As an 'MM1c: Proficiency in Rating 3.9; Mechanical ability 3.9; ability as leader of men 3.8, conduct 4.0 - Authorized for honorable discharge.'[7] It would be another month though, December 10, before Johnnie finally received his honorable discharge. He received a final payment of $180.97; he was authorized to wear the American Theater Medal, Asiatic Pacific Medal with one Star, the Victory Medal and Good Conduct Medal. Johnnie wanted to get back home and back to work at Ford Motor. The following day, Tuesday, December 11, Johnnie was back home in Detroit.[8]

It was late in 1945 or early 1946 when the Kurzyniec Store on Pulaski closed. Tom had saved enough money at Ternstead to move his family into a house that Thomas and Martha Kurzyniec bought outright at 608 Livernois.[9] After the war, Dorothy and Martha Kurzyniec worked in downtown Detroit for about two years. Dorothy worked as a billing clerk for Colliers Publishers. Her office was in a building on Washington Blvd. Dorothy and Martha would meet at a Catholic Church downtown on Washington Blvd for Holy Days services.[10]

Walt did not last long working for WXYZ radio after his on-the-air interview back in June 1945. Doing odd jobs around the radio station was not what he had in mind for a permanent job. Going to work for Ford Motor or General Motors might have provided better opportunities, but wartime production had actually peaked and there was little new hiring. With some of his savings, Walt was able to buy his first car, and in October 1945, he had his first temporary driver's license.[11] Walt signed up with the United States Employment Service, which tried to find jobs for the increasing flood of returning veterans who hoped to re-enter the civilian workforce.[12]

By early 1946, Walt found a job in Highland Park, at one of the many small factories that were tooling up as American – and especially Detroit – industry was making the transition back to a civilian economy. Walt had time to join a bowling league, bowling for the 'Merry-Go-In' team during the 1946-1947 bowling season.[13] The Merry-Go-Inn was a bar located not too far from home, at the corner of Keller and Leigh Streets in Delray.

A good childhood friend of Walt's was John 'Pee Wee' Drabek. The same age as Walt, he was first of his friends to marry after the war. Walt was best man at his wedding in 1946, and young Dorothy Kurzyniec was his date. Pee Wee had a reputation as a very good dancer.[14]

In early 1947, Walt heard about another company that was hiring in Highland Park. Walt thought this might be a good opportunity and he put in an application to work for the Detroit Stamping Company, at 350 Midland. Walt was hired and reported for his first day of work on Tuesday, March 11, 1947. He quickly started learning the ropes. He became friends with a man who started the day before he did – Joe Chroebuck. Detroit Stamping had many jobs for press operators, drill operators, and spot welders. The company had a wide variety of customers, from the auto industry, to the oilfield industry, and they even had their own line of toggle clamps, which the company had patented. Many women worked in the plant. These 'Rosie-the-Riveters' had been hired during the War, but many stayed on to continue working. Walt liked the work and he was good at any job that he was given. He felt at home – this could be the job that he had been looking for and the chance to settle down.[15]

Before the end of 1947, Walt had proposed to Dorothy. They were married in February 1948. Their wedding cost $600 total for 250 people, including food, drinks, the hall, a band, and all. And at that time when you hired a band for a wedding, they also played outside both the bride and the groom's house the morning before the wedding ceremony![16] Their honeymoon was in New York City. They took the time to visit Ed Dunne, who lived in Brooklyn.[17] They wrote a postcard showing the Hotel New Yorker to Leona:

Hi Leona,
How are you? We are fine and we are up on the 35th Floor here. We have seen a lot of nice scenery and are having a swell time.
-Dottie & Lize[18]

Soon Martha Kurzyniec had married young Tony Tranchida, and Stella Kurzyniec married John Besek.[19] Walt's sisters married soon too. Anne to Frank Bogeslaw, Helen to Eddie Czerwinski, and Genevieve to Eddie Mendryga. Life in Detroit was good and life in Delray was filled with joyful people and growing families. Johnnie married. Walt and Dorothy had the first of many children in 1949 and soon moved into a little house of their own at 6888 Penrod in Detroit.

Young Hank was drafted into the US Army during the Korean War. He was stationed in Korea and then Japan. On his return to the U.S., he arrived by ship in Seattle and returned to Michigan by rail.[20] He was stationed at Sault Ste. Marie before his discharge and return to Detroit, where he went to work for Great Lakes Steel.[21] Soon he was driving a beautiful new 1955 Ford Thunderbird and married to an even more beautiful young girl from a Greek family – Gladys. And finally, Leona married Jim Smith – a handsome man of Polish descent from Glasgow, Scotland.

Life was sweet. Life was good.

- 17 -
AFTERTHOUGHT: LINES OF LIFE

The Passing of Decades and Choosing to Remember

What is the meaning of life? What does a single life mean? How and what do we choose to remember of our life and the life-experiences of other?

For two Babinski relatives stranded in Poland when war broke out in 1939, life took different courses. Julian Gebolys had escaped to England in 1939 and helped the war effort by training British and Polish paratroopers. He was fortunate to stay in the U.K. after the war ended. He lived in a rural area of northern England where he raised horses. He died in 1984 in Manchester, England. For decades, no one knew the fate of Capt. Zbigniew Babinski until archives in Moscow opened after the fall of the Soviet Union in the late 1980's. Records there showed that Capt. Babinski had been captured by the Red Army in September 1939 and taken to a prison camp at Katyn, in what is now Belarus. At Katyn, under the direct signed order of Josef Stalin, he was one of some 22,000 Polish officers, teachers, professionals, and intelligentsia executed in April 1940 and buried in secret mass-graves. Both the Babinski and Kurzyniec families lost many other friends and relatives in Poland during the course of the war under the Nazis and its aftermath under the Soviet Union.

What does society choose to remember about an event like World War Two, which was experienced by so many people? The arts provide one way of remembering.

In 1945, a U.S. Army Air Force veteran, Randall Jarrell, published a five-line poem, *The Death of the Ball Turret Gunner*.[1] Jarrell's short poem distilled for the post-war public the essence of what Walt and thousands of other ball turret gunners experienced when they left home for service to the State. His intense words evoke the discomfort and terror that these men endured in their detached dream-like isolation far above the earth. For Walt and the other lucky ones, only Jarrell's last line did not apply.

This poem and the terror of facing death dozens of times in a plexiglass dome miles above the surface of mother earth inspired a June 1976 article[2] in Playboy Magazine by John Irving, which later became a chapter in his 1978 novel, *The World According to Garp*.[3] The theme was revived again in Anna Moench's 2008 play *The Death of the Ball Turret Gunner*.[4]

In 1968, Richard Hooker and W.C. Heinz published a book titled *MASH: A Novel about Three Army Doctors*. This book was based on Hooker's experiences as an army combat surgeon during the Korean War in a Mobile Army Surgical Hospital (MASH). The concept of the MASH originated in World War Two when the Army created PASH units, Portable Army Surgical Hospitals, designed to be transported and staffed by 29 men: four medical officers (three general surgeons and a general surgeon/anesthetist) and 25 enlisted men, including two surgical and 11 medical technicians.[5]

A 307th Bomb Group veteran, Lt. Robert Altman, had served as a B-24 copilot on the Lt. Dale Dennison crew in the 371st Bomb Squadron in 1945. On a mission to bomb the A/A gun positions at Balikpapan on June 18, Dennison's plane was hit by flak shrapnel. One of the gunners, Bill Keel, was hit in the foot. John Horoshak tended to Keel as the shrapnel pelted their plane like buckshot. There were flak holes in the floor, top turret, #1 engine cowling and four in the wing. The windshield in front of Dennison and Altman was shattered.[6]

Lt. Altman flew 50 combat missions. He was well familiar with the PASH on Morotai, and the somewhat chaotic and dangerous life on the island, with Japanese air raids continuing into June 1945, and a thousand or more Japanese troops still just beyond the front lines. When he returned to the U.S., he began working as an industrial filmmaker. He produced 65 documentaries and industrial films by 1957. In 1953, he began directing TV shows, including episodes of *Bonanza, Combat, Surfside 6, Route 66, Maverick,* and *Hawaiian Eye*. In 1968 he was offered a chance by 20th Century Fox to direct Hooker and Heinz' book *MASH*. Released in 1970, *MASH* was a big commercial success, also winning international critical acclaim. Altman used his experiences with the 307th Bomb Group on Morotai to understand and depict the chaotic, dangerous, and some-times funny camp environment shared by the Air Corps in World War Two on Morotai and a MASH in Korea in 1950-1953. Altman's first hand-experience contributed to making *MASH* a critical success.[7]

In the Mediterranean Theater, Joseph Heller served as a B-25 bombardier with the 488th Bombardment Squadron, 340th Bomb Group, 12th Air Force. Heller flew 60 combat missions. Heller, the same as men half way around the world in the 307th Bomb Group, experienced the frustration of not knowing when their tour of combat duty would end and when they would be sent home. In the Mediterranean, first 25 missions was the magic number to go home, then 35 missions, then 40, and so on. He coined a phase: 'Catch-22' to describe the frustration felt by all Air Force combat crews. You can be sent home if you are crazy, but if you want to go home, you are clearly not crazy: Catch-22. In 1961, his novel *Catch-22* was published. In 1970, the same year that *MASH* was released, the film version of Catch-22 was released, again to commercial success and critical acclaim.[8]

These two books and films, *MASH* and *Catch-22*, reminded Americans of the face of combat as experienced by their sons, husbands, and brothers in the 1944-1953 era. The Vietnam War was in progress in 1970. *MASH* and *Catch-22* contributed to the reflection on the insanity of war that began to intensify as a new generation was asked to fight in faraway foreign lands.

One 13th Air Force veteran wrote of his combat experiences decades afterwards:

> *'To be involved in combat requires a feeling, an attitude unexplainable and elusive. The activities are foreign to our natures and to the rules of polite behavior in regular society.*
>
> *'I believe totally and emphatically that those who were at one time or another involved in actual combat cannot, and will not, and should not attempt to express their feelings or relate their specific experiences in such activities. It is just something that is not proper to do.'*[9]

In 1977, Steve Goodman, a Chicago singer and songwriter, wrote and recorded a song titled *My Old Man*.[10] Goodman senior was a U.S. Army Air Corps pilot in the "big war' who flew C-47's with heavy loads of combat cargo. After America 'dropped the bomb', his father was able to return home, marry, and then become, Steve's old man. Goodman recognized that his dad was not only charming, funny, and frugal, but also able to accomplish difficult things in war and peace. Goodman, as many children do, 'tuned-out' his father. But then his dad died at the age of 58. Time passed, but he never forgot the old man. And then, eventually, late at night, thinking of his father, Steve Goodman cried. And he had the courage to tell us not to forget.

So therefor, maybe there is a time in life when remembrances, reflection, and retelling of horrible, but honorable shared experiences is proper.

In December 1974, four Japanese, dressed in World War Two Japanese Army uniforms, were carried deep into the jungle on Morotai. At a particularly remote location, they played a recording of the Japanese national anthem and waved the old Japanese rising sun flag. After a period of time, a frail old man wearing tattered clothes came cautiously out of the jungle. He was Teruo Nakamura, a 57-year-old Japanese soldier who became separated from his unit in August 1945. He never found his company again and never learned the news that Japan had surrendered. He built a small grass hut deep in the rugged mountains and lived off the land for almost 30 years, running, hiding, and thinking that Japan was still at war.[11]

Earlier that same year, in Reno, Nevada, six 307th Bomb Group veterans met in an informal gathering. Carl Whitesell, Dan Caufiel, George Jaffe, Ed Jurkens, M.P. Nelson, and Arthur Downs kept in contact off and on after the end of the war. They met in Reno in 1974 just to reminisce among themselves, but that meeting was the genesis of the 307th Bomb Group (HV) Association.[12]

What a contrast – in one case one lone man thinking the war had not ended for 30 years. In the other case, hundreds of men hardly thinking at all about the war for 30 years, but instead getting on with their lives. But then, when careers are finished, families raised and launched on their own trajectories, and individuals have time for reflection, perhaps it is then proper to reconnect with those who shared past experiences, to recollect, remember, and then pass their stories on for the future.

The Momentum of Life

One 307th Bomb Group veteran reflected on his experience in the war as he travelled east from San Francisco after his return home in 1945:

> *'I had acquired the rudiments of a complex technical skill in the Army, but felt no desire to use that skill, ever again. [Many] young men...left dead on Morotai...would never raise sons or daughters. Their lines of life were gone forever. I could not know what their lives might have been, but I wanted to contribute to the world on their behalf. I did not want that contribution to be more war...In war, not just the evil die and the good survive.'*[13]

What I discovered during the journey to understand my dad's experiences during the war included learning what some of his comrades did with their 'lines of life' after the war.

T/Sgt. Wayne Cooper, a 307th BG gunner, went to school at the University of California for a while, but he really had no idea what he wanted to do or be. His cousin worked for the telephone company and suggested that Wayne work there too. So he did and worked for the telephone company for 36 years before he retired in 1982. Along the way he got married, had a son, a daughter, and three grandsons. After he lost his wife, it took him years 'to come back to the world'.[14]

Lt. William D. Holston, a 307th BG co-pilot moved to California. He worked for the Southern California Edison Company as a powerhouse operator for about 25 years. Then he became a power and hydroelectric power dispatcher. Holston saw a similarity between his power-dispatching job and his career as a B-24 pilot, because of the heavy equipment he was operating, reading instruments and correcting problems.[15]

Robert Manley was a B-24 radio operator. After returning to civilian life, he used the GI Bill to attend a Catholic college in Manhattan run by the Christian Brothers.[16] Thomas Pelle was also a 307th BG radio operator gunner. He had been severely wounded during the October 3, 1944 mission to Balikpapan and had his leg amputated. Pelle recounted meeting someone years later at an Air Force reunion:

> "At the 13th air force reunion in Springfield, Missouri a guy was telling he was in the 42nd bomb group, a B-25 outfit at Cape Sansapor. Well that's where my operation was. Well he was walking past the tent and heard some guy's leg getting sawed off and I said "Yeah it was me!"

Pelle survived and returned to the States in December 1944. He was operated on several more times. After discharge, he went to work for the L&N railroad as a rate agent. In 1958, after the L&N and the Seaboard railroads merged he was transferred to Jacksonville, Florida and then to Louisville, Kentucky, retiring as an Assistant Rate Manager.[17] Thomas Pelle died in 2014.

James V. Walsh, who worked as an armorer and on a plane fueling crew for the 307th BG, came back after the war in 1945 and went to work for a barrel company. In 1951, he quit that job and went to work for Gulf Oil for thirty years, before retiring in 1981. A resident of Fannett, Texas, when he was '87 years young' he remembered that in the Air Force, 'I learned a lot, I traveled a lot. Showed me right from wrong'.[18] James Walsh died in 2015.

Glenn Norwood was a corporal in the 307th BG ground crew. After the war, he finished high school. He had played baseball in high school, and every chance he could while he was in the Air Force. At one point, he thought maybe he could play pro after the war and high school. Instead, he went to college and became a coach, taught for a while, was a principal, and eventually a superintendent. He got his high school degree on the GI Bill, a business degree on the GI Bill, and three masters' degrees on the GI Bill. Norwood believed 'The GI bill changed the world more than the war. Think about it I don't know if I could have gone to college without the GI bill, three kids and my wife got degrees all of that the GI Bill brought. All of it was the best for the country'.[19]

Sgt. Robert Robinson, who spent months cleaning radar set vacuum tubes with q-tips on Morotai, used the GI Bill too. He found a program at Boston University where you could get your undergraduate degree and a law degree in six years instead of seven. He graduated from law school in 1949 and did very well with a job at a law firm. Robinson married, and had two kids. He was able to send his kids to private schools, private colleges, and in essence, the GI Bill made it all possible.[20]

Ed Jurkens' story is interesting, because he was there with the 307th Bomb Group when it was first activated at Ephrata, Washington in 1942, and he saw it through to the war's end. He became a colonel and commanded the 307th when it was reactivated after the war. Later he commanded the Air Force ROTC program at Yale University, then the ROTC programs at all schools in New England, Michigan, Ohio, Kentucky, Virginia, and West Virginia. Jurkens married and had a daughter and was sent to England, assigned as a liaison officer with the RAF in Windsor. When the Korean War broke out, Jurkens headed a program at Lackland AFB near San Antonio to accelerate training for newly activated pilots. He was stationed as base commander at Pusan, Korea, then back to the US with the Strategic Air Command. Then Jurkens had assignments in Japan and Hong Kong, then finally ending his Air Force career at Hickam AFB in Hawaii. During the Vietnam War, he became disillusioned with the American involvement and retired from the Air Force. Jurkens became a volunteer for SCORE, a service corps for retired executives. He traveled across California, Guam, Saipan, and Samoa putting on Score workshops. Later he became a docent at the Waikiki Aquarium.

Then he got involved with the AARP safe driving program as chief trainer for the Hawaiian Islands. 'It keeps me mentally occupied. And, that's about it'.[21]

Jurkens was not with the 424th, but Walt and his crewmates would have known and seen John Vanderpoel, the Squadron Commander for much of their time in the South Pacific. Vanderpoel served as the 424th BS commander from July 12, 1944 until May 19, 1945.[22] After the war, he stayed in the Air Force and served in Japan during the Korean War. His family came out and they lived in a traditional Japanese house with a servant and a driver. Vanderpoel enjoyed skiing in Japan – one time with a Japanese Olympic team skier. After Korea, Vanderpoel was assigned by the Air Force to work on the B-52 project in Seattle with Boeing. He remembered getting a car from Boeing so that he could ski weekends at Mt Baker. Vanderpoel became an avid cyclist – he rode cross-country several times logging more than 200,000 miles by bike. In 1967, when he retired from the Air Force, he partnered with Bob Hall, a wheelchair athlete, to form Hall's Wheels, a company that manufactured wheelchairs for a variety of sports. His company employed nine employees – five wheel-chair disabled and four Mexican-Americans.[23] John Vanderpoel died in 2008 in Concord, Massachusetts at age 89, survived by his wife of 66 years, Joan, son Eric, and four grandchildren.

What about the men that Walt served with? After Jimmie Clark left the eastbound train in Denver that he shared with Cliff Llewellyn and Gene Schreiner in April 1945, I have learned nothing more of his later life. Len Sherman made the move to the South Pacific with the McRae crew in 1944, but he flew few combat missions with Walt. After December 1944, there is no mention of Len Sherman, Carl Appling, or Charles McRae. However, I have met, talked to, or corresponded with many of Walt's other friends or their families.

Boris Hidalgo was the first of Walt's wartime friends I spoke with. He stayed in the Air Force after the South Pacific. He was based in Alaska and then later flew on the Berlin Airlift in Germany. After he left the Air Force, Boris joined the US Customs service in Brownsville, Texas and married his wife Sara. Together they had five children. Boris worked for the US Customs Service for 40 years before retiring. He said that he would not retire until he put all five of his children through college. He ended his career as a Border and Customs Service inspector. On one trip, he had to go up to Vermont. On his way home to Texas, he took a detour via Ontario and stopped in Detroit. Boris tried to contact Walt. He found three Walter Babinski names in the Detroit phone directory and called each one. However, no one answered any of the numbers he called. When I talked to Boris, he said that he was very sorry to learn that 'Babs' had died before he could see him again.[24] In 2009, Boris and Sara moved to Round Rock, Texas to be close to their grandchildren. Boris died in 2014.

In late 2002 as I began my research for this book, I thought I had found a phone number, but not an address for an Edward T. Dunne, close to Ed Dunne's WWII home address in Brooklyn. I called the number a few times, but no one ever answered. I left a message on the voice recorder a couple times, explaining that I was the son of Walter Babinski, who had died but who left me information that led me to seek his old crew mates. A few weeks later, I received a card from Edward Dunne, 651 Vanderbilt, 11218, New York City. It was a religious greeting card with a note written in a shaky hand: 'With deep sympathy in your loss. I will write you soon. –Ed.'[25] That was the only contact I ever had from Ed Dunne, who had safely navigated Walt's crew so many times in California and across the vast South Pacific Ocean.

Don Aubrey completed 45 combat missions, and his gunners were credited with six Jap fighters shot down. After the war, he became an airline pilot. Aubrey married Margaret Coffin (a Northeast Airlines stewardess) from Bangor, Maine and had two sons. They lived in Pompano Beach, Florida.[26] Don Aubrey reflected about his time at war: 'Back in the U.S., I went to work for Northeast Airlines in New England in 1945. I started out on DC3's and flew for 20 years. My last 10 years as a pilot was with Delta airlines, where I retired. It strikes me as funny…in the Air Force; I dropped bombs and killed a lot of people. For that, I got awards and medals. But I was an airline pilot for 30 years, and not one of my passengers ever needed even a Band-Aid. For that, all I got was a plaque when I retired.'[27] Don Aubrey did succeed along with Carl Appling and Charles McRae in bringing his crewmates home safely many times.

Walt flew with Lt. Jim Fielding on just two missions in December 1944. After he was discharged in mid-1945, Jim wanted to be a commercial pilot. He went to many airlines, but with no luck in landing a job. He was told that because he had not gone to college, he could not get a commercial pilot job. The airlines were looking for college-educated pilots to enhance the public's confidence in commercial aviation after the war. Later in 1945, Jim went to the Sikorsky Helicopter company in Connecticut. He pleaded to be taught to fly helicopters. Jim thought this would land him a commercial job flying post war. He convinced the company and was soon trained to fly helicopters, but shortly after VJ-Day the Sikorsky plant shut down and there was no market for civilian helicopter pilots. Later Fielding developed a career in pharmaceutical sales.

Jim Fielding married and raised a large family at their home in Grosse Pointe, Michigan. In 2002, he showed me his leather Air Force jacket, which was in beautiful shape, with the 13th AF patch over 1 breast, and the 424th BS patch over the other. He showed me the inside pockets where emergency equipment and silk maps would be stored in case of ditching. Jim did periodic presentations to fifth grade classes at St. Paul's Catholic Church in Grosse Pointe. He spoke to the children about the role of the 13th Air Force in World War Two and about the experiences of 307th BG crewmen like himself. He would describe the details of a typical strike mission, things like preparing, feelings during the mission, and afterwards. He would show the schoolchildren photos and souvenirs of his time in the South Pacific Theater and also show them his leather bomber jacket - often letting one of the kids put the jacket on. Jim and his wife were also very involved in local community theater work.[28]

Dick Reis was the youngest of 14 kids in a Portuguese-American family growing up in Gustine, California. Later his family moved to San Jose, where Dick moved after the war. They bought a store, which his wife Mina ran.[29] Later Dick and Mina moved to Vancouver, Washington, where their daughter Kitty lived.

Jack Riley returned to his wife Nancy in South Carolina. They lived in Gastonia, North Carolina and had three sons. The two oldest sons enlisted in the Air Force, each serving four years. In 1972, the two older sons were duck hunting in a boat on a nearby river. The boat was swept over a dam and the two boys drowned. Their youngest son Chris was only nine years old at the time. Jack died in 1986 leaving his wife, one son, and four grandchildren. 58 years later, Nancy remembered the cross-country trip with Virgie Mansir and Helen Bretherton to Muroc Air Base in 1944.[30]

Ed 'Jupe' Bretherton stayed in the Air Force after the war. He and his wife Helen moved to Bossier City, Louisiana, near Barksdale Air Force Base. Ed died in 1987 in Shreveport, Louisiana. His wife still lived in Bossier City, with his two sons, Dennis and Bill nearby in Texas. Dennis was one of the two infants that Walt, Cliff, Len Sherman and the other crewmen babysat back in 1944 when the three wives drove out to Muroc Air Base to be with their husbands before they deployed.

After the war, Frank Mansir left the service, but then after a year or so he reenlisted. In the late 1940's or early 1950's the Air Force wanted to assign him to a new U.S. Bomber base in England. Frank and his wife Virgie did not want to go because all their family and friends were in the Ft Worth area. The Air Force offered to move his family and make him a warrant officer as added inducement, but still Frank did not want to go. Eventually he realized that if he didn't go, his military career would end before retirement age. He accepted and moved to the UK, but he did not accept the warrant officer promotion because it was not much more money and it had a fuzzy status in the military.

On his way to the UK, Mansir contacted Don Aubrey and Ed Dunne in New York City and arranged to meet them at a bar to 'relive old times.' However, Frank got cold feet and did not show up at the bar. Year later he regretted the lost chance to see these old comrades. In England, Frank participated in the Berlin Airlift, and later was top NCO at a new B-52 base there. From England, Frank, Virgie and their sons Paul and Don were able to travel across Europe, to Germany, France, and they especially liked Scotland. Paul Mansir was the other infant who made the cross-country trip to Muroc AAB in 1944 so that the three wives could be with their husbands, Frank, Jupe, and Jack.

Frank and his family came back to the U.S. and Frank ended his Air Force career in 1963 after 27-1/2 years of service. When he retired, he was the senior USAF NCO in the Dallas/Ft. Worth area, where he had established his family and had many friends. For many years after retirement, Frank suffered from bone deterioration and emphysema. Virgie got a degree from Texas Christian University and worked as a schoolteacher for many years. She eventually went back and got a Master's Degree. Frank and Virgie bought a small farm about 60 miles outside Fort Worth. During the summer (when school was out), they would live on the farm and during the school year they lived in Fort Worth. They both liked life on the farm.

Their oldest son, Paul Mansir, died in 1965 in Vietnam, when he stepped on a mine while on patrol. He was the first Fort Worth resident to die in service in Vietnam. An American Legion Post in Fort Worth is named for Paul Mansir. Virgie was invited to be honored at many activities at this American Legion Post. The Mansir's second son, Don Mansir, lives in the Fort Worth area. Virgie remembered that Frank gave her a seashell necklace that he made while in the South Pacific, which she treasured. Frank died in 1988.[31]

Gene Schreiner and Walt wrote a few times after the war, but then lost touch with each other as they returned to civilian life. Gene married, worked a few years as a mailman, and then started farming in Colorado. Later he became a rancher in Wyoming and also a State of Wyoming cattle inspector. His son Roger became a CPA in Casper, Wyoming. Gene retired and lived in Overton, Nevada.[32] I met Gene and his son Roger at the 2002 Reunion of the 307th Bombardment Group in Salt Lake City.

I also met and developed a friendship with Cliff Llewellyn and his wife Betty. In 2002, I visited them at their home in Thiensville, Wisconsin, just outside of Milwaukee. Cliff only completed eight years of school. After he returned from the Pacific, Santa Ana, and then Texas, he was finally released from the service. He considered more school, even college with the GI Bill, but in 1945 and 1946 all the schools were full. He went to work and never returned to school.

Cliff met Betty in 1945 and they were married shortly thereafter. Cliff came from a family of seven kids in Winona, Minnesota and Betty was one of 12 kids. She and her family lived all their lives in the Milwaukee area. Cliff went to work first as a maintenance/repair man working for Bendix and Maytag. Cliff said that he was not the best mechanic, but he was good at supervising crews, as he had done in the Air Force. He changed jobs about 12 times, moving from one Bendix or Maytag office to another. He usually got about a $100/month raise when he moved offices, but in retrospect, he felt that those who stayed in Milwaukee and worked for a big brewery actually made more money. He worked for Bendix offices in Milwaukee, Minnesota, and Indianapolis. He also took short trips for Bendix to New York City working at a Bendix display at the 1964 World's Fair, and to Puerto Rico.

Cliff and Betty's two daughters live in Texas. One worked as a school nurse and the other was the office manager for a small company. One daughter had three kids and the oldest had a child of her own, so that Cliff and Betty were great-grandparents. In Thiensville Cliff showed me a dollhouse that he made that was a scale model of their house in Milwaukee, where Cliff, Betty and their family lived for 37 years. The dollhouse was a work of love, very detailed and complete with furnishings and wall coverings. It eventually went to his family in Texas.

Cliff and Betty visited Frank and Virgie Mansir several times in Texas when they were on trips to visit their daughters. Their son, Cliff Junior finished about two and a half years of college, and then quit to start working. He worked as a draftsman for a gas and electric utility. Then he moved to Colorado with a friend and went to work for an electric and phone utility. He married and moved to Aspen, where he and his wife bought an old dorm-style ski lodge. They rebuilt it and operated it as a ski resort. Cliff Senior spent some summer vacations in Aspen, helping Cliff Junior and his wife remodel the lodge. They turned the open dorm areas into private rooms. Cliff and Betty also learned to cross-country ski and would ski in Aspen with Cliff Junior and his wife.

After 11 years, Cliff Junior sold the Aspen ski lodge and moved to Escondido, California and then to Ashland, Oregon. He and his wife buy old houses and remodel them, preserving the historical exterior, but gutting and modernizing the interior. Cliff and Betty enjoyed visiting Cliff in Ashland and liked the southern Oregon area with salmon bakes, float trips, the airshow in Medford, and jet boating on the Rogue River.

Cliff only learned about the 307th Bombardment Group Association because of the efforts of Gene Schreiner. Schreiner had tried to contact many of the old crew. He knew that Cliff was from Winona, Minnesota. Schreiner had a friend in Minnesota who agreed to go to Winona to try to find Cliff or any Llewellyn who might know Cliff's whereabouts. But they had all moved away. Gene's friend went to the American Legion in Winona and found one person there who had known and kept in contact with Ray Llewellyn, Cliff's younger brother who had been in the Army. So Schreiner was able to contact Cliff via his friend in Minnesota and Cliff's brother Ray. That is how he learned of and came to join the 307th BG Association.

When I visited him in Thiensville, Cliff showed me a yellowed list from the IV Bomber Command - 382nd Group - 539th Squadron. The list named each of the men from each of the 13 crews who completed their bomber phase training on March 25, 1944. Cliff's list had been annotated by him to show which of the crews or individual crewmen had been lost, killed, or wounded during the war. Only the McRae Crew and one other went through the war completely unscathed. I told Cliff that Nancy Riley had confirmed this record for me. This fact is one of the things I remember hearing my father say many times.[33] Sometimes luck or fate determines our lines of life.

Walt and Dorothy lived their lives in Detroit and Dearborn, raising 10 children. Walt's family shared his joy when his beloved Tigers won the World Series and the Red Wings won the Stanley

Cup. Walt put in 40 years at Detroit Stamping Company, retiring in 1987. I know that he would have enjoyed meeting his Long Ranger friends again.

I organized the 2006 Reunion of the 307th Bombardment Group Association in Bellevue, Washington. About 75 of the vets and 125 of their family and friends attended. Derek Poppe and Emily Mount conducted videotaped interviews of ten of the vets, which were then submitted to the U.S. Library of Congress Veterans Oral History Project. A woman named Laura Hildebrandt attended to meet and speak with some of the vets. She used some of what she learned to later write her book about one of the Long Ranger Vets.[34] For many years, one of the friends of the Long Rangers has been Dr. Pat Scannon, of the Bent Prop Project. He attended the 2006 reunion in Bellevue and made an important announcement. He and the Bent Prop Project had found the wreck of A/C 453, piloted by Lt. Jack Arnett and lost over Koror, Palau on September 1, 1944. During a ceremony at the reunion, folded American flags that had been flown by the Bent Prop crew over the spot in the ocean where A/C 453 had been lost, and then found, were presented to families of the crewmen, while my son played *Amazing Grace* and *Coming Home* on the bagpipes.[35] Some lines of life end, but for good lives well lived, some people do not forget.

Virgie Mansir recounted to me that for many years her husband Frank was troubled by nightmares related to memories of combat in the South Pacific. Cliff Llewellyn told me that he was troubled by similar nightmares, and my mom told me that dad, Walt, was also troubled for many years by combat related nightmares.

In 2013, I had an opportunity to travel to Djakarta, Indonesia. I spent some extra time and succeeded in traveling from Djakarta via the Celebes, Ternate, and Halmahera, to Morotai. I met warm and friendly people who were eager to talk to Americans. I drove across Halmahera, below the IP for Walt's December 22, 1944 mission to bomb Lolobata and Hate Tabako, which I could see across the bay. I stayed in a small Morotai hotel on a highway that I believe was built over the Wama Airfield runway. I managed to talk my way onto the Indonesian Air Force Base on Morotai and walk out on what had been Pitoe Air Field, where my dad flew half of his combat missions. I visited a new World War Two Museum located on the old 424th Squadron camp site on the Gila Peninsula that honored the Americans, Dutch, and Australians who fought on Morotai. The Museum also honored Teruo Nakamura, the Japanese soldier who lived more than three decades on Morotai, refusing to believe that the war was over. More than anything, I went to Morotai to try to experience the sound of the birds and the waves on the beach, the feel of the hot sun and cool ocean breeze, the color of the trees and flowers, and the clouds scudding across the sky, that my dad would have experienced for his many months on Morotai.

We each have our own single unique line of life. We each have our own daytime experiences, personal evening reflections, and unique and uncontrollable nighttime dreams. Across a lifetime, we each encounter and to some degree interact with many thousands of people. The meaning of life is not our accumulated riches. It is our accumulated interactions with people, our empathy with them, and our understanding of them and of their life-experiences.

In the film 'Saving Private Ryan', an old James Ryan says to his comrades 'I tried to live my life the best that I could. I hope that was enough. I hope that, at least in your eyes, I've earned what all of you have done for me.' Walt and his buddies were a team. They survived because of what they did for each other.

My father and his comrades were brave men to fly vast distances in rickety aircraft to fight a determined enemy on behalf of their country and for all humanity. There is so much that I did not know about him during the 42 years of life that we shared together. Learning more of his life experiences and those of his comrades has enriched my life. He and his buddies played a small part in a big war. They were good men who lived good lives. They each have a big place in my heart. Memories fade fast, but I hope I have helped to keep a little of their past alive.

APPENDIX A: THE BALL TURRET

A ball turret gunner needs a ball turret. There have been many strange contraptions developed throughout the history of warfare, from archers on chariots, to soldiers on elephants; knights clad in heavy armor mounted on massive war-horses, blimps carrying fighter planes, flame throwers, and Molotov cocktails. But none of these infernal devices can compare to a tiny claustrophobic aluminum and Plexiglas capsule that restricted the freedom of movement of the operator who crouched in a semi-fetal position while suspended by a single steel tube miles above the surface of the earth (or ocean) below a big slow aircraft filled with thousands of gallons of highly flammable aviation fuel and a few tons of high-explosive bombs. The operator of this device could not easily wear a parachute or life vest, nor could he extract himself from his capsule without the help of one of his crewmates. Those crewmates depended on him to defend their aircraft from attack across the entire bottom half of the sphere in which they flew with two .50 caliber machine guns situated just inches on either side of his head.

The heavy bomber as a weapon of war emerged during World War One. During the 1920s and 1930s a theory of military strategy emerged that vast fleets of heavy bombers could defeat an enemy, not by crushing its armies in battle, but by destroying its industrial capacity and its people's will to fight by striking its cities hundreds of miles behind the conventional front lines. By the late 1930s, Italy and Germany during the Spanish Civil War and Japan during its invasion of China seemed to prove that bombers could get through to bomb cites and help achieve victory. This success focused each major nation, including Germany, Italy, and Japan, on the problem of how to defend against the heavy bomber. Two means of attacking bombers that evolved during World War One were still the most effective during World War II: anti-aircraft guns and fast fighter planes.

Nothing could effectively defend a warplane from A/A fire from below other than flying higher or destroying the A/A guns themselves. However, if fast fighters threatened bombers, the bombers could be equipped with aerial gunners who could try to shoot the fighters out of the sky. As bombers evolved, almost every major power put a gunner in the nose of the plane, one at the tail, and two in the waist, one defending each side of the aircraft. This is like defending a castle or fortress. But to avoid these gunners, an attacking fighter could dive from above. But another gunner could be placed at the top of the bomber to defend the upper half of the sphere. Each of these gunners could sit or stand to operate their guns. However, the bottom half of the sphere is more difficult to defend against a fast fighter attacking a bomber from below. Germany, England, Japan, and Italy each developed a method of defending a portion of the bottom half of the sphere by having a gunner lie prone on the belly of the airplane firing a single gun.

The ultimate solution to the problem of effectively defending the lower half of the sphere was developed in the early 1940s by the Sperry Corporation of Brooklyn, New York. The Sperry Ball Turret was fitted into B-17 and B-24 bombers to provide bottom-of-the-sphere defense. It was small, but still too large to be fitted into B-25 and B-26 medium bombers. The Sperry Ball Turret was just 44 inches in diameter. Most operators were between 5'-4" and 5'-6" tall. Walter Babinski was well out of this range.

How was the Ball Turret operated? First, the gunner had to see to his guns on the ground. Prior to each mission the gunner or an armorer had to install them in the turret and load belts with about 400 rounds of ammunition for each gun in boxes. During takeoff and landings, the B-24 ball turret was retracted into the belly of the plane with the guns facing aft. The ball turret gunner would find a comfortable position in the belly of the plane for take offs and landings. In general, to man the turret, it would be lowered and the guns depressed, or pointed straight down. This brought the turret hatch in line with the bottom of the fuselage, so the gunner could open it and lower himself inside. He would then slide down and assume a fetal position, with a 13-inch diameter vision panel about 30 inches from his face. Directly in front of his eyes was the computing gun sight. No room to stretch, but otherwise comfortable.[1]

The detailed instructions or operating the ball turret included this warning for Walt:

THE SPERRY BALL TURRET
To become acquainted with the Sperry Ball, get in.

YOU WILL HAVE TO BE CAREFUL! MEN HAVE BEEN KILLED BECAUSE THEY DID NOT KNOW THE CORRECT WAY TO GET IN OR THEY WERE CARELESS ABOUT IT!

The turret ball is heavy and yet delicately balanced. Unless it is locked in place, it may swivel and break a man's leg or snap him almost in two as he attempts to enter. As you get in, take these steps carefully and in the proper order. If you go about it right, getting into the turret is as safe a climbing into a rocking chair.

1. Be sure the elevation hand clutch is in the IN position. Remove the elevation hand crank from its clip on the outside of the turret. Slip it on the elevation hand control shaft.

2. Loosen the elevation hand brake --- a small lever beside the elevation hand control shaft.

3. Keep your left hand firmly on the hand crank. Take the outside elevation power clutch handle from its clip, put it on the outside elevation power clutch shaft, and move it to the OUT position. This disengages the elevation power gearing, allowing free up and down movement of the turret. Unless you keep a tight hold on the hand crank, the weight of the guns will spin the turret down.

4. By turning the elevation hand crank with your left hand, move the turret until the guns are pointing straight down at 90 degrees. This will bring the turret door into position in the floor of the plane so that you can open it.

5. Keep your grip on the hand crank. Unfasten and lift the door. Don't let the door fall back. (It is heavy armor plate) To prevent springing the hinge, never put excess weight on the door.

6. Still keep your grip on the hand crank. Reach down inside the turret, forward and behind the big elevation gear housing on the left, and move the inside elevation power clutch lever up to engage it. (You can't see the lever from the outside, you must be able to know it by touch with your right hand. With practice, you can tell from the position of the lever if you have fully engaged the gears.) You will have to rock the elevation hand crank back and forth as you push the clutch in place. This re-engages the elevation power gearing, locking the turret firmly in gear.

7. Move the outside elevation hand clutch to OUT. Remove the hand crank and outside power clutch handle and replace in their clips.

8. Test the turret by grasping the supporting structure and putting your right foot on the seat. Try to rock the turret with the pressure of your foot. THERE SHOULD BE NO MOVEMENT. If there is movement, the elevation gearing is not engaged.

9. Lower yourself into the turret by grasping and swinging down on the supporting frame. Put your right heel in the right foot rest, then your left heel in the range pedal - - - being careful not to throw your whole weight on the range pedal.

10. Fasten the safety belt. Close and fasten the door. Make sure the latches are securely closed. Then you are ready to go.

OPERATING THE TURRET

1. Make sure the azimuth power clutch is engaged - - - the lever at the upper right should be down.

2. Flip on the main power switch - - - a toggle switch under a wire guard on the junction box beside your left knee.

3. Right beside the power switch are two gun selector switches for firing the guns separately or together. Flip both of them on.

4. Turn on the sight switch directly in front of your nose on the sight. Turn the sight rheostat beside it to adjust the brightness of the light. Never operate the turret under power with the sight switch off.

5. Finally, grasp the two handgrips. Don't worry about safety switches -- the turret has none. But, notice that in azimuth the controls work exactly opposite to all other turrets. To move the turret RIGHT, swing the handgrips to the left; to turn LEFT, swing them to the right. Pull back on them to tilt the turret and guns up, and press forward to turn the turret and guns down. Don't jerk. Steady, smooth tracking is especially important in using the Sperry sight.

The other controls you will use in combat are right at hand. On the top of each handgrip, under your thumbs, are firing buttons, used instead of triggers on the Sperry Ball.

The fuses which protect the Sperry's electrical circuits are in the junction box beside your left knee, where you found the main power switch.

For charging the guns, you will find two handles next to your feet. Reach down to the floor, arms crossed, and pull the handles up carefully until the slack is out of the cables. Then pull them up sharply all the way. Don't ride the cables on the return action; allow them to go back under their own power, still keeping your hands on the handles.

USING THE AUXILIARY UNITS

You will find a demand type regulator with a hose connection for your oxygen mask right under your seat. Alongside your right ear is the flow and pressure gage. The oxygen is drawn from the plane's central tanks in most models. The heated suit plug-in is also under your seat. The interphone jacks for your headphone and throat microphone lead out of the junction box. A trouble light is clamped in a clip just under top rim of the ball to the left of your head. By your right foot is a new azimuth turret position dial, in which a cut-out of your guns rotate upon a clock face and shows your azimuth position at all times. For manual operation of the turret in emergencies, shafts designed for use with removable hand cranks are on the right and left side of the turret -- - right for azimuth, left for elevation. The hand cranks are stored in clips above your head.

TO CHANGE TO MANUAL OPERATION:
1. Turn off the main power switch.
2. Remove the hand cranks from their clips and place them on the shafts.
3. Disengage azimuth and elevation clutches.

GETTING OUT

To get out of the turret, drive it to 90 degrees in elevation, so that the guns point straight down; this will bring the doorway into place in the floor of the bomber. At the same time, drive the turret to 180 degrees in azimuth -- six o'clock on your azimuth turret position dial.

Turn off all switches. Open the door and step out.

Now move the guns up to zero degrees elevation, pointing back toward the tail. First, attach the outside elevation hand clutch to it IN position. Keeping a firm grip on the hand crank -- to keep the turret from moving -- reach inside the turret and move the inside elevation power clutch lever down to disengage it. Close and latch the turret door. Then crank the guns to zero degrees elevation. Hold the elevation crank and remove the outside elevation power clutch handle from its clip, slip it on its shaft, and move the clutch to IN.

Tighten the hand brake, and then return the elevation hand crank and clutch handle to their clips.[2]

Now wasn't that simple? No wonder that it took months of aircraft mechanic, aerial gunnery, and bomber phase training before an aerial gunner, especially a ball turret gunner, was ready to go into combat to defend the lower sphere of his aircraft.

The Sperry ball turret was not a position for either a tall man or anyone who was claustrophobic. The twin .50 caliber machine guns could each fire 600 rounds per minute. If the gunner fired for more than a few seconds, the gun barrels could overheat and deform, causing a jam. Gunners were trained to fire in three-second bursts, but in the heat of battle, they often exceeded this rule. Spent cartridge links and cartridges from all the guns were ejected through a chute and would sometimes rain down on bombers below.

The turret was powered by a 32-volt motor. It could rotate 360 degrees in the horizontal plane and the guns could be depressed from the horizontal to the vertical position. To rotate the turret and elevate or depress the guns, the gunner operated a hydraulic/electric control stick in each hand, on either side of the gun sight. The triggers were also mounted on these controls. When operating the turret, the gunner's hands were located at his sides. Moving the sticks sideways made the turret rotate from side to side. Pushing or pulling the sticks moved the guns up and down. This provided control to track a target at any position within the half sphere below the aircraft.

Shortly after takeoff, enroute to a target, each gunner would fire a short burst from his guns to ensure that they operated properly. Even in the tropics, when flying above 10,000 feet, the gunner had to wear an oxygen mask, with a built-in throat mike and earphones to communicate with the rest of the crew. At very high altitudes, gunners wore an electrically heated flying suit, boots and gloves. The ball turret gunner wore a parachute harness, but there was not enough room in the turret for the parachute, so it was stored in the fuselage. If a ball turret gunner had to bail out he would have to be extricated from the turret by first undoing his electrical, oxygen and safety harnesses, point the guns aft and down, open the armored door and climb back into the aircraft, where he would clip on his parachute.[3]

After the war, a study of bomber crew fatalities would find that actually, the ball turret had been the safest of all the crew positions on a B-17 or a B-24.

The Ball Turret Gunner: This idealized view of a ball turret gunner appeared in a typical magazine ad by a manufacturing company that had switched over to war production work. Walt likely saw depictions like this as he was training as an aircraft mechanic, aerial gunner, and then during bomber phase training at Muroc AAF. *Author's personal collection.*

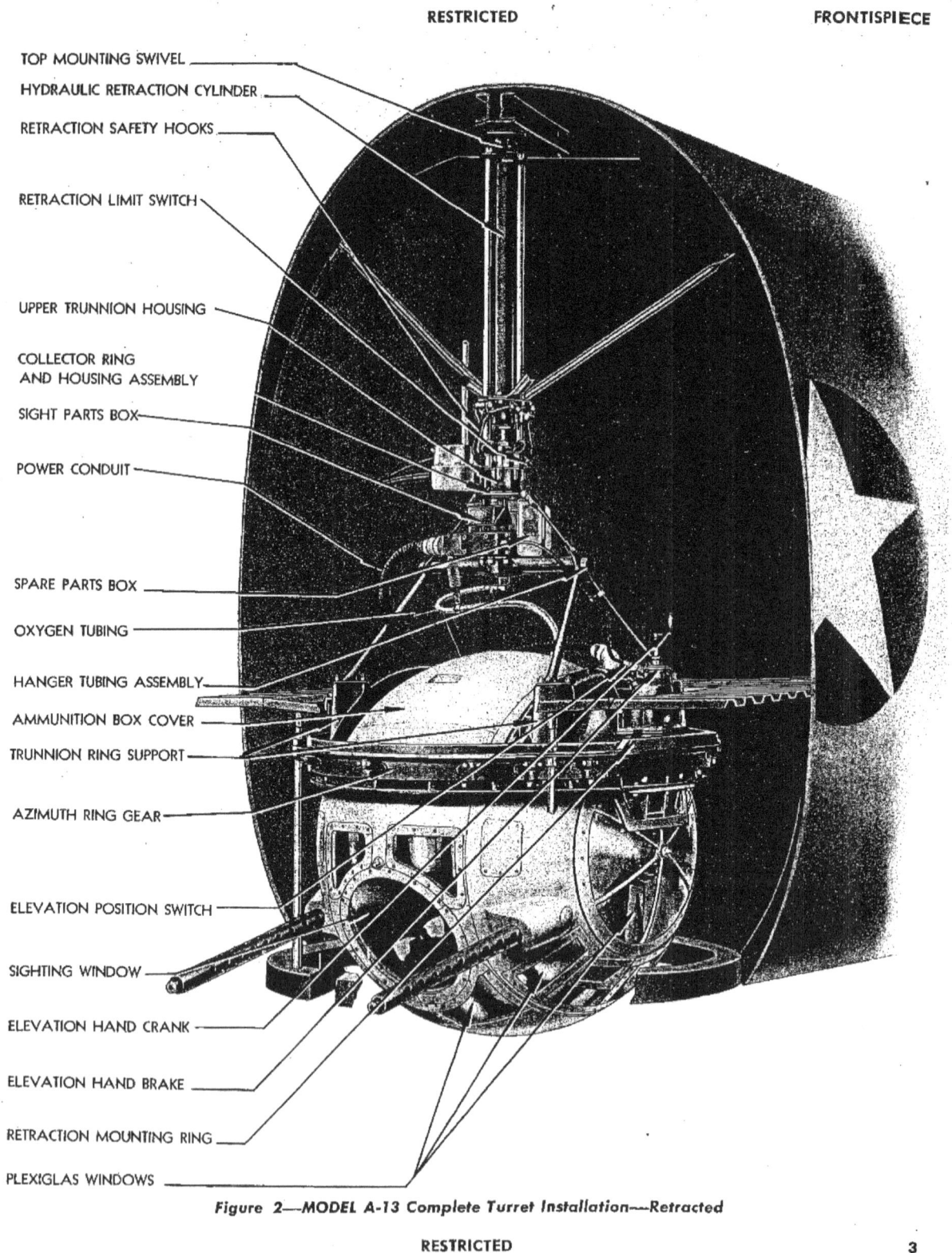

Figure 2—MODEL A-13 Complete Turret Installation—Retracted

The Ball Turret: View of Sperry Ball turret retracted into B-24 fuselage.
Handbook of Operation and Service Instructions for Retractable Lower Ball Turret, Model A-13.

FRONTISPIECE RESTRICTED

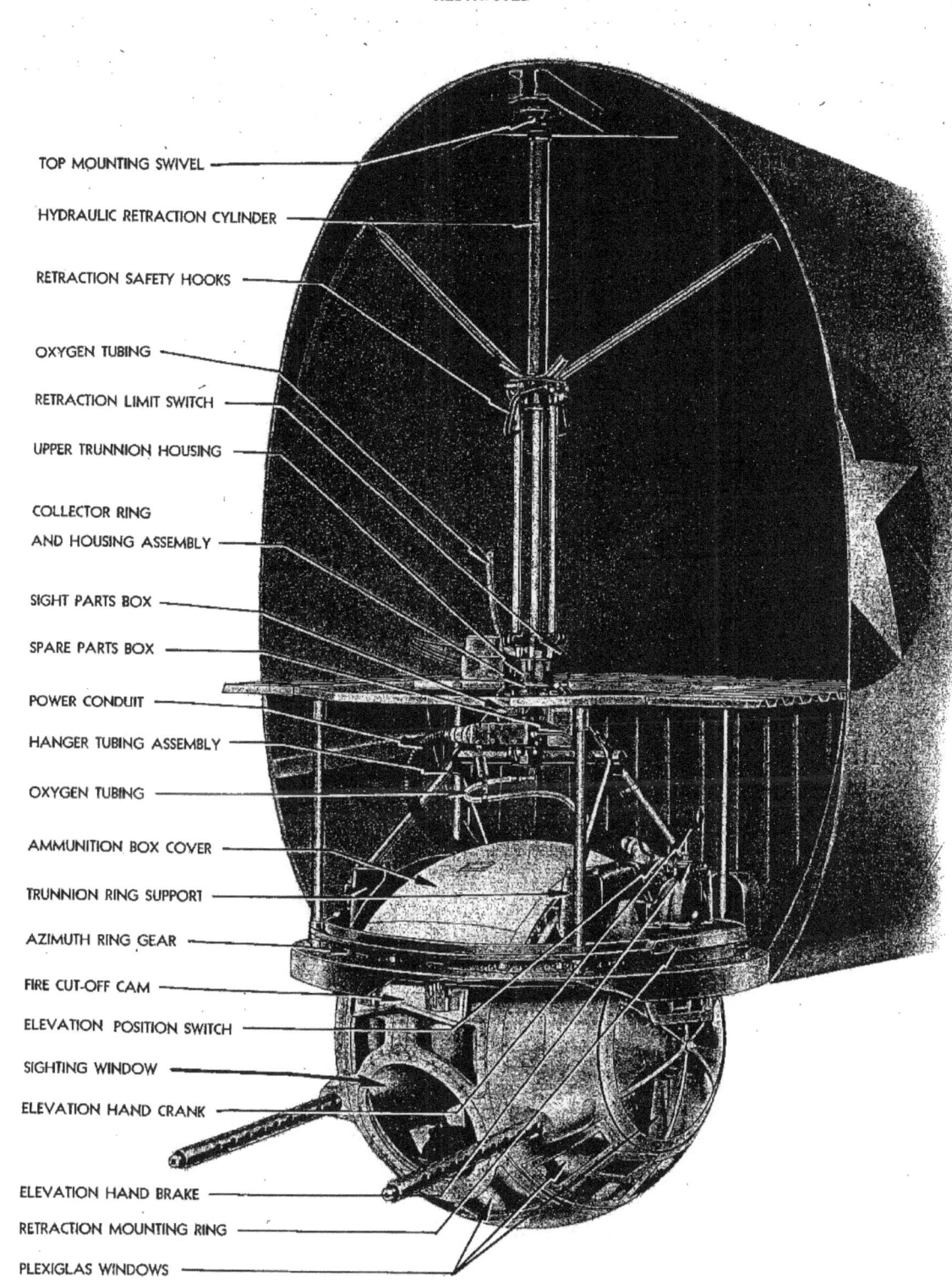

Figure 1—MODEL A-13 Complete Turret Installation—Extended

RESTRICTED

The Ball Turret: View of Sperry Ball turret as lowered below B-24 fuselage.
Handbook of Operation and Service Instructions for Retractable Lower Ball Turret, Model A-13.

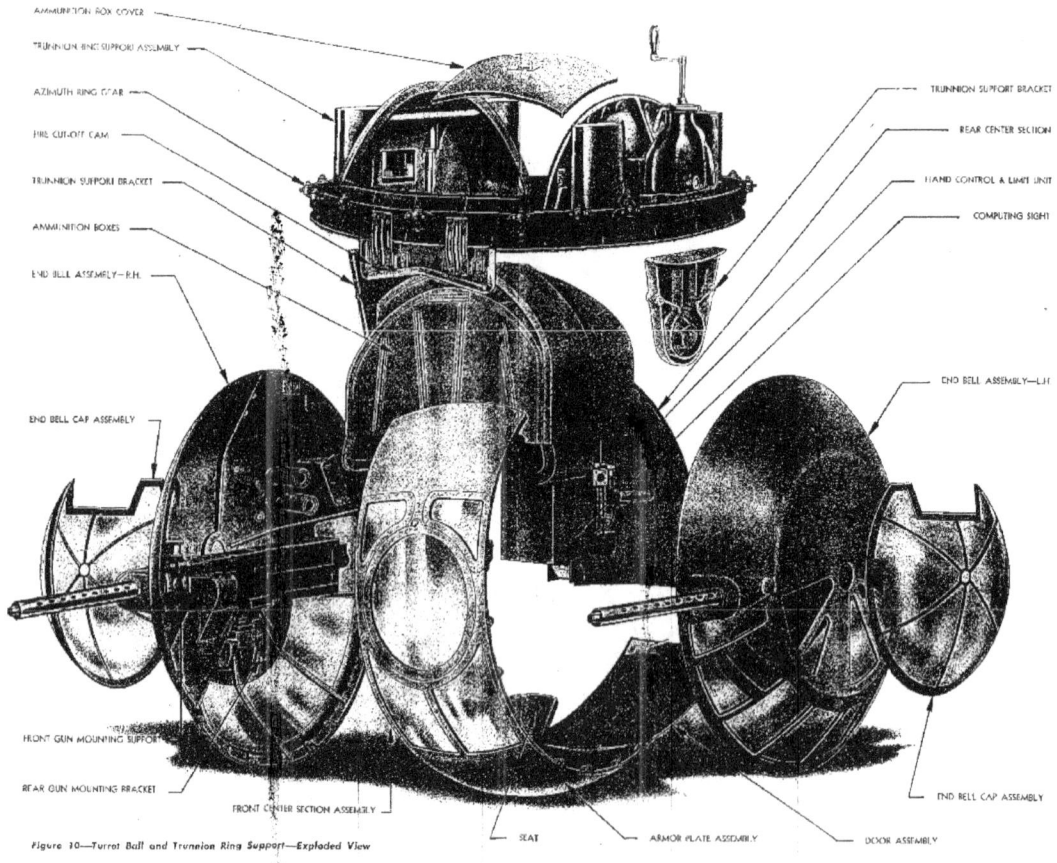

The Ball Turret: Cutaway views of ball turret structural components (above) and gunner seated in turret (below). *Handbook of Operation and Service Instructions for Retractable Lower Ball Turret, Model A-13.*

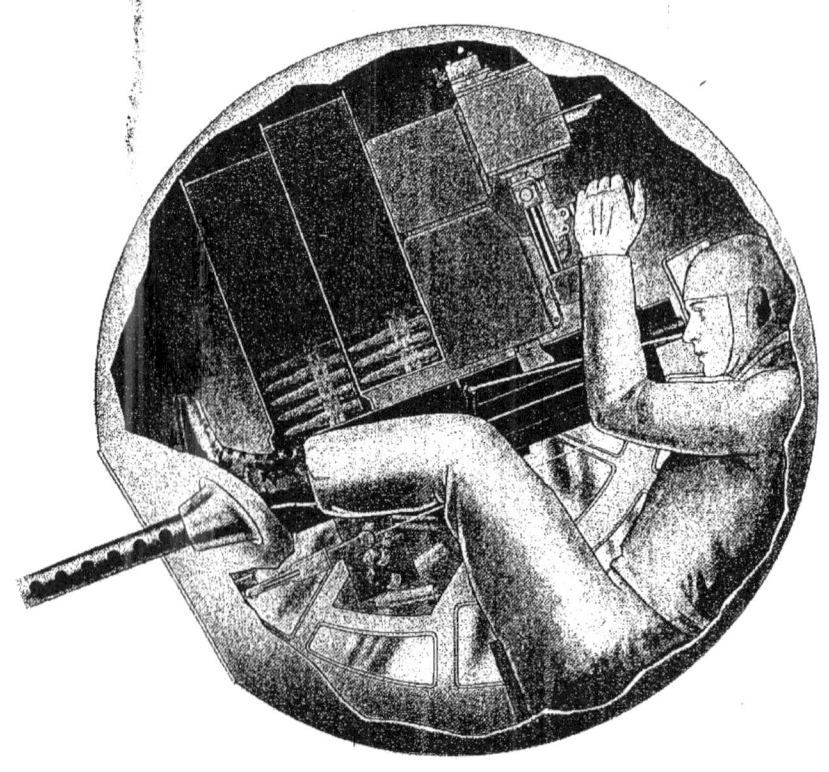

Appendix B: Selected Military Records

App. not Req.

Prepare in Duplicate

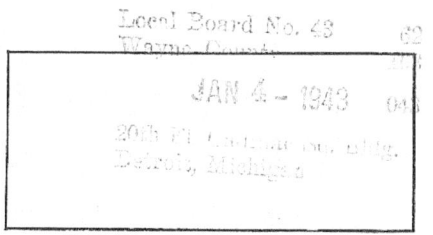

(Local Board Date Stamp With Code)

January 4, 1943

(Date of mailing)

ORDER TO REPORT FOR INDUCTION

The President of the United States,

To   Walter   Anthony   Babinski   Jr.
       (First name)   (Middle name)   (Last name)

Order No. 13551

GREETING:

Having submitted yourself to a local board composed of your neighbors for the purpose of determining your availability for training and service in the armed forces of the United States, you are hereby notified that you have now been selected for training and service in the __Army__
(Army, Navy, Marine Corps)

You will, therefore, report to the local board named above at __3162 East Jefferson Ave. Detroit,__
(Place of reporting)
__Michigan.__

at __7:00 A.M.__ m., on the __14__ day of __January__, 19__43__
(Hour of reporting)

    This local board will furnish transportation to an induction station of the service for which you have been selected. You will there be examined, and, if accepted for training and service, you will then be inducted into the stated branch of the service.

    Persons reporting to the induction station in some instances may be rejected for physical or other reasons. It is well to keep this in mind in arranging your affairs, to prevent any undue hardship if you are rejected at the induction station. If you are employed, you should advise your employer of this notice and of the possibility that you may not be accepted at the induction station. Your employer can then be prepared to replace you if you are accepted, or to continue your employment if you are rejected.

    Willful failure to report promptly to this local board at the hour and on the day named in this notice is a violation of the Selective Training and Service Act of 1940, as amended, and subjects the violator to fine and imprisonment.

    If you are so far removed from your own local board that reporting in compliance with this order will be a serious hardship and you desire to report to a local board in the area of which you are now located, go immediately to that local board and make written request for transfer of your delivery for induction, taking this order with you.

Member or clerk of the local board.

D. S. S. Form 150
(Revised 7-13-42)

Form 150: Walt's order to report for induction, January 14, 1943

TSKF Form No. 86
(Revised Mar. 25, 1943)

TECHNICAL SCHOOL
ARMY AIR FORCES TECHNICAL TRAINING COMMAND
KEESLER FIELD, MISS.   20-8-jw

CERTIFICATE OF PROFICIENCY

Date September 14, 1943

THIS IS TO CERTIFY THAT:

Babinski	Walter	A.
(Surname)	(Christian)	(Middle)

36559500	Pfc.	18th Repl Wing
(Serial No.)	(Rank)	(Org. or Arm)

WAS *GRADUATED* FROM THE

B-24 AIRPLANE MECHANICS COURSE

ON September 14, 1943

SUBJECTS COVERED	HOURS	GRADES
Airplane Mechanics' Tools	48	74
Air Forces Fundamentals	48	80
Airplane Structures	48	82
Airplane Hydraulic Systems	48	70
Airplane Engines	48	79
Airplane Electrical Systems	48	84
Airplane Fuel Systems	48	81
Airplane Instruments	48	78
Airplane Propellers	36	78
Airplane Engine Operation	72	82
Airplane Inspection I	48	80
Airplane Inspection II	48	79
Airplane Inspection III	48	78
Graduation Field Test	42	88
TOTAL HOURS AND AVERAGE GRADE	678	79.6

Subjects covered and grades made are as listed.

95 - 100   Superior
90 - 94   Excellent
80 - 89   Very Satisfactory
70 - 79   Satisfactory
Below 70   Unsatisfactory

For the DIRECTOR OF TRAINING: L. C. EULBERG,
Capt., Air Corps,
Personnel Officer.

Form 86: 9/14/43: Walt's grades in the Aircraft Mechanics Course, September 14, 1943

S/Sgt Babinski

RESTRICTED

HEADQUARTERS
THIRTEENTH AIR FORCE
APO #719

GENERAL ORDERS)   1 December 1944.
                 :
NUMBER   190)   E X T R A C T

OFFICIAL CREDIT FOR THE DESTRUCTION OF ENEMY AIRCRAFT BY BOMBER CREWS

1. The following named officers and enlisted men, Air Corps, United States Army, are officially credited with the destruction of one enemy fighter type aircraft in aerial combat over Negros Island, Philippines, at 1053/I on 6 November 1944:

2nd Lt	CARL D. APPLING	O-747190
2nd Lt	DONALD W. AUBREY	O-760761
2nd Lt	EDWARD T. DUNNE	O-699148
2nd Lt	JIMMIE W. CLARK	O-701565
S/Sgt	Edward R. Bretherton	39330042
S/Sgt	Boris V. Hidalgo	18201062
T/Sgt	Raymond E. Schreiner	17088524
S/Sgt	Walter A. Babinski	36559500
S/Sgt	Francis M. Manzir	20808391
S/Sgt	Jack B. Riley	14062227

This crew was part of a formation of B-24's that was intercepted by a formation of Japanese fighters. Sergeant Riley, tail gunner, fired on a Zeke attacking from 6 o'clock and the enemy aircraft began smoking and went into a dive. It was observed to crash into the sea.

* * * * * * *

By command of Major General STREETT:

JOHN M. STERLING
Colonel GSC
Chief of Staff

OFFICIAL:

/s/ Peyton G. Nevitt
/t/ PEYTON G. NEVITT
Colonel AGD
Adjutant General

A TRUE EXTRACT COPY:

David Davis
DAVID J. DAVIS,
Captain, Air Corps,
Adjutant.

RESTRICTED

GO 190: Appling crew credit for one Japanese Zeke destroyed on November 6, 1944

424TH BOMBARDMENT SQUADRON (H)
307th Bombardment Group (H)
APO # 719

Date 25 February 1945

SUBJECT: Battle Participation Credit - Awards & Decorations.

TO: Commanding Officer, of S Sgt Walter A. Babinski, 36559500.

1. S Sgt Walter A. Babinski, a former member of this organization transferred to your command per _____, is authorized to wear Battle Stars on the Asiatic-Pacific Campaign ribbon for participation in the campaigns listed below:

CAMPAIGN	AUTHORITY
New Guinea	Cir 195, Par 7B, Sec I, War Department 1944
Mandated Islands	Cir 195, Par 7B, Sec I, War Department 1944
Phillipine Islands	Cir 195, Par 7B, Sec I, War Department 1944
Phillipine Liberation Ribbon	Hq, Commonwealth of the Phillipine Army 12-20-44
Bismarck Archipelago	Cir 195, Par 7B, Sec I, War Department 1944

2. Subject Enlisted Man further authorized to wear the below listed awards:
BOLC (AM)   G.O. #138   dd 17 Nov 44   5 Sorties   13th Air Force

3. Subject Enlisted Man has the following awards upon which action has not been completed:

Air Medal	Letter 307th Bomb Gp (H)	100 Opn Hrs.	dd. 31 Aug 44
BOLC (AM)	Letter 307th Bomb Gp (H)	100 Opn Hrs.	dd. 30 Nov 44
BOLC (AM)	Letter 307th Bomb Gp (H)	100 Opn Hrs.	dd. 20 Dec 44
BOLC (AM)	Letter 307th Bomb Gp (H)	100 Opn Hrs.	dd. 5 Feb 45
Unit Presidential Citation	Letter	13th Bomber Command	dd. 3 Dec 44
Unit Presidential Citation	Letter	13th Bomber Command	dd. 6 Dec 44

4. Request acknowledgement of receipt by indorsement hereon.

For the Commanding Officer:

DAVID J. DAVIS,
Captain, Air Corps,
Adjutant.

BPCA&D: 2/25/45: Walt's Battle Participation Credit Awards & Decorations

424TH BOMBARDMENT SQUADRON
307th Bombardment Group (H)
Office of the Operations Officer

NAME BABINSKI, WALTER A. **RANK** S Sgt
DUTY Asst. Rad. Opr. **ASN** 36559500
DATE DEPARTED FOR FOREIGN DUTY 8 Apr 44 **ARRIVED SOPAC** 17 Apr 44

FORMULA	STRIKES	MONTHS OVERSEAS	SP FLYING TIME
	10	3	100

DATE	SEARCH	STRIKE	COMBAT	OTHER	TOTAL	TARGET AND REMARKS
20 May 44		8:00				Strike Rabaul
27		8:15				Strike Rabaul
May 30				12:30		
				6:15		709-324
3 Jun		9:00				Strike Truk
7		6:35				Strike Truk
9		6:00				Strike Truk
11		8:25				Strike Truk
17		8:45				Strike Truk
19			1:30			Strike Truk (Returned)
23			5:15			Strike Yap (Returned)
26		12:35				Strike Yap
30		11:40				Strike Noemfoor
3 Jul		12:15				Strike Yap
7		10:00				Strike Yap
13		13:00				Strike Yap
6 Aug		12:30				Strike Yap
10		12:25				Strike Yap
19				3:50		719-Wakde
29		9:05				Strike Palau
3 Oct			6:10			Strike Balikpapan (Returned)
10		16:40				Strike Balikpapan
18		14:55				Strike Balikpapan
24		18:40				Strike - Search -- Shipping
28				5:15		719-Morotai-719
2 Nov		13:50				Strike Shipping -P.I.
6		15:55				Strike Alicante
10				3:15		Noemfoor-Morotai
13		9:20				Strike Fabrica
15		10:00				Strike LaCarlota
18		10:15				Strike Tarakan
21		10:30				Strike Lumbia
23		10:30				Strike Bacolod
25		9:45				Strike Fabrica
27		10:00				Strike Malogo
29		11:05				Strike Puerta Princesa
1 Dec		10:00				Strike Bacolod
3		11:00				Strike Celebes
5		4:10				Strike Halmahera
8		9:25				Strike LaCarlota
11		9:45				Strike Panay
14		9:50				Strike Malogo
17		11:30				Strike Borneo
22		2:15				Strike Hatetabako
25		8:50				Strike Sandakan
29		11:45				Strike Search
12 Jan 45		13:35				Strike Grace Park A/d
13		11:40				Strike Nielson A/d
22		12:20				Strike Fabrica
24 Dec 44		10:50				Strike Puerta Princesa
				1:00	481:50	

FDFR: Walt's 424th Bomb Squadron Combat Credit report used to calculate 'points'

MISSION	DATE	PILOT	PLACE	TARGET	FIGHTERS	FLACK	TIME	TOTAL TIME
1	5/20	Appling	New Britain	Rabaul Tobera Airstrip			8:00	8:00
2	6/29	"	"	"			8:15	16:15
3	6/3	Appling	Truk	Etten Island Airstrip			9:00	25:15
4	7	"	"	"			6:35	31:50
5	9	"	"	Dublon Town			6:00	37:50
6	11	"	"	"			8:25	46:15
7	17	"	"	"			8:45	55:00
8	19	"	"	Turnback			1:30	56:30
9	23	"	"	"			12:35	69:05
10	8/26	Ardett	Yap	Airstrip			5:15	74:20
11	9/30	"	"	"			11:40	86:00
12	10/3	Appling	Noemfoor	"			12:15	98:15
13	11/7	"	Yap	Yap Town			10:00	108:15
14	12/13	"	Sorol	Radio Tower			13:00	121:15
15	13/8	Ardett	Yap	Airstrip			12:30	133:45
16	14/10	"	Palau	Yap Town			12:25	146:10
17	15/29	Cooksen	Borneo	Koror Town			9:05	155:15
18	16/10	McRae	"	Turnback			6:10	161:25
19	17/18	"	"	Balikpapan Oil Refineries			16:40	178:05
20	18/24	Appling	Celebes	Shipping Prowl Manado			14:55	193:00
21	19/2	"	Philippines	Mindanao Sea Task Force			18:40	211:40
22	20/6	"	"	Negros Island Alicante			13:50	224:30
23	21/31	"	"	"			15:55	240:25
24	22/15	"	"	Fabrica			9:20	249:45
25	23/18	"	Borneo	Balikpapan Oil Refineries			10:00	259:45
26	24/21	"	Philippines	La Carlota			10:15	270:00
27	25/23	"	"	Mindanao Island Lumbia			10:30	280:30
28	26/25	"	"	Negros Island Bacolod			10:30	291:00
29	27/27	"	"	Fabrica			9:45	300:45
30	28/29	"	"	Malogo			10:00	310:45
31	29/7	Appling	"	Palawan Island Puerto Princesa			11:05	321:50
32	30/3	"	"	Negros Island Bacolod			10:00	331:50
33	31/5	Celebes	Mampoeng Airstrip				11:00	342:50
34	32/8	"	Halmahera	Lolobata Airstrip			4:10	347:00
35	33/11	"	Philippines	Negros Island - La Carlota Iloilo			9:25	356:25
				Panay Island Iloilo			9:45	366:10

MISSION	DATE	PILOT	PLACE	TARGET	FIGHTERS	FLACK	TIME	TOTAL
34	14	Aubrey	Philippines	Negros Island Malogo			9:50	376:00
35	17	"	Borneo	Jesselton Airstrip			11:30	387:30
36	22	"	Halmahera	Hatebako			2:15	389:45
37	24	Michel	Philippines	Palawan Puerto Princessa			10:50	400:35
38	25	Aubrey	Borneo	Sandakan Airstrip			8:50	409:25
39	29	"	Philippines	Shipping Prowl Cape Montequirat			11:45	421:10
40	12	Hughes	"	Luzon Grace Park			11:40	432:50
41	13	"	"	Nielson Field			13:35	446:25
42	22	Rade	Guild	Negros Island Fabrica			12:20	458:45

May 49 12:36

May 30-49 6:15 645
6-44 355
9-44 325   100
March 3-49 6:15 600
9 415
16 155   330
18 605
6-44 440
Aug 17-49 3:50 630
28-44 5:15 655
Ext 3:15 1:00

Feb 13-45

Feb 5-44 500
9 415
16 155
18 605
25 300
26 200
28 330
29 430   390
Jan 17-44 200
19

PFMR: Walt's personal flying and mission record

```
Jan. 14, 1943-- Inducted into United States Army
Jan. 21, 1943-- Arrived at Camp Custer, Battle Creek, Mich.
Feb.  1, 1943-- Arrived at St. Petersburgh, Fla.
Feb.  5, 1943-- Assigned to United States Army Air Force
Feb.  8, 1943-- Arrived at Clearwater, Fla.
Mar.  3, 1943-- Arrived at Salt Lake City, Utah
Mar. 23, 1943-- Arrived at Harvard AAB, Harvard, Neb.
Apr. 10, 1943-- Promoted to rank of Private First Class
Apr. 15, 1943-- Arrived at Keesler Field, Biloxl, Miss.
May   5, 1943-- Started Airplane Mechanichs Course
Sep. 13, 1943-- Graduated from Airplane Mechanichs Course
Oct. 13, 1943-- Arrived at Laredo AAF, Laredo, Texas
Oct. 25, 1943-- Started Ariel Gunnery Course
Nov. 27, 1943-- Arrived at Eagle Pass AAF, Eagle Pass, Texas
Dec.  4, 1943-- Arrived back to Laredo AAF
Dec. 12, 1943-- Graduated from Ariel Gunnery Course,. Presented Ariel
                Gunners Wings. Promoted to rank of Seargent.
Dec. 17, 1943-- Arrived at home on 12 day furlough
Jan.  3, 1944-- Arrived at Hammer AAF, Fresno, Cal.
Jan. 15, 1944-- Arrived at Muroc AAF, Muroc, Cal.
Jan. 19, 1944-- Started Overseas Combat Training Course
Mar. 27, 1944-- Arrived at Hamilton AAF, San Francisco, Cal.
Mar. 30, 1944-- Arrived at Fairfield-Suisun AAF, Fairfield, Cal.
Apr.  4, 1944-- Arrived back to Hamilton AAF
Apr.  6, 1944-- Left the United States For Overseas by A.T.C.
Apr.  7, 1944-- Arrived at Hickam Field, Hawaii
Apr. 10, 1944-- Arrived at Canton Island, Phoenix Group
Apr. 11, 1944-- Arrived at Ndeni Island, Santa Cruz Group. Arrived at
                Guadalcanal, Solomon Islands
May. 17, 1944-- Flew first combat mission to Rabaul, New Ireland
May. 30, 1944-- Arrived at Negros, Admiralty Islands. Flew missions
                from there to, Truk, Yap, and Sorol in the Caroline
                Islands, to Noemfoor, Schouten, Islands
Aug. 19, 1944-- Arrived at Wakde Island, New Guinea, Flew missions
                from there to, Palau Islands
Sep. 12, 1944-- Arrived at Nadzab, New Guinea. Promoted to Staff Seargent
                Arrived at Rockhampton, Australia
Sep. 13, 1944-- Arrived at Sydney, Austrailia On 10 day furlough
Sep. 24, 1944-- Arrived at Rockhampton, Australia
Sep. 25, 1944-- Arrived at Port Moresby, New Guinea
                Arrived at Wakde Island, New Guinea
Sep. 26, 1944-- Arrived at Noemfoor, Schouten Islands, Flew missions from
                there to Balikpapan, Borneo. Halmahera, and Philippines.
Nov. 11, 1944-- Arrived at Morotai, Halmahera. Flew missions from there to
                Tarakan, Borneo. Philippines, And Sulu Sea.
Jan. 30, 1945-- Flew last mission to Manila, Philippines.
Feb. 25, 1945-- Arrived at Biak, Schouten Island, for shipment home.
Apr.  4, 1945-- Arrived at San Francisco, Cal.
Apr.  9, 1945-- Arrived at Fort Sheridan, Chicago, Illinois
Apr. 10, 1945-- Arrived at home for 21 day furlough
May.  6, 1945-- Arrived at Santa Ana AAB, Santa Ana, Cal.
Jun.  2, 1945-- Arrived at Fort Sheridan, Illinois. Presented Air Medal
Jun.  3, 1945-- Honorably Discharged.
```

PB&TR: Walt's Personal Base and Transit Record

```
SECRET                 424TH BOMBARDMENT SQUADRON (H)         SECRET
                       307th Bombardment Group (H)
                                                              A.P.O. # 713
                                                              18 November 1944
                         (Voice)   (CW)                       ATTACK    ATTACK
   PILOT     SHIP #  T/C  CALL LTRS CALL LTRS   TARGET        ALT.      TIME

   VANDERPOEL 1619  0730  RANGER 1A  F48A   PAMOESIAN Tank Farm 10,750'  1235
   APPLING    1276  0731  RANGER 1B  F48B   & Oil Separation              to
   HEILLE     1236  0732  RANGER 1C  F48C       Plant                   1240
   RODGERS    0958  0733  RANGER 1D  F48D
   MITCHELL   1546  0734  RANGER 1E  F48E
   RAUE       1290  0735  RANGER 1F  F48F
   CRAWFORD   1318  0734  RANGER 1G  F48G
```

SQDNS & GP ASSEMBLY:	Over Bum Bum Is. (04-27'N - 118-42'E). 424th at 9,500' circling left at 1124 - 1200.
FIGHTER RENDEZVOUS :	Over TARAKAN - 3 Sqdns (P-47's) over target from 1230-1300.
ROUTE FORMATION :	Individual ships to Bum Bum Is. (04-27'N 118-42'E) Sqdns in trail to target.
ROUTE OUT :	Base to assembly Point (04-27'N 118-42'E) to turning point (03-56'N 117-21'E) to IP (03-22'N 117-31'E) to target (03-18'N - 117-35'E) to base.
PLAN OF ATTACK :	No loss in altitude - turn Left of 42° onto target, IP 03°22'N - 117-31'E) Mag Heading of Bomb Run 118°. Bombing Formation will be Sqdns in trail, echelon down 250 ft. 424th in #1, 371st in #2, 372nd in #3, and 370th in #4. 160 MPH (I) on Bomb Run.
METHOD OF SIGHTING :	Lead Bombardiers in each Sqdn sight for range and deflection all other bombardiers sight for Range only.
BOMB LOAD AND INTERVAL:	7-500# G.P. - 2-500# M76. 1/10 nose 01 tail in GP. Instantaneous on M76. 85 ft. (Nine Stations)
WITHDRAWAL :	Left Breakaway. 165 MPH (I). Formation - Group box will be reformed after leaving target.
ALTERNATE TARGETS :	Secondary: SANDAKAN A/D Runway #1. Last Resort: MAPANGET A/D Runway. Alternate plan of attack as approached; Attack altitude same as primary. Bomb Interval - 85 Ft.
ROUTE BACK :	Direct to Base.

REMARKS:
1. Flash Reports will be sent by Sqdn leaders as soon as possible.
2. Radio Discipline will be maintained according to POP.
3. Ground station Calls - MOROTAI S85; LEYTE GG4; SANSAPOR DS72;
4. IFF to be on east of 124° and off west of 124°.
5. Collective call sight for this Gp will be F48.
6. Fuel transfer in accordance with Cruise Control chart. (3100 gal)
7. Four (4) Sqdns of the 5th Gp will follow 307th.
8. Three Sqdns of P-47's for Escort. CALL LETTERS - FIGHTER LEADER: OUTCAST; LEPER; BEAVER;

FREQUENCIES: Command: 6280   LIAISON: 8455 0600-1900   8045 (S) 0600-1900
                       4475   3664 1900-0600   3476 (S) 1900-0600

VHF "A" Channel (Bomber interplane) "B" Channel (Fighter Interplane) "C" Chann for tower) "D" Channel (Distress and Homing).

RESCUE FACILITIES: 1. Daylight 21 (CW - 6PQ/K) orbits at southern tip SIBUTU IS. (04-40'N 119-30'E) orbit time 1245 - 1330.
2. Daylight 22 (CW - 6PQ/L orbits at SIRCE IS. (02-40'N 125-20'E) orbit time 1515 - Dark.

CHARLES W. McRAE,
1st Lieut., Air Corps,
Asst. Operations Officer.

424th Bomb Squadron Mission Plan: November 18, 1944 Mission to Tarakan, Prepared by Lt. Charles McRae

CONFIDENTIAL

424TH BOMBARDMENT SQUADRON (H) AAF
307TH BOMBARDMENT GROUP (H) AAF

A.P.O. # 719
6 November 1944

NARRATIVE COMBAT REPORT OF MISSION # 307 - 357

A. MISSION NUMBER : 307 - 357; 6 November 1944; 424th Bombardment Squadron (H); Six (6) Baker 24's.

B. TARGET : Four (4) Baker 24's hit secondary target (Alicante dispersal area, NEGROS Island.)
One (1) Baker 24 jettisoned bombs in GUIMARAS STRAITS.
One (1) Baker 24 jettisoned twenty (20) bombs in MINDANAO SEA and landed with remainder of bomb load (10 bombs).

C. TIME & ALTITUDE OF ATTACK : Bombs were dropped over ALICANTE at 1053/I from 12,500 to 12,750 feet true altitude.

D. RESULTS : The results of this mission cannot be rated due to cloud cover. Four (4) planes bombed the dispersal area south of the runway. It is believed that the bombs hit south east of the shop area. It was impossible to record any results.
Enemy losses: One probable destroyed.
Own damage loss: A/C 179 (Lt. Raus, pilot) reports 25 holes in plane. The antenna was shot, holes in No. 2 and 3 prop.
A/C # 951 (Lt. Mulevich, pilot) had No. 3 engine shot out. This plane has not reported into this base at the time this report was written.

E. INTERCEPTION : Twelve (12) to fifteen (15) enemy planes were reported. One (1) HAMP was definitely reported with the rest believed to be Zekes.

F. COMBAT TACTICS : Fifteen (15) to twenty (20) passes were made on our squadron with them all coming from the tail. Level between five (5) and seven (7) o'clock. One (1) intercepter was probably destroyed. This plane came in level at six (6) o'clock on the B-24 flying in B-1 position. Gunners from each plane in the B element got concentrated bursts in on this plane at close range. Corporal Jack Hastach, right waist gunner, in A/C 236, T/Sgt. Jack Miley, tail gunner, in A/C 276 and Corporal James McAllister, tail gunner in A/C 179,

424th Bomb Squadron Mission Report: November 6, 1944 Mission to Alicante, Negros Island, p. 1

CONFIDENTIAL

This base started to smoke badly, passed under A/C 276 and swerved over to the right, and went into a steep vertical dive. It was last seen to go into the clouds at 8000 feet with black smoke pouring out of the fuselage.

G. AA FIRE : Nil.

H. PHOTOGRAPHS : Ten (10) photographs were taken at 1053/I over ALICANTE A/D at 12,530 feet true altitude by A/C 381. One (1) photograph taken at 1147/I over BAY POINTE A/D by A/C 381.

I. ROUTE AND OBSERVATION : Base to FLECHA Point to GUIMARAS Island to target. Route back: Target to MOROTAI to HOMEWOOD.

Observations: One (1) large AK was reported at wharf in BACOLOD Harbor.

J. WEATHER : From base to rendezvous: .8 to a full overcast of alto-stratus at 15000 feet with broken cumulus tops at 9000 feet. Rendezvous to GUIMARAS Island conditions ranged from CAVU to broken cumulus over land masses tops at 9000 feet. GUIMARAS to target broken to overcast with a full overcast reported over the target. Route back: Essentially the same conditions existed to MOROTAI with a .9 cloud coverage of cumulus with tops at 9000 feet and scattered conditions over water. MOROTAI to base: .4 cumulus tops at 7000 feet.

K. REMARKS : A/C 951 (Lt. Belovich, pilot) was attacked by a Zeke from 3 o'clock high at 1043/I. This plane had the No. 3 engine knocked out but remained in formation until 1049/I. At which time the pilot peeled off to the right and descended to an altitude of approximately 8000 feet, where he leveled off and went into a cloud coverage with two (2) P-47's acting as escort. Other planes in the squadron reported that the pilot had the A/C under control and were of the opinion that this plane was proceeding to LEYTE. There was no other indication of trouble except the No. 3 engine being out. A/C 179 (Lt. Reme, pilot) was unable to drop his bombs over the target. The roller on the left bomb bay door had been damaged by enemy fire and the crew was unable to get the door open. This plane jettisoned twenty (20) bombs in the MINDANAO Sea when they were able to get the right bomb bay door open. They landed with the rest of the bomb load at MOROTAI. This damaged plane (A/C 179) was left at MOROTAI, and A/C 958 was brought back to HOMEWOOD by this crew.

- 2 -

CONFIDENTIAL

Two (2) enemy A/C were reported destroyed by P-47's escorts.

L. BOMB LOAD : 4 x 30 x 100# GP bombs over ALICANTE dispersal area.
1 x 30 x 100# GP bombs were jettisoned in SURAGAO Straits.
1 x 20 x 100# GP bombs were jettisoned in MINDANAO Sea.
1 x 10 x 100# GP bombs were returned to MOROTAI.

All bombs dropped over ALICANTE had instantaneous nose fusing and .025 tail fusing.
7500 rounds of 50 calibre ammunition were expended.

M. FORMATION AND BOMBING DATA :
Formation : Modified group box was flown due to weather conditions.
Bomb run heading : 180° - 190°
Length bomb trains : 2900 feet.
Position of plane in formation : The squadron lead plane was in No. 4 position and sighted for range only due to cloud coverage, others dropped on leader. The lead plane used A.F.C.E.
Aiming point
 Range : Intersection of taxiways south east of shop area.
 Deflection : Due to weather conditions it was impossible to sight for deflection.
Target : South dispersal, supply and personnel areas of ALICANTE A/D.

JOHN H. SHIELDS,
2nd. Lt., Air Corps,
Intelligence Officer.

-3-

424th Bomb Squadron Mission Report: November 6, 1944 Mission to Alicante, Negros Island, p. 3

424TH BOMBARDMENT SQUADRON (H) AAF
307TH BOMBARDMENT GROUP (H) AAF

A.O. : 718
6 November 1944

LOADING LIST FOR MISSION, 6 November 1944. (Group # 307 - 357)

SHIP # 723

Lt. McMillen
Lt. Fielding
Lt. Sawyer
Lt. Rineboy
Sgt. Deque
Sgt. Doblekar
Sgt. Cropsem
Sgt. Scott
Sgt. Friedman
Sgt. Coster
Sgt. Walker

SHIP # 961

Lt. Bolovich
Lt. Houghton
Lt. Yuschak
Lt. Prinzenberger
Sgt. Sudzik
Sgt. Sennow
Sgt. Capron
Sgt. Harness
Sgt. Shaw
Sgt. Minier

SHIP # 178

Lt. Guild
Lt. Nobel
Lt. Eichelberger
Lt. Feucht
Sgt. Ross
Sgt. Gentry
Sgt. Roth
Sgt. Reis
Sgt. Lott
Sgt. Aufka

SHIP # 278

Lt. Appling
Lt. Aubrey
Lt. Dumas
Lt. Clark
Sgt. Bratherton
Sgt. Shaughnessy
Sgt. Schreiner
Sgt. Babinski
Sgt. Mansir
Sgt. Biley

SHIP # 238

Lt. Neilio
Lt. Henry
Lt. Lanning
Lt. Patton
Sgt. Cootes
Sgt. Hidalgo
Sgt. Johnson
Sgt. Cooper
Sgt. Wantach
Sgt. Solomon

SHIP # 179

Lt. Rowe
Lt. Quigley
Lt. Birkenfield
Lt. Bates
Sgt. Smith
Sgt. Schmaring
Sgt. Llewellyn
Sgt. Bockman
Sgt. McAllister
Sgt. Williamson

424th Bomb Squadron Mission Report: November 6, 1944 Mission to Alicante, Negros Island, p. 4

BIBLIOGRAPHY

Author's Bibliographical Note
For anyone wanting to learn more about the Thirteenth Air Force and the 307th Bombardment Group, the following items would be a good foundation for additional research. The Thirteenth Air Force Veterans Association (http://www.13afvets.org/) seems to be inactive, but their website still maintains newsletters and other useful information. The Jungle Air Force of WWII website (http://www.entnet.com/~personal/rocketeer/html/13thaaf/13thmain.html) includes some excellent research material. The website of the 307th Bombardment Group Association (http://307bg.net/) contains a wealth of information, including many online photos, mission reports, and other research material. The Missing Aircrew Project (http://www.missingaircrew.com/) provides additional useful information about the Long Rangers.

Published books that are both well-written and give a good flavor of the experiences of B-24 crewmen include Stephen Ambrose's *The Wild Blue*, John Boeman's *Morotai: A Memoir of War*, Sam Britt's *The Long Rangers: A Diary of the 307th Bombardment Group (H)*, and William Edwin Hemphill (editor) *Aerial Gunner from Virginia: The Letters of Don Moody to His Family During 1944*. The book *From Fiji Through the Philippines With the Thirteenth Air Force*, written by Lt. Col. Benjamin Lippincott and others and beautifully illustrated by S/Sgt. Robert A. Laessig provides a unique insight into the life of Jungle Air Force veterans. Wil Hylton's book *Vanished*, tells the complete story of the Jack Arnett crew.

There are many useful first-person reminiscences by Thirteenth Air Force vets that can be accessed via the U.S. Library of Congress Veterans Oral History website (http://www.loc.gov/vets/). Ten of these interviews were collected during the 2006 Reunion of the 307th Bombardment Group Association that I organized in Bellevue, Washington. One of the most useful resources for my work was the first-person reminiscence by Clifford Llewellyn *My Time in the Military*.

Original Source Documents
1930 U.S. Census: 1930 U.S. Census Detailed Enumeration Sheets, U.S. National Archives.
1940 U.S. Census: 1940 U.S. Census Detailed Enumeration Sheets, U.S. National Archives.

Published Books & Articles:
307th Bomb Group (HV) Association: Reunion Book No. 5: Kendall, James and Sterkel, Harry, editors, Orlando, Florida, 1984.
307th Bomb Group (HV) Association: Reunion Book No. 6: Kendall, James and Sterkel, Harry, editors, Milwaukee, Wisconsin, 1986.
307th Bomb Group (HV) Association: Reunion Book No. 7: Kendall, James and Sterkel, Harry, editors, Nashville, Tennessee, 1988.
307th Bomb Group (HV) Association: Reunion Book No. 8: Kendall, James, Sterkel, Harry, and Reeves, John, editors, Las Vegas, Nevada, 1990.
307th Bomb Group (HV) Association: Reunion Book No. 9: Kendall, James, Sterkel, Harry, and Reeves, John, editors, Dayton, Ohio, 1992.
307th Bomb Group (HV) Association: Reunion Book No. 10: Kendall, James, Sterkel, Harry, and Reeves, John, editors, St. Louis, Missouri, 1994.
307th Bomb Group (HV) Association: Reunion Book No. 11: Kendall, James, Sterkel, Harry, and Reeves, John, editors, San Diego, California, 1996.
307th Bomb Group (HV) Association: Reunion Book No. 12: Kendall, James, editor, Hampton, Virginia, 1998.
307th Bomb Group (HV) Association: Reunion Book No. 13: Kendall, James, editor, San Antonio, Texas, 2000.
307th Bomb Group (HV) Association: Reunion Book No. 14: Kendall, James, editor, Salt Lake City, Utah, 2002.
307th Bomb Group (HV) Association: Reunion Book No. 15: Kendall, James, editor, Nashville, Tennessee, 2004.
Alexander, Thomas G., Brief Histories of Three Federal Military Installations in Utah. Reprinted from Utah Historical Quarterly, Vol. 34, No. 2, Spring, 1966.
Ambrose, Stephen E., The Wild Blue New York: Simon & Schuster, 2001.
Army Air Forces Technical School, Daily Schedule: Specialized B-24 Airplane Mechanics Course. Keesler Field, Mississippi, Headquarters, Technical School, February 25, 1943.
Army Air Forces Technical School, Students Notebook: Airplane and Engine Mechanics Course B-24 and B-32. Keesler Field, Mississippi, Headquarters, Technical School, N.D. (author's copy annotated by student Harold C. Anderson).
Articles of War: Washington, D.C.: United States Government Printing Office, 1912.
Automatic Computing Sights Sperry Type K (.50 Calibre): Operating Manual. No date, no location.
Baime, A.J., The Arsenal of Democracy. Boston: Houghton Mifflin Harcourt, 2014.

Baross, James P., <u>The War Years WWII.</u> Photocopy Typed Manuscript, Library of Congress, Veterans History Project, http://lcweb2.loc.gov/diglib/vhp/story/loc.natlib.afc2001001.04316/pageturner?ID=pm0001001&page=1, Riverside, CA, September 1998, Revised November 1998, (accessed June 7, 2009).

<u>Basic Field Manual: Soldier's Handbook.</u> Washington, D.C.: United States Government Printing Office, 1941.

<u>Basic Principles of Aerial Gunnery.</u> Prepared by the Training Department, Army Air Force Flexible Gunnery School, Las Vegas Army Air Field, 1944?

Bergerud, Eric M., <u>Fire in the Sky: The Air War in the South Pacific.</u> Boulder, Colorado, Westview Press, 2000.

Boeman, John, <u>Morotai: A Memoir of War</u> Manhattan, Kansas: Sunflower University Press, 1989.

Bolin, Robert L. Depositor, <u>Photographic Interpretation Handbook, United States Forces: Section 14 Ships and Shipping.</u> DOD Military Intelligence. Paper 17. (1944). Accessed 1/14/2016 at: http://digitalcommons.unl.edu/dodmilintel/17

Boylston, Raymond P., <u>World War II South Pacific Letters.</u> Raleigh, North Carolina, Boylston Enterprises, 1997. (Accessed in the Library of Congress Veterans History Project).

Briggs Manufacturing Company, <u>Handbook of Operation and Service Instructions for Retractable Lower Ball Turret, Model A-13.</u> Published by Authority of Commanding General Army Air Forces, Air Service Command, Patterson Field, Fairfield, Ohio, February 1943.

Britt, Sam S., Jr., <u>The Long Rangers: A Diary of the 307th Bombardment Group (H)</u> Baton Rouge, Louisiana: The Reprint Company, Publishers, (Spartansburg, SC), 1990.

Buckingham, William Frederick, The Establishment and Initial Development of a British Airborne Force, June 1940-January 1942 PhD Thesis, University of Glasgow, 2001.

Charles, Roland W., <u>Troopships of World War II.</u> Washington, DC: The Army Transportation Association, 1947.

Clark, Lyman "Ace" Jr., <u>Missions of Shehasta: A Story of World War II Bomber Aces.</u> Terra Alta, West Virginia: Headline Books Inc., 1992.

Coffey, Frank, <u>Always Home: 50 Years of the USO.</u> Washington: Brassey's (US), Inc., 1991.

Conder, Albert E., Editor, <u>The History of Enlisted Aerial Gunnery 1917-1991.</u> Paducah, Kentucky, Turner Publishing Company, 1994.

Deese, Alma W., <u>St. Petersburg Now and Then.</u> Arcadia Publishing, 2006.

Dowdy, Charles B. Jr., <u>"Charlie" Memoirs of a 13th Air Force Flight Engineer in the Southwest Pacific, 1944-1945.</u> Utica, Kentucky: McDowell Publications, 1988.

Droege, John A., <u>Passenger Terminals and Trains.</u> New York: McGraw-Hill Book Company, 1916.

Farnell, Major John M. The Operations of the 136th Infantry (33d Infantry Division) on Morotai Island, 18 December 1944 to 19 January 1945 New Guinea Campaign. Fort Benning, Georgia, The Infantry School, 1947?

Faulkner, Marcus, <u>War at Sea: A Naval Atlas 1939-1945.</u> Annapolis: Naval Institute Press, 2012.

Griffith, Thomas E., Jr., <u>MacArthur's Airman: General George C. Kenney and the War in the Southwest Pacific.</u> Lawrence, Kansas: University Press of Kansas, 1998.

Gideon, Francis C., <u>Airdromes Guide Southwest Pacific Area:</u> Office of the Assistant Chief of Air Staff, A-3 Headquarters, Far East Air Forces, November 1944.

<u>Gunner's Information File: Flexible Gunnery.</u> No author, Training Aids Division, Office of the Assistant Chief of Air Staff, Training, (Air Forces Manual No. 20), May 1944.

Harvey, Gordon K. and Hamilton, Eugene K., editors, <u>We'll Say Goodbye: Story of the 307th Bombardment Group (HV).</u> Sydney, Australia: F. H. Johnston Publishing Company, 1945.

Hemphill, William Edwin, editor, <u>Aerial Gunner from Virginia: The Letters of Don Moody to His Family during 1944.</u> Richmond, Virginia: Virginia State Library, 1950.

<u>Hickam Field Guide.</u> No author, introduction by Gurr, James W., Colonel, AC Commanding, HQ 1521st AAF Base Unit, Pacific Division, Air Transport Command, Honolulu: 6263 Army Printing Plant, no date (1946?).

Hillenbrand, Laura, <u>Unbroken.</u> New York: Random House, 2010.

Historical Studies Branch, USAF Historical Division, <u>Combat Crew Rotation: World War II and Korean War.</u> Maxwell Airforce Base, Alabama, Aerospace Studies Institute, Air University, January 1968.

Hooker, Richard and Heinz, W.C. <u>MASH: A Novel About Three Army Doctors.</u> New York: William Morrow, 1968.

Hough, Capt. Donald and Arnold, Capt. Elliot, <u>Big Distance</u> New York: Duell, Sloane and Pearce, 1945.

Hylton, Wil S., <u>Vanished.</u> New York: Riverhead Books, 2013.

Jensen, Cecile Wendt, <u>Detroit's Polonia.</u> Charleston, SC, Chicago, IL, Portsmouth, NH, San Francisco, CA: Arcadia Publishing, 2005.

<u>Keesler Field, Mississippi:</u> No author, San Antonio, Texas: Universal Press, 1942.

Kochanski, Halik, <u>The Eagle Unbowed: Poland and the Poles in the Second World War.</u> Cambridge: Harvard University press, 2012.

<u>Laredo Army Air Field:</u> Laredo, Texas, no date, no author, with introduction by Colonel Charles G. Pearcy. (A copy in the Babinski family collection has extensive handwritten annotations by (then) Private Babinski.)

Leiser, Gary, <u>A History of Travis Air Force Base: 1943-1996</u>, Sacramento: Travis Air Force Base Historical Society, 1996.

Lippincott, Lt. Col. Benjamin E., (with Duffey, Capt. Kenneth E., Hays, Capt. C. Blaine Jr., Freiman, Capt. Marvin, & Westlin, S/Sgt. Bertil), Illustrated by Laessig, S/Sgt. Robert A., <u>From Fiji Through the Philippines With the Thirteenth Air Force.</u> San Angelo, Texas: Newsfoto Publishing Co., 1948.

Lucas, J.S., <u>Austro-Hungarian Infantry 1914-1918.</u> London: Almark Publishing, 1973.

Magriel, T/Sgt. Paul, editor, <u>Art and the Soldier</u>. Biloxi, Mississippi: Special Services, Keesler Field, 1943.

Maxwell, Major Reuben J., (with Huntington, Major Haskins and Matthias, M/Sgt. John I.), Illustrated by Sundberg, Sgt. Arthur W., <u>Camp Living.</u> Thirteenth Air Force, 905th Engineer Air Force Headquarters Company, 1945.

Makos, Adam, with Alexander, Larry, <u>A Higher Call</u>, Penguin Group, New York, 2012.

Maleckas, Frank Jr., <u>One 11 Millionth of a War.</u> Elk Rapids, MI, Bookability, Inc., 2000.

Maurer, Maurer, <u>Air Force Combat Units of World War II:</u> Washington, D.C.: Office of Air Force History. 1983.

Miller, Edrick J., Illustrations by Miller, Patricia M., <u>The SAAAB Story: The History of the Santa Ana Army Air Base.</u> Santa Ana, California: Tri-Level Inc., Lithographers, (Sponsored by the Costa Mesa Historical Society), 1989.

Munson, Kenneth, Illustrated by John W. Wood, <u>Fighters and Bombers of World War II:</u> Exeter Books, New York, 1969.

Newnham, Maurice, <u>Prelude to Glory</u>, Sampson Low, Marston, and Co., 1948.

Norris, Dia Vee, editor, <u>Diary of Texas Kate: The Recollection of Bob E. Graves, Radio Operator and Gunner, and his B-24's crew's 40 Missions in the Pacific Theater during World War II.</u> Portland, Oregon, 1994.

Office of History, 81st Training Wing, <u>Legacy of Excellence: The History of Keesler AFB and the 81st Training Wing.</u> Keesler Air Force Base, Mississippi, 1 August 1999.

Office of the Assistant Chief of Air Staff, A-3, Headquarters, Far East Air Forces, <u>Airdromes Guide Southwest Pacific Area:</u> US Army GHQ SWPA, November 1944.

<u>Official Guide of the Railways of the United States:</u> No author, New York, National Railway Publication Co., 1930.

Okumiya, Masatake and Horikoshi, Jiro, with Caiden, Martin, <u>Zero! The Story of Japan's Air War in the Pacific: 1941-45.</u> New York: Ballantine Books, 1956.

Parker, Philip, Editor, <u>The Collins Atlas of Military History.</u> London: Harper Collins, 2004.

Pearce, E.A. and Smith, C.G., <u>The World Weather Guide.</u> London: Hutchison & Co. Publishers Ltd., 1984.

Poremba, David Lee, <u>Detroit: 1930-1969</u>. Charleston, SC: Arcadia Publishing, 1999.

Powell, Murella H., <u>Biloxi: Queen City of the Mississippi Coast.</u> (No citation data)

Ranfranz, Patrick, <u>Coleman Crew, MACR #10023, Lost over Yap Island on 25 June 1944.</u> Shoreview, MN, MissingAirCrew.com, 2006.

Russell, S/Sgt. Gail R., <u>Keesler Field: The War Years 1941–1945</u>. (KTTC Historical Monograph) Keesler Air Force Base, Mississippi: Office of History, Keesler Technical Training Center, 1986.

Rust, Ken C. and Bell, Dana, <u>Thirteenth Air Force Story...in World War II</u> Temple City, California: Historical Aviation Album, 1981.

<u>Service and Instruction Manual: Armament B-24D.</u> (No author), San Diego: Flight and Service Department, Consolidated Aircraft Corporation, 1942.

Sides, Hampton, <u>Ghost Soldiers.</u> New York, Anchor Books, 2001.

Smith, T/Sgt. Paul T., USAF (Ret.), <u>The Pacific Crusaders.</u> Reseda, California: Mojave Books, 1980.

Stauter-Halsted, Keely, <u>The Nation in the Village: The Genesis of Peasant National Identity in Austrian Poland, 1848-1914.</u> Ithaca and London: Cornell University Press, 2001.

Stone, Joel, <u>Detroit: The Arsenal of Democracy.</u> Detroit Historical Society, Undated article, accessed December 19, 2015 at: https://detroithistorical.org/sites/default/files/pdfs/AoD%20Paperv2.pdf.

<u>Story of St. Petersburg.</u> St. Petersburg: P. K. Smith & Co., 1948.

Szawala, Dennis, <u>A Landmark in Delray.</u> Undated (1977?) typewritten manuscript, from the Archives of the Archdiocese of Detroit.

Titler, Dale M., <u>Gulf Tempest: Major Hurricanes and Their Effects on Keesler Technical Training Center</u>. Office of History, Keesler Technical Training Center, ND.

Titler, Dale M. and Murphy, S/Sgt. Gary M., <u>Keesler Field: Inception to Pearl Harbor: 1939 – 1941</u>. Office of History, Keesler Technical Training Center, Keesler Air Force Base, MS, 1981 (4th Printing 1988).

Tourtellot, Arthur B., Editor, (with full-page text by John Dos Passos, <u>Life's Picture History of World War Two.</u> New York: Time Incorporated, 1950.

Ulmer, S. Sidney, <u>Waist Gunner: The Diary of William Davis Parker in World War II.</u> Lexington, Kentucky: Xlibris Corporation, 2000.

United States Army Air Force, <u>The Story Of The Fifth Bombardment Group (Heavy).</u> NP, Hillsborough House, 1946.

United States Holocaust Museum, <u>Holocaust Encyclopedia</u>. See: https://www.ushmm.org/.

U.S. Army, 41st Division. History of the Biak Operation, 15-27 June 1944. U.S. Army 41st Division. 1944.

Weinberg, Gerhard L., A World at Arms: A Global History of World War II. New York: Cambridge University Press, 1994.
Weislow, Saul C., Davis, William P., and Whitesell, Carl F. History of the 307th Bombardment Group (Hv). Salt Lake City, 307th Bomb Group, 1973.
Witts, David A., Forgotten War-Forgiven Guilt. Las Cruces, New Mexico: Yucca Tree Press, 2003.
Wolff, Larry, The Idea of Galicia: History and Fantasy in Habsburg Political Culture. Stanford, California, Stanford University Press, 2010.
Woods, Wiley O., Jr., Diary of Wiley O. Woods, Jr., Navigator, 320th Bomb Squadron, 90th Bomb Group, November 26, 1944 to November 1, 1945, SWPA. Photocopy Manuscript, Library of Congress, Veterans History Project, http://lcweb2.loc.gov/diglib/vhp/story/loc.natlib.afc2001001.05895/pageturner?ID=pm0001001, (accessed June 5, 2009).
Writers' Program of the Work Projects Administration, Alabama: A Guide to the Deep South. (Sponsored by the Alabama State Planning Commission), New York: Hastings House Publishers, 1941.
Writers' Program of the Work Projects Administration, California: A Guide to the Golden State. New York: Hastings House Publishers, 1939.
Writers' Program of the Work Projects Administration, Colorado: A Guide to the Highest State. New York: Hastings House Publishers, 1941.
Writers' Program of the Work Projects Administration, Florida: A Guide to the Southernmost State. (Sponsored by the Florida Department of Public Information), New York: Oxford University Press, 1939.
Writers' Program of the Work Projects Administration, Georgia: A Guide to its Towns and Countryside. Atlanta: Tupper & Love, 1940.
Writers' Program of the Work Projects Administration, Kansas: A Guide to the Sunflower State. (Sponsored by the State Department of Education), New York: The Viking Press, 1939.
Writers' Program of the Work Projects Administration, Louisiana: A Guide to the State. (Sponsored by the Louisiana Library Commission at Baton Rouge), New York: Hastings House Publishers, 1941.
Writers' Program of the Work Projects Administration, Michigan: A Guide to the Wolverine State. (Sponsored by the Michigan State Administrative Board), New York: Oxford University Press, 1941.
Writers' Program of the Work Projects Administration, Mississippi: A Guide to the Magnolia State. (Sponsored by the Mississippi Advertising Commission), New York: The Viking Press, 1943.
Writers' Program of the Work Projects Administration, Missouri: A Guide to the "Show Me" State. (Sponsored by the Missouri State Highway Department), New York: Duell, Sloan and Pearce, 1941.
Writers' Program of the Work Projects Administration, Nebraska: A Guide to the Cornhusker State. (Sponsored by the Nebraska State Historical Society), New York: Hastings House Publishers, 1939.
Writers' Program of the Work Projects Administration, San Francisco: The Bay and Its Cities. (Sponsored by the City and County of San Francisco), New York: Hastings House Publishers, 1940.
Writers' Program of the Work Projects Administration, Tennessee: A Guide to the State. (Sponsored by the Department of Conservation, Division of Information), New York: The Viking Press, 1939.
Writers' Program of the Work Projects Administration, Utah: A Guide to the State. (Sponsored by the Utah Institute of Fine Arts), New York: Hastings House Publishers, 1941.
Writers' Program of the Work Projects Administration, Texas: A Guide to the Lone Star State. (Bureau of Research in the Social Sciences of the University of Texas), New York: Hastings House Publishers, 1940.

Brochures & Miscellaneous Paper Items:
Harvey, Gordon K. and Hamilton, Eugene K., Long Rangers Combat History. (Hand drawn and typed sheet), February 13, 1945.
History of Harvard Army Air Field. (No Author), no date, 11 page typewritten manuscript, obtained from the Adams County Historical Society, Hastings, Nebraska, in 2004.
Harvard Air Base: 1942-1946. (No Author), no date, 9 pages type and hand written manuscript and photos, obtained from the Adams County Historical Society, Hastings, Nebraska, in 2004.
Moore, W.O.(jg) Edgar J. Song of the Thirteenth Airforce, Thirteenth Air Force, W.O. Edgar J. Moore, 1945.
The Samsonite Story. (No Author), no location, no date, (Samsonite Corporation Historical Brochure).
75th Anniversary: The Fort Harrison. Clearwater: The Fort Harrison Hotel, 2002.

Contemporary Articles:
[Air Force 5/44 BRA] Bachman, Capt. Lawrence P., "Blocking Rabaul by Air" Air Force, the Official Service Journal of the U.S. Army Air Forces, May 1944
[Air Force 8/44 MFK] Jolly, Master Sgt. Arthur, "Maintenance for Keeps" Air Force, the Official Service Journal of the U.S. Army Air Forces, August 1944
[Air Force 5/44 BCTSP] Jones, Lieut. Col. H. E., "Bomber Crew Training in the South Pacific" Air Force, the Official Service Journal of the U.S. Army Air Forces, May 1944
[Air Force 3/45 Snoopers] Air Force Overseas Staff Correspondents, Illustrated by Lt. Norman F. Todhunter, "Snoopers," Air Force, the Official Service Journal of the U.S. Army Air Force, March, 1945.

[Air Force 5/45 DF] Marshall, Lt. Henry W., "Dodging Flak," <u>Air Force, the Official Service Journal of the U.S. Army Air Force</u>, May, 1945.
[Air Force X/YY DoC] Hardman, Maj. Thomas C., "Drop on Corregidor," <u>Air Force, the Official Service Journal of the U.S. Army Air Force</u>, no date.
[Air Force PS] _____, "Puzzle Solved," <u>Air Force, the Official Service Journal of the U.S. Army Air Force</u>, no date.
[Air Force SG] _____, "Shell Game," <u>Air Force, the Official Service Journal of the U.S. Army Air Force</u>, no date
[Air Force SWP] _____, "Southwest Pacific," <u>Air Force, the Official Service Journal of the U.S. Army Air Force</u>, no date.
[Air Force MB] Reynolds, Lt. Richard S., Illustrated by Cpl. Louis S. Glanzman, "Mission to Balikpapan," <u>Air Force, the Official Service Journal of the U.S. Army Air Force</u>, no date.
[Air Force F13th] Johansen, Maj. Herbert O., "The Fighting 13th," <u>Air Force, the Official Service Journal of the U.S. Army Air Force</u>, no date.
[BGFMR:NBJS] _____, "Borneo Gets First Mass Raid: New Blows on Jap Shipping," Misc. newspaper article, no date (1944?)
[Biloxi Herald] Copies of <u>Biloxi Gulfport Daily Herald</u>, published in Biloxi, Mississippi. April 16 thru June 10, 1943.
[Detroit News] <u>War...in Headlines from the Detroit News.</u> No publication details, but apparently published by the Detroit News .
[EIO] _____, "East Indies Oil," Misc magazine article, no date (1944?)
[FTC Review: Ghost Sails On] Edmonds, M/Sgt. Ivy G., "The Ghost Sails On," <u>Flight Test Center Review</u>, (Edwards Air Force Base) Feb. 1960, Vol. 2, Number 8
[Keesler Field News] Copies of <u>The Keesler Field News</u> base newspaper, dated 9/16/1943.
[Laredo Times] Copies of <u>Laredo Times</u>, published in Laredo, Texas. October 13 thru December 12, 1943.
[Mechanical Brains] "Mechanical Brains," <u>Life Magazine</u>, January 24, 1944, pp. 66-72.
[Mess Talks] February 12, 1944 issue of <u>Mess Talks</u>, No. 3, published by Headquarters Army Air Field, Muroc, Office of the Base Mess Supervisor, California.
[MOY] Keighley, Larry, "Mission Over Yap," <u>Saturday Evening Post</u>, no date (1944).
[Muroc Mirage] Copies of <u>Muroc Mirage</u> base newspaper, dated 1/15/44, 1/22/44, and 3/11/44.
[St. Petersburg Times] Copies of <u>St. Petersburg Times</u>, published in St. Petersburg, Florida, dated 8/12/39 and 5/15/60.
[8JSDNBY 7/23/44] _____, "8 Japs Shot Down in New Blows at Yap," Misc. newspaper article, (Dateline 'Allied Headquarters, New Guinea), July 23, 1944)
[30B] _____, "Thirty for Bong" Misc newspaper article, no date (1944?)

Contemporary Personal Correspondence (all from Babinski Family collection, unless otherwise noted):
[Walter to JJB 7/30/35] Post card from 'Walter' to J Babinski, (Pere Marquette Depot @ Petosky Mich.: Mrs. AG Cook, Petosky, Mich.,) postmark 7/30/35, message regarding location.
[Walter to JJB 8/6/35] Post card from 'Walter' to John Babinski, (Pere Marquette Depot @ Petosky Mich.: Mrs. AG Cook, Petosky, Mich.,) postmark 8/6/35, message regarding location.
[Walter to JJB, 8/12/36] Postcard from 'Walter' to John Babinski (Monument at Sault Ste. Marie, Mich.), postmarked 8/12/36, Sault Ste. Marie, Mich., message regarding 'came over to the Soo to spend the day...having swell time...'
[JJB to Mr. & Mrs. B., 5/8/37] Postcard from John Babinski to Mr. & Mrs. Babinski (Rifle River at Omer, Mich.), postmarked 5/8/37, Omer, Mich., message regarding '...this is where I spend my weekends...'
[Walter to JJB, 7/26/37] Postcard from 'Walter' to John Babinski (Alanson, Mich.), postmarked 7/26/37, Alanson, Mich., message regarding '...fishing is swell...but haven't tried ourselves yet...'
[Walter to JJB, 7/29/37] Postcard from 'Walter' to John Babinski (Grand Hotel, Mackinaw Island, Mich.), postmarked 7/29/37, Mackinaw Island, Mich., message regarding '...nice place for a rest...beautiful scenery and weather fine...'
[W&S to JJB, 6/25/38] Postcard from Stella & 'Walter' to 'Tornado John' Babinski (Rock Garden, Michigan City, Ind.) postmarked 6/25/38, Michigan City, Ind., message regarding '...having wonderful time...lots of beautiful women and lots of altes lager to get tight on...'
[JJB to Mr. & Mrs. B. 6/24/40] Postcard from John Babinski to Mr. & Mrs. Babinski (John Tobias Cottage @ Norway Cove, Lake Avalon, Hillman, Mich.), postmarked 6/24/40, Hillman, Mich., message regarding '....its raining but quiet and nice...'
[JJB to WAB 8/24/42] Postcard from John Babinski to Walter Babinski (Great Lakes Naval Training Station Mess Hall), postmarked 8/24/42, Great Lakes, Ill.), message regarding '...this is where we get our chow...very clean, but no beer...'
[JJB to WAB 10/9/42] Postcard from John Babinski to Walter Babinski (Great Lakes Naval Training Station Drill Hall), postmarked 10/9/42, Great Lakes, Ill.), message regarding '...everything alright...warm, but mornings cold...this is drill hall where we have mass...'

[WAB to GB 1/29/43] Postcard from Walter Babinski to G. Babinski (St. Mary's Church, St. Petersburg, Fla.) postmarked 1/29/43, St. Petersburg, Fla., message regarding '...walking in this town is like being a millionaire...how rich place looks...glad I'm in AAC Tech Sq. School...'

[WAB to LB 1/29/43] Postcard from Walter Babinski to L. Babinski (Recreation Pier, St. Petersburg, Fla.) postmarked 1/29/43, St. Petersburg, Fla., message regarding '...hope you are alright...you could swim a lot here...'

[JSJ to Mrs. TK 2/1/43] Postcard from Joseph S. Jekielek to Mrs. Thomas Kurzyniec (AC-7 and Scout Bombers with 'Keep 'Em Flying' caption) postmarked 2/1/43, Miami, Fla., message regarding '...Store closed yet?....nice here....I'll be a soldier yet.'

[OJ to KB 2/1/43] Postcard from O. Justyn to K. Babinski (Buffalo, NY Church Sanctuary, w/ card text in Polish), postmarked 2/1/43, Buffalo, NY, message written in Polish. (O. Justyn was a NY relative (cousin) of KB – per DFB.)

[WAB to Mr. & Mrs. B., 2/2/43] Postcard from Walter Babinski to Mr. & Mrs. Babinski (Waving color 48 star US Flag), postmarked 2/2/43, St. Petersburg, Fla., message regarding '...I had an exam and past...going to school to be an airplane mechanic or armorer...after...chance to be a gunner on an airplane...' (D. F. Babinski collection).

[JJB to Mr. & Mrs. B 2/7/43] Postcard from John Babinski to Mr. & Mrs. Babinski (Rainbow Bridge, Utah) postmarked 2/7/43, Albuquerque, New Mexico, message regarding '...am fine...still travelling.'

[JJB to WAB 2/7/43] Postcard from John Babinski to Walter Babinski ('Keep-em-Flying') postmarked 2/7/43, Albuquerque, New Mexico, message regarding '...heading west...seen some fine country...'

[WAB to Mr. & Mrs. B 2/9/43] Postcard from Walter Babinski to Mr. & Mrs. Babinski (Clearwater, Fla., St. Celia's Church), postmarked 2/9/43, Clearwater Florida (ret. address Clearwater, Fla.), message written in Polish, signed 'Wladek'.

[WAB to GB 2/9/43] Postcard from Walter Babinski to G. Babinski (Clearwater, Fla., Hotel Fort Harrison), postmarked 2/9/43, Clearwater, Fla. (ret. address Clearwater, Fla.), message regarding '...got the letters...glad you wrote...see a lot of things I never seen....that's my hotel I live in now...show it to the family...'

[WAB to HB 2/14/43] Postcard from Walter Babinski to H. Babinski (Lake Wales, Fla., Singing Tower), postmarked 2/15/43, Clearwater, Fla. (dated 2/14/43, w/ ret. Address Clearwater, Fla.), message regarding '...received your letters...don't have much time to write...'

[WAB to LB 2/14/43] Postcard from Walter Babinski to L. Babinski (Florida Palms), postmarked 2/15/43, Clearwater, Fla. (dated 2/14/43, w/ ret. Address Clearwater, Fla.), message regarding '...hope you are taking care of yourself...hope you like the emblem I sent you and Henry...'

[WAB to GB 3/3/43] Postcard from Walter Babinski to G. Babinski (Bingham, Utah Copper Mine), postmarked 3/3/43, Salt Lake City, Utah (ret. Address Salt Lake City Army Air Base), message regarding '...couldn't write because I was on a train for 4 days. Busy now but will write more.'

[WAB to LB 3/3/43] Postcard from Walter Babinski to L. Babinski (Salt Lake City, Mormon Temple), postmarked 3/3/43, Salt Lake City, Utah (ret. Address Salt Lake City Army Air Base), message regarding '...I couldn't write because I was on a train....will write more soon.'

[WAB to GB 3/23/43] Postcard from Walter Babinski to G. Babinski (D&RGW RR view of Gore Canyon), postmarked 3/23/43 Harvard, Neb. (ret. Address Harvard AAB), message regarding '...how are you getting along in school and in work? See that the package gets here, will you...'

[WAB to GB 4/8/43] Postcard from Walter Babinski to G. Babinski (Hastings, Neb., Aerial View), postmarked 4/8/43, Harvard, Neb. (ret address Harvard AAB), message regarding '...thank you for the song sheets you sent me...when I asked, I didn't know I would get that many.'

[WAB to HB 4/8/43] Postcard from Walter Babinski to H. Babinski, (Hastings, Neb., Elm Ave), postmarked 4/8/43, Harvard, Neb. (ret address Harvard AAB), message regarding '...am doing alright...how did the Red Wings make out...beat Boston for championship...?'

[WAB to LB 4/8/43] Postcard from Walter Babinski to L. Babinski (Hastings, Fountain in Highland Park), postmarked 4/8/43 Harvard, Neb (Harvard AAB ret address), message regarding '...weather is nice outside here...I will send more pictures home soon.'

[WAB to GB 4/14/43] Post card from Walter Babinski to G. Babinski, (St Louis, Mo. - Old Court House), postmarked St Louis Mo, 4/14/43 (ret address Harvard AAB), message regarding '...here in St Louis for 8 hours, at USO now, having swell time.'

[WAB to HB 4/14/43] Postcard from Walter Babinski to H. Babinski, (St. Joseph, Mo, Felix Street), postmarked St. Louis, MO 4/14/43 (ret address Harvard AAB), message regarding '...on my way to school...having quite a time on the train.'

[WAB to Mr. & Mrs. B, 4/14/43] Postcard from Walter Babinski to Mr. & Mrs. Babinski (Univ. of Nebraska Student Union Bldg.), postmarked St. Louis, Mo., 4/14/43, message regarding 'having a swell time on the train, may get there by Wednesday night.'

[WAB to LB, 4/14/43] Postcard from Walter Babinski to L Babinski (Kansas City Union Station), postmarked St. Louis, Mo., 4/14/43, message re: 'how are you? I'm on the train going to school.'

[WAB to HB, 4/15/43] Postcard from Walter Babinski to H Babinski (Guv's Mansion, Montgomery Ala.), postmarked Montgomery, Ala., 4/15/43, message regarding 'having a swell time, will reach camp this evening.

[EN to WAB, 4/22/43] Postcard from Pvt. Eddie Nycz to Walter Babinski (Jefferson Barracks, St. Louis), postmarked Jefferson Barracks, Mo., 4/22/43 addressed to Harvard AFB, re-postmarked Harvard Neb., 4/30/43 as forwarded on to Keesler AFB, message regarding 'how are you, I'm in the same outfit as you, we had a GI party, we sleep 6 men to a hut, good luck to you.'

[JK to WAB, 4/22/43] Postcard from Pfc. Joseph Kaszyca to Walter Babinski (Balboa Park, San Diego), postmarked San Diego, Calif., 4/22/43, message regarding 'Lize is a Pfc now, soon he'll be Cpl or Sgt, or he may be Lt., well you're in a tough place again – we had a swell time, -'Clark.

[WAB to LB, 4/23/43] Postcard from Walter Babinski to L Babinski (Airplane Flying Over Biloxi), postmarked Keesler Field, 4/23/43, message regarding 'plane in pix is what I'll fix, my camp is seen below in pix, show ma & pa.'

[WAB to HB, 4/23/43] Postcard from Walter Babinski to H Babinski (Mechanics work on B-24 @ Keesler), postmarked Keesler Field, 4/23/43, message regarding 'this is what I'm going to learn to do – fix big planes - show ma & pa.'

[WAB to LB, 4/XX/43] Postcard from Walter Babinski to L Babinski (Keesler Field Arriving Soldiers & Trucks), postmarked Keesler Field, 4/XX/43 (Precise date unclear), message regarding 'hope you had a nice Easter, thanks for the card.

[WAB to GB, 4/XX/43] Postcard from Walter Babinski to G Babinski (Keesler Field Chow Line), postmarked Keesler Field, 4/XX/43 (Precise date unclear), message regarding 'hope you had a nice Easter, mine was swell.'

[WAB to HB, 4/XX/43] Postcard from Walter Babinski to H Babinski (Keesler Field Main Gate), postmarked Keesler Field, 4/XX/43 (Precise date unclear), message regarding 'hope you had a nice Easter, I did.

[WAB to GB, 5/2/43] Postcard from Walter Babinski to G Babinski (Keesler Field Assembly Room), postmarked Keesler Field, 5/2/43, message regarding letter requesting package.

['Maggie to WAB, 7/1/43] Postcard from 'Maggie' to Walter Babinski (Dayton Art Institute), postmarked Dayton, Ohio, 7/1/43, message regarding on vacation, picking raspberries, what's for dinner, am busy, will write again, love 'Maggie.'

[Ange to WAB, 7/7/43] Postcard from Ange to Walter Babinski (Rose Garden, Hamilton, Ont.), postmark Hamilton, Ont., 7/7/43, message regarding on vacation and why I didn't write. (Ange was WAB cousin – per DFB)

[SK to WAB, 7/21/43] Postcard from Stella Kurzyniec to Walter Babinski (Canoe @ Standish Mich.), postmarked Standish, Mich., 7/21/43, message regarding having a good time.

['Babe' to WAB, 8/2/43] Postcard from 'Babe' to Walter Babinski (Niagara Falls), postmarked Niagara Falls, NY, 8/2/43, message regarding enjoying NY. ('Babe' was Joe Lafata – WAB Detroit friend – per DFB)

[DK to GB, 8/2/43] Postcard from Dorothy Kurzyniec to G Babinski (Detroit Beach), postmarked Monroe, Mich., 8/2/43, message regarding 'thanks for the rackets…here till Thursday…I still like you.

[WAB to Mr. & Mrs. Babinski, 10/12/43] Postcard from Walter Babinski to Mr. & Mrs. Babinski (Gen. Houston Monument, Houston, Tex.), postmarked Houston, Tex., 10/12/43, message regarding in Houston, en route to where I don't know…will write more.

[WAB to GB, 10/16/43] Postcard from Walter Babinski to G Babinski (Picking southwest cotton), postmarked Laredo Texas Gunnery School 10/16/43, message regarding best camp with good food.

[WAB to GB, 1/6/44] Postcard from Walter Babinski to G Babinski (Fresno – Aerial View), postmarked Fresno, Cal., 1/6/44, message regarding 2 days in Cal and seen no sun – hoping to get pictures from you.

[MK to WAB 2/2/44] Post Card from Martha Kurzyniec @ Chicago, Ill., to Walter Babinski @ Muroc AFB, Cal., (Chicago – Morrison Hotel), postmarked Chicago, 2/2/44, message regarding visit to Chicago.

[AB to GB, 8/15/44] Postcard from Ann Babinski to G Babinski, (St. Mary's Academy, Monroe, Mich.), postmark Monroe, Mich. 8/15/44, message regarding 'how are you – out riding much?'

[JJB to Mr. & Mrs. W Babinski, 1/22/45] Postcard from John J Babinski (MM1/C) to Mr. & Mrs. Walter Babinski, (Broadway @ 6th Street, Los Angeles), return address Terminal Island, San Pedro, Cal., postmarked US Navy, 1/22/45, message re: "…I'm still waiting for assignment."

[DFB & WAB to LB, 2/11/48] Post card from Dorothy Babinski and Walter Babinski to Leona Babinski, (Hotel New Yorker, New York City), postmarked NYC 2/11/48, message regarding location & activity.

[JJB to G Babinski ND] Post card from John J Babinski, APO Pool Terminal Island, San Pedro Cal, to Miss Gene Babinski Detroit, (Sunset & Vine, Hollywood Cal.: Western Publishing & Novelty #835, Los Angeles, Cal, ND) message regarding weather.

[JJB to L Babinski ND] Post card from John J Babinski to L Babinski, (Fisherman's Wharf, San Francisco, Cal.: WM Smith #87, San Francisco, Cal, ND) message regarding palm & orange trees, etc.

[WAB to DFK V-Mail from Admiralties] Undated V-Mail from S/Sgt Walter A. Babinski, 307th BG, 424th BS, Admiralty Islands, via APO 718 to Dorothy F. Kurzyniec, Detroit Michigan.

[WAB to DFK V-Mail 'Soldiers Dream'] Undated V-Mail from S/Sgt Walter A. Babinski, 307th BG, 424th BS, via APO 718 to Dorothy F. Kurzyniec, Detroit Michigan, with message regarding 'Soldier's Dream.'

[WAB to MK V-Mail 'Japs Have Surrendered'] Undated V-Mail from S/Sgt Walter A. Babinski, 307th BG, 424th BS, via APO 718 to Martha Kurzyniec, Detroit Michigan, with message regarding 'Japs Have Surrendered.'
[JEP to Mr. & Mrs. Kelly, 4/12/43] V-Mail dated 4/12/43 from Pvt. Joseph E. Prus, 67th General Hospital via APO 511 to Mr. & Mrs. Kelly, Detroit Michigan. 'Kelly' was a Delray nickname for the entire Kurzyniec family.
[JEP to Mr. & Mrs. Kelly, 11/3/43] V-Mail dated 11/3/43 from Pvt. Joseph E. Prus, 67th General Hospital via APO 511 to Mr. & Mrs. Kelly, Detroit Michigan. V-Mail says 'Seasons Greetings from Great Britain.'
[JS to Mr. & Mrs. Kurzyniec, 12/10/43] V-Mail dated 12/10/43 from Cpl. Joseph Stoklosa, Co. 'C' 3rd Tank Batt., In the Field, via Fleet Post Office, San Francisco, Calif.

Contemporary Postcards (blank) (all from Babinski Family collection, unless otherwise noted):
[PC: Biloxi: Historic Back Bay]
[PC: Biloxi: Beauvoir]
[PC: Biloxi: Keesler: General Review]
[PC: Biloxi: Keesler: Technical School]
[PC: Chicago, Ill.: Stevens Hotel]
[PC: Detroit: Naval Armory]
[PC: Evansville]
[PC: Hastings, Neb.: Masonic Temple]
[PC: Hastings, Neb.: State Hospital]
[PC: Hawaii: Fern Forest]
[PC: Hawaii: Kalapana Beach]
[PC: Hawaii: Lei Seller]
[PC: Hawaii: Bird of Paradise]
[PC: Hawaii: Rainbow Falls]
[PC: Hawaii: Pineapple]
[PC: Hawaii: Pineapple Fields]
[PC: Hawaii: Dawn Silhouette]
[PC: Houston, Tex.: Airport]
[PC: Houston, Tex.: Chamber of Commerce Building]
[PC: Kansas City, Mo.: Airport]
[PC: Kansas City, Mo.: Fern Valley Park: The Scout]
[PC: Kansas City, Mo.: Union Station at Night]
[PC: Laredo, Tex.: Ft. McIntosh]
[PC: Laredo, Tex.: Jarvis Plaza]
[PC: Laredo, Tex.: Laredo Army Air Field]
[PC: Lincoln, Neb.: State Capitol]
[PC: Miami: Army Airforces Training Command]
[PC: Miami: Indian Creek]
[PC: Miami: Jockeyclub]
[PC: Monterrey, Mexico: Cathedral & Saddle Back Mountain]
[PC: Montgomery, Ala.: State Capitol]
[PC: Nuevo Laredo, Mexico: Juarez Plaza]
[PC: New Orleans, La.: Pontchartrain Beach]
[PC: Salt Lake City: Greetings From UTAH]
[PC: Salt Lake City, Ut.: Seagull Monument]
[PC: St. Joseph, Mo.: Free Bridge]
[PC: St. Joseph, Mo.: Lovers Lane]
[PC: St. Louis, Mo.: Municipal Auditorium]
[PC: St. Louis, Mo.: Reservoir Park]
[PC: St. Louis, Mo.: War Memorial Building
[PC: St. Petersburg, Fla.: Yacht Basin]
[PC: Sydney, NSW, Australia: Olympic Pool at Luna Park, w/ notation on rear 'WAB, 1944']

Maps:
[AAF Chart C-41] AAF Cloth Chart – <u>Philippine Series No. C-41, Mindoro Island</u>, Washington, D.C.: Aeronautical Chart Services, April 1944.
[AAF Chart C-42] AAF Cloth Chart – <u>Philippine Series No. C-42, Samar Island</u>, Washington, D.C.: Aeronautical Chart Services, April 1944.
<u>Advancing the Bomber Line – Hollandia to Morotai.</u> Southwest Pacific News Map, Vol. 3, No. 10 – 9 Mar. 1945, Information & Education Section, USAFFE.
<u>Australia and New Guinea.</u> Sydney: H. E. C. Robinson Pty. Ltd., no date (assume purchased by WAB 1944 during furlough to Sydney)

Finschhafen to Hollandia – On the Road Back. Southwest Pacific News Map, Vol. 3, No. 3 – 19 Jan. 1945, Information & Education Section, USAFFE.

Morotai: Maps and Plans. Various maps and plans drafted by the 1/2 Australian CRE Army Topo Survey Company, of the Morotai base area, including: 1) 1/2 General Area, 2) 10/1 General Map of Staging Area (February 18, 1945), 3) 10/4 Staging Maintenance Area (February 21, 1945), 4) 13/2 Docks Area (February 26, 1945), 5) 14/5 Initial Maintenance Area, and 6) 15/1, 2, 3, 4 Staging Area Showing Camp (March 7, 1945), from the Australian National Archives, Australia War Museum, Access File AWM54 469/5/36.

Muroc Army Air Field. No author, no date, no publication information (Assume 1942-1945 era. Map was provided by R.L. Puffer, Edwards AFB, in response to request for information about WW2 era Muroc AAF).

Pacific Ocean. Sydney: H. E. C. Robinson Pty. Ltd., no date (assume purchased by WAB during 1944 furlough to Sydney)

United States Map. Denver, Colo.: Hotchkiss Map Co., no date (assume purchased by WAB 1943 or 1944 – shows major US rail routes, with WAB routes during training and after return to US during 1943-45 indicated)

Darley, James, M. (Chief Cartographer), Southeast Asia and Pacific Islands from the Indies and the Philippines to the Solomons. Washington, D.C.: National Geographic Society, October 1944.

Garver, John B. (Chief Cartographer), World War II: Asia and the Pacific. Washington, D.C.: National Geographic Society, December 1991.

Garver, John B. (Chief Cartographer), World War II: Europe and North Africa. Washington, D.C.: National Geographic Society, December 1991.

Advancing the Bomber Line – Hollandia to Morotai. Southwest Pacific News Map, Vol. 3, No. 10 – 9 Mar. 1945, Information & Education Section, USAFFE.

Personal Reminiscences:

[Llewellyn: My Time in the Military] Llewellyn, Clifford J., My Time in the Military Thiensville, Wisconsin: Privately printed, 1997.

[Llewellyn NCRM] T/Sgt. Clifford J. Llewellyn, Narrative Combat Report of Mission: 424th Bombardment Squadron, Office of the Operations Officer.

[Paul Journal] Paul, Richard A., 1st Lt., 13th Army Air Force, RICHARD PAUL JOURNAL, XIII "Jungle: Air Force Heavy Bombardment Squadrons: 1942-1945", in XIII Bombardment Groups (H) web site: http://www.5thbomberbarons.com/html/richardpaul.html

[Roth Interview] Roth, Richard, An Interview with Richard Roth. Rutgers Oral History Archives of World War II, Interview conducted by Sandra Stewart Holyoak in Somerset, New Jersey, June 9, 1999.

[Stebbins War History] George Hobart "Hobbie" Stebbins Jr War History Type written history, Author is Stebbins son or daughter (?), no date.

[Tuck's Diary] Sgt. Harley Tuck's Wartime Diary Transcribed by Harley Tuck Jr., ND (covers period 1943-1944), 447th Bomb Group Association, accessed July 7, 2016 at: http://www.447bg.com/Harley%20Tuck.pdf.

Oral History Interviews with 307th Bomb Group Veterans:

[Ayala Interview: August 25, 2006] U.S. Library of Congress Interview (2006 Reunion of the 307th Bomb Group Association, Bellevue, Washington) with S.J. 'Ike' Ayala, August 25, 2006.

[Coggins Interview: August 25, 2006] U.S. Library of Congress Interview (2006 Reunion of the 307th Bomb Group Association, Bellevue, Washington) with Harry E. Coggins, August 25, 2006.

[Cooper Interview: August 24, 2006] U.S. Library of Congress Interview (2006 Reunion of the 307th Bomb Group Association, Bellevue, Washington) with Wayne Cooper, August 24, 2006.

[Holston Interview: August 25, 2006] U.S. Library of Congress Interview (2006 Reunion of the 307th Bomb Group Association, Bellevue, Washington) with William D. Holston, August 25, 2006.

[Jurkens Interview: August 25, 2006] U.S. Library of Congress Interview (2006 Reunion of the 307th Bomb Group Association, Bellevue, Washington) with Edward A. Jurkens, August 25, 2006.

[Manley Interview: August 24, 2006] U.S. Library of Congress Interview (2006 Reunion of the 307th Bomb Group Association, Bellevue, Washington) with William Manley, August 24, 2006.

[Norwood Interview: August 24, 2006] U.S. Library of Congress Interview (2006 Reunion of the 307th Bomb Group Association, Bellevue, Washington) with Glenn Norwood, August 24, 2006.

[Pelle Interview: August 24, 2006] U.S. Library of Congress Interview (2006 Reunion of the 307th Bomb Group Association, Bellevue, Washington) with Thomas W. Pelle, August 24, 2006.

[Robinson Interview: August 25, 2006] U.S. Library of Congress Interview (2006 Reunion of the 307th Bomb Group Association, Bellevue, Washington) with Robert Robinson, August 25, 2006.

[Walsh Interview: August 24, 2006] U.S. Library of Congress Interview (2006 Reunion of the 307th Bomb Group Association, Bellevue, Washington) with James V. Walsh, August 24, 2006.

Personal Interviews:

[Aubrey Interview: August 15, 2002] Personal Interview with Donald W. Aubrey, August 15, 2002.

[H. Babinski Interview: August 27, 2002] Personal Interview with Henry Babinski, August 27, 2002.
[D.F. Babinski Interview: August 25, 2002] Personal Interview with Dorothy F. Babinski, August 25, 2002.
[D.F. Babinski Interview: August 27, 2002] Personal Interview with Dorothy F. Babinski, August 27, 2002.
[D.F. Babinski Interview: August 29, 2002] Personal Interview with Dorothy F. Babinski, August 29, 2002.
[D.F. Babinski Interview: November 27, 2002] Personal Interview with Dorothy F. Babinski, November 27, 2002.
[D.F. Babinski Interview: August 21, 2004] Personal Interview with Dorothy F. Babinski, August 21, 2004.
[D.F. Babinski Interview: April 9, 2005] Personal Interview with Dorothy F. Babinski, April 9, 2005.
[D.F. Babinski Interview: October 2005] Personal Interview with Dorothy F. Babinski, October 2005.
[D.F. Babinski Interview: January 21, 2013] Personal Interview with Dorothy F. Babinski, January 21, 2013.
[D.F. Babinski Interview: February 8, 2015] Personal Interview with Dorothy F. Babinski, February 8, 2015.
[D.F. Babinski Interview: May 30, 2016] Personal Interview with Dorothy F. Babinski, May 30, 2016.
[Fielding Interview: August 26, 2002] Personal Interview with James Fielding, August 26, 2002.
[V Mansir Interview: April 12, 2003] Personal Interview with Mrs. Virgie Mansir, April 12, 2003.
[Hidalgo Interview: April 14, 2002] Personal Interview with Boris Hidalgo, April 14, 2002.
[C. Llewellyn Interview: October 15, 2002] Personal Interview with Cliff Llewellyn, October 15, 2002.
[C. Llewellyn Interview: October 27, 2002] Personal Interview with Cliff Llewellyn, October 27, 2002.
[G. Mendryga Interview: November 21, 2003] Personal interview with Genevieve Mendryga, November 21, 2003.
[M. Powell Interview: November 22, 2002] Personal Interview with Murella H. Powell, November 22, 2002
[W. Pruett Interview: August 30, 2002] Personal Interview with William R. Pruett, August 30, 2002.
[Reis Interview: August 20, 2004] Personal Interview with Richard J. (Dick, Junior) Reis, August 20-21, 2004.
[Schreiner Interview: August 30, 2002] Personal Interview with Ray (Gene) Schreiner, August 30, 2002.
[J. Vanderpoel Interview: August 29, 2002] Personal Interview with John Vanderpoel, August 29, 2002.
[J. Vanderpoel Interview: August 19, 2004] Personal Interview with John Vanderpoel, August 19, 2004.
[J. Vanderpoel Interview: August 27, 2006] Personal Interview with John Vanderpoel, August 27, 2006.

Personal Correspondence:
[Adrian Letter: 9/13/2004] Email reply from Adrian, September 13, 2004, via www.armyairforces.com forum.
[C. Babinski Letter: 4/24/02] Letter from Christine Babinski, April 24, 2002.
[H. Babinski Letter: 5/8/02] Letter from Hank Babinski, May 8, 2002.
[Bolce Letter: 5/14/03] Email from Harold Bolce, May 14, 2003, via www.armyairforces.com forum.
[Boudreaux Letter: 1/22/03] Letter from Edmond Boudreaux, Mississippi Coast Historical & Genealogical Society, January 22, 2003.
[Bretherton Letter: August 10, 2002} Email from Dennis Bretherton (son of Jupe Bretherton) dated August 10, 2002.
[Burris Letter: 1/2/04] Email reply re: Harvard AFB, Nebraska, from Steve Burris, January 2, 2004, via www.armyairforces.com forum.
[Burris Letter: 11/12/03] Email reply re: Clearwater AAF, Florida, from Steve Burris, November 12, 2003, via www.armyairforces.com forum.
[Dunne Letter: 10/3/2002] Note from Edward T. Dunne, Brooklyn, NY, No date or postmark but received October 3, 2002.
[Fawcett Letter: 6/24/02] Letter from Bill Fawcett, June 24, 2002.
[Fielding Letter: August 19, 2002] Letter from James Fielding postmarked August 19, 2002.
[Guild Letter: April 13, 2002] Letter from Alan C. Guild, April 13, 2002.
[Guild Letter, 4/24/02] Email from Alan C. Guild, April 24, 2002.
[Hefferan Email, 7/25/15] Email from James Hefferan, 'Gilowice', July 25, 2015.
[Llewellyn Letter: 4/29/02] Letter from Cliff Llewellyn, April 29, 2002.
[Llewellyn Letter: 12/12/03] Letter from Cliff Llewellyn, Received December 12, 2003.
[McDonald Letter: 4/22/02] Letter from Mary McDonald, April 22, 2002.
[McDonald Letter: 5/3/02] Email from Mary McDonald, May 3, 2002.
[H. Mitchell Letter: 6/22/02] Letter from Harold Mitchell, June 22, 2002.
[Peters Letter: 1/29/03] Letter from James S. Peters, Sr., M/Sgt., USAF (Ret.), January 29, 2003, via www.armyairforces.com forum.
[Peters Letter: 1/31/03] Letter from James S. Peters, Sr., M/Sgt., USAF (Ret.), January 31, 2003, via www.armyairforces.com forum.
[Price Letter: 3/23/02] Email from Harlan Price, March 23, 2002.
[K. Reis Letter: 5/1/02] Undated letter from Kitty Reis, Received May 1, 2002.
[R. Reis Letter: May 1, 2002] Undated letter from Richard Reis, Received May 1, 2002.
[R. Reis Letter: 6/17/02] Letter from Richard Reis, June 17, 2002.
[N. Riley Letter: 10/15/02] Letter from Mrs. Nancy Riley, 10/15/02]
[Schreiner Letter: 5/16/02] Undated letter from R. E. 'Gene' Schreiner, received May 16, 2002.
[Vanderpoel Letter: 12/10/03] Email from John Vanderpoel, December 10, 2003.

Unit and Base Military Records:

[Edwards AFB WW2] _____ "History of Edwards AFB – (The World War II Era)." United States Air Force Fact Sheet 92-20, Air Force Flight Test Center, Office of Public Affairs, Edwards AFB, CA, (1992?).

[Keesler Air Force Base Guide: History of Keesler] "History of Keesler: Keesler Field to Keesler Air Force Base" from Keesler Air Force Base Guide, no author, no date (Xeroxed copy provided from Mississippi Coast Historical & Genealogical Society collections).

[Keesler Air Force Base Telephone Directory: History: Keesler Technical Training Center] "History: Keesler Technical Training Center" from Keesler Air Force Base Telephone Directory, no author, no date (Xeroxed copy provided from Mississippi Coast Historical & Genealogical Society collections).

[Muroc Army Airfield: 1942-1945] _____ "Historical Summarization of Muroc Army Airfield: 23 January 1942 – 2 September 1945." Seven fragmentary typewritten pages reproduced from microfilm, proved by Air Force History Support Office, Bolling AFB, DC.

[MAAF Chronology] Muroc Army Air Field Chronology, 1/3/44 thru 5/4/44, pp. 30-32, no author, no date, no publisher listed (provided by R.L. Puffer, Edwards AFB Historian, in response to request for Muroc AAF history during 1944).

[307th BG History] "History of the 307th Bombardment Group (HV), 13th AAF, 1944-1945." Frame citations from microfilm roll from Air Force History Support Office, Bolling AFB, D.C.

[424th BS History] "History of the 424th Bombardment Squadron (HV), 307th Bombardment Group, 13th AAF, 1944-1945." Frame citations from microfilm roll A0610 and A0611, from Air Force History Support Office, Bolling AFB, D.C.

[424th BS Mission Plan] Individual Mission Plans for the 424th Bombardment Squadron (HV). From the 307th Bombardment Group Association Archival Holdings (http://www.307bg.net/).

BS History: 12/1/43 – 12/20/43] "History of the 539th Bombardment Squadron (H) AAF from 1 December 1943 to 20 December 1943." Begins at microfilm image 1293, from Air Force History Support Office, Bolling AFB, D.C.

[539th BS History: 12/21/43 – 1/31/44] "History of the 539th Bombardment Squadron (H) AAF from 21 December 1943 to 31 January 1944." Begins at microfilm image 1296, from Air Force History Support Office, Bolling AFB, D.C.

[539th BS History: 2/1/44 – 2/29/44] "History of the 539th Bombardment Squadron (H) AAF from 1 February 1944 to 29 February 1944." Begins at microfilm image 1299, from Air Force History Support Office, Bolling AFB, D.C.

[539th BS History: 3/1/44 – 3/31/44] "History of the 539th Bombardment Squadron (H) AAF from 1 March 1944 to 31 March 1944." Begins at microfilm image 1327, from Air Force History Support Office, Bolling AFB, D.C.

Military Records from Babinski Family Collection:

These records were collected before, during, and after WW2 and retained by S/Sgt. Walter A. Babinski in the Babinski family homes in Detroit and Dearborn, Michigan.

[Arnold Letter] Arnold, Hap A., Commanding General, Army Air Forces, Certificate of Appreciation for War Service to Walter Babinski, undated.

[BPCA&D: 2/25/45] Battle Participation Credit – Awards & Decorations letter, S/Sgt. Walter A. Babinski, 307th BG, 424th BS, February 25, 1945, signed by Capt. David J. Davis.

[DSS Form 2] DSS Form 2 Registration Certificate, Walter A. Babinski, Local Board 43, Wayne, Michigan, June 30, 1942.

[DSS Form 150] DSS Form 150, Order to Report for Induction, Walter A. Babinski, Local Board 43, Wayne County, Michigan, dated January 4, 1943.

[FDFR] Foreign Duty Flight Record, 424th Bombardment Squadron, 307th Bombardment Group (H), Office of the Operations Officer, Babinski, Walter A., (undated, but covers period from 4/6/44 thru 1/22/45, signed by Capt. Carl S. Looker.

[Form 1: 2/25/45] AAF Form #1 Flight Report Operations, APO 920, 403rd Transport Command Group, 13th Transport Squadron, February 25, 1945.

[Form 2: 5/8/45] AAFRS4 Form 2-72 Certification of entitlement to wear medals and ribbons for S-Sgt. Walter A Babinski, dated 8 May 1945 at Santa Ana Army Airforce Base, signed by Major James A. Smith Jr.

[Form 5: 1/44] AAF Form #5, IV Bomber Command, 382nd Group, 539th Squadron, Muroc, Calif., Individual Flight Record, Babinski, Walter A., January 1944, signed by Capt. Paul Pestel.

[Form 5: 2/44] AAF Form #5, IV Bomber Command, 382nd Group, 539th Squadron, Muroc, Calif., Individual Flight Record, Babinski, Walter A., February 1944, signed by Capt. Paul Pestel.

[Form 5: 3/44] AAF Form #5, IV Bomber Command, 382nd Group, 539th Squadron, Muroc, Calif., Individual Flight Record, Babinski, Walter A., March 1944, signed by Capt. Herbert E. Lindhe.

[Form 5: 4/44] AAF Form #5, XIII Bomber Command, Individual Flight Record, Babinski, Walter A., April 1944, signed by Capt. Charles N. Kulman.

[Form 5: 5/44] AAF Form #5, XIII Bomber Command, Individual Flight Record, Babinski, Walter A., May 1944, signed by Capt. Charles N. Kulman.

[Form 8-117] W.D. Form 8-117, Immunization Register, Walter A. Babinski, no location, dated from 1/22/43 thru 1/3/45.
[Form 28] Form 28, Soldier's Individual Pay Record, S/Sgt. Walter A. Babinski, no location, various dates 1943-1945.
[Form 29-5: 1/22/43] W.D., A.G.O Form 29-5, War Savings Bond, Class A Pay Reservation Application, Pvt. Walter A. Babinski, Fort Custer, Michigan, January 22, 1943.
[Form 57] Form 57, Notice of Classification, Walter A. Babinski, Local Board 43, Wayne County, Michigan, June 6, 1945.
[Form 86: 9/14/43] TSKF Form 86, Certificate of Proficiency, B-24 Airplane Mechanics Course, Keesler Field, Mississippi, Pfc. Walter A. Babinski, September 14, 1943, signed by Capt. L. C. Eulberg.
[Form 121] Form 121, Individual Issue Record, S/Sgt. Walter A. Babinski, no date, no location.
[Form 206] AAF Form 206, Army Air Forces Pilot and Crew Member Physical Record Card, Sgt. Walter A. Babinski, signed by flight surgeon, Muroc, Calif., 1/5/44.
[Form230] WD ODB Form 230, Pay Allotment Notice, 539th Bomb Sq., AAF Muroc, Calif., Sgt. Walter A. Babinski, April 4, 1944.
[Form 812] Form 812 Office of Dependency Benefits, no date.
[Form 350: 1/22/43] VA Insurance Form 350, Application for National Service Life Insurance, Pvt. Walter A. Babinski, Fort Custer, Michigan, January 22, 1943, signed by Lt. S. F. Brower.
[GO5] Headquarters 307th Bombardment Group (HV), APO 719, General Orders 5, February 15, 1945.
[GO10] Headquarters Far East Air Forces, APO 925, General Orders 10, January 2, 1945.
[GO138] HQ Thirteenth Air Force, APO 719, General Orders 138, November 17, 1944.
[GO190] HQ Thirteenth Air Force, APO 719, General Orders 190, December 1, 1944.
[GO216] Headquarters Far East Air Forces, APO 925, General Orders 216, February 5, 1945.
[GO389] Headquarters Far East Air Forces, APO 925, General Orders 389, March 17, 1945.
[IR] Immunization Register, Walter A. Babinski, no location, dated from 1/22/43 thru 1/3/45, signed by Capt. Charles E. Cook.
[NCRM] Narrative Combat Report of Mission: 424th Bombardment Squadron (Format: NCRM: Mission # :Date = NCRM:307-338:10/3/44)
[O8: 1/15/44] 539th Bombardment Squadron (H) AAF, Muroc Army Air Field, California, Order 8, dated January 15, 1944.
[PFMR] Personal Flying Mission Record, Unsigned handwritten log, Walter A. Babinski, undated, but covering period 5/20/44 thru 1/22/45 (with addendum covering 2/44 thru 2/45).
[PB&TR] Personal Base and Transit Record, unsigned typewritten log, Walter A. Babinski, undated, but covering period 1/14/43 thru 6/3/45.
[Sanders Letter] Sanders, Brigadier General Richard C., Undated letter to "Former Members of the AAF Team, Headquarters, AAF Personnel Distribution Command, Louisville, Kentucky.
[SO5: 4/10/43] Headquarters, Army Air Base, Harvard Nebraska, Special Orders 5, April 10, 1943.
[SO11] Headquarters U.S. Army Recruiting & Induction District, Special Orders 11, Walter A. Babinski, Detroit, Michigan, dated January 14, 1943.
[SO49] HQ Far East Air Forces, APO 925, Special Orders 49, February 18, 1945
[SO56] HQ 307th Bombardment Group (H), APO 324, Special Orders 56, June 3, 1944
[SO64] 307th BG, APO 324, Special Orders 64, June 20, 1944.
[SO99] Army Service Forces – Sixth Service Command, Fort Sheridan, Illinois, Special Orders 99, April 9, 1945.
[SO125] Headquarters Santa Ana Army Air Base, Santa Ana, California, Special Orders 125, May 25, 1945.
[SO297: 12/14/43] Headquarters Army Air Forces Flexible Gunnery School, Laredo Army Air Field, Laredo, Texas, Special Orders 297, dated December 14, 1943.
[Streett Letter, 12/7/44] Streett, St. Clair, Major General, Commendation Letter to S/Sgt. W. A. Babinski, APO 719, December 7, 1944.
[Suggestions for Returning Combat Crew Members] Suggestions for Returning Combat Crew Members, Headquarters Thirteenth Air Force, APO #719. Undated. (Attached to SO49)
[VIC] Army Air Forces Veteran Identity Card, S Sgt. Walter Babinski, dated May 18, 1946, signed by H. A. Arnold, Commanding General.

S/Sgt. Walter A. Babinski Records from the National Personnel Records Center:
These record copies were obtained in 1993 by the author from the National Archives and Records Administration, National Personnel Records Center (NPRC) in St. Louis, Missouri. The vast majority of the copies show fire damage from a major 1973 fire at the NPRC storage facility. Records fragments are cited as [NPRC Fragment].
[Form 371] Army Form 371, Final Payment Roll, Sgt. Walter A. Babinski, Fort Sheridan, Illinois, 1/3/45.

Machinist Mate First Class John J. Babinski Records from the National Personnel Records Center:
These record copies were obtained in 2002 by the author from the National Archives and Records Administration, National Personnel Records Center (NPRC) in St. Louis, Missouri. All references to this material are cited as [JJB:Item].

John J. Pearce Material:
This material is comprised of various training logs, training maintenance records, aircraft instructional information, and notebooks, all related to Private John J. Pearce, while assigned to the 397th T.S.S., Fl. A., Barracks 8G, at Keesler Air Force Base, Biloxi, Mississippi, in 1943. Acquired from Mr. Aaron Bittner, North Carolina, 2002, via eBay auction. Pearce was the father of a friend of Bittner. Material from this collection notated as: Pearce: Maintenance Inspection Record, Squadron Airplane No. 13246, etc.

Photographs:
[Babinski Collection] Photos from Babinski family collection.
[Biloxi Public Library] Photos and postcards from Biloxi Public Library, Local History & Genealogy Department Collection.
[Truk] Two contemporary newspaper article photos of Truk Harbor & Dublon Island Seaplane Base under USAAF attack.
[Llewellyn Collection] Photos from Cliff Llewellyn Collection.
[307th BG Association Collection] Photos from the archives of the 307th Bombardment Group (HV) archives.
[Bretherton Collection] Photos from Dennis Bretherton Collection.
[Fronczak Collection] Photos preserved in the Fronczak Room of the E. H. Butler Library at Buffalo State University.
Film and Video:
[WW2 at Muroc] Johnson, Frederick A., Writer and Director: World War II at Muroc. Produced by Air Force Flight Test Center History Office, Edwards Air Force Base, California, Dr. James O. Young, Chief Historian, August 1995.

Internet Web Sites:
[53rd WRS] 53rd Weather Reconnaissance Squadron "Hurricane Hinters". United States Air Force Reserve, 2002. (www.hurricanehunters.com/fac53.htm)
[Abandoned Airfields] Freeman, Paul, Abandoned & Little-Known Airfields. 2004, (http://members.tripod.com/airfields_freeman/) .
[Babinski, Wladyslaw] Babinski, Wladyslaw. Ellis Island Passenger Record, (www.ellisisland.com).
[Baseball Almanac] The Baseball Almanac. 2004, (www.baseball-almanac.com).
[Battle of Morotai] Wikipedia: https://en.wikipedia.org/wiki/Battle_of_Morotai (December 31, 2015).
[Beard] Beard, Dick, Lindbergh Flew With Us (307th Bomb Group). Charles Lindbergh, http://www.charleslindbergh.com/history/b24.asp (6/3/2009), El Reno, OK
[Canton] Resture, Jane, Jane's Kiribati Home Page: Canton. 2001, (http://www.janeresture.com/kiribati_phoenix_group/canton.htm).
[CCUSAAF] COMBAT CHRONOLOGY OF THE US ARMY AIR FORCES: 1. ftp.rutgers.edu in directory pub/wwii/usaf; 2. byrd.mu.wvnet.edu (129.71.32.152) in pub/history/military/airforce/wwii_chronology;
[Combat Box] Combat Box, Wikipedia, http://en.wikipedia.org/wiki/Combat_box, accessed December 15, 2013.
[DAT] Carleton, Major William H., USAAF, History of the Directorate of Air Transport Allied Air Force South West Pacific Area and the 322nd Troop Carrier Wing. Australia @ War, 2007, www.ozatwar.com.
[Days of Infamy] Days of Infamy. American RadioWorks, Minnesota Public Radio and NPR News, September 2002, http://www.americanradioworks.org/features/daysofinfamy/index.html.
[Detroit Hockey.Net] Detroit Hockey.net. The Definitive Detroit Red Wings Resource, 2003, (www.detroithockey.net).
[Edwards Air Force Base] Historic California Posts: Edwards Air Force Base. California State Military Museum, 2004. (www.militarymuseum.org/edwardsafbase.html).
[Fort Mason] Wikipedia: https://en.wikipedia.org/wiki/Fort_Mason, accessed May 29, 2016.
[Fort McDowell] Wikipedia: https://en.wikipedia.org/wiki/Angel_Island_(California)#Fort_McDowell, accessed May 29, 2016.
[Frank M. Coxe] Wikipedia: https://en.wikipedia.org/wiki/USAT_General_Frank_M._Coxe, accessed May 29, 2016.
[Fresno Air National Guard Base] Historic California Posts: Fresno Air National Guard Base – Hammer Army Air Field. California State Military Museum, 2004. (www.militarymuseum.org/fresnoangbase.html).
[Hamilton Air Force Base] Historic California Posts: Hamilton Air Force Base. California State Military Museum, 2004. (www.militarymuseum.org/hamiltonafbase.html).

[Handbook of Texas Online: LAFB] Leatherwood, Art, "Laredo Air Force Base" <u>The Handbook of Texas Online.</u> The Texas State Historical Association, 2002. (www.tsha.utexas.edu/handbook/online/articles/view/LL/qbl2.html)

[Handbook of Texas Online: EPAFB] Leatherwood, Art, "Eagle Pass Air Force Base" <u>The Handbook of Texas Online.</u> The Texas State Historical Association, 2002. (www.tsha.utexas.edu/handbook/online/articles/view/EE/qbe12.html)

[Hester Lake B-24] Jordan, Don, <u>The Hester Lake B-24E</u>. 2001, (http://djordan.cyberlink.com/hesterlakeb24.html).

[Hickam Field: Sand Dunes to Sonic Booms] <u>Hickam Field: From Sand Dunes to Sonic Booms</u>. US National Park Service, ND. (http://www.cr.nps.gov/nr/travel/aviation/hic.htm).

[Historical Hurricane Tracks] <u>NOAA Historical Hurricane Data</u>. National Oceanographic and Atmospheric Administration, 2002. (http://hurricane.csc.noaa.gov/website/hhtm)

[History of the Rouge] <u>History of the Rouge</u>. The Henry Ford, 2004. (http://www.hfmgv.org/rouge/history.asp).

[M/S Orinoco] "M/S Orinoco" <u>Shiplover WebSite</u>, 2003, (http://shiplover2.virtualave.net/germany/orinoco.html).

[JDA&SM: History of Travis AFB] <u>Origin and History of Travis AFB</u>. Jimmy Doolittle Air & Space Museum, Travis Airforce Base, 2002 (www.travis.af.mil).

[Joseph Heller] <u>Joseph Heller</u>. Wikipedia, https://en.wikipedia.org/wiki/Joseph_Heller, accessed April 10, 2016.

[My Military Service] Parmer, Hoyt, <u>My Military Service</u>. 2004. (http://aaron.smith.free.fr/stories/my_military_service.html).

[Pacific Wrecks] <u>Pacific Wrecks Database.</u> , 2005, (www.pacificwrecks.com).

[Portable Surgical Hospital] <u>Portable Surgical Hospital</u>, Wikipedia, https://en.wikipedia.org/wiki/Portable_Surgical_Hospital, consulted April 10, 2016.

[Robert Altman] <u>Robert Altman</u>, Wikipedia, https://en.wikipedia.org/wiki/Robert_Altman, accessed April 10, 2016.

[Salt Lake City US Army Air Base] Balls, Jami, "Salt Lake City US Army Air Base" from <u>History of Kearns</u>. 2002 Kearns Community Website, (www.kearns-utah.org)

[Ships List] <u>The Ships List.</u> 2004, (http://www.theshipslist.com).

[St. Mary's Church] Rajtar, Steve, "St. Mary's Roman Catholic Church," <u>St. Petersburg Historical Trail.</u> 1999, (http://www.geocities.com/yosemite/rapids/8428/hikeplans/st_petersburg/planstpete.html).

[USAF Web Site Biography: D. L. Harlow] "Chief Master Sergeant Of The Air Force Donald L. Harlow" <u>United States Air Force Web Site Biography.</u> (http://www.af.mil/news/biographies/harlow_dl.html).

[US Army WW2 Dog Tags] Steinert, David, <u>US Army WW2 Dog Tags.</u> (http://home.att.net~steinert/ us army ww2 dog tags.htm), 2000, submitted by Alain Batens.

Articles and other published material:

[Aubrey Home] 'Lt. Don Aubrey Home From Pacific' Undated, unattributed newspaper article, appears late-1945, local Norwich (?), Connecticut Newspaper.

[Aubrey Pilot] 'Don Aubrey Jet Pilot for N. E. Airlines' Undated, unattributed newspaper article, appears mid-1960, local Norwich (?), Connecticut Newspaper.

[Fielding Letter] Fielding, Jim, 'Mission Logs Worth a Read: Jimmy Fielding Letter' <u>307th BG (HV) Newsletter</u>: June 2009, pp. 10-11.

[Johnson: 'Greetings from Earl Johnson' 307BGN: April 2004] Johnson, Earl, 'Greetings from Earl Johnson,' <u>307th BG (HV) Newsletter</u>: April 2004, p. 2.

[Kendal: 'Bacolod City, Negros Island' 307BGN: October 2002] Kendal, James, 'Bacolod City, Negros Island, <u>307th BG (HV) Newsletter</u>: October 2002, p. 6.

[Kendal: 'T/Sgt. Nelson Allan Knowlton,' 307BGN: September 2001] Kendal, James, 'T/Sgt. Nelson Allan Knowlton, <u>307th BG (HV) Newsletter</u>: September 2001, pp. pp. 5-6.

[Kendal: 'Current Requests,' 307BGN: December 2000] Kendal, James, 'Current Requests, <u>307th BG (HV) Newsletter</u>: December 2000, pg. 4.

[Messina: New History Data Reveals Bungling] Messina, Mauro J., 'New History Data Reveals Bungling, and a Far Different Pacific War than the War We Know,' The 13th Air Force Veterans Association, No date.

[Messina: Rip Van Winkle of the Pacific War Awakens] Messina, Mauro J., 'The Rip Van Winkle of the Pacific War Awakens to High Honors After Half a Century of Deep Sleep,' The 13th Air Force Veterans Association, No date.

[Randell, Jared] Randell, Jared, The Death of the Ball Turret Gunner. 1945.

Miscellaneous Babinski Family Collection Items:

[ABC Card] <u>American Bowling Congress Membership Card:</u> Merry-Go-Inn Team, 1946-1947 Season

<u>Aircraft Recognition Note Book:</u> Inscribed on P. 1: "Pfc. Walter B. Detroit" with 30 hand drawn aircraft recognition pictures, notebook is 7" x 4" published in Atlanta, GA.

[DESTACO: 2/3/48] Detroit Stamping Company: Leave of Absence letter, Walter Babinski, dated February 3, 1948.

[DESTACO: 4/21/49] Detroit Stamping Company: Statement of Earnings letter, Walter Babinski, dated April 21, 1949.

[DESTACO: 12/11/86] Detroit Stamping Company, Employment Service letter, Walter Babinski, dated December 11, 1986.

[Form 130] OPA Form R-130, War Ration Book No. 3, Martha Josephine Kurzyniec, no date.

[Form 145] OPA Form R-145, War Ration Book No. 4, Thomas Kurzyniec, no date.

[Form 506] United States Employment Service Form 506, Applicant Identification Card, Walter A. Babinski, Detroit, dated 8/21 thru 10/4/(45?).

[Earnings 1942] Earnings report for 1942, Walter Babinski # 304, handwritten sheet (from Samsonite Corporation?).

[Keesler Field Newspaper: 8/12/43] Keesler Field, Miss., Newspaper issue August 12, 1943 (fragment).

[MLCC:MJK] Michigan Liquor Control Commission card issued to Martha Josephine Kurzyniec, no date.

[MLCC:TK] Michigan Liquor Control Commission card issued to Thomas Kurzyniec, no date.

<u>Military Record K.u.K. Regiment 56: Ladislaus Babinski:</u> Military pass and record book, dated 1903.

[Mop on the B-24] Engram, Pfc. Julius E., "Where is the Mop on the B-24" Song Sheet, Keesler Field, Miss: Reproduction Unit, dated 5-19-43.

[NK B24 BT Ad] Nash Kelvinator Advertisement: "I've Got a Front Row Seat" B-24 Ball Turret, no date.

[Slagle Interview] Interview Script titled "Onterview {sic} Number three" Walter Babinski, Slagle & Koppitz interviewers, no location, but assume Detroit, no date, but assume 1945.

[TOP 10/8/45] Temporary Operators Permit, State of Michigan, Issued to Walter Anthony Babinski, Detroit, October 8, 1945 (reisued January 8, 1946.

[UAW: 1/28/87] United Auto Workers, Application for UAW Journeyman Card, Walter A. Babinski, dated 1-28-87.

[WUT: 3/20/43] Western Union Telegram from Walter Babinski, Hastings, Neb, to Mr. & Mrs. Babinski, Detroit, dated 3/20/43.

[WUT: 1/16/44] Western Union Telegram from Sgt. Walter Babinski, Muroc, Calif., to Mr. & Mrs. Babinski, Detroit, dated 1/16/44.

END NOTES

Chapter 1: Beginnings: We Love This Splendid Land

[1] Most of the information in this chapter is based on various Babinski family genealogical records, as well family reminiscences, interviews, letters, and personal recollections.

[2] Wolff, The Idea of Galicia

[3] Military Record K.u.K. Regiment 56: Ladislaus Babinski. Ladislaus' civilian occupation is noted as Handlesmann (farm harvest hand).

[4] Lucas, Austro-Hungarian Infantry 1914-1918, pp. 16-19.

[5] Military Record K.u.K. Regiment 56: Ladislaus Babinski. There was a family belief that Ladislaus served in the Austrian Cavalry, although there is no indication of this in his military records.

[6] Ship List. The Samland had been built in Camden, New Jersey for the Atlantic Transport Steamship Line as the SS Mississippi. She was launched on December 15, 1902 and registered at 7,913 gross tons, 490.4 feet long, with a 58.2 foot beam, and could make 14 knots. The Mississippi had accommodations for 1,900 third class passengers and started out sailing from London to Baltimore. In 1903 she was sold to the Red Star Line of Belgium and began sailing on the Antwerp to New York route. Her name was changed to the Samland on July 24, 1906, just three months before Ladislaus sailed from Antwerp to New York. She was later placed in service between Hamburg, Antwerp, Quebec, and Montreal, and later still made passages between Europe and Australia. During World War One she sailed as a cargo only transport between New York and England or Rotterdam. She resumed the Antwerp to New York passenger service route in 1919 until 1931. She was scrapped in 1931.

[7] Ellis Island records indicate that Wladyslaw (Ladislaus) Babinski's place of residence was Ropczyce, Austria. This must have been a misspelling of Roczyny. His age was recorded as 24 and marital status single.

[8] Szawala, A Landmark in Delray, p. 2.

[9] The 1930 US Census indicated that Walter Babinski's age was 44 and that he had been first married at the age of 24. Therefore the date of his first marriage must have been 1910.

[10] The Detroit Polk Directory listed his residence at 30 Copeland and his occupation as 'laborer' (1910), 'butcher' (1912), 'machinist hand' (1913), and residing at 9050 Pulaski as 'machinist' (1921).

[11] This information annotated on an old Babinski family genealogical chart.

[12] Ship List. The Lapland was built by Harland & Wolff in Belfast for the Red Star Line of Belgium. She was launched on June 27, 1908 and registered at 17,540 gross tons, 605.8 feet long, with a 70.4 foot beam, and could make 17 knots. She had accommodations for 400 first class, 450 second class, and 1,500 third class passengers. Until the outbreak of World War One she sailed on the Antwerp to New York route. After the outbreak of the war in 1914 she was chartered by the Cunard Line, sailing between Liverpool and New York. In April 1917 she struck a mine, but was able to reach Liverpool where she was repaired. In 1919 or 1920 she was refitted to 18,565 gross tons with accommodations for 389 first class, 448 second class, and 1200 third class passengers, and then resumed service for the Red Star Line between Antwerp and New York. She made her last passage to New York in 1932, then commenced London to Mediterranean cruises for one season. In October 1933 she was sold to Japanese owners and scrapped in Osaka the following year.

[13] D.F. Babinski Interview: August 29, 2002. My mother told me that her father (Thomas Kurzyniec) said that Cora was a remarkably beautiful woman and a great dancer. Thomas said "When Cora danced her feet never touched the ground. I remember how beautiful she looked at her wedding to Walter."

[14] The Detroit Polk Directory lists Walter Babinski's residence at 9050 Pulaski and his occupation as 'auto worker' (1923-1924), 'painter' (1924-1925), and 'laborer' (1925-1926, 1926-1927, and 1927-1928).

[15] 1930 US Census

[16] 1930 US Census. The store was valued at $6,000 in 1930.

[17] D.F. Babinski Interview: February 8, 2015

[18] Hefferan Email 7/25/15. Recounting 2015 interview with Kurzyniec family in Gilowice.

[19] Thomas told the author about eating at the Trzesniewski Buffet in Vienna when his regiment was stationed there. The author ate there when he first visited Vienna in 1969.

[20] Hefferan Email 7/25/15. Recounting 2015 interview with Kurzyniec family in Gilowice.

[21] Ship List. The Lutzow was a relatively new ship, having been launched on December 17, 1907 in Bremen, Germany for the North German Lloyd Steamship Line. The Lutzow, registered at 8,818 gross tons, 462.3 feet long, and with a 57.6 foot beam, could make 14 knots. She had accommodations for 104 first class, 104 second class, and 1,700 third class passengers. Thomas sailed on the third and final Bremen to New York voyage of the Lutzow, after which she went into service between Bremen and the Far East via the Suez Canal. In August 1914, after the start of the First World War, she was captured by the British Royal Navy and renamed the Huntsend. In 1924 she was returned to the North German Lloyd Line and was again renamed the Lutzow. She sailed on the Bremen, Halifax, New York route until 1932. She was scrapped in 1933. Additional information courtesy of Mr. Edwin Drechsel.

[22] Michigan, A Guide to the Wolverine State, p. 248. The bridge was rebuilt in 1923.

[23] The Detroit Polk Directory (1921-1922 and 1922-1923) lists Thomas Kurzyniec residing at 8620 W. Jefferson with the occupation of 'confectioner.'

[24] D.F. Babinski Interview: February 8, 2015

[25] Bertha's year of birth in 1875 is based on an undated family geneology. However, Martha's birth certificate lists her mother's age in 1907 as 20, implying a year of birth of 1886 or 1887. This birth certificate was issued in 1943, based on information from Martha's maternal aunt, Josephine Anderson. Bertha's age at the time of Martha's birth listed on this birth certificate is most likely in error. If Bertha was 20 years old in 1907, she would have been only 8 or 9 years old when married to John in 1895. Also, while not unheard of, it is unlikely that Bertha would have given birth to five children by age 20.

[26] Although their marriage license stated that Thomas was 35 and Martha 18.

[27] Jensen, Detroit's Polonia, p. 50.

[28] Jensen, Detroit's Polonia, p. 41. The boat to Put-in-Bay would leave Detroit at 10:00am and bring the partiers back at about 10:45pm. They advertised 'Entertainment galore, Afloat and Ashore.'

[29] Detroit Polk Directory (1925-1926 and 1927-1928). Thomas' occupation was listed as 'axelworker.'

30 *D.F. Babinski Interview: August 27, 2002*
31 *Carbonworks got its name from the Michigan Carbon Works Company, which began in 1873 in Detroit. It was established to turn dried buffalo bones from the American Prairies into industrial products, such as fertilizer, glues, neats foot oil, and filter material. Michigan carbon Works was the largest employer in Detroit, with millions of tons of the bones delivered by rail each year, until about 1895 when the Buffalo was finally near extinction. The company was successful though in finding alternate sources of raw material to remain in operation. Their factory was located on about 100 acres of land along the Rouge River west of Delray Proper. Successor companies to Michigan Carbon Works remain in operation to this day, although their plant has been moved to suburban Melvindale.*
32 *D.F. Babinski Interview: August 27, 2002*
33 *Jensen, Detroit's Polonia, pp. 4, 7-8.*
34 *Jensen, Detroit's Polonia, p. 79.*
35 *H. Babinski Interview: August 27, 2002. The author remembers bathing this way in the house on Pulaski in the early 1950's.*
36 *The shocking visual contrast of walking through that back fence gate, from the neat, quiet yard and into the chaos of rail lines and empty trash-strewn fields beyond is one of the most vivid childhood memories of the author.*
37 *H. Babinski Letter: May 8, 2002*
38 *D.F Babinski Interview and map sketch, April 9, 2005.*
39 *The author remembers the Babinski family Christmas tree being lit with candles as late as the early 1950's.*
40 *H. Babinski Letter: May 8, 2002*
41 *Poremba, Detroit: 1930-1969, p. 22.*
42 *D.F. Babinski Interview: August 27, 2002*
43 *JJB, Form 609.*
44 *Walt was only 13 years old in 1935. It is not clear if he travelled with his ma and pa, older sisters, or family friends.*
45 *Walter to JJB 7/30/35*
46 *Walter to JJB 8/6/35*
47 *Walter to JJB, 8/12/36*
48 *JJB to Mr. & Mrs. B., 5/8/37.*
49 *Walter to JJB, 7/26/37*
50 *Walter to JJB, 7/29/37*
51 *W&S to JJB, 6/25/38*

Chapter 2:   The World in the 1920s and 1930s: Era of Isms
1 *Many fortunes had been made earlier in the century by lumbermen who had cut down 40,000 square miles of Michigan forests. By the end of the century the forests (sadly) where gone but many of these men did not want to leave Michigan. They provided a ready supply of local capital for inustrail development in Detroit.*
2 *The author worked as a truck driver and warehouse man during 1971, 1972, and 1973 in Detroit with Paul Brooks, an elderly ex-US Army Paratrooper, who had gone to the Soviet Union to live in the 1930's. Mr. Brooks recounted to the author how many Americans felt deeply that the Soviet Communist system was superior to American democracy and capitalism, and was bound to succeed.*
3 *The United States was an 'imperial power' too, with possessions in the Philippine Islands, Samoa, Guam, Puerto Rico, the Virgin Island, and elsewhere. Most Americans though rationalized that the basis for the American possessions was different from the evil European empires.*
4 *These Melanesian and Micronesian islands included Yap, Truk, Palau, and others of which we will hear more about later.*
5 *Detroit News, September 1, 1939.*
6 *Capt. Babinski had been born May 13, 1896. Surviving the initial German attack, he would be captured by the Red Army on September 18 after they joined the Nazi onslaught to dismember Poland.*
7 *Julian Gebolys was born on April 2, 1911 in Auschwitz, in the Austrian Province of Galicia (later Brzezinka Oświęcim, Poland).*
8 *Detroit News, September 29, 1939.*

Chapter 3:   At Home 1939-1942: A Tremendous Shock
1 *D.F. Babinski Interview, January 21, 2013.*
2 *U.S. Census 1940*
3 *U.S. Census 1940*
4 *The Samsonite Story, p. 2.*
5 *The Samsonite Story pp. 2-7. Later, during WWII, the Ecorse plant was converted to war production, producing casings for bangalore torpedoes, metal parts for incendiary bombs, and tank parts. In 1964, Shwayder closed the Detroit plant, moving its manufacturing to a huge new factory in Murfreesboro, Tennessee. The following year the company name itself was changed to the Samsonite Corporation.*
6 *G. Mendryga Interview: November 21, 2003. Another benefit must have been a folding card table and chairs set. As a child, I remember Walt and my uncles playing cards on that table in the Babinski house on Pulaski.*
7 *D.F. Babinski Interview, November 27, 2002. However, Gennie Mendryga reported that Walt's 'Lize' nickname was because he liked to eat pork steaks, and in Polish pork is 'shliza' (G. Mendryga Interview: November 21, 2003).*
8 *D.F. Babinski Interview, August 25, 2002.*
9 *U.S. Census 1940*
10 *D.F. Babinski Interview, January 21, 2013.*
11 *D.F. Babinski Interview, April 9, 2005.*
12 *JJB to Mr. & Mrs. B. 6/24/40*
13 *The Detroit News, April 9, 1940.*
14 *The Detroit News, May 10, 1940.*
15 *The Detroit News, May 31, 1940.*
16 *The Detroit News, June 10, 1940.*
17 *The Detroit News, June 23, 1940.*
18 *The Detroit News, August 9, 1940.*
19 *The Detrot News, September 16, 1940.*
20 *The Detroit News, May 28, 1941.*
21 *The Detroit News, June 22, 1941.*

22 *Poremba, Detroit: 1930-1969, p. 45.*
23 *Szawala, A Landmark in Delray and Letter from Archbishop Edward Mooney to Archbishop Amleto Cicognani, dated July 8, 1944.*
24 *The Detroit News, August 14, 1941.*
25 *The Red Wings would win that game 3-2.*
26 *Days of Infamy.*
27 *D.F. Babinski Interview, April 9, 2005.*
28 *Days of Infamy.*
29 *Llewellyn, My Time in the Military, p. 3*
30 *The Detroit News, December 8, 1941.*
31 *The Detroit News, December 10, December 11, December 22, December 25, 1941.*
32 *The Detroit News, January 2, January 23, January 26, 1942*
33 *The Detroit News, April 18, 1942.*
34 *The Detroit News, June 6, 1942.*
35 *The Detroit News, July 2, 1942.*
36 *DSS Form 2.*
37 *The Detroit News, August 11, 1942.*
38 *JJB Form 609.*
39 *JJB: Form 16-26357*
40 *JJB: PT&BR.*
41 *JJB to WAB, August 24, 1942.*
42 *JJB: Form 6 and JJB:PT&BR. Johnnie completed basic training on September 18, 1942. He started his Mariners Mechanics Training course on October 6, 1942. He was rated as Seaman First Class.*
43 *JJB to Mr. WB, October 9, 1942. Johnnie listed his address as: W-10-1 Barracks 405 L.S., Group 3, U.S.N.T.S., Service School, Great Lakes, Ill.*
44 *The Detroit News, November 8, 1942.*
45 *The Detroit News, November 15, 1942.*
46 *Earnings, 1942.*

Chapter 4:   Induction and Basic Training 1943: Confident Faith
1 *DSS Form 150.*
2 *JJB: Form 16-9733*
3 *Articles of War, 110.*
4 *SO11, PB&TR, Form 28.*
5 *SO11*
6 *Fort Custer began as Camp Custer in 1917, where over 100,000 soldiers were trained over two years. Named for General George Armstrong Custer, after WWI it was used for Army officer training and for the Covilian Conservation Corps. In 1940 it was designated Fort Custer, with 16,005 Acres, and quarters for 1,279 officers and 27,553 enlisted men. During WWII, over 300,000 men would be receive processing and training there. Wikipedia 'Fort Custer Training Center' (https://en.wikipedia.org/wiki/Fort_Custer_Training_Center) accessed January 12, 2016.*
7 *Michigan: A Guide to the Wolverine State, p. 404.*
8 *Hemphill, Aerial Gunner from Virginia, p. 5.*
9 *Llewellyn, My Time in the Military, p. 6.*
10 *Hemphill, Aerial Gunner from Virginia, p. 9.*
11 *Hemphill, Aerial Gunner from Virginia, p. 6.*
12 *Form 29-5: 1/22/43.*
13 *Form 350: 1/22/43.*
14 *Form 8-117.*
15 *Basic Field Manual: Soldier's Handbook, p. iii.*
16 *Llewellyn, My Time in the Military, p. 9.*
17 *United States Map. Walt penciled in all of his wartime rail journeys on this detailed railroad map.*
18 *Pearce & Smith, World Weather Guide, pp. 135 & 138.*
19 *St. Petersburg Times, 'Razing of Beverly Hotel,' May 15, 1960. The Beverly Hotel was torn down in the early 1960's. The site is occupied today by the Florida International Museum. Nearby are at least two other St. Petersburg hotels that Walt would have noticed in 1943: the Detroit Hotel and the Hotel De Leon (author visit, 2004).*
20 *Florida: A Guide to the Southernmost State, pp. 259-262.*
21 *Story of St. Petersburg, pp 192-193.*
22 *Florida: A Guide to the Southernmost State, p. 264. The St. Petersburg Pier was rebuilt many years after the war as a city redevelopment project. The east end of the pier is still occupied by an amusement building similar to the casino that Pvt. Babinski visited in 1943 (author visit, 2004).*
23 *St. Petersburg Story, p. 151.*
24 *WAB to GB 1/29/43*
25 *WAB to LB 1/29/43*
26 *Florida: A Guide to the Southernmost State, pp. 259-261.*
27 *St. Mary's Church. St. Mary's church still stands today, appearing in good repair, in an area of redevelopment (2004 author visit).*
28 *PB&TR*
29 *Baross, The War Years WWII, p. 6.*
30 *Titler & Murphy, Keesler Field: Inception to Pearl Harbor, 1939-1941, p. 52.*
31 *'WAB to Mr. & Mrs. B, 2/2/43*
32 *Burris Letter: 11/12/03*
33 *Russell: Keesler Field: The War Years 1941-1945, p. 29 & 31.*
34 *Llewellyn, My Time in the Military, p. 8.*
35 *Hemphill, Aerial Gunner from Virginia, p. 10.*
36 *Basic Field Manual: Soldier's Handbook, pp. 10-15.*
37 *Basic Field Manual: Soldier's Handbook, pp. 16-21, 26-29.*

[38] Burris Letter: 11/12/03.
[39] PBTR.
[40] The Hotel Fort Harrison had been built in 1925 for Ransom E. Olds, of Oldsmobile and Reo in Michigan. Wikipedia: 'Fort Harrison Hotel' (https://en.wikipedia.org/wiki/Fort_Harrison_Hotel), accessed January 12, 2016.
[41] Form 8-117.
[42] Florida: A Guide to the Southernmost State, pp. 424-425.
[43] 75th Anniversary: The Fort Harrison. The Hotel Fort Harrison was acquired by the Church of Scientology in the 1970's and is used as the church's international headquarters and as a retreat center. The Church of Scientology is constructing an addition to the hotel in a similar architectural style across Fort Harrison Drive. The addition will house Church operational headquarters, freeing the original hotel structure for more usage as a retreat center. The hotel and grounds are carefully guarded, but the public is taken on guided tours of the hotel and grounds on Sundays (author visit, 2004).
[44] WAB to Mr. & Mrs. B 2/9/43.
[45] WAB to GB 2/9/43.
[46] JJB to WAB 2/7/43.
[47] JJB: Form 4-6111, JJB: Form NTSGL-S26-10-7-42, JJB: Form 16-9733. Johnnie's best test scores were in boilermaking (100%), shop theory (97%), and machine tools (96%). He scored only a 50% in welding and shipfitting. In metal working, his 84.84% placed him only 446th out of a class of 488. On February 5, 1943 he was ordered to report to the U.S. Navy Receiving Ship in San Francisco for shipment to the South Pacific, for duty where 'his specialized training may be further utilized.' (JJB: Form 9).
[48] St. Cecilia's Church is still in good repair and seems to serve a large congregation (author visit, 2004).
[49] Florida: A Guide to the Southernmost State, P. 498.
[50] WAB to HB 2/14/43.
[51] WAB to LB 2/14/43.
[52] Hemphill, Aerial Gunner from Virginia, p. 19.
[53] Coffey: Always Home, p. 3.
[54] Coffey: Always Home, pp. 5-6.
[55] Hemphill, Aerial Gunner from Virginia, pp. 57-61.
[56] Basic Field Manual: Soldier's Handbook, pp. 138-150.
[57] Form 29-5: 1/22/43.
[58] PB&TR.
[59] Georgia: A Guide to its Towns and Countryside, pp. 277-279.
[60] Coffey: Always Home, p. 6-8.
[61] Alabama: A Guide to the Deep South, pp. 220-228.
[62] Alabama: A Guide to the Deep South, p. 4.
[63] Alabama: A Guide to the Deep South, pp. 164-176.
[64] Alabama: A Guide to the Deep South, p. 5.
[65] Tennessee: A Guide to the State, pp. 206-229.
[66] Kansas: A Guide to the Sunflower Sate, pp. 192-195.
[67] Kansas: A Guide to the Sunflower State, pp. 198-204.
[68] Colorado: A Guide to the Highest State, p. 296.
[69] Colorado: A Guide to the Highest State, pp. 179-183.
[70] Llewellyn, My Time in the Military, p. 9.
[71] Colorado: A Guide to the Highest State, p. 301.
[72] Utah: A Guide to the State, pp. 225-226.
[73] PB&TR.
[74] WAB to GB 3/3/43.
[75] WAB to LB 3/3/43.
[76] Baross, The War Years WWII, p. 16.
[77] See SO5: 4/10/43 which indicated that a copy of PFC. Babinski's orders were to be sent to the 18th Replacement Wing in Salt Lake City. Also, Walt's Aircraft Mechanic's Proficiency record (Form 86: 9/14/43) from Keesler Field on Sept. 14, 1943 indicated that he was attached to the 18th Replacement Wing.
[78] Salt Lake City US Army Air Base.
[79] Alexander: Brief History of Three Federal Military Installations in Utah, p. 126.
[80] Pearce & Smith, World Weather Guide, p. 150.
[81] Utah: A Guide to the State, pp. 234-238.
[82] Utah: A Guide to the State, p. 228.
[83] Salt Lake City US Army Air Base.
[84] Utah: A Guide to the State, pp. 321-322.
[85] Colorado, A Guide to the Highest State, p. 164.
[86] Nebraska: A Guide to the Cornhusker State, pp. 352-353.
[87] Nebraska: A Guide to the Cornhusker State, p. 169. The CB&Q station is now used as Amtrak's Hastings, NB station.
[88] WUT: 3/20/43.
[89] Nebraska: A Guide to the Cornhusker State, pp. 169-171.
[90] Harvard Air Base: 1942-1946.
[91] History: Harvard Army Air Field, pp. 7-8..
[92] PB&TR.
[93] WAB to GB: 3/23/43.
[94] Burris Letter: 1/2/2004.
[95] Harvard Air Base: 1942-1946.
[96] Nebraska: A Guide to the Cornhusker State, pp. 169-175.
[97] History: Harvard Army Air Field, pp. 3-4.
[98] WAB to GB: 4/8/43.
[99] WAB to HB: 4/8/43. (The Red Wings lost to the Boston Bruins on April 8, tying the Stanley Cup Finals series 2-2)
[100] WAB to LB: 4/8/43.

[101] Baross, *The War Years WWII*, p. 8.
[102] SO5: 4/10/43.

Chapter 5:   Airplane Mechanics School 1943: The Mop on the B-24

[1] WAB to Mr. & Mrs. B 4/14/43. This postcard to Walt's parents and the two other postcards to Hank and Leona sent the same day were all postmarked at 12:30 p.m. in St. Louis, indicating they were written on the train and mailed as soon as Walt arrived in St. Louis.
[2] WAB to HB 4/14/43.
[3] WAB to LB 4/14/43.
[4] *Missouri: A Guide to the "Show-Me" State*, pp. 295-316.
[5] WAB to GB 4/14/43.
[6] PC: Evansville.
[7] WAB to HB 4/15/43. This postcard and the other four Walt sent the previous day continued to list his return address as Harvard AAB. Any mail sent to Walt at Harvard would be forwarded to his next base.
[8] *Mississippi: A Guide to the Magnolia State*: pp. 285 & 165. This station is still used by Amtrak as of 2004.
[9] PB&TR. In April, average temperatures in Biloxi range from a high of 76 to a night time low of 60.
[10] Titler & Murphy, *Keesler Field: Inception to Pearl Harbor, 1939-1941*, p. 35.
[11] Titler & Murphy, *Keesler Field: Inception to Pearl Harbor, 1939-1941*, p. 1.
[12] Titler & Murphy, *Keesler Field: Inception to Pearl Harbor, 1939-1941*, pp. 3-23.
[13] Titler & Murphy, *Keesler Field: Inception to Pearl Harbor, 1939-1941*, pp. 21-35.
[14] Titler & Murphy, *Keesler Field: Inception to Pearl Harbor, 1939-1941*, pp. 52-53.
[15] *Mississippi: A Guide to the Magnolia State*, pp.165, 168-175.
[16] Russell, *Keesler Field: The War Years, 1941-1945*, p. 112.
[17] *Biloxi Daily Herald*, 4/16/43, "Corporal's Ratings for Specialists," and "Keesler Expected to Get 'Hit Kit' Within Few Days."
[18] The record of this transfer has not been found. The first reference by Walt to the 396th TSS was in postcards sent to his family on 4/23/43. See: WAB to LB, 4/23/43 and WAB to HB, 4/23/43.
[19] Russell, *Keesler Field: The War Years, 1941-1945*, p. 4.
[20] *Biloxi Daily Herald*, 5/1/43, 'Jewish Soldiers Attend Seder Supper.'
[21] Pearce & Smith, *World Weather Guide*, p. 146.
[22] *Biloxi Daily Herald*, 4/23/43, 'Service at Sunrise Set at Oak Park.'
[23] Russell, *Keesler Field: The War Years, 1941-1945*, p. 5.
[24] *Biloxi Herald*, 4/22/43, 'These Soldiers Eat Stew in Swing-Time.'
[25] Russell, *Keesler Field: The War Years, 1941-1945*, pp. 5-6.
[26] Russell, *Keesler Field: The War Years, 1941-1945*, p. 20.
[27] Russell, *Keesler Field: The War Years, 1941-1945*, p. 29.
[28] Russell, *Keesler Field: The War Years, 1941-1945*, pp. 73-75.
[29] Russell, *Keesler Field: The War Years, 1941-1945*, p. 29.
[30] Russell, *Keesler Field: The War Years, 1941-1945*, pp. 34-35.
[31] *Mississippi: A Guide to the Magnolia State*, pp. 165-168.
[32] M. Powell Interview: 11/22/2002. After the war Biloxi's wild ways continued. By 1952 a Federal study determined that 15% of Biloxi's economy was related to gambling. The Air Force told Biloxi civic leaders they had to choose between gambling and Keesler AFB -. they could not keep both. Biloxi choose Keesler, but gambling just went underground. Today casinos are legal in Biloxi, as they are in many parts of the country. Another image of Biloxi's wild side can be seen in the Neil Simon film, *Biloxi Blues*.
[33] *Biloxi Daily Herald*, 4/19/43.
[34] *Biloxi Daily Herald*, 5/1/43.
[35] WAB to LB 4/23/43.
[36] WAB to HB 4/23/43.
[37] JK to WAB 4/22/43. ' (Joe Kaszyca, known as 'Clark,' was a friend of Walt's from Detroit. Kaszyca later hung himself – per D.F. Babinski interview 11./27/43.)
[38] WAB to JB 4/xx/43; WAB to LB 4/xx/43; WAB to HB 4/xx/43.
[39] *Biloxi Herald*, 5/1/43.
[40] JJB PT&BR, JJB: Form 9.
[41] WAB to GB 5/2/43.
[42] EN to WAB 4/22/43.
[43] Russell, *Keesler Field: The War Years, 1941-1945*, p. 35.
[44] Russell, *Keesler Field: The War Years, 1941-1945*, p.67.
[45] Russell, *Keesler Field: The War Years, 1941-1945*, p. 68.
[46] PB&TR.
[47] *Biloxi Daily Herald* 6/4/43, "AM School Goes on 2 Shifts; BTC Expands Schedule."
[48] *Daily Schedule: Specialized B-24 Airplane Mechanics Course*, p.1.
[49] Pearce: Personal Lecture Notebook.
[50] Form 86
[51] Russell, *Keesler Field: The War Years, 1941-1945*, p. 35.
[52] *Daily Schedule: Specialized B-24 Airplane Mechanics Course*, p. 1.
[53] *Biloxi Daily Herald*, 5/14/43, "New Setup on Graduates of AM School."
[54] At this point, Ford produced B-24's were not in operation in large numbers.
[55] Pearce: Personal Lecture Notebook.
[56] *Biloxi Daily Herald*, 5/21/43, "Keesler Negro Soldier Modern Tower of Babel," "Pfc. Doo Jam Ng Now is Learning Art of Mechanics."
[57] Russell, *Keesler Field: The War Years, 1941-1945*, p.67.
[58] *Daily Schedule: Specialized B-24 Airplane Mechanics Course*, p. 2.
[59] Munson, *Fighters and Bombers of World War II*, p. 250.
[60] Engram, *Where is the Mop on the B-24*.
[61] *Daily Schedule: Specialized B-24 Airplane Mechanics Course*, p. 2.
[62] Pearce: *Aircraft Mechanics Course Notebook*.

63 *Biloxi Daily Herald, 6/10/43.*
64 *Temperatures in Biloxi would now average daytime highs of 90 and night time lows of about 75 for the next three months.*
65 *Daily Schedule: Specialized B-24 Airplane Mechanics Course, pp. 2-3.*
66 *The Pratt & Whitney R-1830 Twin Wasp was an American aircraft engine widely used in the 1930s and 1940s. Produced by Pratt & Whitney, it was a two-row, 14-cylinder, air-cooled radial design. It displaced 1,830 cu in (30.0 L) and its bore and stroke were both 5.5 in (140 mm). A total of 173,618 R-1830 engines were built, and from their use in two of the most-produced aircraft ever built, the B-24 bomber and DC-3 transport, more Twin Wasps may have been built than any other aviation piston engine in history. (Wikepedpia)*
67 *Daily Schedule: Specialized B-24 Airplane Mechanics Course, p. 3.*
68 *Form 206; Form 8-117.*
69 *Daily Schedule: Specialized B-24 Airplane Mechanics Course, p. 4.*
70 *JJB: Form 4-C.*
71 *Maggie to WAB 7/1/43.*
72 *Daily Schedule: Specialized B-24 Airplane Mechanics Course, p. 4.*
73 *Ange to WAB 7/7/43.*
74 *Daily Schedule: Specialized B-24 Airplane Mechanics Course, pp. 4-5.*
75 *Pearce Material - Ford Airplane School, Army Air Forces, Propeller Job Sheets #1-A (Installation and removal) and #2-D (Blade and Barrel Disassembly and Assembly).*
76 *Pearce & Smith, World Weather Guide, p. 146.*
77 *Historical Hurricane Tracks.*
78 *53rd WRS.*
79 *Daily Schedule: Specialized B-24 Airplane Mechanics Course, p. 5.*
80 *SK to WAB 7/21/43.*
81 *Daily Schedule: Specialized B-24 Airplane Mechanics Course, pp. 5-6.*
82 *Babe to WAB 8/2/43. ("Babe" – was Joe "Babe" Lafata, a Detroit friend of Walt's. Lafata later played professional baseball for the New York Giants from 1947 until 1949, and also for the Minneapolis Millers in 1948 and 1950. – per D.F. Babinski interview 11/27/2002 and major-league baseball records.)*
83 *Keesler Field News, 8/12/43.*
84 *Pearce Material – Ford Airplane School, Army Air Forces, Electrical – Phase III, Instruments 100B, 7/6/43.*
85 *Daily Schedule: Specialized B-24 Airplane Mechanics Course, pp. 6-7.*
86 *Daily Schedule: Specialized B-24 Airplane Mechanics Course, p. 6.*
87 *Daily Schedule: Specialized B-24 Airplane Mechanics Course, p. 7.*
88 *Russell, Keesler Field: The War Years, 1941-1945, p.67.*
89 *Pearce Material Ford Airplane School, Army Air Forces –Engine Mechanics, Fuel System Inspection, 7/22/43.*
90 *Keesler News, 9/16/1943.*
91 *Form 86.*
92 *Historical Hurricane Tracks.*

Chapter 6:   Aerial Gunnery Training 1943: Faith, Trust, and My Duty

1 *Mississippi: A Guide to the Magnolia State, pp. 194-198; 285-302.*
2 *Louisiana: A Guide to the State, pp. 316 & 383.*
3 *Louisiana: A Guide to the State, pp. 199-202.*
4 *Texas: A Guide to the Lonestar State, pp. 287-288.*
5 *WAB to Mr. & Mrs. Babinski, 10/12/43.*
6 *Texas: A Guide to the Lonestar State, pp. 596-597.*
7 *Texas: A Guide to the Lonestar State, pp. 450-451.*
8 *Texas: A Guide to the Lonestar State, pp. 306-313.*
9 *Texas: A Guide to the Lonestar State, pp. 311-312.*
10 *PC: Laredo, Tex.: Ft. McIntosh*
11 *Opened on 23 September 1942, Laredo AAF was part of the Eastern Flying Training Command under the Army Air Forces Training Command at Fort Worth AAF, Texas. The 2d Aerial Gunnery Training Group of seven squadrons (1021st - 1027th Gunnery Training) taught aerial gunnery to new cadets primarily for B-17 Flying Fortress and B-24 Liberator duty, training on the .50 and .30 caliber Browning machine guns and their firing platforms. Classes in aerial gunnery were ended in September 1945 with the end of the war. Wikipedia: 'Laredo Air Force Base' (https://en.wikipedia.org/wiki/Laredo_Air_Force_Base), accessed January 12, 2016.*
12 *Handbook of Texas Online: LAFB. Laredo Army Air Field was deactivated in late 1945 and the City of Laredo began using it as a municipal airport in 1950. It was later reactivated by the USAF as a basic flight training facility for jet pilots from the US as well as 24 foreign countries.*
13 *PB&TR. When Walt arrived in Laredo, average daytime highs were about 86. Within a month though, highs would get down into the high 60's with nighttime lows below 50 degrees.*
14 *Clark, Missions of Shehasta, p. 11.*
15 *PB&TR*
16 *Laredo Army Airfield (W. Babinski handwritten notations).*
17 *Pearce & Smith, World Weather Guide, pp. 125 &144.*
18 *WAB to GB 10/16/44*
19 *Laredo Times, October 13, 1943.*
20 *Laredo Times, October 21, 1943.*
21 *PB&TR*
22 *SO297*
23 *Laredo Army Air Field*
24 *Dowdy, Charlie, p. 299.*
25 *Hemphill, Aerial Gunner from Virginia, p. 118.*
26 *Hemphill, Aerial Gunner from Virginia, pp. 115, 122, 145.*
27 *Basic Principles of Aerial Gunnery, p. 0.*
28 *Basic Principles of Aerial Gunnery, pp. 1-9.*
29 *Basic Principles of Aerial Gunnery, p. 19.*

30. *Basic Principles of Aerial Gunnery*, p. 51.
31. *Basic Principles of Aerial Gunnery*, p. 201.
32. *Aircraft Recognition Notebook.*
33. Clark, *Missions of Shehasta*, p. 11.
34. Clark, *Missions of Shehasta*, pp. 11-12.
35. Ambrose, *The Wild Blue*, pp. 158-159. Ambrose reports the experience of a 743rd Squadron B-24 tail gunner, Sgt. Louie Hansen, who discovered while testing his guns on a mission over Germany, that the cocking levers had been put in backwards after the last mission and would not fire. While wearing gloves due to the cold and with his goggles streamed over due to the strain, he disassembled the cocking mechanism on each gun and reassembled them properly while in his cramped turret. Potentially Hansen's life and the survival of his crewmates depended on his ability to fire his guns in defense against enemy aircraft.
36. *Laredo Times*, October 24, 1943.
37. *Laredo Army Airfield* (W. Babinski handwritten notations). The author determined the exact street location of the Laredo USO, by examining digital aerial photographs of Laredo available on the U.S. Geological Survey web site and identifying the USO structure depicted in the Laredo Army Airfield booklet. The photo of the USO in the booklet also shows railroad tracks running parallel to the street in front of the building, which can be identified as the Tex-Mex Railroad from the Laredo City Map in *Texas: A Guide to the Lone Star State*, p. 311.
38. *Laredo Times*, October 24, 1943.
39. Hemphill, *Aerial Gunner from Virginia*, pp. 106-108, 114.
40. *Basic Principles of Aerial Gunerry*, p. 527.
41. *Laredo Times*, November 1, 2, & 9, 1943.
42. *Laredo Army Airfield* (W. Babinski handwritten notations).
43. PC: Laredo, Tex.: Jarvis Plaza
44. *Texas: A Guide to the Lonestar State*, p. 306.
45. *Texas: A Guide to the Lonestar State*, pp. 309-310.
46. *Texas: A Guide to the Lonestar State*, p. 312.
47. *Texas: A Guide to the Lonestar State*, pp. 312-313.
48. *Laredo Times*, November 15, 1943.
49. *Laredo Army Airfield* (W. Babinski handwritten notations).
50. Clark, *Missions of Shehasta*, p. 12.
51. Llewellyn, *My Time in the Military*, p. 12., and Clark, *Missions of Shehasta*, p. 12.
52. *Laredo Army Airfield* (W. Babinski handwritten notations).
53. *Basic Principles of Aerial Gunnery*, p. 52.
54. *Basic Principles of Aerial Gunnery*, p. 101.
55. *Basic Principles of Aerial Gunnery*, p. 105.
56. *Basic Principles of Aerial Gunnery*, pp. 110-111.
57. *Basic Principles of Aerial Gunnery*, p. 115, p. 122.
58. Peters Letter: 1/29/2003
59. Peters Letter: 1/31/2003
60. *Basic Principles of Aerial Gunnery*, pp. 601-602.
61. Clark, *Missions of Shehasta*, p. 11.
62. Clark, *Missions of Shehasta*, pp. 12-13.
63. *Laredo Times*, November 24, 1943.
64. *Laredo Times*, November 28 & 29, 1943.
65. *Texas: A Guide to the Lonestar State*, p. 324.
66. *Texas: A Guide to the Lonestar State*, pp. 507-508.
67. Handbook of Texas Online: EPAFB. All training was discontinued at Eagle Pass Army Air Field by 1945.
68. PB&TR
69. USAF Web Site Biography: D. L. Harlow. Later, Harlow received a BS in Business Administration and also graduated from the Strategic Air Command Noncommissioned Officer Academy. In 1969 he was appointed Chief Master Sergeant of the Air Force, the highest non-commissioned rank in the USAF.
70. Llewellyn, *My Time in the Military*, p. 12.
71. Peters Letter: 1/31/2003
72. Clark, *Missions of Shehasta*, p. 14.
73. PB&TR
74. PB&TR
75. Conder, *The History of Enlisted Aerial Gunnery 1917-1991*, p. 3.
76. *Laredo Times*, December 12, 1943.
77. SO297
78. Ambrose, *The Wild Blue*, p. 65.
79. H. Babinski Letter: May 8, 2002
80. Sgt. Adam Sordyl would fight on Saipan, Iwo Jima, and Okinawa, before returning to Detroit after the war.
81. DF Babinski Interview, August 2004.
82. JJB: PB&TR. This was the 134th Naval Depot in Brisbane. The transfer had occurred November 25, 1943. While assigned to Task Force 78, his superiors had given him good fitness ratings, but not as good as John thought he should have had. His ratings for both June and November 1943 were: Proficiency in Rating – 3.5, Mechanical Ability – 3.6, Leadership – 3.3, Conduct – 4.0. Having worked at Ford Motor for almost 10 years, Johnnie felt he was being held back by biased ratings.
83. D.F. Babinski Interview, October 2005.

Chapter 7: Bomber Crew Phase Training 1944: Goodbye Golden Gate

1. *Missouri: A Guide to the Show Me State*, pp. 241-255.
2. Llewellyn, *My Time in the Military*, p. 9.
3. *California: A Guide to the Golden State*, pp. 188-191.
4. Fresno Air National Guard Base.
5. Llewellyn, *My Time in the Military*, p. 9.

[6] PB&TR
[7] WAB to GB 1/6/44.
[8] MAAF Chronololgy
[9] My Military Service; The Hester Lake B-24E.
[10] Form 206; Form 8-117
[11] WAB to GB 1/6/44
[12] California, A Guide to the Golden State, p. 606.
[13] PB&TR. When Walt arrived at Muroc, average high temperatures were less than 60 degrees and at night it often got down below freezing.
[14] Air Force Combat Units of World War II, pp. 269-270, 645.: The IVth Air Force was one of the major Army Air Corps training commands in the continental U.S. during the war. The 382nd Bombardment Group (Heavy) had been activated as an operational training unit on November 3, 1942 at Salt Lake City AAB, equipped with B-24's. After moving to Davis-Montham Field in Arizona and Pocatello AAF in Idaho during 1943, the 382nd Group had moved to Muroc AAF on December 6, 1943, where it would remain until March 31 of that year when it was inactivated. While at Muroc, the 382nd Group was commanded by Lt. Col. George E. Glober. The 382nd Group was later reactivated again in 1944 as a B-29 equipped combat unit and transferred to the Pacific Theater in July 1945, but inactivated on January 6, 1946, having never gone into combat. The 539th Bombardment Squadron (Heavy) was activated as a B-24 training unit on November 3, 1942, but not manned until January 23, 1943. During its existence, it was a component of the 382nd Bombardment Group. The 539th Squadron was inactivated on March 31, 1944.
[15] 539th BS History 12/1/43-12/20/43; 539th BS History 12/21/43-1/31/44;
[16] O8:1/15/44.
[17] WUT: 1/16/44.
[18] WW2 at Muroc.
[19] Edwards AFB WW2, Edwards Air Force Base
[20] Muroc Army Air Field: 1942-1943. Units at Muroc included the 338th (Colored) Aviation Squadron, 1021st (Colored) Quartermaster Platoon, and 1313th (Colored) Guard Squadron, attesting to the segregated nature of much of the service during WW2.
[21] WW2 at Muroc. Although Walt and other crews never realized it at the time, Muroc in late 1943 and early 1944 specialized in the training of replacement bomber crews intended for the Pacific Theater of combat.
[22] Clark, Missions of Shehasta, p. 20
[23] Muroc Army Air Field Map
[24] Cliff told me he did his aerial gunnery training at March AFB near Las Vegas, but this conflicts with his written narrative. C. Llewellyn Interview: October 27, 2002
[25] Llewellyn, My Time in the Military, p. 13.
[26] Llewellyn, My Time in the Military, p. 14.
[27] C. Llewellyn Interview: October 27, 2002
[28] C. Llewellyn Interview: October 27, 2002
[29] 'Shoots Down Zero' clipping from unknown local newspaper dated 10/8/1943, from Llewellyn Family collection.
[30] Muroc Mirage, 1/15/44, 'General Arnold Pays a Visit to Field at Muroc,' 'General and Mrs. Lynd at Muroc,' 'Post Theater to Get USO Camp Show,' 'Muroc Teams Hold First, Third Places,' and 'Lives of AAF Pilots Safer Within Armor.'
[31] Form 206; Form 8-117.
[32] Aubrey Interview, 8/15/2002.
[33] Llewellyn, My Time in the Military p. 14. Throughout his narrative of his time in the U.S.A.A.C. in WW2, Llewellyn refers to the crew 430 bombardier as Jim Lynch. I believe this is in error. The bombardier of the McRae (later Appling) crew was Jimmie W. Clark (see crew 430 graduation photo in: 539th BS History: 3/1/44 – 3/31/44).
[34] 539th BS History, 2/1/44-2/29/44.
[35] Paul Journal, p. 6.
[36] C. Llewellyn Interview: October 27, 2002
[37] 539th BS History, 12/21/43-1/31/44. The 539th historian reported: 'Amid great rejoicing, it was announced that the Flight Line Latrine was finally completed and ready for active service.'
[38] Ambrose, The Wild Blue, p. 95.
[39] Form 5: 1/44.
[40] Form 5: 1/44. For January 1944 Sgt. Babinski was credited with a total of 6:30 flying time, including 3:00 night flying.
[41] MAAF Chronology, p, 30.
[42] Ambrose, The Wild Blue, pp. 80-81.
[43] Hemphill, Aerial Gunner from Virginia, pp. 165 & 185.
[44] Muroc Mirage, 1/22/44. 'Fourth Air Force Highly Praised,' 'Red Cross Aids,' 'New Chaplain,' 'G.I. Prisoners Aid M.P. Garding Them,' 'Stellar Lockheed Team Here Tonight,' "Jack Benny Will Broadcast From Here Tomorrow,' 'USO Show Will Come Monday.'
[45] World War 2 at Muroc.
[46] Edwards Air Force Base.
[47] WW2 at Muroc.
[48] Llewellyn, My Time in the Military, p. 15.
[49] Edwards Air Force Base. After the war Barnes' Happy Valley Riding Ranch continued to expand into a true dude ranch, with a horse corral and barns, dance hall, motel and swimming pool, and even her own airstrip with tower and hangars. Hollywood personalities would fly into her airstrip to spend a few days in the desert, and the club continued to be popular with base pilots and officers. After Edwards (as Muroc was renamed) Air Force Base began to specialize in aircraft testing, one of the most famous patrons of the Happy Valley Riding Ranch was Chuck Yeager, before and after he first broke the sound barrier.
[50] 539th BS History, 2/1/44-2/29/44.
[51] 539th BS History, 2/1/44-2/29/44.
[52] 539th BS History, 2/1/44-2/29/44.
[53] Llewellyn, My Time in the Military, p. 15.
[54] Form 5: 2/44.
[55] Paul Journal, p. 6.
[56] FTC Review: Ghost Sails On. The Muroc Maru was constructed of a wooden framework with wiremesh screen 'sides and decks.' It was constructed at a total cost of $35,819. It was actually visible in the distance from a highway near the base boundary, and

motorists spread stories that it was an actual captured Japanese warship, towed out into the desert by the Air Corps. The Muroc Maru was demolished in 1950 during expansion of the east-west runways at Edwards Air Force Base.

57 MK to WAB 2/2/44
58 539th BS History, 2/1/44-2/29/44.
59 Llewellyn, My Time in the Military, p. 15.
60 Form 5: 2/44.
61 Photo from Babinski family collection. The author has seen copies of this same photo from the Llewellyn and Bretherton family collections.
62 California, A Guide to the Golden State, pp. 192, 546-547.
63 Hollywood Canteen
64 Based on Babinski Family collection photo from De Valle's Studio, dated 2/10/44.
65 California, A Guide to the Golden State, pp. 192-196.
66 Grauman's Chinese Theater.
67 Coffey, Always Home, pp. 6-7.
68 California, A Guide to the Golden State, p. 225.
69 Babinski family collection photo.
70 V. Mansir Interview, April 12, 2003. Virgie Mansir recounted in this interview that Lt. McRae was very friendly to the crewmen and their wives.
71 MAAF Chronology, p. 30.
72 539th BS History, 2/1/44-2/29/44.
73 Ambrose, The Wild Blue, p. 86.
74 Basic Principles of Aerial Gunnery, pp. 530-539.
75 W. Pruett Interview: August 30, 2002
76 Form 5: 2/44.
77 MAAF Chronology, p. 30.
78 539th BS History, 2/1/44-2/29/44.
79 Boeman, Morotai: A Memoir of War, p. 35.
80 Llewellyn, My Time in the Military, p. 15.
81 Boeman, Morotai, p. 30, describes this method of gunnery practice.
82 W. Pruett Interview: August 30, 2002
83 Form 5: 2/44.
84 V Mansir Interview: 4/12/2003
85 MAAF Chronology, p. 31. The XB-32 was being developed as insurance if the B-29 program failed. Lt. Col. Osmond J. Ritland would conduct eight test flights of the XB-32 over Muroc over the next couple weeks.
86 539th BS History, 2/1/44-2/29/44.
87 Boeman, Morotai: A Chronicle of War, p. 31.
88 Form 5: 2/44.
89 Ambrose, pp. 99-100.
90 MAAF Chronology, p. 31. These tests were of two different concrete filled bomb facsimiles. The two shapes were code named Thin Man and Fat Man. 24 models were drop-tested for flight stability, ballistics characteristics, fusing equipment, and suitability of the B-29 as a delivery aircraft. The test results at Muroc led to the final design of the shape of the first operational atomic bombs.
91 Form 5: 2/44.
92 Llewellyn, My Time in the Military, p. 16. A posting by Mr. John Dill of Westlake, Ohio on the www.armyairfoces.com forum dated November 22, 2000 indicated that the plane Lorian Llewellyn was lost on was a B-24-D, S/N 41-23938, piloted by Lt. Charles Hopkins, with co-pilot Lt. John Muncy, navigator Lt. William P. Rowe, Jr., bombardier Lt. Frank Washburn, flight engineer T/Sgt. Lorian Llewellyn, radio operator Sgt. Eugene Gurzenda, and gunners Sgt. Alan Hibbert, Sgt. Orvel Estes, and Sgt. Sam Belfiore. The plane was reported over the target, Wotje Island, at 0230 on 1/20/44, and last heard from at 0315. An undated newspaper clipping from the Llewellyn family collection ('Lorian Llewellyn Presumed Dead, Parents Informed') indicated that the War Department informed the Llewellyn family that Lorian was officially declared dead, two years after he was first reported missing. T/Sgt. Lorian A. Llewellyn is memorialized at an Air Force memorial located in Honolulu, Hawaii.
93 Form 5: 2/44
94 539th BS History, 3/44.
95 V. Mansir Interview, 4/12/2003.
96 The Earl Carroll Theater/Supper Club
97 Aquarius Theater. In the 1960's the Earl Carroll Theater was used as a venue for rock concerts. In 1968 it was redesigned and reopened as the Aquarius Theater.
98 The date of Walt's visit to the Earl Carroll Theater is based on the date of the souvenir photo showing Walt, Len Sherman, and another unknown man at the club (Babinski Family Collection).
99 Clark, Missions of Shehasta, pp. 32-34.
100 Clark, Missions of Shehasta, pp. 27-28 & 56.
101 Form 5: 3/44.
102 539th BS History, 3/1/44-3/31/44.
103 WW2 at Muroc. The men training at Muroc did not know which theater they would be assigned to after they completed phase training. However, unbeknownst to the men, the stationary Sperry ball turret training machines were special tropical climate models.
104 Clark, Missions of Shehasta, pp. 44-47.
105 539th BS History, 3/1/44-3/31/44.
106 Form 5: 3/44.
107 Fielding Interview: August 26, 2002. Dunne developed a reputation with pilots as an excellent navigator.
108 Form 5: 3/44.
109 539th BS History, 3/1/44-3/31/44.
110 Muroc Mirage, 3/11/44.
111 Date based on Llewellyn Familiy Collection photo of Walt Babinski and Clif Llewellyn at Muroc and dated 3/11/44.
112 V. Mansir Interview, 4/12/2003.

[113] *MAAF Chronology*, pp. 31-32. This was the beginning of a two month study conducted by Lt. Col. W. Randolph Lovelace, Chief of the Aero Medical Laboratory.
[114] *539th BS History*, 3/1/44-3/31/44.
[115] *539th BS History*, 3/44.
[116] During World War II, military arircraft were assigned a seven-digit serial number, but only the last three digits were painted on the side for identification. Despite the fact that there were many duplicate three-digit numbers throughout the Air Force, it was common practice to use only the last three digits for reporting purposes.
[117] Photo from Babinski Family collection.
[118] *539th BS History*, 3/44.
[119] Form 5: 3/44.
[120] Form 5: 3/44. This ended Walt's training flying at Muroc as part of the 539th HBC training program. He totaled 70 hours and 50 minutes of training flying time, and about 160 hours of ground school instruction time while at the base.
[121] Paul Journal.
[122] Form 230.
[123] V. Mansir Interview, 4/12/2003. The women drove east to drop off Helen and Dennis Bretherton with family, while Virgie and Paul returned to North Carolina with Nancy to stay a while with the Riley family.
[124] *539th BS History* 3/44. Two days later 15 additional crews were ordered out for shipment overseas, followed by ten crews on the 29th and eight crews on the 30th of March. At the time the 539th Squadron was deactivated on March 31, 1944, eight additional crews were still being held at Muroc to complete required flying hours or for other reasons.
[125] Form 5: 3/44
[126] *Air Force Combat Units*, p. 645.
[127] Ambrose, *The Wild Blue*, p. 102.
[128] PB&TR, Form 5: 3/44.
[129] *San Francisco: The Bay and Its Cities*, p. 445
[130] Hamilton Air Force Base.
[131] Hemphill, *Aerial Gunner from Virginia*, pp. 267 & 277.
[132] *San Francisco: The Bay and Its Cities*, pp. 436-445. Decades later on a visit to California, Walt reminisced with the author about the bus ride from Hamilton Field to San Francisco, and the precise route to the Golden Gate Bridge.
[133] Hemphill, *Aerial Gunner from Virginia*, p. 282.
[134] PB&TR
[135] Leiser, *A History of Travis Air Force Base*, p. 6.
[136] Leiser, *History of Travis Air Force Base*, pp. 1-7. After World War Two, Fairfield-Suisun was renamed Travis Air Force Base and continues in operation today as one of the major facilities of the Air Force's Military Air Transport Service (MATS).
[137] Boeman, *Morotai: A memoir of War*, pp. 37-38.
[138] Walt's dog tags were inscribed on three lines: The first was his name: W A BABINSKI. The second line listed his US Army Air Corps serial number, 365595000, and the codes: T43 44 a. This code indicated that he had received his first tetanus inoculation in 1943 with a tetanus toxoid booster in 1944, and his blood type 'A.' On the last line was a single letter 'C,' which indicated that he was a Catholic. By 1943 US Army 'dog tags' had been standardized to be issued in sets of two. They were to be worn on two metal necklaces – the first 28" long and the second suspended from a 6" necklace suspended from the first. The dog tag itself was stainless steel, 2" by 1-1/8" is size with a small notch on the outer left side and a 1/8" dia hole on the right side for the necklace. Serial number beginning with the digit 3, indicated the man was drafted. Serial numbers in the range 36,000,000 – 36,999,999 indicated the man had been drafted from the Second Army, Sixth Army Corps area which included Illinois, Michigan, and Wisconsin, with Headquarters at Fort Sheridan, Illinois. US Army regulations stated that dog tags were considered part of the uniform and men were to wear dog tags at all times, except when required for personal hygiene. In case of battlefield death, regulations stated that one of the two tags was to be interred with the deceased, and the second attached to the grave marker by means of nails through the hole and notch, for identification until a permanent grave marker was provided. Source: *US Army WW2 Dog Tags*.
[139] Llewellyn, *My Time in the Military*, p. 17.
[140] Form 28
[141] PB&TR
[142] Form 28
[143] Leiser, *A History of Travis Air Force Base*, p. 6.
[144] The distance from Hamilton Field to Hickam Field is approximately 2,399 miles.

Chapter 8: Southwest Pacific 1941-1944: How Long and What Cost?
[1] Rust and Bell, *Thiteenth Air Force Story*, pp. 4-5, 15-16.
[2] Lippincott, *From Fiji through the Philippines with the 13th Air Force*, p. 91.
[3] 307th Bomb Group (HV) Association: Reunion Book No. 12, "Short History of the Thirteenth Air Force," pp. 66-68.

Chapter 9: Over the Isles We Fly: Guadalcanal & the 13th Air Force
[1] From Moore, *Song of the Thirteenth Air Force*.
[2] PB&TR. Temperatures at Fairfield-Suisun at this time of year ranged from highs of about 68 to lows just under 50.
[3] Boeman, *Morotai*, p. 55.
[4] Boeman, *Morotai*, p. 56.
[5] *Hickam Field Guide*, pp. 9-10. Temperatures at Hickam in April would range from an average of 82 during the day to about 67 at night. There is little rain on Oahu in April.
[6] *Hickam Field: Sand Dunes to Sonic Booms*
[7] *Hickam Field Guide*, pp. 12-15, 21-23.
[8] NPRC-30: An IBM 'Morning Report Locater Card' recorded Walt as being at Hickam AFB on April 8, 1944.
[9] Norris, *Texas Kate*, May 31, 1944.
[10] Llewellyn, *My Time in the Military*, p. 21.
[11] Roth Interview.
[12] Coffey, *Always Home*, p. 6. The Honolulu USO used as much as a ton of bananas and 250 gallons of ice cream each day to meet the demand for their banana splits.

[13] *Walt and Jack apparently did not know it at the time, but the rest of the McRae crew had arrived at Hickam Field late the previous day (April 8) after the long flight from Fairfiled-Suisun. The enlisted men on the crew were exhausted, because they had spent a long, raucous night in Sacramento, arriving back at Fairfield-Suisun late, and then finding out that they were on alert to leave for Hawaii in just a few hours. They had arrived at Hickam in B-24J 075 at 6:08 p.m. on Saturday, while Walt and Jack were still in Honolulu. On Sunday, they apparently left for Honolulu while Jack and Walt were still finishing breakfast after Easter mass. Llewellyn, My Time in the Military, pp. 17-20.*

[14] *Hickam Field Guide, pp. 9-12.*

[15] *PB&TR*

[16] *Canton. The island is now part of the Republic of Kiribati. Average high temperatures in April when Walarrive are about 90 degrees.*

[17] *Llewellyn, My Time in the Military, pp. 21-22.*

[18] *PB&TR states that the next destination for Walt was Ndeni Island in the Santa Cruz Group. But the author has been able to find very little information about Ndeni. Before and during World War Two it was part of the British Solomon Islands. No reference has been found to a U.S. base or airfield located on the island. The sparsely populated island is now called Nendo and forms part of the nation of the Solomon Islands. Another possibility is that the airplane did not land at Ndeni, but simply used it as a way-point, continuing on directly to Guadalcanal. Flying from Canton to Ndeni would be 1,633 miles, and then a further 250 miles to Guadalcanal. The author believes that the PB&TR is in error. Walt bought a map of the Pacific in Australia. It shows Ndeni in the Santa Cruz Islands, but Nandi on Fiji is not shown. It is likely that Walt or whoever typed the PB&TR looked at the map, saw the name Ndeni, and confused it with Nandi. Note that many contemporary narratives refer to flying from Canton to Guadalcanal via Nandi, Fiji (For example,, Cooper Interview: August 24, 2006) The Pacific War Online Encyclopedia states that an attempt to build an airfield at Ndeni was abandoned due to difficult terrain and a the presence of a vicious strain of malaria. (http://pwencycl.kgbudge.com/N/d/Ndeni.htm).*

[19] *Lippincott, From Fiji Through the Philippines, p. 1. United States Army Air Field Nandi had been originally started in 1939 by the New Zealand government. Fiji was a British colony, but New Zealand had become active in Pacific Ocean defense planning. In 1942 the USAAF expanded the base and stationed a medium bomber squadron there for defense against a possible Japanese attack. Many Thirteenth Air Force units had been based on Fiji, until moving up the the Solomon Islands in 1943.*

[20] *PB&TR. There is some error in the record of dates that cannot be resolved. His Personal Base and Transit Record indicates that Sgt. Babinski arrived at Canton on April 10 and at Nandi (or Ndeni) and Guadalcanal on April 11. However, when he traveled westward from Canton to Ndeni, Sgt. Babinski crossed the International Date Line, thereby loosing a day. If he left Canton on the 11th, his arrival at Nandi and Guadalcanal must have been on the 12th. If he arrived at Nandi and Guadalcanal on the 11th, his departure from Canton must have been on the 10th.*

[21] *Paul Journal, p. 7-8.*

[22] *Form 5: 4/44.*

[23] *NPRC-31. A Morning Report Locator Card confirmed that Sgt. Babinski was assigned to the XIII Bomber Command at Guadalcanal on April 15, 1944.*

[24] *Hemphill, Aerial Gunner From Virginia, pp. 293-303. These are excerpts written during October 1944 by Corporal Don Moody, aerial gunner with the 307th Bomb Group, to his family in Virginia. Mail to men assigned to the 13th Air Force was addressed to A.P.O. 719 c/o Postmaster, San Francisco, California.*

[25] *Lippincott, From Fiji Through the Philippines, p. 45.*

[26] *Jurkens Interview: August 25, 2006*

[27] *Harvey & Hamilton, We'll Say Goodbye, pp. 32-48.*

[28] *Aubrey Interview, 8/15/2002.*

[29] *Llewellyn, My Time in the Military, pp. 21-22.*

[30] *Boeman, Morotai, p. 73.*

[31] *Boeman, Morotai, p. 74.*

[32] *Britt, The Long Rangers, p. 75.*

[33] *Air Force 5/44 BRA, pp. 2-5.*

[34] *Air Force 5/44: BCTSP, p. 1.*

[35] *Johnson: 'Greetings from Earl Johnson' 307BGN: April 2004*

[36] *Boeman, Morotai, pp. 75-76.*

[37] *Air Force 5/44: BCTSP, p. 2.*

[38] *Adrian Letter, 9/13/2004.*

[39] *Britt, The Long Rangers, p. 86.*

[40] *Britt, The Long Rangers, p. 92.*

[41] *Air Force 5/44: BCTSP, p. 2.*

[42] *Harvey & Hamilton, We'll Say Goodbye, p. 62.*

[43] *Form 5: 4/44. Walt's XIII Bomber Command individual flight record report for April 1944 stated 'No time accomplished at this station for the month of April.'*

[44] *Llewellyn, My Time in the Military, p. 24.*

[45] *Smith, The Pacific Crusaders, pp. 108-109.*

[46] *Llewellyn Interview, 10/27/2002.*

[47] *Air Force 5/44: BCTSP, p. 2.*

[48] *Form 5: 5/44.*

[49] *Air Force 5/44: BCTSP, pp. 2-3.*

[50] *Britt, The Long Rangers, pp. 97-101.*

[51] *Air Force 5/44: BCTSP, p. 3.*

[52] *Form 5: 5/44.*

[53] *Llewellyn, My Time in the Military, p. 23.*

[54] *Form 5: 5/44.*

[55] *Llewellyn, My Time in the Military, p. 23.*

[56] *FDFR*

[57] *Aubrey Interview: August 15, 2002. Gregory 'Pappy' Boyington was the flamboyant leader of a US Marine Corsair Fighter Squadron known as the 'Black Sheep Squadron.'*

[58] *Form 5: 5/44.*

59 Britt, *The Long Rangers*, p. 102.

Chapter 10: Los Negros: The 307th Bombardment Group (Heavy)

1 Britt, *The Long Rangers*, pp. 2-4.
2 *Air Force Combat Units of WW II*, pp. 181-182.
3 424th BS History, 3/25/44, Roll A0610, pp. 0608-0609.
4 424th BS History, 6/5/44, Roll A0610, p. 0628.
5 *Finschhafen to Hollandia – On the Road Back*.
6 Lippincott, *From Fiji Through the Philippines*, p.91.
7 Lippincott, *From Fiji Through the Philippines*, p.93.
8 Gideon, *Airdromes Guide Southwest Pacific Area*, p. 27.
9 Harvey & Hamilton, *We'll Say Goodbye*, pp. 62-63.
10 424th BS History, Roll A0610, pp. 0629-0630.
11 307th BG History, pp. 1594-1595.
12 FDFR.
13 Baross, *The War Years WWII*, p. 28.
14 424th BS History, 6/5/44, Roll a0610, pp. 0628-0629.
15 Based on Llewellyn, *My Time in the Military*, p. 24.
16 SO56
17 424th BS History, Roll A0610, pp. 0637-0638.
18 Britt, *The Long Rangers*, p. 104. The pilot was Lt. John R. Sawyer, killed instantly by 20 mm cannon fire. His navigator, Lt. Phillip J. Heimlick, was wounded but died shortly after the plane returned to Los Negros, piloted by the copilot, Lt. William E. Way, who was also wounded.
19 307th BG Consolidated Mission Report 307-274, June 3, 1944.
20 Weinberg, *A World at Arms*, pp. 648-649.
21 Tourtellot, *Life's Picture History of World War Two*, p. 210.
22 Britt, *The Long Rangers*, p. 106; Clark, *Missions of Shehasta*, p. 157.
23 H. Babinski Letter: May 8, 2002
24 424th BS History, Roll A0610, p. 0638.
25 307th BG Consolidated Mission Report 307-274, June 3, 1944.
26 Reunion Book 7, p. 111.
27 307th BG Consolidated History – June 1 – June 30, 1944, dated July 4, 1944.
28 307th BG Consolidated Mission Report 307-275, June 7, 1944.
29 424th BS History, Roll A0610, p. 0638, CCUSAAF, FDFR. Both Walt's and Clif Llewellyn's Foreign Duty Flight Records indicate a flight of 6:35 with credit for a strike against Truk, despite the fact that the 424th BS history indicates that all their planes turned back due to the bad weather. PFMR indicates that the intended target was Eten Island, while Clif Llewellyn's records indicate that the target was Dublon Town.
30 307th BG Consolidated Mission Report 307-277, June 9, 1944.
31 FDFR; 424th BS History, Roll A0610, p. 0638.
32 Ulmer, *Waist Gunner*, p. 54.
33 Barros, *The War Years*, p. 65.
34 Britt, *The Long Rangers*, p. 102.
35 Boylston, *WWII South Pacific Letters*, p. 40.
36 Llewellyn, *My Time in the Military*, pp. 23-24.
37 Lippincott, *From Fiji Through the Philippines*, p. 97.
38 FDFR, 424th BS History, Roll A0610, p. 638. FDFR indicates the original Dublon Town target, but the 424th BS History describes the change to the secondary target due to clouds.
39 Volchko seemed to fly with Lt. Jordan and Lt. Randolph's crew frequently, based on later records in 424th BS History.
40 Ulmer, *Waist Gunner*, p. 46.
41 307th BG Consolidated Mission Report 307-285, June 17, 1944.
42 307th BG Consolidated Mission Report 307-285, June 17, 1944.
43 424th BS History, Roll A0610, p. 0639; Britt, *The Long Rangers*, pp. 103 & 107; FDFR.
44 PFMR
45 FDFR
46 FDFR, PFMR, 424th BS History, Roll A0610, p. 0639.
47 SO64, NPRC Fragments 15 & 17.
48 Ranfranz, *Coleman Crew*.
49 307th BG Consolidated Mission Report 307-290, June 23, 1944.
50 FDFR, PFMR, 424th BS History, Roll A0610, p. 0640.
51 Britt, *The Long Rangers*, p. 109.
52 MOY
53 Ulmer, *Waist Gunner*, p. 56.
54 Britt, *The Long Rangers*, pp. 109-110.
55 FDFR, PFMR, 424th BS History, Roll A0610, pp. 0640-0641.
56 MOY
57 Britt, *The Long Rangers*, p. 112.
58 307th BG Consolidated Mission Report 307-296, June 30, 1944.
59 307th BG Consolidated Mission Report 307-296, June 30, 1944.
60 Britt, *The Long Rangers*, pp111-112, FDFR, PFMR, 424th BS History, Roll A0610, p. 0641. There is some discrepancy between Britt and the official 424th BS History. The 424th BS history states that there was no opposition, while Britt describes the scores by Kaestner's crew and the hit on Balovitch's plane.
61 424th BS History, Roll A0610, p. 0643.
62 307th BG Consolidated History – June 1 – June 30, 1944, dated July 4, 1944.
63 Boeman, *Morotai*, pp. 184-185.

64 Britt, *The Long Rangers*, p. 114.
65 424th BS History, Roll A0610, p. 0654.
66 307th BG Consolidated Mission Report 307-298, July 3, 1944.
67 424th BS History, Roll A0610, page. 0655.
68 Britt, *The Long Rangers*, pp. 112-113.
69 FDFR. The date of this photo is assumed from the evidence that the crew flew in A/C 540 the day before.
70 FDFR. 424th BS History, Roll A0610, page 0656.
71 Britt, *The Long Rangers*, pp. 114-115.
72 424th BS History, Roll A0610, p. 0662.
73 Britt, *The Long Rangers*, p. 112.
74 For A/A fire reported as 'heavy, slight, innacurate'; 'heavy' indicated that the fire was from heavy – or large caliber A/A guns; 'slight' indicated that there was only a slight rate of A/A fire, and 'innacurate' indicated that the A/A fire was off-target. This three-part pattern of reporting A/A fire was a standard in 424th BS mission reports.
75 424th BS History, Roll A0610, p. 0657.
76 Britt, *The Long Rangers*, pp. 116-118.
77 FDFR.
78 Britt, *The Long Rangers*, pp. 119-120.
79 Based on Babinski family reminiscences. The exact date of Walt's injury is not certain, but the mid-July timeframe is likely, considering the 3-1/2 weeks with no combat missions that occurred during this period.
80 Coggins Interview: August 25, 2006
81 WAB to DFK V-Mail from Admiralties.
82 424th BS History, Roll A0610, p. 0661-0662.
83 Harvey & Hamilton, *We'll Say Goodbye*, p. 67.
84 NPRC-13.
85 424th BS History, Roll A0610, p. 0665.
86 424th BS History, Roll A0610, P. 0674.
87 Hylton, *Vanished*, pp. 174-175.
88 J. Vanderpoel Interview: August 19, 2004. Hepfer later became an Air Force Major General. He ended his career commanding the Ballistic Missile Office, Air Force Systems Command, Norton Air Force Base, Calif. He retired from the Air Force in 1980 and died in 1997.
89 424th BS History, August 6, 1944.
90 Hidalgo Interview: April 14, 2002
91 Schreiner letter 5/16/02
92 424th BS History, Roll A0610, pp. 0705-0706.
93 424th BS history, 8/6/44.
94 Dowdie, *Charlie*, pp. 345-346.
95 PFMR.
96 424th BS History, Roll A0610, pp. 0705-0706.
97 Campora was remembered as a 'fat dago' and not very likable by other crewmembers (Schreiner Interview: August 30, 2002).
98 424th BS History, Roll A0610, p. 0713. Lt. Edmund G. Ginnann later questioned Hidalgo and other crewmen at length about this radio transmission. No one else over Yap that day reported hearing it, although it was agreed that there was heavy VHF radio traffic during the bomb run and many garbled messages.
99 PFMR.
100 424th BS History, 8/10/44.
101 Britt, *The Long Rangers*, p. 123.
102 NPRC-17.

Chapter 11: Wakde: Thanks for the Memories?
1 Pacific Wrecks, 'Wakde' - The P47's were from the 348th Fighter Group.
2 FDFR.
3 Gideon, Francis C., *Airdromes Guide Southwest Pacific Area*, p. 74.
4 *Advancing the Bomber Line – Hollandia to Morotai*
5 Britt, *The Long Rangers*, p. 124.
6 Baross, *The War Years WWII*, p. 33.
7 Llewellyn, *My Time in the Military*, FDFR.
8 424th BS History, Roll A0610, p. 0765.
9 Baross, *The War Years WWII*, p. 44.
10 Based on photos and map depicted in Reunion Book #10, pp. 105-107.
11 Baross, *The War Years*, p. 65.
12 Britt, *The Long Rangers*, pp. 125-126.
13 307th BS History, A0610, p. 0679.
14 Reunion Book 10, p. 105.
15 Llewellyn, *My Time in the Military*, p. 25.
16 Harvey, *We'll Say Goodbye*, pp. 80-82. Bob Hope had begun his USO tours in 1942, taking Langford, Romano, and Colona to U.S. Bases in Alaska. In 1943 Hope, Langford, Romano, and Clark Gable toured England, Africa, Sicily, and Iceland. His 1944 Pacific tour would include Hawaii, Eniwetok Atoll, Milne Bay, the Treasury Islands, many stops in and along New Guinea, and Biak. Judith Anderson was knighted by Queen Elizabeth in 1960. Jerry Colonna accompanied Hope on many more USO tours. He died in 1986. Frances Langford made movies into the 1950's and many TV appearances into the late 1980's. She died in 2005. Bob Hope continued to take his USO tour to entertain the troops for many more decades.
17 Ulmer, *Waist Gunner*, pp. 75-76.
18 Pelle Interview: August 24, 2006
19 Reunion Book 13, p. 93.
20 Wright Diary, 8/25/44.
21 Britt, *The Long Rangers*, pp. 127-128.

[22] *Reunion Book 7, p. 151.*
[23] *Hylton, Vanished, p. 169.*
[24] The description of preparations for this mission is taken from Llewellyn, *My Time in the Military*, pp. 26-27. Llewellyn describes this this mission for August 28 or September 1 stating that Walt was on this crew. This is in error. However I have utilized his description for the August 29 mission.
[25] *Llewellyn, My Time in the Military, p. 29.*
[26] *424th BS History, 8/29/44.*
[27] *FDFR*
[28] *424th BS History, image 1010.*
[29] *Hometowns from National WWII Memorial Database.*
[30] *R. Reis Letter: May 1, 2002.* This was Richard (Dick) Reis, who Walt would fly with later in December.
[31] *424th BS History, Roll A0610, p. 0754.*
[32] *Tuck's Diary, March 8, 1944.*
[33] *Britt, The Long Rangers, p.p. 130-131.*
[34] *424th BS History, Roll A0610, p. 0679.*
[35] *Baross, The War Years, p. 65.*
[36] *PB&TR.*
[37] *424th BS History, Roll A0610, pp. 753 & 755.*
[38] *424th BS History, Roll A0610, p. 752.* Through September 1944, the 424th Squadron included one C-47A, in addition to its complement of B24 D's and J's. At the end of September, the C-47 was transferred to the XIII Bomber Command.
[39] *Boeman, Morotai, p. 72.*
[40] *Paul Journal, p. 8*
[41] *Boeman, Morotai, p. 73.*
[42] Cliff Llewellyn said that Walt '…liked to play cards for relaxation….but he never 'got in too deep. Walt would take $5 or $10 or $15 to a game and put it on the table. If he lost it – OK, he would quit. If he won though, after an hour or so, Walt would take his original $15 (or whatever he started with) and put that back in his pocket and never touch that again. Then he would continue playing…either winning and continuing to play, or losing what he had left on the table (then quitting, with the original $15 safe). Walt was prudent with money that way.' *C. Llewellyn Interview: October 27, 2002*
[43] *DAT, p. 9.*
[44] *Roth Interview, p. 9.*
[45] *PB&TR.*
[46] *Lippincott, With the Thirteenth Air Force, p. 137.* 13th Air Force personnel used Sydney for air crew rest leave beginning in July 1944. The program ended in March 1945.
[47] *Paul, pp. 12-13.*
[48] *JJB: Form 9*
[49] *C. Babinski Letter, 4/24/02.*
[50] *Paul, pp. 12-13.*
[51] *Wikipedia, Luna Park, see:* https://en.wikipedia.org/wiki/Luna_Park_Sydney, accessed July 5, 2016. The influx of servicemen drew prostitutes to the area and large-scale brawls were a common occurrence - usually between Australian home defence troops and American sailors on shore leave.
[52] *Lippincott, With the Thirteenth Air Force, p. 141.*
[53] *JJB: PT&BR.*
[54] *Lippincott, With the Thirteenth Air Force, pp. 109-111.*
[55] *JJB: Form 16-26357.* John had received an excellent rating during his most recent fitness report – with '…Proficiency in Rating 3.9; Mechanical ability 3.9; ability as leader of men 3.8, conduct 4.0' (JJB: Form 4-6111).
[56] The Polish Contribution to this effort, known as Operation Market-Garden, was the 1st (Polish) Independent Parachute Brigade. Lt. Julian Gebolys, Cora Babinski's cousin, had helped train this Brigade in parachute techniques at RAF Ringway near Manchester, and then at Upper Largo, in Scotland.
[57] *PB&TR*

Chapter 12: Noemfoor: The Damned 13th Air Force Strikes Again

[1] By the end of September, the 424th Squadron C-47 had been transferred to XIII Bomber Command. *424th BS History, Roll A0610, p. 0752.*
[2] *424th BS History, Roll A0610, p. 0753.*
[3] *Llewellyn NCRM.*
[4] *Llewellyn NCRM*
[5] *PB&TR*
[6] *Britt, The Long Rangers, pp. 132-33.*
[7] *Gideon, Airdromes Guide Southwest Pacific Area, p. 83.*
[8] *Hylton, Vanished, pp. 73-74.*
[9] *Hylton, Vanished, pp. 76-78.*
[10] *Fielding Interview: August 26, 2002.* According to Jim Fielding, good competent pilots like him could always spot 'poor pilots' – men they would not want to fly with – and they could always figure out some way to avoid flying with these losers if they were worried about their competence. But the enlisted men in the rest of the crew could not. The enlisted men "…just had to put their blind faith in the two guys doing the flying."
[11] *Britt, The Long Rangers, p. 133.*
[12] *Dowdy, Charlie, p. 55.*
[13] *424th BS History, image 1010.*
[14] *Britt, The Long Rangers, p. 135.*
[15] *Griffith, MacArthur's Airman, pp. 180-183.*
[16] *Reprinted in* <u>307th Bomb Group (HV) Association: Reunion Book No. 11</u>, pp. 92-96.
[17] *Britt, The Long Rangers, p. 134.*
[18] *Llewellyn NCRM*
[19] *Llewellyn, My Time in the Military, pp. 35-36.*

[20] *30 Sept. 1944, reproduced in Reunion Book No. 12, p. 73.*
[21] *Britt, The Long Rangers, pp. 137-138.*
[22] *424th BS History, Roll A0610, p. 0918.*
[23] *424th BS History, Roll A0610, p. 0919.*
[24] *Britt, The Long Rangers, p. 137.*
[25] *424th BS History, Roll A0610, p. 0917.*
[26] *Pelle Interview: August 24, 2006*
[27] *Britt, The Long Rangers, p. 148.*
[28] *307th Bomb Group History, October 1944, frame 0078.*
[29] *Griffith, MacArthur's Airman, p. 183.*
[30] *John Vanderpoel told me that Coorssen had 'lost his nerve and considered him to be a coward.' J. Vanderpoel Interview: August 19, 2004*
[31] *424th BS History, Roll A0610, p. 0907.*
[32] *Combat Box.*
[33] *Griffith, MacArthur's Airman, p. 184.*
[34] *Britt, The Long Rangers, p. 149.*
[35] *307th Bomb Group History, October 1944, frame 0078.*
[36] *307th BG History Mission Plan 10 October, p. 4, image 0092. The 424th BS planes were instructed to take off at 65 second intervals, beginning at 3:42 am.*
[37] *424th BS History, Roll A0610, p. 0934.*
[38] *424th BS History, Roll A0610, p. 0932-34.*
[39] *In April 2013 the author visited Morotai. Wama airstrip is now the main highway through Daruba, the capitol city of Morotai. Ptoe airstrip is now an Indonesian Air Force base.*
[40] *Llewellyn, My Time in the Military, p. B.8.*
[41] *424th BS History, Roll A0610, p. 0934 reports McRea's detour with A/C 273 from Balikpapan to Morotai on 10/10/44 and their return to Noemfoor the following day. Walt's records (FDFR and PFMR) do not mention this detour.*
[42] *307th Bomb Group History, October 1944, frame 0078.*
[43] *Dowdy, Charlie, pp. 54-57.*
[44] *Britt, The Long Rangers, p. 151.*
[45] *424th BS History, Roll A0610, pp. 0943-0946*
[46] *Britt, The Long Rangers, p. 152.*
[47] *Faulkner, War at Sea, pp. 242-243.*
[48] *Britt, The Long Rangers, p. 152.*
[49] *Aerial Gunner from Virginia, p. 306.*
[50] *Britt, The Long Rangers, pp. 152-153.*
[51] *Faulkner, War at Sea, pp. 242-243.*
[52] *Britt, The Long Rangers, p. 153.*
[53] *424th BS History, Roll A0610, pp. 0943-0946, and Schreiner Interview: August 30, 2002 for the entire mission of A/C 951 on October 24, 1944. Some details between these two sources do not agree. The 424th BS History does not mention the attack on the fighter strip, which Schreiner described as being on Borneo. But he was quite certain that the mission date was October 24, 1944, because his records confirm the 18 hour 40 minute duration of the flight. Both sources agree that 951 returned to Morotai, not Noemfoor.*
[54] *Llewellyn, My Time in the Military, p. B.8.*
[55] *Britt, The Long Rangers, pp. 155-157.*
[56] *424th BS History, image 1010.*
[57] *424th BS History, Roll A0610, p. 0913.*
[58] *FDFR*
[59] *FDFR, PFMR, 424th BS History, images 1019-1027.*
[60] *Dowdy, Charlie, p. 361.*
[61] *424th BS History, images 1026-1030. Jack Riley and the entire crew was later credited with the kill [GO 190]. Note that GO 190 lists a different composition for this crew, than the 424th BS Mission History. GO 190 lists the crew as: Appling, Aubrey, Dunne, Clark, Bretherton, Hidalgo, Schreiner, Babinski, Mansir, and Riley.*
[62] *FDFR, PFMR*
[63] *Dowdy, Charlie, p. 121.*
[64] *424th BS Mission History, image 1012.*
[65] *FDFR, PFMR give the date as November 10. But Walt's PB&TR gives the arrival date as November 11. 424th BS History, image 1011 states that the squadron move was completed on November 10.*

Chapter 13: Back Home 1943-1945: Beyond a Wide and Spacious Sea

[1] *The Detroit News, January 27, 1943.*
[2] *The Detroit News, May 23, 1943.*
[3] *The Detroit News, September 8, 1943.*
[4] *The Detroit News, Decmebr 8, 1943.*
[5] *Baime, Arsenal of Democracy, pp. 258-266.*
[6] *DK to GB, 8/2/43*
[7] *G. Mendryga Interview: November 21, 2003*
[8] *AB to GB, 8/15/44*
[9] *G. Mendryga Interview: November 21, 2003*
[10] *MOY*
[11] *EIO*
[12] *Air Force: MB*
[13] *G. Mendryga Interview: November 21, 2003*
[14] *V Mansir Interview: April 12, 2003*
[15] *JSJ to Mrs. TTTK, 2/1/43.*
[16] *JEP to Mr. & Mrs. Kelly, 4/12/43*

17 JEP to Mr. & Mrs. Kelly, 11/3/43. Joe Prus must have had little confidence in speedy delivery of this 'Seaso's Greetings' V-mail, as it was written at the beginning of November.
18 JS to Mr. & Mrs. Kurzyniec, 12/10/43. In December 1943 the USMC 3rd Tank Battalion was still involved in the Battle of Bougainville, back in the Solomon Islands. This campaign would drag in to 1945, although 3rd Tank Battalion would be redeployed to support the invasions of Guam and Iwo Jima.
19 WAB to MK V-Mail 'Japs Have Surrendered'
20 WAB to DFK V-Mail 'Soldiers Dream'
21 The Detroit News, June 6, 1944.
22 The Detroit News, June 15, 1944.
23 The Detroit News, July 21, 1944.
24 The Detroit News, September 21, 1944.
25 The Detroit News, October 20, 1944.
26 JJB: Form 16-26357
27 JJB: PT&BR
28 JJB: Form 9
29 JJB to Mr. & Mrs. W Babinski, 1/22/45
30 JJB to G Babinski ND
31 H. Babinski Interview: August 27, 2002
32 Buckingham, Establishment of a Polish Airborne Force, pp. 246-247.
33 Kochanski, Halik, The Eagle Unbowed, p. 227.
34 D.F. Babinski Interview: August 27, 2002. After about seven years, conditions had improved and Tony and Martha Skrabut took Norman back. This devastated Thomas, who had grown to love Norman as his own son.
35 D.F. Babinski Interview: February 8, 2015
36 Stone, Detroit: The Arsenal of Democracy, pp. 14-15.
37 Form R-130
38 Form R-145
39 General information about ration books from Genealogy,com website (http://www.genealogytoday.com/guide/war-ration-books.html) accessed 12/26/2015). A question is, why did Thomas and Martha's ration books from late 1943 survive? Perhaps it was because the Kurzyniec store on Pulaski had access to many basic products acquired wholesale, that then could be used outside the ration process.
40 MLCC: TK and MLCC:MJK
41 D. F. Babinski interview, January 13, 2017.
42 Stone, Detroit: The Arsenal of Democracy, pp. 11-12.

Chapter 14:   Morotai: With Courage and Devotion to Duty
1 Or was it Saturday, November 11. PFMR and FDFR state November 10, but PB&TR gives the date as November 11.
2 Farnell, Operations on Morotai Island, pp.8-9.
3 Gideon, Francis C., Airdromes Guide Southwest Pacific Area, pp. 89-91.
4 Battle of Morotai.
5 Britt, The Long Rangers, p. 160.
6 Camp Living.
7 Llewellyn: My Time in the Military, p. 40.
8 J. Vanderpoel Interview: August 27, 2006
9 Thirteenth Air Force Story...in World War II
10 424th Bomb Squadron Mission Plan, November 18, 1944. This document was signed by Lt. McRae as Assistant Operations Officer.
11 Brit, The Long Rangers, p. 161.
12 Battle of Morotai.
13 Morotai: A Memoir of War
14 Fielding Interview: August 26, 2002
15 Combat Crew Rotation, pp. 1-2.
16 Combat Crew Rotation, pp. 5-6.
17 Combat Crew Rotation, p. 7.
18 Combat Crew Rotation, p. 8. This was the formula for bomber crews. The formula for fighter pilots was slightly different.
19 See FDRF.
20 Combat Crew Rotation, p. 14.
21 Combat Crew Rotation, pp. 8-18.
22 The author in his conversations with 307th BG vets encountered a bewildering number of theories and explanations of the rotation policy.
23 Paraphrased from Combat Crew Rotation, pp. 20-21.
24 Llewellyn, My Time in the Military, p. B.8.
25 Baross, The War Years WWII, pp. 45-47.
26 Schreiner Interview: August 30, 2002
27 Llewellyn, My Time in the Military, p. B.8.
28 424th BS History, image 1037.
29 Llewellyn, My Time in the Military, p. B.8.
30 Llewellyn, My Time in the Military.
31 424th BS History, image 1047-1049.
32 Llewellyn, My Time in the Military, p. B.9.
33 GO 138.
34 McRae was the author of the plan for this mission. See Appendix 2.
35 Lippincott, From Fiji Through the Philippines With the Thirteenth Air Force, p. 161.
36 424th BS History, Images 1010-1017. The 424th Mission report for 11/18/44 is missing, so the exact composition of the crew is uncertain except for Appling and Babinski. Other details of the mission are inferred from the 372nd BS History, Images 521-524, and Britt, The Long Rangers, pp. 168-169.

37 *Dowdy, Charlie, pp. 108-113.*
38 *Llewellyn, My Time in the Military, p. 37*
39 *Photographic Interpretation Handbook, United States Forces: Section 14 Ships and Shipping, p. 14.04.A. 'Fox Tare Dog' refers to a system of code names to help aerial observers like Walt classify ships sighted on a consistent basis. For example, code name Fox Tare Dog referred to a small freighter between 700 and 1300 gross tons, two or three masts with the number two mast being located against the deck house, and a normal cruising speed of 10 knots.*
40 *424th BG Mission History, images 1063-1065.*
41 *Battle of Morotai.*
42 *Rust, Thirteenth Air Force Story, p. 38.*
43 *Llewellyn, My Time in the Military, p. B.9.*
44 *424th Mission History, images 1069-1071.*
45 *Farnell, Operations on Morotai, p. 6.*
46 *Llewellyn, My Time in the Military, p. B.9. That night. at 0300 on November 23, 1944, four Jap planes raided Wama and the surrounding tent area – a PB4Y and two Aussie Beaufighters were destroyed, five Beaufighters damaged, three men were killed and 11 wounded.*
47 *Llewellyn, My Time in the Military, pp. 37-38 & p. B.9.*
48 *Llewellyn, My Time in the Military, p. B.9.*
49 *Dowdie, Charlie, p. 115.*
50 *Dowdie, Charlie, p. 222.*
51 *424th Squadron Mission History, images 1076-1078.*
52 *Llewellyn, My Time in the Military, p. B.9.*
53 *Dowdy in Charlie p. 116, refers to White's plane being hit, but he states that they were attacked by 15 to 20 Zekes and Jacks, but neither the 424th BS Mission History nor Britt, The Long Rangers, refer to this aerial attack.*
54 *424th Squadron Mission History, images 1084-1086.*
55 *Llewellyn, My Time in the Military, p. B.10.*
56 *Dowdy, Charlie, p. 117.*
57 *North Islet is now known as Tubbataha Reef, a Philippines nature preserve.*
58 *424th Squadron Mission History, images 1090-1092.*
59 *Llewellyn, My Time in the Military, p. B.10.*
60 *424th BS History, image 1010.*
61 *J. Vanderpoel Interview: August 19, 2004*
62 *424th BG History, image 1012-1013.*
63 *424th BG History, images 1016-1017. The November Summary History Report stated that the 424th started November with 12 B24's, then lost two and gained two, but had a total of 14 B-24's by the end of November. This must be an error. The December History states that the Squadron began with 12 A/C.*
64 *Britt, The Long Rangers, p. 173.*
65 *C. Llewellyn Interview: October 27, 2002*
66 *Aubrey Interview: August 15, 2002*
67 *Schreiner Interview: August 30, 2002*
68 *C. Llewellyn Interview: October 27, 2002. Clif couldn't remember who the crewman was –not him, because he was the only one in the crew who did not smoke.*
69 *C. Llewellyn Interview: October 27, 2002*
70 *Boylston, WWII South Pacific Letters, p. 51. Between September 15, 1944 and February 1, 1945 – 82 Jap air raids destroyed 42 U.S. planes on Morotai and damaged 33.*
71 *Britt, The Long Rangers, p. 160.*
72 *D.F. Babinski Interview: August 21, 2004*
73 *C. Llewellyn Interview: October 27, 2002*
74 *Reis Interview: August 20, 2004*
75 *Britt, The Long Rangers, p. 175.*
76 *Llewellyn, My Time in the Military, p. 40.*
77 *Llewellyn, My Time in the Military, p. 40.*
78 *Reis Interview: August 20, 2004*
79 *307th Bomb Group (HV) Association: Reunion Book No. 14, p. 91.*
80 *Fielding Interview: August 26, 2002*
81 *Norwood Interview: August 24, 2006*
82 *Robinson Interview: August 25, 2006*
83 *Walsh Interview: August 24, 2006*
84 *Fielding Interview: August 26, 2002*
85 *J. Vanderpoel Interview: August 29, 2002*
86 *Reis Interview: August 20, 2004. Dick Reis told the author about the Guild crew . One day back at March field during their bomber phase training, Guild called all the enlisted men into the officers quarters and told them that every man was going to get a nickname and that there would be no saluting or calling officers Lt., etc. Guild became 'Smokey', Vern Michel was 'Sam' (from 'Marryin Sam in Lil Abner – because he was 26 years old with a wife and kid), Michel was Guild's co-pilot, Reis was the youngest man in the crew – hence he was named 'Junior.*
87 *J. Vanderpoel Interview: August 19, 2004*
88 *Guild Letter, 4/24/02*
89 *Personal reminiscence from Walter Babinski, recounted to the author.*
90 *Ayala Interview: August 25, 2006.*
91 *Britt, The Long Rangers, pp. 173-174.*
92 *Coggins Interview: August 25, 2006*
93 *424th BS Mission History, images 1207-1209.*
94 *Britt, The Long Rangers, p. 176.*
95 *424th BS Mission History, images 1207-1209.*

96 *424th BS Mission History, image 1196, and Britt, The Long Rangers, pp. 176-177.*
97 *GO190.*
98 *Ulmer, Waist Gunner, p. 104.*
99 *GO10*
100 *Dowdy, Charlie, p. 119.*
101 *424th BS Mission History, images 1217-1220.*
102 *424th BS Mission History, images 1225-1229.*
103 *424th BS Mission History, image 1217.*
104 *424th BS Mission History, image 1225*
105 *Llewellyn, My Time in the Military, p. B.10.*
106 *Streett Letter, 12/7/44. Each man, including Walt, had a personally addressed copy of this letter put in his file.*
107 *Llewellyn, My Time in the Military, p. B.10.*
108 *424th BS Mission History, images 1235-1237. Dowdie, Charlie, pp. 126 & 223 gives conflicting information on this mission. He states that there was a Japanese fighter attack over La Carlota, but this is not mentioned by the official history or by Britt.*
109 *424th BS Mission History, images 1267-1269.*
110 *Appling's return is supposed because the December 11 mission is the last he is recorded on.*
111 *Rust, Thirteenth Air Force Story, p. 4. The 419th Night Fighter Squadron was formed in November 1943. In April 1944 it was reequipped with P-61's. In December 1944 the newly formed 550th Night Fighter Squadron was added, also with P-61's.*
112 *Llewellyn, My Tie in the Military, p. 39.*
113 *R. Reis Letter: May 1, 2002*
114 *Brit, The Long Rangers, pp. 162-163.*
115 *Brit, The Long Rangers, pp. 164-165. Sgt. Augsberger would be picked up by another PB-Y Catalina and returned to Morotai on January 21, 1945.*
116 *424th BS Mission History, images 1197-1198.*
117 *Fielding Interview, August 26, 2002.*
118 *Faulkner, War at Sea, p. 256. What Walt saw was part of the U.S. invasion task force headed to Mindoro.*
119 *Bolin, Photographic Interpretation Handbook, p. 14.04A. This ship was between 8,000 and 10,000 GT.*
120 *424th BS Mission History, images 1257-1260.*
121 *Llewellyn, My Time in the Military, p. B.10.*
122 *Llewellyn, My Time in the Military, p. B.10.*
123 *424th BS Mission History, images 1269-1271.*
124 *Britt, The Long Rangers, p. 181.*
125 *Llewellyn, My Time in the Military, p. B.10.*
126 *See Brit, The Long Ranger, pp. 143-145 and Vanderpoel Interview: August 19, 2004. During the October 3 Balikpapan attack, White was wounded by 20mm shrapnel which broke his left wrist and shredded his legs. But his pilot and copilot had it worse. The pilot, Lt. James Wheeler, was wounded in the right arm and had blood oozing over his face. His copilot, Lt. Frank N. Stephany, was seriously wounded around the left cheek and left arm. The shrapnel had been fired by a Japanese kamikaze pilot who crashed his plane into another B24 after his attack on Wheeler's A/C. Stephany went into shock as White and another man dressed his wounds. Then Wheeler's right arm became paralyzed and he could no longer control the plane. He had White, the navigator, get in the copilot seat to help fly the plane. At this point the #1 engine cut out due to low oil pressure. White determined that they could not make it back to Noemfoor, so they decided to try for Morotai where there was a short fighter strip and a crash strip that they might use. As they approached Morotai they determined that their radio was not working, so they made one pass over the fighter strip and fired flares to indicate that there were wounded on board. They then made a 180° turn and with Wheeler and White at the controls, and the flight engineer, Sgt. Byrum working the flaps, they set the plane down successfully. In my conversation with John Vanderpoel, he had the highest praise and respect for Lt. John A. White IV.*
127 *424th BS Mission History, images 1283-1285.*
128 *Llewellyn, My Time in the Military, p. B.10.*
129 *Guild Letter: April 13, 2002*
130 *Location of this islet, noted on the mission report, is uncertain.*
131 *R. Reis Letter: May 1, 2002. The next day (12/25) they were picked up by PBY and returned safely to Morotai. They were believed to be a 370th or 371st crew. One of the rescued men later attended the 307th BG reunions and maintained that their lives had surely been saved by the crew that day (K. Reis Letter: May 1, 2002). Walt also recounted the story of this rescue to Babinski family (C. Babinski letter April 24, 2002.)*
132 *424th BS Mission History, images 1291-1293.*
133 *Brit, The Long Rangers, p. 182.*
134 *Ulmer, Waist Gunner, pp. 108-109.*
135 *Llewellyn, My Time in the Military, p. B.10.*
136 *In January 1945, the Japanese began to move these prisoners by forced march through the jungle. Only a small handful lived to complete what became known as the Sandakan Death March. See: https://en.wikipedia.org/wiki/Sandakan_Death_Marches.*
137 *424th BS Mission History, images 1294-1296.*
138 *424th BS Mission History, images 1201, 1202, 1291-1296.*
139 *Llewellyn, My Time in the Military, p. B.10.*
140 *Farnell, Operations on Morotai, pp. 17-18.*
141 *Llewellyn, My Time in the Military, p. 39.*
142 *Farnell, Operations on Morotai, p. 14.*
143 *Farnell, Operations on Morotai, p. 18.*
144 *Llewellyn, My Time in the Military, p. B10.*
145 *Llewellyn, My Time in the Military, p. 39.*
146 *Llewellyn, My Time in the Military, p. B10.*
147 *Ulmer, Waist Gunner, p. 111.*
148 *PFMR, FDFR. I have found no 424th Mission History images for this specific mission, although it is listed on the summary December report, 424th Mission History, image 1200. The only other crew member I am certain of is Don Aubrey. I assume that Jimmie Clark, Jack Riley, and Frank Mansir were on this mission because they flew together with Walt and Aubrey on December 25.*

[149] Rust, Thirteenth Air Force Story, p. 64.
[150] 424th BS Mission History, images 1195-1202.
[151] Llewellyn, My Time in the Military, p. B34.
[152] 424th BS Mission History, image 1459.
[153] Britt, The Long Rangers, p. 184.
[154] Farnell, Operations on Morotai, p. 22.
[155] Farnell, Operations on Morotai, pp. 25-29.
[156] GO10, NPRC-29.
[157] Form 8-117, IR, Form 28, NPRC-12.
[158] Llewellyn, My Time in the Military, p. B10.
[159] Farnell, Operations on Morotai, p. 28.
[160] Farnell, Operations on Morotai, p. 27-28.
[161] Farnell, Operations on Morotai, pp. 30-31. Note that elsewhere Farnell reports the Japanese losses as 1,004.
[162] Rust, Thirteeth Air Force Story, p. 64.
[163] Llewellyn, My Time in the Military, p. B11.
[164] J. Vanderpoel Interview: August 19, 2004, Hughes would later be promoted by John Vanderpoel to be 424th BS Operations Officer.
[165] Britt, The Long Rangers, pp. 185-188. Unfortunately, six days previous, on January 8, a 370th BS B-24 piloted by Lt. John . Lucey was shot down by A/A fire over Santo Tomas.
[166] There is a problem with reconciling some of Walt's January missions. PFMT and FDFR indicate this mission on January 12 to Grace Park with Lt. Hughes. But there is no 424th BS Mission History available for verification and details of this mission. Furthermore the 424th BS Mission History summary of missions for January (image 1463) does not indicate a mission to Grace Park. I have chosen to assume that PFMT and FDFR are correct.
[167] Llewellyn, My Time in the Military, pp. 39-40. Clif had much more SP flying time than most men. As a radio operator, he flew many inter-island supply hops and also flew as radio operator when a new aircraft was received and needed to be checked out before going into combat service.
[168] PFMR and FDFR place the date of this mission as January 13, 1945, but the 424th Mission Report, images 1519-1521 clearly state that this mission was on January 14 and confirms Walt's participation. PFMR and FDFR state that the target for the mission was Nielsen Airfield, but the 424th BS Mission report indicates that the primary target was Grace A/D, while the optional targets were, Nielsen A/D, Batangas Airfield and Cavite Naval Base in Manila Bay. For my narrative I assume that the mission was on January 14 and the intended target was Grace Park A/D.
[169] Fawcett Letter: 6/24/02.
[170] Dowdy, Charlie, p. 371.
[171] Dowdy, Charlie, p. 109.
[172] Pacific Wrecks web site, consulted April 17, 2016, see: http://www.pacificwrecks.com/airfields/philippines/nielson/index.html.
[173] 424th BS Mission History, images 1519-1523.
[174] Vanderpoel Letter, 12/10/03. On his first mission, on October 14, as Raue was taking off on Noemfoer, his plane hit some rocks and kicked up a log at the end of the runway because he had failed to lock his turbo-supercharger fully forward and his manifold pressure wasset for 91 octane fuel, so that full take-off power could not be achieved. He managed to get the plane into the air, but after an hour his crew could not fully open the bomb bay doors because of the damage from the log, so Raue aborted and returned to base. Of that take-ff on October 14, John Vanderpoel wrote to the author: 'I was down at the runway that morning, but not flying, and I knew he was in trouble as he went by the tower because of the lack of noise. I really didn't think he would get off the ground' (also in Brit, The Long Rangers, p. 151). Another time, on November 8, Raue's plane was hit by A/A fire which damaged the bomb bay doors. His crew was finally able to get the right side doors open to jettison half his bomb load, but Raue returned to Morotai with the other half of the bomb load aboard (Brit, p. 164).
[175] The Story Of The Fifth Bombardment Group, Luketz was later promoted to Major.
[176] J. Vanderpoel Interview: August 19, 2004
[177] 424th BS Mission History, images 1554-1557.
[178] NPRC-13.
[179] Ulmer, Waist Gunner, pp. 115-117.
[180] 'South Pacific' flying included transfer or supply flights from base to base in theater, but not flying time to or from rest leave in Australia.
[181] Cooper Interview: August 24, 2006
[182] PFMR
[183] FDFR
[184] Tuesday, January 30, 1945 was one of these slowly unfolding days. Walt's PB&TR states on this date 'Flew last Mission to Manila, Philippines.' There was a 424th BS mission to attack Canacao Point A/D in Manila bay that day, but this mission does not appear on Walt's PBFR or FDFR.
[185] 424th BS Mission History, image 1457.
[186] GO 216. The official criteria are: - "The Air Medal is awarded to any person who, while serving in any capacity in or with the armed forces of the United States, shall have distinguished himself by meritorious achievement while participating in aerial flight". - "Awards may be made to recognize single acts of merit or heroism or for meritorious service". - "Award of the Air Medal is primarily intended to recognize those personnel who are on current crew member or non-crew member flying status which requires them to participate in aerial flight on a regular and frequent basis in the performance of their primary duties. Any further awarded Air Medal came in the form of an Bronze Oak Leaf Cluster.
[187] This one hour flight is listed on Walt's FDFR, although the purpose for it is conjecture on its likely purpose by the author, based on the 424th BS Mission History, image 1758.
[188] Hemphill, Aeial Gunner, p. 336.
[189] Llewellyn, My Time in the Military, p. 41.
[190] 424th BS Mission History, images 1753-1761.
[191] GO5. The Good Conduct medal is awarded to any active-duty enlisted member of the United States military who completes three consecutive years of "honorable and faithful service". During times of war, the Good Conduct Medal may be awarded for one year of faithful service.
[192] SO49.

193 *Suggestions for Returning Combat Crew Members*
194 *FDFR.*
195 *Form 1: 2/25/45*
196 *The author in his youtn heard this statement from Walt many times. Cliff Llewellyn confirmed this fact and showed the author the list that he had kept from Muroc that showed the other crews and crewmembers who were casulatiesof war.*
197 *Battle of Morotai.*

Chapter 15:   1945: The Old Malfunction and the Return Home

1 *Form 1: 2/25/45*
2 *41st Division, History of the Biak Operation.*
3 *Gideon, Airdromes Guide Southwest Pacific Area, pp. 75-79.*
4 *Boeman, Morotai, pp. 98-100. Boeman's description of the transient area and experience on Biak is from his time there just before Walt arrived. Boeman left for Morotai on 2/26/45, the day after Walt arrived.*
5 *Llewellyn, My Time in the Military, pp. 41-42.*
6 *NPC-17.*
7 *Paul, p. 14*
8 *Llewellyn, My Time in the Military, pp. 41-42.*
9 *Holocaust Encyclopedia, 'Seeking Refuge in Cuba, 1939', (see https://www.ushmm.org/wlc/en/article.php?ModuleId=10007330) accessed May 15, 2016. On May 27, 1939 the Orinoco left Hamburg bound for Cuba with 200 Jewish refugees fleeing Germany. But at sea, the Captain was informed by radio that the Jewish passengers would not be allowed to land in Cuba without visas. The Orinoco put in to Cherburg, France to await instructions. Some of the refugees jumped into the water, to avoid being returned to Germany. Neither France nor England would accept the refugees. The American Ambassador in London received assuuarnces from German authorities that no harm would befall the Jewish passengers if they returned. The Orinoco then retruned to Germany, but the fate of the refugees is unkown to this day.*
10 *Charles, Troopships of World War Two, p. 51.*
11 *Llewellyn, My Time in the Military, p. 42.*
12 *Reis Interview: August 20, 2004.*
13 *PB&TR.*
14 *Llewellyn, My Time in the Military, p. 42.*
15 *Baross, The War Years, p. 71.*
16 *Llewellyn, My Time in the Military, pp. 42-43.*
17 *Reis Interview: August 20, 2004*
18 *Fielding Interview: August 26, 2002*
19 *Baross, The War Years, p. 71.*
20 *Fielding Interview: August 26, 2002*
21 *Llewellyn, My Time in the Military, p. 43.*
22 *Schreiner Interview: August 30, 2002*
23 *Fielding Interview: August 26, 2002*
24 *Barros, The War Years, p. 72.*
25 *Fort Mason: Wikipedia article. The Civil War had prompted the construction of a set of coastal defense batteries located inside the Golden Gate. Initially these defenses were built as temporary wartime structures rather than permanent fortifications and one of these was constructed in 1864 at Point San Jose, as the location of Upper Fort Mason was then known. The fort was named Fort Mason in 1882, after Richard Barnes Mason, a former military governor of California. The piers and sheds of Lower Fort Mason were originally built from 1912 to warehouse army supplies and provide docking space for army transport ships. During World War II, Fort Mason became the headquarters of the San Francisco Port of Embarkation, controlling a network of shipping facilities that spread across the Bay Area. Over the years of the war, 1,647,174 passengers and 23,589,472 measured tons moved from the port into the Pacific. This total represents two-thirds of all troops sent into the Pacific and more than one-half of all Army cargo moved through West Coast ports. The highest passenger count was logged in August 1945 when 93,986 outbound passengers were loaded.*
26 *PB&TR*
27 *Llewellyn, My Time in the Military, p. 44.*
28 *Baroos, The War Years, p. 72.*
29 *Frank M. Coxe, Wikipedia article: The General Frank M. Coxe was a steam ferry built for the United States Army to provide transportation services among several military facilities which ring California's San Francisco Bay. The 144-foot ship was launched in 1922. The Coxe was an active military vessel on San Francisco Bay from the 1922 to 1947, being decommissioned and sold for surplus in 1947 after the end of World War II.*
30 *Fort McDowell, Wikipedia article. In the later 19th century, the army designated the entire Angel Island as "Fort McDowell" and developed further facilities there, including what is now called the East Garrison or Fort McDowell. During World War II the need for troops in the Pacific far exceeded prior needs. The facilities on Angel Island were expanded and further processing was done at Fort Mason in San Francisco. Fort McDowell was used as a detention station for Japanese, German and Italian immigrant residents of Hawaii arrested as potential fifth columnists (despite a lack of supporting evidence or access to due process). These internees were later transferred to inland Department of Justice and Army camps. Japanese and German prisoners of war were also held on the island, supplanting immigration needs, which were curtailed during the war years.*
31 *NPRC-6 documents Walt's physical on Angel Island.*
32 *Fielding Interview: August 26, 2002. Jim Fielding told me that once he was released from processing at Angel Island, he went into San Francisco for dinner. He was in his best uniform, but he still looked shabby, compared to the stylish San Franciscans. He went into the Mark Hopkins Hotel and got a table in the dining room. He felt self conscious, but told the waitress, all he wanted was the biggest steak dinner they had. He also ordered pie with ice cream for dessert, beer, coffee, the works. When he asked for his bill, he was told the dinner was 'on the house' for a returning service man. 'Welcome back to the USA' she said.*
33 *Llewellyn, My Time in the Military, pp. 44-45.*
34 *Baroos, The War Years, p. 72.*
35 *Llewellyn, My Time in the Military, pp. 44-45. Llewellyn states that Walt travelled through Omoha, but Walt's war time rail travel map does not show a route through Omaha.*
36 *SO99.*
37 *PB&TR.*

[38] *Overseas American Cemetaries, ID=759728*
[39] *SO99.*
[40] *Tourtellot, Life's Picture History of World War II, p. 280.*
[41] *JJB Form 9. The Pine Island (AV-12) was officially commissioned into the Navy at San Pedro, California on April 26, 1945.*
[42] *JJB to L Babinski ND*
[43] *This song was taken from a Disney-produced propaganda animated film staring Donald Duck.*
[44] *D.F. Babinski Interview: August 21, 2004*
[45] *Miller, The SAAAB Story.*
[46] *PB&TR*
[47] *Santa Ana Army Air Base, Wikipedia, accessed June 12, 2016, see: https://en.wikipedia.org/wiki/Santa_Ana_Army_Air_Base.*
[48] *C. Llewellyn Interview: October 27, 2002*
[49] *Los Angeles Times, May 7, 1945. Germany surrendered on May 8, local time.*
[50] *Form 2: 5/8/45*
[51] *Llewellyn, My Time in the Military, p. 46.*
[52] *NPRC-17*
[53] *VIC.*
[54] *NPRC-7*
[55] *SO125*
[56] *PB&TR*
[57] *PB&TR, Form 371, NPRC-8, NPRC-10*
[58] *G. Mendryga Interview: November 21, 2003]*
[59] *Johnny Slagle was a senior staff announcer for WXYZ Radio. After WWII he left to work in New York City but returned to Detroit in 1947. [Motor City Radio Flashbacks website: http://www.mcrfb.com/?p=22337 accessed 12/26/2015]*
[60] *Slagle Interview. The author has interviewed many 307th BG vets regarding the 424th Squadron B24 nicknamed 'Malfunction.' Although many of them remember 'Malfunction' no photo of this plane and its wolf emblem have been found. The author verified with D.F. Babinski (12/25/15) that this was a radio interview that was broadcast in Detroit shortly after Walt returned home following his discharge.*
[61] *D.F. Babiknski Interview, 5/30/2016.*

Chapter 16: New Beginnings: Dreams of Life
[1] *Llewellyn, My Time in the Military, p. 48.*
[2] *JJB: Navpers-601*
[3] *The Detroit News, July 22, 1945.*
[4] *McDonald Letter: May 3, 2002 and Babinski family reminscences. The author and his brothers and sisters all knew the story of the box of 1920s and 1930s era baseball cards. Walt concluded that at some point, his ma threw them out while cleaning the attic. He never seemed angry about this, but regreted the valuable collection that had been lost.*
[5] *JJB Navpers 601.*
[6] *JJB Navers 601.*
[7] *JJB: Form 4-6111.*
[8] *JJB: Notice of Separation*
[9] *D.F. Babinski Interview: April 9, 2005.*
[10] *D.F. Babinski Interview: August 27, 2002*
[11] *TOP issued October 8, 1945 and reisued January 8, 1946.*
[12] *Form 506.*
[13] *ABC Card.*
[14] *D.F. Babinski Interview: August 29, 2002. Later in life Pee Wee had heart bypass surgery, but within a week of being released from the hospital, he was out on the dance floor again.....all his friends and acquaintances were afraid he'd 'keel over' but he didn't and lived happily many years after until his death in 1997.*
[15] *D.F. Babinski Interview, May 30, 2016.*
[16] *D.F. Babinski Interview: August 25, 2002*
[17] *D.F. Babinski Interview: August 21, 2004*
[18] *DFB & WAB to LB, 2/11/48*
[19] *D.F. Babinski Interview: August 21, 2004. John Besek, five years younger than Walt had been in the USN during the war – stationed in Pennsylvania – but never went overseas.*
[20] *H. Babinski Interview: August 27, 2002*
[21] *D.F. Babinski Interview: August 21, 2004*

Chapter 17: Afterthought: Lines of Life
[1] *Randall Jarrell, "The Death of the Ball Turret Gunner" from The Complete Poems. 1980 by Randall Jarrell.*
[2] *Irving, John, The Ball Turret Gunner, Playboy Magazine, June 1976.*
[3] *Irving, John, The World According to Garp, E.P. Button, 1978.*
[4] *Moench, Anna Ouyang, The Death of the Ball Turret Gunner, performed at the 2008 New York Fringe Festival.*
[5] *Wikipedia, Potable Surgical Hospital.*
[6] *Britt, The Long Rangers, p. 234.*
[7] *Robert Altman. MASH, 1970.*
[8] *Joseph Heller. Catch-22, 1970.*
[9] *Baross, The War Years WWII, p. c.*
[10] *Goodman, Steve, My Old Man, 1977.*
[11] *Seattle Post-Intelligencer, May 28, 2005. Nakamura wasborn onJapanese ruled Formosa (now Taiwan) in 1918. He served in the Japanese Army from 1943. He died in Taiwan in 1979.*
[12] *Reunion Book 11, p. 3.*
[13] *Boeman, Morotai, p. 270.*
[14] *Cooper Interview: August 24, 2006*

[15] Holston Interview: August 25, 2006
[16] Manley Interview: August 24, 2006
[17] Pelle Interview: August 24, 2006
[18] Walsh Interview: August 24, 2006
[19] Norwood Interview: August 24, 2006
[20] Robinson Interview: August 25, 2006
[21] Jurkens Interview: August 25, 2006
[22] Dowdy, Charlie, p. 320.
[23] J. Vanderpoel Interview: August 19, 2004
[24] Hidalgo Interview: April 14, 2002.
[25] Dunne Letter: 10/3/2002
[26] Aubrey Home.
[27] Aubrey Interview: August 15, 2002
[28] Fielding Interview: August 26, 2002
[29] Reis Interview: August 20, 2004
[30] N. Riley Letter, October 15, 2002.
[31] V Mansir Interview: April 12, 2003
[32] Schreiner Interview: August 30, 2002
[33] C. Llewellyn Interview: October 15, 2002
[34] Hillenbrand, Laura, Unbroken, is based on the life and experiences of Louis Zamperini.
[35] Hylton, Wil S., Vanished, describes the Arnett crew, their loss on 9/1/1944, and Scannon's efforts to find the plane with the help of information from 307th Bomb Group Historian, Jim Kendall.

Appendix A: The Ball Turret
[1] William (Bill) E. Correll Interview, see: http://www.southernoregonwarbirds.org/b24b.html#corel
[2] Bolce Letter: 5/14/03
[3] William (Bill) E. Correll Interview, see: http://www.southernoregonwarbirds.org/b24b.html#corel

www.ingramcontent.com/pod-product-compliance
Lightning Source LLC
Chambersburg PA
CBHW080419230426
43662CB00015B/2148